Developments in Mathematics

Volume 80

Aims and Scope

The **Developments in Mathematics** (DEVM) book series is devoted to publishing well-written monographs within the broad spectrum of pure and applied mathematics. Ideally, each book should be self-contained and fairly comprehensive in treating a particular subject. Topics in the forefront of mathematical research that present new results and/or a unique and engaging approach with a potential relationship to other fields are most welcome. High-quality edited volumes conveying current state-of-the-art research will occasionally also be considered for publication. The DEVM series appeals to a variety of audiences including researchers, postdocs, and advanced graduate students.

Alexander J. Zaslavski

Approximate Fixed Points of Nonexpansive Mappings

Springer

Alexander J. Zaslavski
Department of Mathematics
Technion – Israel Institute of Technology
Haifa, Haifa, Israel

ISSN 1389-2177 ISSN 2197-795X (electronic)
Developments in Mathematics
ISBN 978-3-031-70709-4 ISBN 978-3-031-70710-0 (eBook)
https://doi.org/10.1007/978-3-031-70710-0

This Springer imprint is published by the registered company Springer Nature Switzerland AG
The registered company address is: Gewerbestrasse 11, 6330 Cham, Switzerland

Preface

In this book we study iteration processes associated with a given nonlinear mapping which generate its approximate fixed point and in some cases converge to a fixed point of the mapping. We consider various classes of nonlinear single-valued as well as set-valued mappings including nonexpansive mappings, monotone nonexpansive mappings, uniformly locally nonexpansive mappings, mappings in metric spaces with a graph, and many other nonlinear mappings. We also study iteration processes under the presence of computational errors which always present in practice.

The fixed point theory of nonlinear operators has been a rapidly growing area of research. The starting point of this theory is Banach's classical theorem [15] concerning the existence of a unique fixed point for a strict contraction. The main goals of this theory is to show the existence of a fixed point for a given nonlinear mapping and to construct an iterative process which generates approximate fixed point and in some cases converge to a fixed point of the mapping. The fixed point theory contains the study of various classes of nonlinear mappings. This theory plays an important role in the study of variational inequalities, monotone operators, feasibility problems, optimization theory, proximal point methods, superiorization and their applications in engineering, medical, and the natural sciences.

The monograph contains 13 chapters. Chapter 1 is an introduction. In Chap. 2 we study asymptotic regularity for Halpern-type iterations and for iterations of alternative iterative methods generated by a nonexpansive self-mapping of a star-shape set in a metric space with a hyperbolic structure. Linear rates of this asymptotic regularity is computed. Such results are known in the literature under the assumption that the nonexpansive mapping possesses a fixed point. In Chap. 2 we prove our results under a weaker assumption assuming only the existence of a bounded orbit of the nonexpansive mapping. It is shown that in the space of nonexpansive mappings equipped with some natural topology there exists an open everywhere dense set such that for each its element all orbits are bounded. We study asymptotic regularity for single-valued and set-valued mappings and also under the presence of computational errors. Most of results of this chapter are new.

In Chaps. 3, 4, and 5 we extend the theory developed in Chap. 2 for classes of monotone nonexpansive operators, uniformly locally nonexpansive operators, and nonexpansive operators in spaces with graphs, respectively. All the results of these chapters are new. In Chap. 6 we discuss the viscosity approximation method for solving of fixed-point problems for nonexpansive mappings in a Hilbert space and for finding a common element of the set of solutions of an equilibrium problem and the set of fixed points of a nonexpansive mapping in a Hilbert space. We show the uniform convergence of inexact iterates generated by the viscosity approximation method.

In Chap. 7 we consider a common fixed point problem with a finite family of nonexpansive self-mappings of a convex subset of a metric space with the hyperbolic structure. For Halpern-Mann iterations we prove a quantitative asymptotic regularity result for the common fixed point problem under assumption that this problem has approximate solutions. We also obtain two asymptotic regularity results for Krasnoselskii-Mann iterations. All the results of Sects. 7.1– 7.9 are new.

In Chap. 8 we study contraction mappings of Perov type in a generalized complete metric space. We show that these mappings possess a unique fixed point which attracts all (inexact) iterates of the mapping uniformly on bounded sets.

Chapter 9 is devoted to the study of cyclical mappings acting in a complete metric space. We study the asymptotic behavior of (inexact) iterates of cyclical mapping of contractive type. Using the Baire generic approach it is shown that a typical cyclical mapping is of contractive type. More precisely, we consider a complete metric space of cyclical nonexpansive mappings acting on a union of two sets in a complete hyperbolic space. If these sets have a nonempty intersection, then using Baire category approach, we show that for most mappings their iterates converge to a unique fixed point belonging to the intersection of the sets. If these two sets have an empty intersection, then using Baire category approach, we show for most mappings the distance between iterates converges to the distance between these two sets.

In Chap. 10 we analyze monotone nonexpansive mappings. First we consider three classes of monotone contractive mappings defined on a complete metric space, and for each mapping in one of these classes, we establish the existence a unique fixed point, which attracts all iterates. Then it is shown that the convergence of iterates of a monotone nonexpansive mapping is stable in the presence of summable computational errors. We consider monotone nonexpansive self-mappings of a bounded closed subset of a Banach space and using the Baire category approach, we show that for most of monotone nonexpansive mappings the fixed point problem is well-posed.

In Chap. 11 we study uniformly locally contractive mappings. We show that a uniformly locally contractive mapping has a fixed point, its fixed point problem is well posed and that inexact iterates of such a mapping converge to its unique fixed point, uniformly on bounded sets. Using the porosity notion, we also show that most uniformly locally nonexpansive mappings are, in fact, uniformly locally contractive. We extend these results to uniformly locally contractive non-self mappings defined on a closed subset of the metric space.

In Chap. 12 we study iteration processes associated with a given set-valued mapping which generate its approximate fixed point and in some cases converge to a fixed point of the mapping. We consider these processes under the presence of computational errors, which always present in practice. Our results are proved for various classes of set-valued mappings including strict contractions and the Mizoguchi-Takahashi class of mappings. We consider spaces of nonexpansive mappings, acting on a closed subset of a complete metric space, which is equipped with the topology of uniform convergence on bounded sets. We show the existence of an open and everywhere dense subset in this space such that every its element possesses an approximate fixed point, which is stable under small perturbations.

Chapter 13 is devoted to the study of nonexpansive mappings in spaces with graphs. We consider single-valued operators as well as set-valued maps. Our goals are to show the existence of a fixed point and construct iteration processes associated with a given mapping which generate its approximate fixed point and in some cases converge to a fixed point of the mapping. Our results are proved for various classes of mappings of contractive type.

In this book we analyze the behavior of iterative processes associated with various classes of fixed point problems. Our results can be useful in numerical solutions of fixed point problems. It should be mentioned that in practice it is impossible to obtain an exact solution of a fixed point problem. Therefore, approximate solution of these problems are of great importance. In many cases an exact solution does not exist while approximate solutions are available and can be obtain after a finite number of iterations. In practice one can do a certain number of iterations and needs to know what an approximate solutions can be obtained in the presence of computational errors. Many results of the book answer this question. Therefore, the author believes that this book will be useful for researchers working on numerical solutions of fixed point problems.

Rishon LeZion, Israel Alexander J. Zaslavski
March 30, 2024

Contents

Chapter 1
Introduction

In this book, we study iteration processes associated with a given nonlinear mapping which generate its approximate fixed point and in some cases converge to a fixed point of the mapping. We consider various classes of nonlinear single-valued as well as set-valued mappings including nonexpanive mappings, monotone nonexpansive mappings, uniformly locally nonexpansive mappings, mappings in metric spaces with a graph, and many other nonlinear mappings. We also study iteration processes under the presence of computational errors which always present in practice. Some of these iteration processes are discussed is this chapter.

1.1 Notation

The fixed point theory of nonlinear operators has been a rapidly growing area of research. The starting point of this theory is Banach's classical theorem [15] concerning the existence of a unique fixed point for a strict contraction. The main goals on this theory is to show the existence of a fixed point for a given nonlinear mapping and to construct an iterative process which generates approximate fixed point and in some cases converge to a fixed point of the mapping [1, 29, 34, 64–66, 70, 84–86, 93, 95, 100, 102, 104, 151, 160, 165, 168, 171, 186, 188, 189, 191, 193, 194, 209, 214, 217, 235]. The fixed point theory contains the study of various classes of nonlinear single-valued [2–4, 10, 17, 31, 49, 50, 69, 76, 80, 103, 108, 110, 111, 117, 122, 123, 128, 130, 131, 133, 140, 143, 148, 157–159] as well as set-valued mappings [11, 62, 74, 129, 132, 137, 144, 145, 150, 152, 167, 169, 185, 195, 203, 221] including nonexpanive mappings [14, 16, 18, 20, 23, 24, 26, 30, 32, 47, 51, 54–57, 61, 63, 67, 71, 73, 75, 107, 112, 114, 115, 118–121, 125, 141, 142, 156, 161, 164, 166, 179, 187, 204, 207, 219], monotone nonexpansive mappings [7, 8, 77, 170, 172, 173, 218], locally nonexpansive mappings [25, 155, 175–178], mappings in metric spaces with a graph [92, 138, 174, 184], and many other nonlinear mappings [96–98,

A. J. Zaslavski, *Approximate Fixed Points of Nonexpansive Mappings*,
Developments in Mathematics 80, https://doi.org/10.1007/978-3-031-70710-0_1

101, 109, 116, 146, 147, 149, 198, 202, 215, 216, 220, 222, 223]. It plays an important role in the study of variational inequalities [22, 35–37, 39–41, 60, 68, 82, 83, 87–90, 94, 105, 106, 113, 182, 183, 190, 196], monotone operators [19, 180, 211, 212], feasibility problems [6, 9, 27, 28, 33, 42–44, 46, 48, 58, 59, 78, 81, 124, 127, 134–136, 139, 163, 192, 199, 201, 206, 208, 213], optimization theory [13, 22, 53], proximal point methods [5, 12, 38, 72, 91, 126, 205, 211, 212], superiorization [45, 52], and their applications in engineering, medical, and the natural sciences.

In our book, we use the following notation. Let (X, d) be a metric space. For each $x \in X$ and each set $E \subset X$ put

$$d(x, E) = \inf\{d(x, y) : \ y \in E\}.$$

Note that $d(x, \emptyset) = \infty$, $x \in X$. For every pair of nonempty sets $A, B \subset X$ put

$$H(A, B) = \max\{\sup\{d(x, B) : \ x \in A\}, \ \sup\{d(y, A) : \ y \in B\}\}.$$

Given a mapping $P : X \to X$, we define $P^0 = I$, the identity self-mapping of X, $P^1 = P$, and $P^{i+1} = P \circ P^i$ for all nonnegative integers i.

For each mapping $P : X \to 2^X \setminus \{\emptyset\}$ and each nonempty set $E \subset X$, set

$$P(E) = \cup\{P(x) : \ x \in E\},$$

$P^1 = P$, and $P^{i+1}(E) = P(P^i(E))$ for all integers $i \geq 1$. In some chapters, a metric is denoted by ρ instead of d.

For each $x \in X$ and each $r > 0$ set

$$B_d(x, r) = \{y \in X : \ d(x, y) \leq r\}.$$

If the metric d is understood, then

$$B(x, r) = B_d(x, r), \ x \in X, \ r > 0.$$

For each $x \in R^1$ set

$$\lfloor x \rfloor = \max\{i : \ i \text{ is an integer and } i \leq x\}.$$

Denote by $\mathrm{Card}(A)$ the cardinality of a set A. We assume that the sum over empty set is zero and that the infimum of an empty set is ∞. The set $X \times X$ is equipped with the metric

$$d_1((x_1, x_2), (y_1, y_2)) = d(x_1, y_1) + d(x_2, y_2), \ x_i, y_i \in X, \ i = 1, 2.$$

For each set $A \subset X$ denote by $\mathrm{cl}(A)$ its closure.

1.2 Asymptotic Regularity for Iterations of Nonexpansive Mappings

In Chap. 2 we study asymptotic regularity for Halpern-type iterations and for iterations of alternative iterative methods generated by a nonexpansive self-mapping of a star-shaped set in a metric space with a hyperbolic structure. Linear rates of this asymptotic regularity are computed. Such results are known in the literature under the assumption that the nonexpansive mapping possesses a fixed point. In Chap. 2, we prove our results under a weaker assumption assuming only the existence of a bounded orbit of the nonexpansive mapping. It is shown that in the space of nonexpansive mappings equipped with some natural topology, there exists an open everywhere dense set such that for each its element all orbits are bounded. We study asymptotic regularity for single-valued and set-valued mappings and also under the presence of computational errors.

We say that a sequence $\{a_n\}_{n=0}^{\infty}$ in a metric space (X, d) is asymptotically regular if $\lim_{n\to\infty} d(a_n, a_{n+1}) = 0$. If, furthermore, T is a self-mapping of X, then $\{a_n\}_{n=0}^{\infty}$ is said to be T-asymptotically regular if $\lim_{n\to\infty} d(a_n, T(a_n)) = 0$.

Recently, H. Cheval and L. Leustean [56] computed linear rates of asymptotic regularity for a viscosity version of Halpern-type iterations generated by a nonexpansive self-mapping of a metric space assuming that it has a fixed point. In our recent work [232] we showed that their result is valid without this assumption. It is enough to assume the existence of a bounded orbit of the nonexpansive mapping. It should be mentioned that the viscosity version of Halpern-type iterations was introduced by Xu [204]. The results of Sect. 2.2 were obtained in [232] under the assumption that the mapping acts on a convex subset of the metric space. In Chap. 2 we show that these results are true when the subset is merely star-shaped. The results of Section 2.4 were obtained in [232] too. All other results of the chapter are new.

A W-space is a structure (X, d, W) where (X, d) is a metric space and $W : X \times X \times [0, 1] \to X$. We think of $W(x, y, \lambda)$ as a convex combination of the points $x, y \in X$ with the coefficients $1 - \lambda$, λ and use the notation

$$W(x, y, \lambda) := (1 - \lambda)x \oplus \lambda y, \quad x, y \in X, \ \lambda \in [0, 1].$$

A W-space (X, d, W) is called a W-hyperbolic space [110] if for all $x, y, w, z \in X$ and all $\lambda, \tilde{\lambda} \in [0, 1]$ the following properties hold:

(W1) $d(z, (1 - \lambda)x \oplus \lambda y) \le (1 - \lambda)d(z, x) + \lambda d(z, y)$;
(W2) $d((1 - \lambda)x \oplus \lambda y), (1 - \tilde{\lambda})x \oplus \tilde{\lambda} y) \le |\lambda - \tilde{\lambda}|d(x, y)$;
(W3) $(1 - \lambda)x \oplus \lambda y = \lambda y \oplus (1 - \lambda)x$;
(W4) $d((1 - \lambda)x \oplus \lambda z, (1 - \lambda)y \oplus \lambda w) \le (1 - \lambda)d(x, y) + \lambda d(z, w)$.

It is obvious that convex subsets of normed spaces are W-hyperbolic spaces. Other examples of W-hyperbolic spaces are hyperbolic spaces [168], Busemann spaces [10], and $CAT(0)$ spaces [111].

In the sequel, we assume that (X, d, W) is a W-hyperbolic space which is denoted by X for simplicity.

A nonempty set $C \subset X$ is convex if for each $x, y \in C$ and each $\lambda \in [0, 1]$,

$$(1 - \lambda)x \oplus \lambda y \in C.$$

A nonempty set $C \subset X$ is called star-shaped, and $x_* \in C$ is its star if for each $x \in C$ and each $\lambda \in [0, 1]$,

$$(1 - \lambda)x \oplus \lambda x_* \in C.$$

If C is star-shaped, we denote by $\mathrm{St}(C)$ the sets of all stars of C.

Assume that C is a nonempty star-shaped subset of X, a number $\rho \in [0, 1)$ is given, a mapping $T : C \to C$ satisfies the equation

$$d(T(x), T(y)) \leq d(x, y), \ x, y \in C$$

and that a strict contraction $f : C \to \mathrm{St}(C)$ satisfies the equation

$$d(f(x), f(y)) \leq \rho d(x, y), \ x, y \in C.$$

Assume that a sequence $\{\alpha_n\}_{n=0}^{\infty} \subset (0, 1)$ is given

$$x_0 = x \in C$$

and that for every nonnegative integer n,

$$x_{n+1} = (1 - \alpha_n) f(x_n) \oplus \alpha_n T(x_n).$$

Since the inclusion $f(C) \subset \mathrm{St}(C)$ is true, the sequence of iterates $\{x_n\}_{n=0}^{\infty}$ is well-defined. Let

$$\theta \in C \text{ and } M_0 > 0.$$

We use the following assumption.

(A) There exists a point $p \in C \cap B(\theta, M_0)$ such that for each nonnegative integer i, we have

$$d(T^i(p), \theta) \leq M_0.$$

We study the asymptotic regularity of the sequence of iterates $\{x_n\}_{n=0}^{\infty}$. It was introduced in [204], in the setting of Banach spaces, and called as a viscosity version of the Halpern iteration. Note that if the mapping f is a constant, then we obtain the well-known Halpern iterations.

The following theorem is our first asymptotic regularity result in Chap. 2.

Theorem 1.1 *Let (A) hold,* $J = 2\lfloor(1-\rho)^{-1}\rfloor+1$ *and for every nonnegative integer* n,

$$\alpha_n = 1 - 2(1-\rho)^{-1}(n+J)^{-1}.$$

Then for every nonnegative integer n,

$$d(x_n, x_{n+1}) \le 2MJ(1-\rho)^{-1}(n+J)^{-1}$$

and

$$d(x_n, T(x_n)) \le 2M(J+2)(1-\rho)^{-1}(n+J)^{-1}.$$

Note that in [232] this result was proved when the set C is convex.

1.3 Generic Results

We assume that

$$\theta \in \mathrm{St}(C).$$

Evidently, assumption (A) holds if the set C is bounded. In Sect. 2.4, we prove the following result.

Proposition 1.2 *Assume that the set* C *is bounded. Then for every positive number* ϵ *there exists a point* $p_\epsilon \in C$ *for which*

$$d(p_\epsilon, T(p_\epsilon)) \le \epsilon.$$

Denote by $\mathcal{M}$ the set of all mappings $A : C \to C$ such that

$$d(A(x), A(y)) \le d(x, y)$$

for all $x, y \in C$ equipped with the uniformity determined by the base

$$\mathcal{U}(n) = \{(A, B) \in \mathcal{M} \times \mathcal{M} :$$

$$d(A(x), B(x)) \le n^{-1},\ x \in B(\theta, M) \cap C\},\ n = 1, 2, \ldots.$$

The space $\mathcal{M}$ with this uniformity is metrizable and complete [168].

The following result shows that for most mappings in $\mathcal{M}$ assumption (A) holds.

Proposition 1.3 *There exists an open everywhere dense set* $\mathcal{F} \subset \mathcal{M}$ *such that for every mapping* $T \in \mathcal{F}$*, assumption (A) holds.*

Note that these results were obtained in [232] under the assumption that the mapping acts on a convex subset of the metric space.

In Chap. 2 we prove an analog of Proposition 1.3 for set-valued mappings. Assume that the set C is closed and denote by $S(C)$ the set of all nonempty bounded closed subsets of $S(C)$. It is known [168] that $(S(C), H)$ is a complete metric space.

We assume that

$$\theta \in \mathrm{St}(C).$$

Denote by $\mathcal{M}$ the set of all mappings $T : C \to S(C)$ such that

$$H(T(x), T(y)) \le d(x, y),\ x, y \in C.$$

The space $\mathcal{M}$ is equipped with the uniformity determined by the base

$$\mathcal{U}(m) = \{(T_1, T_2) \in \mathcal{M} :$$

$$H(T_1(x), T_2(x)) \le m^{-1},\ x \in B(\theta, m)\},\ m = 1, 2, \dots.$$

The space $\mathcal{M}$ with this uniformity is metrizable and complete [168].

The following results is proved in Sect. 2.7.

Theorem 1.4 *There exists an open everywhere dense set $\mathcal{F} \subset \mathcal{M}$ such that for each $S \in \mathcal{F}$ there exist $M > 0$ and an open neighborhood V of S in $\mathcal{F}$ such that for each $T \in V$,*

$$T(B(\theta, M) \cap C) \subset B(\theta, M) \cap C.$$

1.4 Halpern-Type Iterations of Set-Valued Mappings

In Sect. 2.8 we study the asymptotic regularity property for Halpern-type iterations of set-valued mappings.

We assume that (X, d, W) is a W-hyperbolic space which is denoted by X for simplicity and that C is a nonempty star-shaped subset of X. Let $\rho \in [0, 1)$, $T : C \to 2^C \setminus \{\emptyset\}$, $T(z)$ be bounded for each $z \in C$,

$$H(T(x), T(y)) \le d(x, y),\ x, y \in C$$

and that $f : C \to \mathrm{St}(C)$ satisfy

$$d(f(x), f(y)) \le \rho d(x, y),\ x, y \in C.$$

We consider Halpern-type iterations of T which are defined below.

Assume that $\{\alpha_n\}_{n=0}^{\infty} \subset (0, 1)$, $\{r_n\}_{n=0}^{\infty} \subset (0, 1]$,

$$x_0 = x \in C, \ \ y_0 \in T(x_0)$$

and that for each integer $n \geq 0$,

$$y_n \in T(x_n),$$

$$x_{n+1} = (1 - \alpha_n) f(x_n) \oplus \alpha_n y_n,$$

$$d(y_n, y_{n+1}) \leq H(T(x_n), T(x_{n+1})) + r_n \leq d(x_n, x_{n+1}) + r_n.$$

Let $\theta \in C$,

$$M_0 > H(\{\theta\}, T(\theta)\}$$

and

$$M > (1 - \rho)^{-1}(2M_0 + d(x, \theta) + d(f(\theta), \theta) + 2 \sup\{d(\theta), \eta) : \ \eta \in T(\theta)\}).$$

We continue to assume that assumption (A) holds and recall it now.
(A) There exists $p \in C \cap B(\theta, M_0)$ such that for each integer $n \geq 0$,

$$T^n(p) \subset B(\theta, M_0).$$

In Sect. 2.8 we will prove the following result.

Theorem 1.5 *Let $J = 2\lfloor(1 - \rho)^{-1}\rfloor + 1$ and for each integer $n \geq 0$,*

$$\alpha_n = 1 - 2(1 - \rho)^{-1}(n + J)^{-1}.$$

Then for each integer $n \geq 0$,

$$d(x_n, x_{n+1}) \leq (2M + 1)J(1 - \rho)^{-1}(n + J)^{-1}$$

and

$$d(x_n, T(x_n)) \leq (2M + 1)(J + 2)(1 - \rho)^{-1}(n + J)^{-1}.$$

In Chap. 2 we extend the results obtained above for an alternative iterative method and for the case when small computation errors are taken into account. In Chaps. 3–5, we extend the theory developed in Chap. 2 for classes of monotone nonexpansive operators, uniformly locally nonexpansive operators and nonexpansive operators in spaces with graphs, respectively. In Chap. 6 we study the viscosity approximation method for solving of fixed-point problems for nonexpansive mappings in a Hilbert space and for finding a common element of the set of solutions of an equilibrium problem and the set of fixed points of a nonexpansive mapping in a Hilbert space. We show the uniform convergence of inexact iterates generated by the viscosity approximation method.

1.5 Krasnoselskii-Mann Iterations for a Common Fixed Point Problem

In Chap. 7 we consider a common fixed point problem with a finite family of nonexpansive self-mappings of a convex subset of a metric space with the hyperbolic structure. For Halpern-Mann iterations we prove a quantitative asymptotic regularity result for the common fixed point problem under assumption that this problem has approximate solutions. This result generalize recent results of [121] obtained for a family consisting only of two operators which have a common fixed point. We also obtain two asymptotic regularity results for Krasnoselskii-Mann iterations. Here we present one of them.

Let $X = (X, d, W)$ be a W-hyperbolic space. We assume that the W-hyperbolic space X has a structure (X, η), where $\eta : (0, \infty) \times (0, 2] \to (0, 1]$ is a so-called modulus of uniform convexity such that the following assumptions hold:

(B1) For each $r > 0$, each $\epsilon \in (0, 2]$ and each $x, y, a \in X$, if

$$d(x, a),\ d(y, a) \leq r \text{ and } d(x, y) \geq \epsilon r,$$

then

$$d(2^{-1}x + 2^{-1}y, a) \leq (1 - \eta(r, \epsilon))r.$$

(B2) For each $\epsilon \in (0, 2]$ and each numbers s, r satisfying $0 < r \leq s$,

$$\eta(s, \epsilon) \leq \eta(r, s).$$

It turns out that this class of spaces is an appropriate setting for obtaining quantitative results on the asymptotic behavior of the Mann iteration for nonexpansive mappings (see [112, 118, 119]), as well as of the Picard iteration for firmly nonexpansive mappings [10]. It contains uniformly convex normed spaces and CAT(0) space.

Let

$$\theta \in C,\ M_0 > 0.$$

Assume that m is a natural number, for $i = 1, \ldots, m$, $T_i : C \to C$ and that the following assumption holds.

For each $\delta \in (0, 1)$ there exists $p_\delta \in B(\theta, M_0) \cap C$ such that for each $i \in \{1, \dots, m\}$ and each $x \in C$,

$$d(T_i(x), p_\delta) \le d(x, p_\delta) + \delta.$$

Let $\delta \in (0, 1)$ and p_δ be as guaranteed by our assumption. Clearly, for $i = 1, \dots, m$,

$$d(T_i(p_\delta), p_\delta) \le \delta.$$

Assume that $\bar{N} \ge m$ is a natural number. Denote by $\mathcal{R}$ the set of all mappings $r : \{0, 1, \dots\} \to \{1, \dots, m\}$ such that for each integer $k \ge 0$,

$$\{1, \dots, m\} \subset \{r(i) : i \in \{k, \dots, k + \bar{N} - 1\}\}.$$

Assume that

$$r \in \mathcal{R}, \ \{\alpha_i\}_{i=0}^{\infty} \subset (0, 1),$$

$$x_0 = x \in B(\theta, M_0) \cap C$$

and that for each integer $n \ge 0$,

$$x_{n+1} = (1 - \alpha_n)x_n \oplus \alpha_n T_{r(n)}(x_n).$$

We prove the following result obtained.

Theorem 1.6 *Assume that*

$$\sum_{n=0}^{\infty} \min\{\alpha_i(1 - \alpha_i) : i \in \{n\bar{N}, \dots, (n + 1)\bar{N} - 1\}\} = \infty.$$

Let $\epsilon \in (0, 1)$, $\gamma = \bar{N}^{-1}\epsilon$ *and a natural number* Q *satisfy*

$$\sum_{n=0}^{Q} \min\{\alpha_i(1 - \alpha_i) : i \in \{n\bar{N}, \dots, (n + 1)\bar{N} - 1\}\}$$

$$> 16M_0\gamma^{-1}\eta(4M_0 + 1, \gamma(4M_0 + 1)^{-1})^{-1}.$$

Then there exists an integer $n \in \{0, \dots, Q\}$ *such that for each*

$$i \in \{n\bar{N}, \dots, (n + 1)\bar{N} - 1\},$$

we have

$$d(x_i, T_{r(i)}(x_i)) \le \epsilon \bar{N}^{-1},$$

for each $i, j \in \{n\bar{N}, \dots, (n+1)\bar{N}\}$,

$$d(x_i, x_j) \le \epsilon$$

and for each $i \in \{n\bar{N}, \dots, (n+1)\bar{N}\}$ *and each* $s \in \{1, \dots, m\}$,

$$B(x_i, \epsilon) \cap \{z \in C : d(T_s(z), z) \le \epsilon \bar{N}^{-1}\} \ne \emptyset.$$

1.6 Perov Contraction Mappings

In Chap. 8, we study contraction mappings of Perov type in a generalized complete metric space. We show that these mappings possess a unique fixed point which attracts all (inexact) iterates of the mapping uniformly on bounded sets. Here we present some results of Chap. 8.

Let R^n be an n-dimensional Euclidean space. In other words,

$$R^n = \{x = (x_1, \dots, x_n) : x_i \in R^1, \ i = 1, \dots, n\}.$$

Let

$$R^n_+ = \{x = (x_1, \dots, x_n) \in R^n : x_i \ge 0, \ i = 1, \dots, n\}$$

and $e = (1, 1, \dots, 1) \in R^n$. We say that $x = (x_1, \dots, x_n), \ y = (y_1, \dots, y_n) \in R^n$ satisfy $x \le y$ if $x_i \le y_i$ for all $i = 1, \dots, n$. For each $x = (x_1, \dots, x_n) \in R^n$ set

$$\|x\|_1 = \sum_{i=1}^{n} |x_i|.$$

Assume that X is a nonempty set and a function $d : X \times X \to R^n_+$ satisfies for each $x, y, z \in X$,

$$d(x, y) = 0 \text{ if and only if } x = y,$$

$$d(x, y) = d(y, x),$$

$$d(x, z) \le d(x, y) + d(y, z).$$

The pair (X, d) is called a generalized metric space, and d is called a generalized metric [146, 147, 149, 198].

For all $x, y \in X$ set

$$d(x, y) = (d_1(x, y), \ldots, d_n(x, y)).$$

Fix $\theta \in X$. For each $x \in X$ and each $r \in R^n_+$ set

$$B(x, r) = \{y \in X : d(x, y) \le r\}.$$

We say that a sequence $\{x_i\}_{i=0}^{\infty} \subset X$ converges to $x_* \in X$ if

$$\lim_{i \to \infty} \|d(x_i, x_*)\|_1 = 0.$$

We say that $\{x_i\}_{i=0}^{\infty} \subset X$ is a Cauchy sequence if for each $\epsilon > 0$ there exists a natural number $n(\epsilon)$ such that for each pair of integers $p, m \ge n(\epsilon)$,

$$\|d(x_p, x_m)\|_1 \le \epsilon.$$

The generalized metric space (X, d) is complete if every Cauchy sequence converges.

For all $x, y \in X$ set

$$\tilde{d}(x, y) = \|d(x, y)\|_1 = d_1(x, y) + \cdots + d_n(x, y).$$

Clearly, $(X, \tilde{d})$ is a metric space, and convergence in (X, d) is equivalent to the convergence in $(X, \tilde{d})$.

We assume that the metric space $(X, \tilde{d})$ is complete.

Assume that $A : R^n_+ \to R^n_+$ and that the following properties hold:

(i) $A(0) = 0$ and A is continuous at zero;
(ii) for each $z_1, z_2 \in R^n_+$ satisfying $0 \le z_1 \le z_2$,

$$A(z_1) \le A(z_2);$$

(iii) for each $z_1, z_2 \in R^n_+$,

$$A(z_1 + z_2) \le A(z_1 + z_2);$$

(iv) $A^k(e) \to 0$ as $k \to \infty$;
(v) $A(\lambda z) = \lambda A(z)$ for each $\lambda \ge 0$ and each $z \in R^n_+$.

Assume that K is a nonempty closed subset of X and $T : K \to X$ satisfies for each $x, y \in K$,

$$d(T(x), T(y)) \le A(d(x, y)).$$

It is natural to call the mapping T as a generalized contraction [156, 168]. This class of mappings was introduced in [146]. A particular case when A is a linear mapping was introduced in [147].

In Chap. 8 we prove the following results obtained in [226].

Theorem 1.7 *Assume that for each $\epsilon > 0$ there exists $x_\epsilon \in K$ such that*

$$d(x_\epsilon, T(x_\epsilon)) \leq \epsilon e.$$

Then the following assertions hold:

1. *There exists a unique point $x_T \in K$ such that $T(x_T) = x_T$.*
2. *For each $\epsilon > 0$ there exists $\delta > 0$ such that if $x \in K$ satisfies $d(x, T(x)) \leq \delta e$, then $d(x, x_T) \leq \epsilon e$.*

Theorem 1.8 *Assume that $c > 0$ and that for each integer $m \geq 1$ there exists $\{x_i^{(m)}\}_{i=0}^m \subset K$ such that*

$$d(x_0^{(m)}, x_1^{(m)}) \leq ce$$

and that for each integer $i \in \{0, \dots, m-1\}$,

$$d(x_{i+1}^{(m)}, T(x_i^{(m)})) \leq m^{-1}e.$$

Then there exists a unique point $x_T \in K$ such that $T(x_T) = x_T$.

Theorem 1.9 *Assume that $x_T \in K$ satisfies $T(x_T) = x_T$ and $c, \epsilon > 0$. Then there exist $\delta \in (0, \epsilon)$ and a natural number n_0 such that for each integer $m \geq n_0$ and each sequence $\{x_i\}_{i=0}^m \subset K$ which satisfies*

$$d(x_0, x_T) \leq ce,$$

$$d(x_{i+1}, T(x_i)) \leq \delta,\ i = 0, \dots, m-1$$

the inequality

$$d(x_i, x_T) \leq \epsilon e$$

holds for all integers $i = n_0, \dots, m$.

Theorem 1.10 *Assume that $x_T \in K$ satisfies $T(x_T) = x_T$ and $\epsilon > 0$. Then there exists $\delta \in (0, \epsilon)$ such that for each integer $k \geq 1$ and each sequence $\{x_i\}_{i=0}^k \subset K$ which satisfies*

$$d(x_0, x_T) \leq \delta e,$$

$$d(x_{i+1}, T(x_i)) \leq \delta,\ i = 0, \dots, k-1$$

the inequality

$$d(x_i, x_T) \leq \epsilon e$$

holds for all integers $i = 0, \ldots, k$.

1.7 Cyclical Mappings

In Chap. 9, we study cyclical mappings acting in a complete metric space. We study the asymptotic behavior of (inexact) iterates of cyclical mapping of contractive type. Using the Baire generic approach, it is shown that a typical cyclical mapping is of contractive type. More precisely, we consider a complete metric space of cyclical nonexpansive mappings acting on a union of two sets in a complete hyperbolic space. If these sets have a nonempty intersection, then using Baire category approach, we show that for most mappings their iterates converge to a unique fixed point belonging to the intersection of the sets. If these two sets have an empty intersection, then using Baire category approach, we show for most mappings, the distance between iterates converges to the distance between these two sets.

Here we present one of these generic results which was obtained in [223].

Let (X, ρ, W) be a complete W-hyperbolic space, A and B be nonempty, bounded and closed subsets of X such that

$$A \cap B = \emptyset$$

and let

$$x_A \in A, \; x_B \in B$$

satisfy

$$\rho(x_A, x_B) = \rho(A, B) := \inf\{\rho(a, b) : \; a \in A, \; b \in B\}.)$$

Assume that for each $a \in A$, each $b \in B$ and each $\gamma \in (0, 1)$,

$$\gamma x_A \oplus (1 - \gamma)a \in A$$

$$\gamma x_B \oplus (1 - \gamma)b \in B.$$

Denote by $\mathcal{M}$ the set of all mapping

$$T : A \cup B \to A \cup B$$

such that

$$T(A) \subset B,\ T(B) \subset A$$

and

$$\rho(T(x), T(y)) \le \rho(x, y) \text{ for all } x \in A \text{ and all } y \in B.$$

For each pair $T_1, T_2 \in \mathcal{M}$, we set

$$d(T_1, T_2) = \sup\{\rho(T_1(x), T_2(x)) :\ x \in A \cup B\}.$$

Clearly, $(\mathcal{M}, d)$ is a complete metric space. Set

$$D = \sup\{\rho(a, b) :\ a, b \in A \cup B\}.$$

We will prove the following result obtained in [223].

Theorem 1.11 *There exists a set $\mathcal{F} \subset \mathcal{M}$ which is a countable intersection of open everywhere dense sets in $(\mathcal{M}, d)$ such that the following assertion holds.*

For each $S \in \mathcal{F}$ and each $\epsilon > 0$, there exist a natural number n_0 and a positive number δ such that for each sequence $\{x_i\}_{i=0}^{\infty} \subset A \cup B$ satisfying for each integer $i \ge 0$

$$\textit{if } x_i \in A, \textit{ then } x_{i+1} \in B \textit{ and } \textit{if } x_i \in B \textit{ then } x_{i+1} \in A$$

and

$$\rho(S(x_i), x_{i+1}) \le \delta$$

the inequality

$$\rho(x_n, x_{n+1}) \le \rho(A, B) + \epsilon$$

holds for all integers $n \ge n_0$.

This theorem easily implies the following result.

Theorem 1.12 *Assume that for each $\epsilon > 0$, there exists $\delta > 0$ such that for each $x \in A$ and each $y \in B$ satisfying*

$$\rho(x, y) \le \rho(A, B) + \delta$$

there exist $\widehat{x} \in A$ and $\widehat{y} \in B$ such that

$$\rho(\widehat{x}, \widehat{y}) = \rho(A, B)$$

and

$$\max\{\rho(x,\widehat{x}),\ \rho(y,\widehat{y})\} \le \epsilon.$$

Let a set $\mathcal{F}$ be as guaranteed by Theorem 1.11. Then the following assertion holds.

For each $S \in \mathcal{F}$ and each $\epsilon > 0$, there exist a natural number n_0 and a positive number δ such that for each sequence $\{x_i\}_{i=0}^{\infty} \subset A \cup B$ satisfying for each integer $i \ge 0$

$$\textit{if } x_i \in A,\ \textit{then } x_{i+1} \in B \textit{ and } \textit{if } x_i \in B \textit{ then } x_{i+1} \in A$$

and

$$\rho(S(x_i), x_{i+1}) \le \delta$$

and for each integer $n \ge n_0$, there exist $\widehat{x},\ \widehat{y} \in A \cup B$ such that

$$\rho(\widehat{x},\widehat{y}) = \rho(A,B),$$

$$\max\{\rho(x_n,\widehat{x}),\ \rho(x_{n+1},\widehat{y})\} \le \epsilon$$

if $x_n \in A$, then $\widehat{x} \in A$, $\widehat{y} \in B$ and and if $x_n \in B$, then $\widehat{x} \in B$, $\widehat{y} \in A$.

1.8 Monotone Nonexpansive Mappings

In Chap. 10 we study monotone nonexpansive mappings. First we consider three classes of monotone contractive mappings defined on a complete metric space, and for each mapping in one of these classes, we establish the existence of a unique fixed point which attracts all iterates. Then it is shown that the convergence of iterates of a monotone nonexpansive mapping is stable in the presence of summable computational errors. We consider monotone nonexpansive self-mappings of a bounded closed subset of a Banach space, and using the Baire category approach, we show that for most of monotone nonexpansive mappings, the fixed point problem is well-posed.

Some of these results are presented here.

Let (X, ρ) be a complete metric space equipped with a partial order $\le$. That is, for all points $x, y, z \in X$, we have

$$x \le x,$$

$$\text{if } x \le y,\ y \le x,\ \text{then } x = y,$$

and

$$\text{if } x \le y, \ \ y \le z, \ \text{ then } x \le z.$$

We also assume that

$$\{(x, y) \in X \times X : \ x \le y\}$$

is a closed subset of $X \times X$.

Let K be a nonempty closed subset of X which is not a singleton. Let $x_K \in K$ and assume that at least one of the following relations holds:

$$x_K \le x \text{ for all } x \in K$$

or

$$x \le x_K \text{ for all } x \in K.$$

Suppose that $T : K \to X$ and that the graph of T

$$\text{graph}(T) = \{(x, T(x)) : \ x \in K\}$$

is a closed subset of $X \times X$,

$$T^i(x_K) \in K \text{ for all integers } i \ge 1$$

and

$$T(x) \le T(y) \text{ for all } x, y \in K \text{ such that } x \le y.$$

In Chap. 10, we prove three theorems obtained in [170] regarding the existence of a unique fixed point of such a mapping T under three different contractivity assumptions. In the first result, we use contractivity in the sense of Rakotch [156], and the second is in the spirit of Boyd and Wong [23], while in the third result T is a contractive mapping in the sense of Matkowski [128].

Theorem 1.13 *Assume that* $\phi : [0, \infty) \to [0, 1]$ *is a decreasing function,*

$$\phi(t) < 1 \textit{ for all } t > 0$$

and that for all $x, y \in K$ *satisfying* $x \le y$*, we have*

$$\rho(T(x), T(y)) \le \phi(\rho(x, y))\rho(x, y).$$

Then the sequence $\{T^i(x_K)\}_{i=1}^{\infty}$ *converges,* $\lim_{i \to \infty} T^i(x_K)$ *is the unique fixed point of the mapping* T *and the following assertion holds.*

Let M and ϵ be positive. Then there exists a natural number n_0 such that if

$$x \in K \cap B(x_K, M),$$

$n > n_0$ is an integer and $T^n(x)$ is defined, then

$$\rho(T^i(x), \lim_{j \to \infty} T^j(x_K)) \le \epsilon$$

for all $i = n_0 + 1, \ldots, n$.

Theorem 1.14 *Assume that the function $\phi : [0, \infty) \to [0, \infty)$ is upper semicontinuous,*

$$\phi(t) < t \text{ for all } t > 0$$

and that for all $x, y \in K$ satisfying $x \le y$, we have

$$\rho(T(x), T(y)) \le \phi(\rho(x, y)).$$

Then the sequence $\{T^i(x_K)\}_{i=1}^{\infty}$ converges, $\lim_{i \to \infty} T^i(x_K)$ is the unique fixed point of the mapping T, and the following assertion holds.

Let M and ϵ be positive. Then there exists a natural number n_0 such that if

$$x \in K \cap B(x_K, M),$$

$n > n_0$ is an integer and $T^n(x)$ is defined, then

$$\rho(T^i(x), \lim_{j \to \infty} T^j(x_K)) \le \epsilon$$

for all $i = n_0 + 1, \ldots, n$.

Theorem 1.15 *Assume that $\phi : [0, \infty) \to [0, \infty)$ is an increasing function,*

$$\lim_{n \to \infty} \phi^n(t) = 0 \text{ for all } t > 0$$

and that for all $x, y \in K$ satisfying $x \le y$, we have

$$\rho(T(x), T(y)) \le \phi(\rho(x, y)).$$

Then the sequence $\{T^i(x_K)\}_{i=1}^{\infty}$ converges, $\lim_{i \to \infty} T^i(x_K)$ is the unique fixed point of the mapping T, and the following assertion holds.

Let M and ϵ be positive. Then there exists a natural number n_0 such that if

$$x \in K \cap B(x_K, M),$$

$n > n_0$ is an integer and $T^n(x)$ is defined, then

$$\rho(T^i(x), \lim_{j\to\infty} T^j(x_K)) \leq \epsilon$$

for all $i = n_0 + 1, \ldots, n$.

1.9 Uniformly Locally Contractive Mappings

In Chap. 11 we study uniformly locally contractive mappings. We show that a uniformly locally contractive mapping has a fixed point, its fixed point problem is well posed, and inexact iterates of such a mapping converge to its unique fixed point, uniformly on bounded sets. Using the porosity notion, we also show that most uniformly locally nonexpansive mappings are, in fact, uniformly locally contractive. We extend these results to uniformly locally contractive non-self-mappings defined on a closed subset of the metric space. Here we present some of these results.

Assume that (X, ρ) is a complete metric space, $\Delta > 0$, and that the following assumption holds.

(A) For each $x, y \in X$, there exist an integer $q \geq 1$ and points $x_i \in X$, $i = 0, \ldots, q$, such that

$$x_0 = x,\ x_q = y,\ \rho(x_i, x_{i+1}) \leq \Delta,\ i \in \{0, \ldots, q\} \setminus \{q\}.$$

For each $x, y \in X$, define

$$\rho_1(x, y) := \inf\{\sum_{i=0}^{q-1} \rho(x_i, x_{i+1}) :\ q \geq 1 \text{ is an integer},$$

$$x_i \in X,\ i = 0, \ldots, q,\ x_0 = x,\ x_q = y,\ \rho(x_i, x_{i+1}) \leq \Delta,\ i \in \{0, \ldots, q\} \setminus \{q\}\}.$$

It follows from assumption (A) that for each $x, y, z \in X$, $\rho_1(x, y)$ is finite,

$$\rho(x, y) \leq \rho_1(x, y) < \infty,$$

$$\text{if } \rho(x, y) \leq \Delta, \text{ then } \rho_1(x, y) = \rho(x, y),$$

$$\rho_1(x, y) = \rho_1(y, x),$$

$$\rho_1(x, x) = 0,$$

$$\text{if } \rho_1(x, y) = 0, \text{ then } \rho(x, y) = 0 \text{ and } x = y,$$

and

$$\rho_1(x, z) \le \rho_1(x, y) + \rho_1(y, z).$$

Clearly, ρ_1 is a metric on X.

Let $T : X \to X$ be a self-mapping of X.

We assume that for each $x, y \in X$ satisfying $\rho(x, y) \le \Delta$, the inequality

$$\rho(T(x), T(y)) \le \rho(x, y)$$

holds. It is not difficult to see that for each $x, y \in X$,

$$\rho_1(T(x), T(y)) \le \rho_1(x, y).$$

Such a mapping T is said to be uniformly locally nonexpansive.

We remark in passing that the smaller class of uniformly local (strict) contractions was introduced in [75], while the larger class of locally nonexpansive mappings was studied in [25]. The work [75] also contains an example of a uniformly local (strict) contraction which is not nonexpansive.

Assume that $\phi : [0, \Delta] \to [0, 1]$ is a decreasing function

$$\phi(t) < 1 \text{ for all } t \in (0, \Delta],$$

and for each $x, y \in X$ satisfying $\rho(x, y) \le \Delta$, we have

$$\rho(T(x), T(y)) \le \phi(\rho(x, y))\rho(x, y).)$$

In Chap. 11 we prove the following theorem obtained in [175]. The first of them shows the well-posedness of the fixed point problem for the mapping T.

Theorem 1.16 *Given $\epsilon > 0$, there exists $\delta > 0$ such that for each $x, y \in X$ satisfying*

$$\max\{\rho(T(x), x),\ \rho(T(y), y)\} \le \delta,$$

the inequality

$$\rho(x, y) \le \epsilon$$

holds.

Theorem 1.17 *Let $\epsilon, M > 0$. Then there exist a number $\delta \in (0, \epsilon)$ and an integer $n_0 \ge 1$ such that for each integer $n > n_0$ and each finite sequence $\{x_i\}_{i=0}^{n} \subset X$ which satisfies*

$$\rho_1(x_0, x_1) \le M$$

and

$$\rho(x_{i+1}, T(x_i)) \le \delta, \ i = 0, \dots, n-1,$$

the inequality

$$\rho(x_{i+1}, x_i) \le \epsilon$$

holds for all $i = n_0, \dots, n-1$.

These two theorems imply that there exists a unique $x_* \in K$ such that

$$T(x_*) = x_*.$$

Note that the existence of the unique fixed point was obtained in [155] (see also [158]).

These two theorems also easily imply the following convergence result for inexact iterates of the mapping T.

Theorem 1.18 *Let the point* $x_* \in X$ *satisfy*

$$T(x_*) = x_*$$

and let $\epsilon, M > 0$. *Then there exist a number* $\delta > 0$ *and an integer* $n_0 \ge 1$ *such that for each integer* $n > n_0$ *and each finite sequence* $\{x_i\}_{i=0}^n \subset X$ *which satisfies*

$$\rho_1(x_0, x_1) \le M$$

and

$$\rho(x_{i+1}, T(x_i)) \le \delta, \ i = 0, \dots, n-1,$$

the inequality

$$\rho(x_i, x_*) \le \epsilon$$

holds for all $i = n_0, \dots, n-1$.

1.10 Set-Valued Mappings

In Chap. 12 we study iteration processes associated with a given set-valued mapping which generate its approximate fixed point and in some cases converge to a fixed point of the mapping. We consider these processes under the presence of computational errors which always present in practice. Our results are proved for various classes of set-valued mappings including strict contractions and the Mizoguchi-Takahashi class of mappings. We consider spaces of nonexpansive mappings, acting on a closed subset of a complete metric space, which is equipped with the topology of uniform convergence on bounded sets. We show the existence of an open and everywhere dense subset in this space such that all its elements possess an approximate fixed point, which is stable under small perturbations. Here we present some of results of Chap. 12.

Let (X, ρ) be a complete metric space. For each $x \in X$ and each $r > 0$, set

$$B(x,r) = \{y \in X : \rho(x, y) \le r\}, \; B^0(x,r) = \{y \in X : \rho(x, y) < r\}.$$

For each $x \in X$ and each $A \subset X$, set

$$\rho(x, A) = \inf\{\rho(x, y) : \; y \in A\}.$$

For each pair of nonempty sets $A, B \subset X$, set

$$H(A, B) = \max\{\sup\{\rho(x, B) : \; x \in A\}, \; \sup\{\rho(y, A) : \; y \in B\}\}.$$

Assume that $T : X \to 2^X \setminus \{\emptyset\}$, $T(x)$ is closed for each $x \in X$, $c \in [0, 1)$ and that for each $x, y \in X$,

$$H(T(x), T(y)) \le c\rho(x, y).$$

We are interested to solve the problem

$$\text{Find } z \in X \text{ such that } z \in T(z).$$

In practice we can obtain only an approximate solution of this problem $z \in X$ such that $\rho(z, T(z))$ is small.

More precisely, in order to meet this goal, we use the following algorithm.

Initialization: select an arbitrary point $x_0 \in X$ and a small positive constant δ.

Iterative step: given a current iteration point x_n, calculate the next iteration point x_{n+1} such that

$$B(x_{n+1}, \delta) \cap \{\xi \in T(x_n) :$$

$$\rho(\xi, x_n) \le \delta + \rho(x_n, T(x_n))\} \ne \emptyset.$$

Clearly, at the each iterative step, x_{n+1} is an approximate solution of the problem

$$\rho(\xi, x_n) \to \min$$

$$\text{such that } \xi \in T(x_n).$$

Since the space X is not compact a solution of the problem above does not exist in general.

In Chap. 12 we prove the following results obtained in [62] where iterative schemes for approximating fixed points of closed-valued strict contractions in metric spaces were considered.

Theorem 1.19 *Assume that* $x_0 \in X$, $\{\epsilon_i\}_{i=0}^{\infty} \subset (0, \infty)$, $\sum_{i=0}^{\infty} \epsilon_i < \infty$ *and that for each integer* $i \geq 0$,

$$x_{i+1} \in T(x_i), \; \rho(x_i, x_{i+1}) \leq \rho(x_i, T(x_i)) + \epsilon_i.$$

Then $\{x_i\}_{i=0}^{\infty}$ *converges to a fixed point of* T.

Theorem 1.20 *Let* $\epsilon > 0$ *be given and let*

$$\delta \in (0, 4^{-1}(1-c)\epsilon].$$

Then if $x \in X$ *and* $\rho(x, T(x)) < \delta$, *then there is* $\bar{x} \in X$ *such that* $\bar{x} \in T(\bar{x})$ *and* $\rho(x, \bar{x}) \leq \epsilon$.

Theorem 1.21 *Fix* $\theta \in X$. *Let* $\epsilon \in (0, 1]$ *and* $M > 0$ *be given, a positive number* δ *satisfy*

$$\delta < 2^{-1}\epsilon(1-c]$$

and an integer $n_0 \geq 1$ *satisfy*

$$c^{n_0}(2M + 1 + \rho(\theta, T(\theta)) < \epsilon/2.$$

Then for each sequence $\{x_i\}_{i=0}^{\infty} \subset X$ *such that* $\rho(x_0, \theta) \leq M$ *and such that for each integer* $n \geq 0$,

$$x_{n+1} \in T(x_n) \text{ and } \rho(x_{n+1}, x_n) \leq \delta + \rho(x_n, T(x_n)),$$

we have

$$\rho(x_{n+1}, x_n) < \epsilon \text{ for all integers } n \geq n_0.$$

Theorems 1.21 and 1.22 imply the following additional result.

Theorem 1.22 *Let* $\epsilon \in (0, 1)$ *and* $M > 0$ *be given, a positive number* δ *satisfy*

$$\delta \leq 8^{-1}(1-c)^2\epsilon$$

and a natural number n *satisfy*

$$c^{n_0}(2M + 1 + \rho(\theta, T(\theta)) < 8^{-1}(1-c)\epsilon.$$

Then for each sequence $\{x_i\}_{i=0}^{\infty} \subset X$ *which satisfies*

$$\rho(x_0, \theta) \le M,\ x_{n+1} \in T(x_n) \text{ and } \rho(x_n, x_{n+1}) \le \rho(x_n, T(x_n)) + \delta$$

for all integers $n \ge 0$*, for each integer* $n \ge n_0$*, there is a point* $y \in X$ *such that* $y \in T(y)$ *and* $\rho(y, x_n) < \epsilon$*.*

The following example shows that Theorem 12.22 cannot be improved in the sense that the fixed point y, the existence of which is guaranteed by the theorem, is not, in general, the same for all integers $n \ge n_0$.

Example 1.23 Let $X = [0, 1]$, $\rho(x, y) = |x - y|$ and $T(x) = [0, 1]$ for all $x \in [0, 1]$. Let $\delta > 0$ be given. Choose a natural number k such that $1/k < \delta$. Put

$$x_0 = 0,\ x_i = i/k,\ i = 1, \dots, k,$$

$$x_{i+k} = 1 - i/k,\ i = 0, \dots, k,$$

and for all integers $p \ge 0$ and any $i \in \{0, \dots, 2k\}$, put

$$x_{2pk+i} = x_i.$$

Then $\{x_i\}_{i=0}^{\infty} \subset X$ and for any integer $i \ge 0$, we have

$$x_{i+1} \in T(x_i) \text{ and } |x_i - x_{i+1}| \le k^{-1} < \delta.$$

On the other hand, for all $x \in X$ and any integer $p \ge 0$,

$$\max\{|x - x_i| : i = 2kp, \dots, 2pk + 2k\} \ge 1/2.$$

Note that in the results stated above at any iterative step t, we have $x_{t+1} \in T(x_t)$. In Chap. 12, we generalize them in the case when x_{t+1} is just sufficiently close to $T(x_t)$. These generalizations are interesting and important since computational errors always occur in practice.

1.11 Nonexpansive Mappings in Spaces with Graphs

In Chap. 13 we study nonexpansive mappings in spaces with graphs. We consider single-valued operators as well as set-valued maps. Our goals are to show the existence of a fixed point and construct iteration processes associated with a given mapping which generate its approximate fixed point and in some cases converge to a fixed point of the mapping. Our results are proved for various classes of mappings of contractive type. Here we discuss some of these results.

Let (X, ρ) be a complete metric space and let G be a (directed) graph. Let $V(G)$ be the set of its vertices and let $E(G)$ be the set of its edges. We identify the graph G with the pair $(V(G), E(G))$.

Denote by $\mathcal{M}_{ne}$ the set of all mappings $T : X \to X$ such that for each $x, y \in X$ satisfying $(x, y) \in E(G)$, we have

$$(T(x), T(y)) \in E(G) \text{ and } \rho(T(x), T(y)) \le \rho(x, y).$$

A mapping $T \in \mathcal{M}_{ne}$ is called G-nonexpansive. If $T \in \mathcal{M}_{ne}$, $\alpha \in (0, 1)$ and for each $x, y \in X$ satisfying $(x, y) \in E(G)$, we have

$$\rho(T(x), T(y)) \le \alpha\rho(x, y),$$

then T is called a G-strict contraction.

Fix $\theta \in X$. For each $x \in X$ and each $r > 0$, set

$$B_\rho(x, r) := \{y \in X : \ \rho(x, y) \le r\}.$$

We may assume without loss of generality that if $x, y \in X$ satisfies $(x, y) \in E(G)$, then

$$(y, x) \in E(G).$$

We assume that the following assumption holds.

(P) For each $x, y \in X$, there exist an integer $q \ge 1$ and points $x_i \in X$, $i = 0, \dots, q$, such that

$$x_0 = x, \ x_q = y, \ (x_i, x_{i+1}) \in E(G), \ i = 0, \dots, q - 1.$$

Thus, $V(G) = X$ and the graph G is connected.

Let $T \in \mathcal{M}_{ne}$ be a G-strict contraction. It is known [92] that under certain mild assumptions, the mapping T has a unique fixed point which attracts all the iterates of T. We begin Chap. 13 with presenting the results of [174] where it was provided a very simple proof of this fact by using a certain metric on X which is defined below. It [174] it was also shown uniform convergence of the iterates of T on bounded subsets of X and that this convergence is stable under small perturbations of these iterates.

For each $x, y \in X$, define

$$\rho_1(x, y) := \inf\{\sum_{i=0}^{q-1} \rho(x_i, x_{i+1}) : \ q \ge 1 \text{ is an integer},$$

$$x_i \in X, \ i = 0, \dots, q, \ x_0 = x, \ x_q = y, \ (x_i, x_{i+1}) \in E(G), \ i = 0, \dots, q - 1\}.$$

It is easy to see that for each $x, y, z \in X$, $\rho_1(x, y)$ is finite,

$$\rho_1(x, y) \geq \rho(x, y),$$

$$\rho_1(x, y) = \rho_1(y, x),$$

$$\rho_1(x, z) \leq \rho_1(x, y) + \rho_1(y, z),$$

and if $\rho_1(x, y) = 0$, then $x = y$. However, ρ_1 is a metric only if $(x, x) \in E(G)$ for all $x \in X$. This pseudometric ρ_1 plays an important role in our study.

For each $x \in X$ and each $r > 0$, we set

$$B_{\rho_1}(x, r) := \{y \in X : \ \rho_1(x, y) \leq r\}.$$

In Chap. 13 we prove the following results obtained in [174].

Theorem 1.24 *Let $T \in \mathcal{M}_{ne}$ be a G-strict contraction, $\alpha \in (0, 1)$ and assume that for each $x, y \in X$ satisfying $(x, y) \in E(G)$,*

$$(T(x), T(y)) \in E(G) \text{ and } \rho(T(x), T(y)) \leq \alpha\rho(x, y).$$

Then for each $x, y \in X$, we have

$$\rho_1(T(x), T(y)) \leq \alpha\rho_1(x, y).$$

If T is continuous as a self-mapping of (X, ρ), then there exists a unique point $x_* \in X$ satisfying $T(x_*) = x_*$ and for each $x \in X$,

$$\lim_{i\to\infty} T^i(x) = x_* \text{ in } (X, \rho).$$

Theorem 1.25 *Let all the assumptions of Theorem 1.24 hold, let the mapping T be continuous as a self-mapping of (X, ρ) and let $x_* \in X$ be as guaranteed by Theorem 1.24 and satisfy*

$$x_* = T(x_*).$$

Suppose that the following assumption holds.

(A1) For each $M_0 > 0$, there exists $M_1 > 0$ such that for each point $x \in B_\rho(\theta, M_0)$, we have

$$\rho_1(x, \theta) \leq M_1.$$

Then $T^i(x) \to x_*$ as $i \to \infty$ in (X, ρ), uniformly on all bounded subsets of (X, ρ).

Theorem 1.26 *Let all the assumptions of Theorem 1.25 hold and let $x_* \in X$ be as guaranteed by Theorem 1.24 and satisfy*

$$x_* = T(x_*).$$

Suppose that T is uniformly continuous and bounded on bounded sets as a self-mapping of (X, ρ) and let $\epsilon, M > 0$ be given. Then there exist $\delta > 0$ and a natural number n_0 such that for each integer $n \geq n_0$ and every finite sequence $\{x_i\}_{i=0}^n \subset X$ satisfying

$$x_0 \in B_\rho(\theta, M)$$

and

$$\rho(x_{i+1}, T(x_i)) \leq \delta, \ i = 0, \dots, n-1,$$

the inequality $\rho(x_i, x_*) \leq \epsilon$ holds for all $i = n_0, \dots, n$.

Denote by $\mathcal{M}$ the set of all mappings $S \in \mathcal{M}_{ne}$ which are uniformly continuous and bounded on bounded sets as self-mappings of (X, ρ). We equip the space $\mathcal{M}$ with the uniformity which has the base

$$\mathcal{E}(\epsilon, M) = \{(S_1, S_2) \in \mathcal{M} \times \mathcal{M} : \ \rho(S_1(x), S_2(x)) \leq \epsilon \text{ for all } x \in B(\theta, M)\},$$

where $M, \epsilon > 0$. It is not difficult to see that the uniform space $\mathcal{M}$ is metrizable and complete (by a metric d).

Theorem 1.27 *Let all the assumptions of Theorem 1.25 hold and let $x_* \in X$ be as guaranteed by Theorem 1.24 and satisfy*

$$x_* = T(x_*).$$

Suppose that $T \in \mathcal{M}$ and that $\epsilon, M > 0$ are given. Then there exist a neighborhood $\mathcal{U}$ of T in $\mathcal{M}$ and a natural number n_0 such that for each $C \in \mathcal{U}$, each $x \in B_\rho(\theta, M)$ and each integer $n \geq n_0$, we have

$$\rho(C^n(x), x_*) \leq \epsilon.$$

Chapter 2
Asymptotic Regularity for Iterations of Nonexpansive Mappings

In this chapter, we study asymptotic regularity for Halpern-type iterations and for iterations of alternative iterative methods generated by a nonexpansive self-mapping of a star-shaped set in a metric space with a hyperbolic structure. Linear rates of this asymptotic regularity are computed. Such results are known in the literature under the assumption that the nonexpansive mapping possesses a fixed point. In this chapter, we prove our results under a weaker assumption assuming only the existence of a bounded orbit of the nonexpansive mapping. It is shown that in the space of nonexpansive mappings equipped with some natural topology, there exists an open everywhere dense set such that for each its element all orbits are bounded. We study asymptotic regularity for single-valued and set-valued mapping and also under the presence of computational errors.

2.1 Preliminaries

We say that a sequence $\{a_n\}_{n=0}^{\infty}$ in a metric space (X, d) is asymptotically regular if $\lim_{n\to\infty} d(a_n, a_{n+1}) = 0$. If, furthermore, T is a self-mapping of X, then $\{a_n\}_{n=0}^{\infty}$ is said to be T-asymptotically regular if $\lim_{n\to\infty} d(a_n, T(a_n)) = 0$.

Recently, H. Cheval and L. Leustean [56] computed linear rates of asymptotic regularity for a viscosity version of Halpern-type iterations generated by a nonexpansive self-mapping of a metric space assuming that it has a fixed point. In our recent work [232], we showed that their result is valid without this assumption. It is enough to assume the existence of a bounded orbit of the nonexpansive mapping. It should be mentioned that the viscosity version of Halpern-type iterations was introduced by Xu [204]. The results of Sect. 2.2 were obtained in [232] under the assumption that the mapping acts on a convex subset of the metric space. In this chapter, we show that these results are true when the subset is merely star-shaped.

A. J. Zaslavski, *Approximate Fixed Points of Nonexpansive Mappings*,
Developments in Mathematics 80, https://doi.org/10.1007/978-3-031-70710-0_2

The results of Sect. 2.4 were obtained in [232] too. All other results of the chapter are new.

The proofs of the asymptotic regularity results of [56] are based on two estimations for $d(x_n, x_{n+1})$, where $\{x_n\}_{n=0}^{\infty}$ is the sequence of iterates. For the first estimation, the existence of the fixed point is not needed, while for the second one it was used. Reading the work [56], the author realized that for the second estimation this assumption can be replaced by a weaker assumption. It is enough to assume the existence of a bounded orbit of the mapping. Of course in this case all the orbits of the mapping are bounded. The realization of this idea was done in [232]. The results of [232] essentially enlarged the class of operators possessing the asymptotic regularity property. This work became a starting point of our research on the asymptotic regularity, and its results are presented in this chapter. In [232] and here we study certain iterative processes associated with a given fixed point problem. The goal is to obtain analogs of the results obtained in our present research for other types of iterative processes including processes associated with set-valued operators and processes when computational errors are taken into account.

A W-space is a structure (X, d, W) where (X, d) is a metric space and $W : X \times X \times [0, 1] \to X$. We think of $W(x, y, \lambda)$ as a convex combinations of the points $x, y \in X$ with the coefficients $1 - \lambda,\ \lambda$ and use the notation

$$W(x, y, \lambda) := (1 - \lambda)x \oplus \lambda y,\ \ x, y \in X,\ \ \lambda \in [0, 1].$$

A W-space (X, d, W) is called a W-hyperbolic space [110] if for all $x, y, w, z \in X$ and all $\lambda, \tilde{\lambda} \in [0, 1]$ the following properties hold:

(W1) $d(z, (1 - \lambda)x \oplus \lambda y) \le (1 - \lambda)d(z, x) + \lambda d(z, y)$;
(W2) $d((1 - \lambda)x \oplus \lambda y), (1 - \tilde{\lambda})x \oplus \tilde{\lambda} y) \le |\lambda - \tilde{\lambda}| d(x, y)$;
(W3) $(1 - \lambda)x \oplus \lambda y = \lambda y \oplus (1 - \lambda)x$;
(W4) $d((1 - \lambda)x \oplus \lambda z, (1 - \lambda)y \oplus \lambda w) \le (1 - \lambda)d(x, y) + \lambda d(z, w)$.

It is obvious that convex subsets of normed spaces are W-hyperbolic spaces. Other examples of W-hyperbolic spaces are hyperbolic spaces [168], Busemann spaces [10], and $CAT(0)$ spaces [111].

In the sequel, we assume that (X, d, W) is a W-hyperbolic space which is denoted by X for simplicity.

A nonempty set $C \subset X$ is convex if for each $x, y \in C$ and each $\lambda \in [0, 1]$,

$$(1 - \lambda)x \oplus \lambda y \in C.$$

A nonempty set $C \subset X$ is called star-shaped and $x_* \in C$ is its star if for each $x \in C$ and each $\lambda \in [0, 1]$,

$$(1 - \lambda)x \oplus \lambda x_* \in C.$$

If C is star-shaped, we denote by $\mathrm{St}(C)$ the sets of all stars of C.

For each $x \in X$, each nonempty set $E \subset X$ and each $\lambda \in [0, 1]$, set

$$\lambda x \oplus (1-\lambda)E = \{\lambda x \oplus (1-\lambda)y : \ y \in E\}.$$

The following result is well-known in the literature [55, 187].

Proposition 2.1 *Let $x, y, w, z \in X$ and let $\lambda, \tilde{\lambda} \in [0, 1]$. Then*

$$d(x, (1-\lambda)x \oplus \lambda y) \le \lambda d(x, y), \ d(y, (1-\lambda)x \oplus \lambda y) \le (1-\lambda)d(x, y) \quad (2.1)$$

and

$$d((1-\lambda)x \oplus \lambda z, (1-\tilde{\lambda})y \oplus \tilde{\lambda}w)$$

$$\le (1-\lambda)d(x, y) + \lambda d(z, w) + |\lambda - \tilde{\lambda}|d(y, w). \quad (2.2)$$

Proof Equation (2.1) follows from (W1). Let us prove (2.2). Properties (W4) and (W2) imply that

$$d((1-\lambda)x \oplus \lambda z, (1-\tilde{\lambda})y \oplus \tilde{\lambda}w)$$

$$\le d((1-\lambda)x \oplus \lambda z, (1-\lambda)y \oplus \lambda w)$$

$$+ d((1-\lambda)y \oplus \lambda w, (1-\tilde{\lambda})y \oplus \tilde{\lambda}w)$$

$$\le (1-\lambda)d(x, y) + \lambda d(z, w) + |\lambda - \tilde{\lambda}|d(y, w).$$

Thus Eq. (2.2) holds and Proposition 2.1 is proved.

In the sequel we use the following lemma obtained in [121] which is a slight reformulation of Lemma 3 of [181]. It is proved by induction.

Lemma 2.2 *Let $L > 0$, $J \ge N \ge 2$, $\gamma \in (0, 1]$, $\{c_n\}_{n=0}^{\infty} \subset (0, L]$ and let*

$$a_n = N(\gamma(n+J))^{-1}, \ n = 0, 1, \ldots.$$

Assume that $\{s_n\}_{n=0}^{\infty} \subset [0, \infty)$, $s_0 \le L$ and that for each integer $n \ge 0$,

$$s_{n+1} \le (1 - \gamma a_{n+1})s_n + (a_n - a_{n+1})c_n.$$

Then for each integer $n \ge 0$,

$$s_n \le JL(\gamma(n+J))^{-1}.$$

Proof We prove the lemma by induction. Clearly,

$$S_0 \le JL(\gamma(0+J))^{-1}.$$

Let $n \geq 0$ be an integer and

$$s_n \leq JL(\gamma(n+J))^{-1}.$$

In view of the relations above,

$$s_{n+1} \leq (1-\gamma a_{n+1})JL(\gamma(n+J))^{-1} + (a_n - a_{n+1})c_n$$

$$\leq (1-N(n+1+J)^{-1})JL(\gamma(n+J))^{-1} + L(N(\gamma(n+J))^{-1} - N(\gamma(n+1+J))^{-1})$$

$$= (n+1+JN)LJ(\gamma(n+1+J)(n+J))^{-1} + NL(\gamma(n+1+J)(n+J))^{-1}$$

$$\leq JL(n+1+J-N)(\gamma(n+1+j)(n+J))^{-1} + JL(\gamma(n+1+J)(n+J))^{-1}$$

$$\leq JL(n+2+J-N)(\gamma(n+1+j)(n+J))^{-1} \leq JL(\gamma(n+J+1))^{-1}.$$

This completes the proof of Lemma 2.2.

2.2 Asymptotic Regularity for a Viscosity Version of Halpern-Type Iterations

Assume that C is a nonempty star-shaped subset of X, a number $\rho \in [0, 1)$ is given, a mapping $T : C \to C$ satisfies the equation

$$d(T(x), T(y)) \leq d(x, y), \quad x, y \in C \tag{2.3}$$

and that a strict contraction $f : C \to \mathrm{St}(C)$ satisfies the equation

$$d(f(x), f(y)) \leq \rho d(x, y), \quad x, y \in C. \tag{2.4}$$

Recall that $T^0(x) = x$ for every point $x \in C$. Assume that a sequence $\{\alpha_n\}_{n=0}^{\infty} \subset (0, 1)$ is given,

$$x_0 = x \in C \tag{2.5}$$

and that for every nonnegative integer n,

$$x_{n+1} = (1-\alpha_n)f(x_n) \oplus \alpha_n T(x_n). \tag{2.6}$$

Since the inclusion $f(C) \subset \mathrm{St}(C)$ is true the sequence of iterates $\{x_n\}_{n=0}^{\infty}$ is well-defined. Let

$$\theta \in C \text{ and } M_0 > 0.$$

We study the asymptotic regularity of the sequence of iterates $\{x_n\}_{n=0}^{\infty}$. It was introduced in [204], in the setting of Banach spaces, and called as a viscosity version of the Halpern iteration. Note that if the mapping f is a constant, then we obtain the well-known Halpern iterations.

We use the following assumption.

(A) There exists a point $p \in C \cap B(\theta, M_0)$ such that for each nonnegative integer i, we have

$$d(T^i(p), \theta) \le M_0.$$

Proposition 2.3 *Assume that* $M_1 > 0$ *and that for every natural number* n *there exists a point*

$$p_n \in B(\theta, M_1) \cap C$$

for which

$$T^i(p_n) \in B(\theta, M_1), \ i = 0, \ldots, n. \tag{2.7}$$

Then for every point $v \in C$ *and every natural number* $n \ge 1$,

$$d(T^n(v), \theta) \le 2M_1 + d(v, p).$$

Proof Let $v \in C$ and a natural number $n \ge 1$ be given. There is a point $p_n \in C$ which satisfies (2.7). In view of (2.3) and (2.7), for every integer $i \in \{0, \ldots, n\}$, we have

$$d(T^i(v), T^i(p)) \le d(v, p),$$

$$d(T^i(v), \theta) \le d(T^i(v), T^i(p)) + d(T^i(p), \theta)$$

$$\le d(v, p) + M_1 \le d(v, \theta) + d(\theta, p) + M_1 \le d(v, \theta) + 2M_1.$$

This completes the proof of Proposition 2.3.

Proposition 2.4 *Assume that* $M_1 > 1$ *and that for every positive number* ϵ *there is a point*

$$p_\epsilon \in C \cap B(\theta, M_1 - 1)$$

for which

$$d(p_\epsilon, T(p_\epsilon))) \le \epsilon.$$

Then for every natural number integer n *there is a point*

$$p \in B(\theta, M_1) \cap C$$

such that

$$T^i(p) \in B(\theta, M_1), \ i = 0, \ldots, n.$$

Proof Let $n \geq 1$ be an integer. There exists a point

$$p \in C \cap B(\theta, M_1 - 1) \tag{2.8}$$

for which

$$d(p, T(p)) \leq 1/n. \tag{2.9}$$

In view of (2.3) and (2.9), for every $i \in \{0, \ldots, n-1\}$,

$$d(T^{i+1}(p), T^{i+2}(p)) \leq d(T^i(p), T^{i+1}(p)),$$

$$d(T^{i+1}(p), T^i(p)) \leq d(p, T(p)) \leq n^{-1}.$$

By (2.8) and the equation above, for every natural number $j \in \{1, \ldots, n\}$, we have

$$d(p, T^j(p)) \leq \sum_{i=0}^{j-1} d(T^{i+1}(p), T^i(p)) \leq jn^{-1} \leq 1$$

and

$$d(T^j(p), \theta) \leq d(T^j(p), p) + d(p, \theta) \leq M_1.$$

The completes the proof of Proposition 2.4.

Fix a number

$$M > (1-\rho)^{-1}(2M_0 + d(x_0, \theta) + 2 + 2d(f(\theta), \theta) + d(\theta, T(\theta))). \tag{2.10}$$

The next theorem is our first asymptotic regularity result which shows that our iterative process generates approximate fixed points of T.

Theorem 2.5 *Let (A) hold,* $J = 2\lfloor(1-\rho)^{-1}\rfloor + 1$ *and for every nonnegative integer* n,

$$\alpha_n = 1 - 2(1-\rho)^{-1}(n+J)^{-1}.$$

Then for every nonnegative integer r n,

$$d(x_n, x_{n+1}) \le 2MJ(1-\rho)^{-1}(n+J)^{-1}$$

and

$$d(x_n, T(x_n)) \le 2M(J+2)(1-\rho)^{-1}(n+J)^{-1}.$$

Note that in [232] this result was proved when the set C is convex.

We begin its proof with auxiliary results.

Lemma 2.6 *Let $n \ge 0$ be an integer. Then*

$$d(x_{n+2}, x_{n+1}) \le (1-(1-\rho)(1-\alpha_{n+1}))d(x_{n+1}, x_n)$$
$$+ |\alpha_{n+1} - \alpha_n| d(f(x_n), T(x_n)), \tag{2.11}$$

$$d(x_n, T(x_n)) \le d(x_n, x_{n+1}) + (1-\alpha_n)d(f(x_n), T(x_n)). \tag{2.12}$$

Proof It follows from Proposition 2.1 and Eqs. (2.2)–(2.4) and (2.6) that

$$d(x_{n+2}, x_{n+1})$$
$$= d((1-\alpha_{n+1})f(x_{n+1}) \oplus \alpha_{n+1}T(x_{n+1}), (1-\alpha_n)f(x_n) \oplus \alpha_n T(x_n))$$
$$\le (1-\alpha_{n+1})d(f(x_{n+1}), f(x_n)) + \alpha_{n+1}d(T(x_{n+1}), T(x_n))$$
$$+ |\alpha_{n+1} - \alpha_n| d(f(x_n), T(x_n))$$
$$\le \rho(1-\alpha_{n+1})d(x_{n+1}, x_n) + \alpha_{n+1}d(x_{n+1}, x_n)$$
$$+ |\alpha_{n+1} - \alpha_n| d(f(x_n), T(x_n))$$

and Eq. (2.11) is true.

It follows from Proposition 2.1 and Eqs. (2.3). (2.11) that

$$d(x_n, T(x_n)) \le d(x_n, x_{n+1}) + d(x_{n+1}, T(x_n))$$
$$= d(x_n, x_{n+1}) + d((1-\alpha_n)f(x_n) \oplus \alpha_n T(x_n), T(x_n))$$
$$\le d(x_n, x_{n+1}) + (1-\alpha_n)d(f(x_n), T(x_n))$$

and Eq. (2.12) is true. This completes the proof of Lemma 2.6.

Lemma 2.7 *Let assumption (A) hold and let a point p be as guaranteed by assumption (A). Then for every nonnegative integer n,*

$$d(x_n, T^n(p)) \le M \tag{2.13}$$

and

$$d(x_n, \theta) \le M + M_0. \tag{2.14}$$

Proof We prove Eq. (2.13) by induction. In view of assumption (A) and Eq. (2.10), we have

$$d(x_0, p) \le M_0 + d(\theta, x_0) < M.$$

Assume that Eq. (2.13) is true for a nonnegative integer n. It follows from Proposition 2.1 and Eqs. (2.4), (2.6), (2.9), (2.10), and (2.13) that

$$d(x_{n+1}, T^{n+1}(p)) \le d((1-\alpha_n)f(x_n) \oplus \alpha_n T(x_n), T^{n+1}(p))$$

$$\le (1-\alpha_n)d(f(x_n), T^{n+1}(p)) + \alpha_n d(T(x_n), T^{n+1}(p))$$

$$\le (1-\alpha_n)d(f(x_n), f(T^n(p))) + (1-\alpha_n)d(f(T^n(p)), T^{n+1}(p)) + \alpha_n d(x_n, T^n(p))$$

$$\le (1-\alpha_n)\rho d(x_n, T^n(p)) + \alpha_n d(x_n, T^n(p))$$

$$+ (1-\alpha_n)(d(f(T^n(p)), f(\theta)) + d(f(\theta), \theta) + d(\theta, T^{n+1}(p)))$$

$$\le (1-(1-\alpha_n)(1-\rho))d(x_n, T^n(p)) + (1-\alpha_n)(2M_0 + d(f(\theta), \theta))$$

$$\le M - (1-\alpha_n)(1-\rho)M + (1-\alpha_n)(2M_0 + d(f(\theta), \theta))$$

$$\le M - (1-\alpha_n)((1-\rho)M - 2M_0 - d(f(\theta), \theta)) \le M.$$

Thus we showed that Eq. (2.13) is true for an integer $n+1$ too. Therefore by induction it holds for every nonnegative integer n.

Assumption (A) and Eq. (2.13) imply that Eq. (2.14) is valid for every nonnegative integer n. This completes the proof of Lemma 2.7.

Proposition 2.8 *Let assumption (A) be true and a point p be as guaranteed by assumption (A). Then for every nonnegative integer n, we have*

$$d(x_{n+2}, x_{n+1}) \le (1-(1-\rho)(1-\alpha_{n+1}))d(x_{n+1}, x_n) + 2M|\alpha_{n+1} - \alpha_n|, \tag{2.15}$$

$$d(f(x_n), T(x_n)) \le 2M, \tag{2.16}$$

$$d(x_n, T(x_n)) \le d(x_n, x_{n+1}) + 2M(1-\alpha_n).$$

Proof Let n be a nonnegative integer. It follows from Lemmas 2.6 and 2.7, assumption (A) and Eqs. (2.3), (2.4), and (2.10) that

$$
\begin{aligned}
d(x_{n+2}, x_{n+1}) &\le (1-(1-\rho)(1-\alpha_{n+1}))d(x_{n+1}, x_n) + |\alpha_{n+1}-\alpha_n| d(f(x_n), T(x_n)) \\
&\le (1-(1-\rho)(1-\alpha_{n+1}))d(x_{n+1}, x_n) \\
&\quad + |\alpha_{n+1}-\alpha_n|(d(f(x_n), f(T^n(p))) + d(f(T^n(p)), f(\theta)) \\
&\quad + d(f(\theta), \theta) + d(\theta, T^{n+1}(p)) + d(T^{n+1}(p), T(x_n))) \\
&\le (1-(1-\rho)(1-\alpha_{n+1}))d(x_{n+1}, x_n) \\
&\quad + |\alpha_{n+1}-\alpha_n|(\rho M + \rho M_0 + d(f(\theta), \theta) + M_0 + M) \\
&\le (1-(1-\rho)(1-\alpha_{n+1}))d(x_{n+1}, x_n) + 2|\alpha_{n+1}-\alpha_n| M
\end{aligned}
$$

and Eq. (2.15) is true for every nonnegative integer n.

Let n be a nonnegative integer. It follows from Lemma 2.7 and relations (2.3), (2.4), and (2.10) that

$$
\begin{aligned}
d(f(x_n), T(x_n)) &\le d(f(x_n), f(\theta)) + d(f(\theta), \theta) + d(\theta, T(\theta)) + d(T(\theta), T(x_n)) \\
&\le (1+\rho)d(x_n, \theta) + d(f(\theta), \theta) + d(\theta, T(\theta)) \\
&\le (1+\rho)(M + M_0) + d(f(\theta), \theta) + d(\theta, T(\theta)) \le 2M.
\end{aligned}
$$

Hence Eq. (2.16) is true for every nonnegative integer n. By Lemma 2.6 and relations (2.2) and (2.16), for every nonnegative integer n,

$$
\begin{aligned}
d(x_n, T(x_n)) &\le d(x_n, x_{n+1}) + (1-\alpha_n)d(f(x_n), T(x_n)) \\
&\le d(x_n, x_{n+1}) + 2M(1-\alpha_n).
\end{aligned}
$$

This completes the proof of Proposition 2.8.

Proof of Theorem 2.5

Recall that

$$J = 2\lfloor (1-\rho)^{-1} \rfloor + 1$$

and for every nonnegative integer n, we have

$$\alpha_n = 1 - 2(1-\rho)^{-1}(n+J)^{-1}.$$

For every nonnegative integer n put

$$s_n = d(x_n, x_{n+1}),\ a_n = 1 - \alpha_n,\ c_n = 2M,\ \gamma = 1 - \rho,\ L = 2M,\ N = 2.$$

Lemma 2.7 implies that

$$s_0 = d(x_0, x_1) \leq 2M = L.$$

It follows from Lemma 2.1 and Proposition 2.8 that for every nonnegative integer n, we have

$$d(x_n, x_{n+1}) \leq 2MJ(1 - \rho)^{-1}(n + J)^{-1}$$

and

$$d(x_n, T(x_n)) \leq 2M(J + 2)(1 - \rho)^{-1}(n + J)^{-1}.$$

Theorem 2.5 is proved.

2.3 Alternative Iterative Method

We use the notation, definitions, and assumptions introduced in Sects. 2.1 and 2.2. In particular, we use assumption (A). Let $M_0 > 0$ and a number M satisfy (2.10). Assume that $\{\alpha_n\}_{n=0}^{\infty} \subset (0, 1)$,

$$x_0 = x \in C \tag{2.17}$$

and for each integer $n \geq 0$

$$x_{n+1} = T((1 - \alpha_n)f(x_n) \oplus \alpha_n x_n). \tag{2.18}$$

This iteration was studied in [57] in the setting of Banach spaces. If the mapping f is a constant one obtains, as a particular case, the alternative regularization method, introduced in [207]. We will show that Theorem 2.5 holds in this case too.

Lemma 2.9 *Let (A) hold and p be as guaranteed by (A). Then for each integer $n \geq 0$,*

$$d(x_n, T^n(p)) \leq M, \tag{2.19}$$

$$d(x_n, \theta) \leq M + M_0,$$

$$d(f(x_n), x_n) \leq 2M.$$

Proof In view of (A) and (2.10),

$$d(x_0, p) \le d(x_0, \theta) + d(\theta, p) \le M_0 + d(\theta, x_0) < M.$$

Assume that $n \ge 0$ is an integer and (2.19) holds. By (A), Proposition 2.1 and (2.3), (2.4), (2.10), and (2.18)

$$d(x_{n+1}, T^{n+1}(p)) \le d(T((1-\alpha_n)f(x_n) \oplus \alpha_n x_n), T^{n+1}(p))$$

$$\le d((1-\alpha_n)f(x_n) \oplus \alpha_n x_n, T^n(p))$$

$$\le (1-\alpha_n)d(f(x_n), T^n(p)) + \alpha_n d(x_n, T^n(p))$$

$$\le (1-\alpha_n)d(f(x_n), f(T^n(p))) + (1-\alpha_n)d(f(T^n(p)), T^n(p)) + \alpha_n d(x_n, T^n(p))$$

$$\le (1-\alpha_n)\rho d(x_n, T^n(p)) + \alpha_n d(x_n, T^n(p))$$

$$+ (1-\alpha_n)(d(f(T^n(p)), f(\theta)) + d(f(\theta), \theta) + d(\theta, T^n(p)))$$

$$\le (1-(1-\alpha_n)(1-\rho))d(x_n, T^n(p))$$

$$+ (1-\alpha_n)(\rho d(T^n(p), \theta) + d(f(\theta), \theta) + d(\theta, T^p(p)))$$

$$\le (1-(1-\alpha_n)(1-\rho))M + (1-\alpha_n)(\rho M_0 + d(f(\theta), \theta) + M_0)$$

$$= M - (1-\alpha_n)((1-\rho)M - \rho M_0 - d(f(\theta), \theta) - M_0) \le M.$$

Thus, by induction, we showed that (2.19) holds for all integers $n \ge 0$. By (A) and (2.19), for each integer $n \ge 0$,

$$d(x_n, \theta) \le d(x_n, T^n(p)) + d(T^n(p), \theta) \le M + M_0.$$

By the relation above and (2.10),

$$d(f(x_n), x_n) \le d(f(x_n), f(\theta)) + d(f(\theta), \theta) + d(\theta, x_n)$$

$$\le \rho d(x_n, \theta) + d(\theta, x_n) + d(f(\theta), \theta)$$

$$\le (\rho + 1)(M_0 + M) + d(f(\theta), \theta) \le 2M.$$

Lemma 2.9 is proved.

Lemma 2.10 *For each integer* $n \ge 0$,

$$d(x_{n+2}, x_{n+1}) \le (1-(1-\rho)(1-\alpha_{n+1}))d(x_{n+1}, x_n) + |\alpha_{n+1} - \alpha_n| d(f(x_n), x_n),$$

$$d(x_n, T(x_n)) \le d(x_n, x_{n+1}) + (1 - \alpha_n)d(f(x_n), x_n).$$

Proof Let $n \ge 0$ be an integer. Proposition 2.1 and (2.3), (2.4), and (2.18) imply that

$$d(x_{n+2}, x_{n+1})$$

$$= d(T((1 - \alpha_{n+1})f(x_{n+1}) \oplus \alpha_{n+1}x_{n+1}), T((1 - \alpha_n)f(x_n) \oplus \alpha_n x_n))$$

$$d((1 - \alpha_{n+1})f(x_{n+1}) \oplus \alpha_{n+1}x_{n+1}, (1 - \alpha_n)f(x_n) \oplus \alpha_n x_n)$$

$$\le (1 - \alpha_{n+1})d(f(x_{n+1})f(x_n)) + \alpha_{n+1}d(x_{n+1}, x_n)$$

$$+ |\alpha_{n+1} - \alpha_n|d(f(x_n), x_n)$$

$$\le \rho(1 - \alpha_{n+1})d(x_{n+1}, x_n) + \alpha_{n+1}d(x_{n+1}, x_n) + |\alpha_{n+1} - \alpha_n|d(f(x_n), x_n)$$

$$= (1 - (1 - \rho)(1 - \alpha_{n+1}))d(x_{n+1}, x_n) + |\alpha_{n+1} - \alpha_n|d(f(x_n), x_n).$$

By (2.3) and (2.18) and Proposition 2.1,

$$d(x_n, T(x_n)) \le d(x_n, x_{n+1}) + d(x_{n+1}, T(x_n))$$

$$= d(x_n, x_{n+1}) + d(T((1 - \alpha_n)f(x_n) \oplus \alpha_n x_n), T(x_n))$$

$$\le d(x_n, x_{n+1}) + (1 - \alpha_n)d(f(x_n), x_n).$$

Lemma 2.10 is proved.

Lemmas 2.9 and 2.10 imply the following result

Proposition 2.11 *For each integer $n \ge 0$,*

$$d(x_{n+2}, x_{n+1}) \le (1 - (1 - \rho)(1 - \alpha_{n+1}))d(x_{n+1}, x_n) + 2M|\alpha_{n+1} - \alpha_n|,$$

$$d(x_n, T(x_n)) \le d(x_n, x_{n+1}) + 2M(1 - \alpha_n).$$

The next theorem is our second asymptotic regularity result which shows that our iterative process generates approximate fixed points of T.

Theorem 2.12 *Let (A) hold, $J = 2\lfloor(1 - \rho)^{-1}\rfloor + 1$ and for each integer $n \ge 0$,*

$$\alpha_n = 1 - 2(1 - \rho)^{-1}(n + J)^{-1}.$$

Then for each integer $n \ge 0$,

$$d(x_n, x_{n+1}) \le 2MJ(1 - \rho)^{-1}(n + J)^{-1}$$

and

$$d(x_n, T(x_n)) \le 2M(J+2)(1-\rho)^{-1}(n+J)^{-1}.$$

Proof For each integer $n \ge 0$ set

$$s_n = d(x_n, x_{n+1}),\ a_n = 1-\alpha_n,\ c_n = 2M,\ \gamma = 1-\rho,\ L = 2M,\ N = 2.$$

By Lemma 2.9 and (A),

$$s_0 = d(x_0, x_1) \le 2M = L.$$

Lemma 2.2 and Proposition 2.11 imply Theorem 2.12.

Arguing as in [120] Theorem 2.12 can be deduced from Theorem 2.5. Here we presented its direct proof.

2.4 Generic Results

The asymptotic regularity results of the previous section were proved under assumption (A). In this section, we show that assumption (A) holds for most nonexpansive mappings.

We assume that

$$\theta \in \mathrm{St}(C).$$

Evidently, assumption (A) holds if the set C is bounded.

Proposition 2.13 *Assume that the set C is bounded. Then for every positive number ϵ, there exists a point $p_\epsilon \in C$ for which*

$$d(p_\epsilon, T(p_\epsilon)) \le \epsilon.$$

Proof Set

$$D = \sup\{d(x,y) : x, y \in C\}.$$

Let a number $\gamma \in (0,1)$ be given. For every point $x \in C$ define

$$T_\gamma(x) = (1-\gamma)T(x) \oplus \gamma\theta,\ x \in C.$$

It is clear that for every pair of points $x, y \in C$, we have

$$d(T_\gamma(x), T_\gamma(y)) \le (1-\gamma)d(x,y).$$

Then there exists a unique point $x_{A,\gamma} \in C$ which satisfies

$$T_\gamma(x_{T\gamma}) = x_{T\gamma}.$$

Hence

$$x_{A\gamma} = (1-\gamma)T(x_{T\gamma}) \oplus \gamma\theta.$$

By Proposition 2.1, we have

$$d(x_{T\gamma}, T(x_{T\gamma})) \le \gamma d(\theta, T(x_{T,\gamma})) \le \gamma D \to 0$$

as $\gamma \to 0^+$. This completes the proof of Proposition 2.13.

Denote by $\mathcal{M}$ the set of all mappings $A : C \to C$ such that

$$d(A(x), A(y)) \le d(x, y)$$

for all $x, y \in C$ equipped with the uniformity determined by the base

$$\mathcal{U}(n) = \{(A, B) \in \mathcal{M} \times \mathcal{M} :$$

$$d(A(x), B(x)) \le n^{-1},\ x \in B(\theta, M) \cap C\},\ n = 1, 2, \ldots.$$

The space $\mathcal{M}$ with this uniformity is metrizable and complete [168].

Proposition 2.14 *There exists an open everywhere dense set $\mathcal{F} \subset \mathcal{M}$ such that for every mapping $T \in \mathcal{F}$, assumption (A) holds.*

Proof Let $A \in \mathcal{M}$ and $\gamma \in (0, 1)$ be given. Define

$$A_\gamma(x) = (1-\gamma)A(x) \oplus \gamma\theta,\ x \in C.$$

Evidently, for each pair of points $x, y \in C$, we have

$$d(A_\gamma(x), A_\gamma(y)) \le (1-\gamma)d(x, y)$$

and there exists a unique point $x_{A,\gamma} \in C$ which satisfies

$$A_\gamma(x_{A\gamma}) = x_{A\gamma}.$$

By Proposition 2.1, for every $x \in C$, we have

$$d(A(x), A_\gamma(x)) \le \gamma d(\theta, A(x)).$$

This implies that the set

$$\{A_\gamma : A \in \mathcal{M}, \ \gamma \in (0, 1)\}$$

is everywhere dense in $\mathcal{M}$.

Let $A \in \mathcal{M}$ and $\gamma \in (0, 1)$ be given. There exists an open neighborhood $V(A, \gamma)$ of the mapping A_γ in the space $\mathcal{M}$ such that

$$V(A, \gamma) \subset \{B \in \mathcal{M} : \ d(A_\gamma(x), B(x)) \le 4^{-1}\gamma$$

$$\text{for each } x \in B(x_{A,\gamma}, 1) \cap C\}.$$

Let

$$B \in V(A, \gamma).$$

Then for every point $x \in B(x_{A,\gamma}, 1)$, we have

$$d(x_{A,\gamma}, B(x)) \le d(x_{A\gamma}, A_\gamma(x)) + d(A_\gamma(x), B(x))$$

$$\le (1 - \gamma)d(x, x_{A,\gamma}) + \gamma/4 \le 1 - \gamma + \gamma/4 < 1.$$

Now it is easy to see that assumption (A) holds for every mapping $B \in V(A, \gamma)$ and that

$$\cup \{V(A, \gamma) : A \in \mathcal{M}, \ \gamma \in (0, 1)\}$$

is an open everywhere dense set in the space $\mathcal{M}$. This completes the proof of Proposition 2.14.

2.5 Inexact Halpern-Type Iterations

We use the notation, definitions, and assumptions introduced in Sect. 2.2.

We assume that C is a nonempty star-shaped subset of X, $\rho \in [0, 1)$, $T : C \to C$ satisfies

$$d(T(x), T(y)) \le d(x, y), \ x, y \in C \tag{2.20}$$

and that

$$f : C \to \mathrm{St}(C) \tag{2.21}$$

satisfies

$$d(f(x), f(y)) \le \rho d(x, y), \ x, y \in C. \tag{2.22}$$

Assume that $\{\alpha_n\}_{n=0}^{\infty} \subset (0,1)$, $\{r_n\}_{n=0}^{\infty} \subset [0,\infty)$

$$x_0 = x \in C \tag{2.23}$$

and that for each integer $n \geq 0$

$$x_{n+1} \in B((1-\alpha_n)f(x_n) \oplus \alpha_n T(x_n), r_n) \cap C. \tag{2.24}$$

We study the asymptotic regularity of the sequence of inexact Halpern-type iterations $\{x_n\}_{n=0}^{\infty}$ with the errors $\{r_n\}_{n=0}^{\infty}$.

Let $\theta \in C$, $M_0 > 0$ and

$$M > (1-\rho)^{-1}(2M_0 + d(x,\theta) + 2 + d(\theta, f(\theta)) + d(\theta, T(\theta))). \tag{2.25}$$

We use assumption (A) introduced in Sect. 2.2 and prove the following result which shows that our iterative process generates approximate fixed points of T under the presence of computational errors.

Theorem 2.15 *Let (A) hold, $J = 2\lfloor (1-\rho)^{-1} \rfloor + 1$ and for each integer $n \geq 0$*

$$\alpha_n = 1 - 2(1-\rho)^{-1}(n+J)^{-1},\ r_n = 2(1-\rho)^{-1}(n+J+1)^{-2}.$$

Then for each integer $n \geq 0$,

$$d(x_n, x_{n+1}) \leq (2M+3)J(1-\rho)^{-1}(n+J)^{-1}$$

and

$$d(x_n, T(x_n)) \leq (2M+3)(J+2)(1-\rho)^{-1}(n+J)^{-1}.$$

We begin with the following lemma.

Lemma 2.16 *Let $n \geq 0$ be an integer. Then*

$$d(x_{n+2}, x_{n+1}) \leq (1 - (1-\rho)(1-\alpha_{n+1}))d(x_{n+1}, x_n)$$

$$+ |\alpha_{n+1} - \alpha_n| d(f(x_n), T(x_n)) + r_{n+1} + r_n.$$

Proof Proposition 2.1 and (2.20), (2.21), and (2.24) imply that

$$d(x_{n+2}, x_{n+1})$$

$$\leq d(x_{n+2}, (1-\alpha_{n+1})f(x_{n+1}) \oplus \alpha_{n+1}T(x_{n+1}))$$

$$+ d((1-\alpha_{n+1})f(x_{n+1}) \oplus \alpha_{n+1}T(x_{n+1}), (1-\alpha_n)f(x_n) \oplus \alpha_n T(x_n))$$

$$
\begin{aligned}
&+ d((1-\alpha_n)f(x_n) \oplus \alpha_n T(x_n), x_{n+1}) \\
&\le r_{n+1} + (1-\alpha_{n+1})d(f(x_{n+1}), f(x_n)) + \alpha_{n+1}d(T(x_{n+1}), T(x_n)) \\
&+ |\alpha_{n+1} - \alpha_n| d(f(x_n), T(x_n)) + r_n \\
&\le r_n + r_{n+1} + \rho(1-\alpha_{n+1})d(x_{n+1}, x_n) + \alpha_{n+1}d(x_{n+1}, x_n) \\
&+ |\alpha_{n+1} - \alpha_n| d(f(x_n), T(x_n)) \\
&= r_n + r_{n+1} + (1-(1-\rho)(1-\alpha_{n+1}))d(x_{n+1}, x_n) \\
&+ |\alpha_{n+1} - \alpha_n| d(f(x_n), T(x_n)).
\end{aligned}
$$

Lemma 2.16 is proved.

Lemma 2.17 *Let (A) hold $p \in B(\theta, M_0)$ be as guaranteed by (A). Then for each integer $n \ge 0$,*

$$d(x_i, T^i(p)) \le M + \sum\{r_j : j \in \{0, \dots, i\} \setminus \{i\}\}, \tag{2.26}$$

$$d(x_n, \theta) \le M + M_0 + \sum\{r_j : j \in \{0, \dots, i\} \setminus \{i\}\}. \tag{2.27}$$

Proof We prove the lemma by induction. By (A) and (2.25),

$$d(x_0, p) \le d(x_0, \theta) + d(\theta, p) \le M_0 + d(\theta, x_0) < M.$$

Assume that (2.26) holds for an integer $i \ge 0$. By Proposition 2.1 and (2.20), (2.22), and (2.24)–(2.26),

$$
\begin{aligned}
&d(x_{i+1}, T^{i+1}(p)) \\
&\le d(x_{i+1}, (1-\alpha_i)f(x_i) \oplus \alpha_i T(x_i)) \\
&+ d((1-\alpha_i)f(x_i) \oplus \alpha_i T(x_i), T^{i+1}(p)) \\
&\le r_i + (1-\alpha_i)d(f(x_i), T^{i+1}(p)) + \alpha_i d(T(x_i), T^{i+1}(p)) \\
&\le r_i + (1-\alpha_i)d(f(x_i), f(T^i(p))) + (1-\alpha_i)d(f(T^i(p)), T^{i+1}(p)) + \alpha_i d(x_i, T^i(p)) \\
&\le r_i + (1-\alpha_i)\rho d(x_i, T^i(p)) + \alpha_i d(x_i, T^i(p)) \\
&+ (1-\alpha_i)(d(f(T^i(p)), f(\theta)) + d(f(\theta), \theta) + d(\theta, T^{i+1}(p))) \\
&\le r_i + (1-(1-\alpha_i)(1-\rho))d(x_i, T^i(p)) + (1-\alpha_i)(2M_0 + d(f(\theta), \theta))
\end{aligned}
$$

$$\leq r_i + M + \sum\{r_j : j \in \{0, \dots, i\} \setminus \{i\}\} - (1 - \alpha_i)(1 - \rho)M$$

$$+ (1 - \alpha_i)(2M_0 + d(f(\theta), \theta))$$

$$\leq M + \sum_{j=0}^{i} r_j - (1 - \alpha_i)((1 - \rho)M - 2M_0 - d(f(\theta), \theta)) \leq M + \sum_{j=0}^{i} r_j.$$

Thus we showed that (2.26) holds for $i + 1$ too. Therefore by induction, it holds for each integer $i \geq 0$. (A) and (2.26) imply that (2.27) holds for each integer $n \geq 0$. Lemma 2.17 is proved.

Proposition 2.18 *Let (A) hold. Then for each integer $n \geq 0$,*

$$d(x_{n+2}, x_{n+1})$$

$$\leq r_n + r_{n+1} + (1 - (1 - \rho)(1 - \alpha_{n+1}))d(x_{n+1}, x_n) + |\alpha_{n+1} - \alpha_n|(2M + 2\sum_{j=0}^{n} r_j), \tag{2.28}$$

$$d(f(x_n), T(x_n)) \leq 2M + 2\sum_{j=0}^{n} r_j, \tag{2.29}$$

$$d(x_n, T(x_n)) \leq d(x_n, x_{n+1}) + (1 - \alpha_n)(2M + 2\sum_{j=0}^{n} r_j).$$

Proof Let $p \in B(\theta, M_0) \cap C$ be as guaranteed by (A). Thus for each integer $i \geq 0$,

$$T^i(p) \in B(\theta, M_0). \tag{2.30}$$

Lemma 2.17, assumption (A) and (2.30) imply that for each integer $i \geq 0$,

$$d(x_i, T^i(p)) \leq M + \sum\{r_j : j \in \{0, \dots, i\} \setminus \{i\}\}, \tag{2.31}$$

$$d(x_i, \theta) \leq M + M_0 + \sum\{r_j : j \in \{0, \dots, i\} \setminus \{i\}\}. \tag{2.32}$$

Let $n \geq 0$ be an integer. Lemma 2.16 and Eqs. (2.20), (2.22), (2.25), (2.30), and (2.31) imply that

$$d(x_{n+2}, x_{n+1})$$

$$\leq (1 - (1 - \rho)(1 - \alpha_{n+1}))d(x_{n+1}, x_n) + |\alpha_{n+1} - \alpha_n|d(f(x_n), T(x_n)) + r_{n+1} + r_n$$

$$\leq r_{n+1} + r_n + (1 - (1 - \rho)(1 - \alpha_{n+1}))d(x_{n+1}, x_n)$$

$$+ |\alpha_{n+1} - \alpha_n|(d(f(x_n), f(T^n(p))) + d(f(T^n(p)), f(\theta)) + d(f(\theta), \theta)$$

$$+ d(\theta, T^{p+1}(p)) + d(T^{n+1}(p), T(x_n)))$$

$$\leq r_{n+1} + r_n + (1 - (1 - \rho)(1 - \alpha_{n+1}))d(x_{n+1}, x_n)$$

$$+ |\alpha_{n+1} - \alpha_n|(\rho d(x_n, T^n(p)) + \rho M_0 + d(f(\theta), \theta) + M_0 + d(T^n(p), x_n)))$$

$$\leq r_{n+1} + r_n + (1 - (1 - \rho)(1 - \alpha_{n+1}))d(x_{n+1}, x_n)$$

$$+ |\alpha_{n+1} - \alpha_n|((\rho + 1)(M + \sum_{j=0}^{n} r_j) + (1 + \rho)M_0 + d(f(\theta), \theta) + (1 + \rho)M_0)$$

$$\leq r_{n+1} + r_n + (1 - (1 - \rho)(1 - \alpha_{n+1}))d(x_{n+1}, x_n)$$

$$+ |\alpha_{n+1} - \alpha_n|(2M + 2\sum_{j=0}^{n} r_j)$$

and (2.28) holds for each integer $n \geq 0$.

Lemma 2.17 and Eqs. (2.20), (2.22), and (2.25) imply that

$$d(f(x_n), T(x_n)) \leq d(f(x_n), f(\theta)) + d(f(\theta), \theta) + d(\theta, T(\theta)) + d(T(\theta), T(x_n))$$

$$\leq (1 + \rho)d(x_n, \theta) + d(f(\theta), \theta) + d(\theta, T(\theta))$$

$$\leq (1 + \rho)(M + M_0 + \sum_{j=0}^{n} r_j) + d(f(\theta), \theta) + d(\theta, T(\theta))$$

$$\leq 2M + 2\sum_{j=0}^{n} r_j$$

and (2.29) holds.

Proposition 2.1 and (2.24) and (2.29) imply that

$$d(x_n, T(x_n)) \leq d(x_n, x_{n+1}) + d(x_{n+1}, T(x_n))$$

$$= d(x_n, x_{n+1}) + d(x_{n+1}, (1 - \alpha_n)f(x_n) \oplus \alpha_n T(x_n))$$

$$+ d((1 - \alpha_n)f(x_n) \oplus \alpha_n T(x_n), T(x_n))$$

$$\leq d(x_n, x_{n+1}) + (1 - \alpha_n)d(f(x_n), T(x_n)) + r_n$$

$$\leq d(x_n, x_{n+1}) + (1 - \alpha_n)(2M + 2\sum_{j=0}^{n} r_j).$$

Proposition 2.18 is proved.

Proof of Theorem 2.15 Let p be as guaranteed by (A). Assumption (A) and (2.25) and (2.26) imply that

$$d(x_0, x_1) \leq d(x_0, \theta) + d(\theta, T(p)) + d(T(p), x_1)$$

$$\leq d(x_0, \theta) + M_0 + M + \sum_{j=0}^{\infty} r_j \leq 2M + \sum_{j=0}^{\infty} r_j.$$

Recall that

$$J = 2\lfloor(1 - \rho)^{-1}\rfloor + 1.$$

Set $N = 2$ and for each integer $n \geq 0$ set

$$s_n = d(x_n, x_{n+1}),\ a_n = 1-\alpha_n,\ c_n = L = 2M+2+\sum_{j=0}^{\infty} r_j,\ \gamma = 1-\rho. \quad (2.33)$$

Clearly,

$$\sum_{j=0}^{\infty} r_j \leq 2(1 - \rho)^{-1} J^{-1}.$$

Proposition 2.18 implies that for each integer $n \geq 0$,

$$s_{n+1} \leq (1 - \gamma a_{n+1})s_n + |a_{n+1} - a_n|(2M + 2\sum_{j=0}^{\infty} r_j) + r_n + r_{n+1}. \quad (2.34)$$

By (2.33), for each integer $n \geq 0$,

$$|a_n - a_{n+1}| = 2(1 - \rho)^{-1}((n + J)^{-1} - (n + J + 1)^{-1})$$

$$= 2(1 - \rho)^{-1}(n + J)^{-1}(n + J + 1)^{-1},$$

$$r_n + r_{n+1} = 2(1 - \rho)^{-1}((n + J + 1)^{-2} - (n + J + 2)^{-2}) \leq 2|a_n - a_{n+1}|. \quad (2.35)$$

In view of (2.34) and (2.35), for each integer $n \geq 0$,

$$s_{n+1} \leq (1 - \gamma a_{n+1})s_n + |a_{n+1} - a_n|(2M + 2\sum_{j=0}^{\infty} r_j) + 2). \tag{2.36}$$

Lemma 2.2 and (2.36) imply that for each integer $n \geq 0$,

$$d(x_n, x_{n+1}) \leq (2M + 2 + \sum_{j=0}^{\infty} r_j)J(\gamma(n + J))^{-1}$$

$$\leq (2M + 3)J(1 - \rho)^{-1}(n + J)^{-1}$$

and together with Proposition 2.18, this implies

$$d(x_n, T(x_n)) \leq (2M + 3)J(1 - \rho)^{-1}(n + J)^{-1} + (1 - \alpha_n)(2M + 3)$$

$$\leq (2M + 3)(J + 2)(1 - \rho)^{-1}(n + J)^{-1}.$$

Theorem 2.15 is proved.

2.6 Alternative Iterative Method with Errors

We use the notation, definitions, and assumptions introduced in Sect. 2.5. In particular, we use assumption (A). Let p be as guaranteed by assumption (A).

Let $M_0 > 0$ and a number M satisfy (2.25). Assume that $\{\alpha_n\}_{n=0}^{\infty} \subset (0, 1)$, $\{r_n\}_{n=0}^{\infty} \subset [0, \infty)$,

$$x_0 = x \in C$$

and that for each integer $n \geq 0$,

$$x_{n+1} \in B(T((1 - \alpha_n)f(x_n) \oplus \alpha_n x_n), r_n). \tag{2.37}$$

We will prove the following result which shows that our iterative process generates approximate fixed points of T under the presence of computational errors.

Theorem 2.19 *Let* $J = 2\lfloor(1 - \rho)^{-1}\rfloor + 1$ *and for each integer* $n \geq 0$,

$$\alpha_n = 1 - 2(1 - \rho)^{-1}(n + J)^{-1}, \tag{2.38}$$

$$r_n = 2(1 - \rho)^{-1}(n + J + 1)^{-2}. \tag{2.39}$$

Then for each integer $n \ge 0$,

$$d(x_n, x_{n+1}) \le (2M+3)J(1-\rho)^{-1}(n+J)^{-1}$$

and

$$d(x_n, T(x_n)) \le (2M+3)(J+2)(1-\rho)^{-1}(n+J)^{-1}.$$

We proceed with the following auxiliary results.

Lemma 2.20 *For each integer* $n \ge 0$,

$$d(x_{n+2}, x_{n+1})$$

$$\le (1-(1-\rho)(1-\alpha_{n+1}))d(x_{n+1}, x_n) + |\alpha_{n+1} - \alpha_n| d(f(x_n), x_n) + r_n + r_{n+1}.$$

Proof Let $n \ge 0$ be an integer. Proposition 2.1 and (2.20), (2.22), and (2.37) imply that

$$d(x_{n+2}, x_{n+1})$$

$$\le d(x_{n+2}, T((1-\alpha_{n+1})f(x_{n+1}) \oplus \alpha_{n+1}x_{n+1}))$$

$$+ d(T((1-\alpha_{n+1})f(x_{n+1}) \oplus \alpha_{n+1}x_{n+1}), T((1-\alpha_n)f(x_n) \oplus \alpha_n x_n))$$

$$+ d(T((1-\alpha_n)f(x_n) \oplus \alpha_n x_n), x_{n+1})$$

$$\le r_{n+1} + d((1-\alpha_{n+1})f(x_{n+1}) \oplus \alpha_{n+1}x_{n+1}, (1-\alpha_n)f(x_n) \oplus \alpha_n x_n) + r_n$$

$$\le r_{n+1} + r_n + (1-\alpha_{n+1})d(f(x_{n+1})f(x_n)) + \alpha_{n+1}d(x_{n+1}, x_n)$$

$$+ |\alpha_{n+1} - \alpha_n| d(f(x_n), x_n)$$

$$\le r_{n+1} + r_n + \rho(1-\alpha_{n+1})d(x_{n+1}, x_n)$$

$$+ \alpha_{n+1}d(x_{n+1}, x_n) + |\alpha_{n+1} - \alpha_n| d(f(x_n), x_n)$$

$$\le r_{n+1} + r_n + (1-(1-\rho)(1-\alpha_{n+1}))d(x_{n+1}, x_n) + |\alpha_{n+1} - \alpha_n| d(f(x_n), x_n).$$

Lemma 2.20 is proved.

Lemma 2.21 *For each integer* $i \ge 0$,

$$d(x_i, T^i(p)) \le M + \sum\{r_j : j \in \{0, \dots, i\} \setminus \{i\}\}, \tag{2.40}$$

$$d(x_n, \theta) \leq M + M_0 + \sum\{r_j : j \in \{0, \dots, i\} \setminus \{i\}\}. \tag{2.41}$$

Proof We prove the lemma by induction. By (A) and (2.25),

$$d(x_0, p) \leq d(x_0, \theta) + d(\theta, p) \leq M_0 + d(\theta, x_0) < M.$$

Assume that (2.40) holds for an integer $i \geq 0$. By Proposition 2.1 and (2.20), (2.22), (2.37), and (2.40),

$$d(x_{i+1}, T^{i+1}(p))$$

$$\leq d(x_{i+1}, T((1-\alpha_i)f(x_i) \oplus \alpha_i x_i))$$

$$+ d(T(1-\alpha_i)f(x_i) \oplus \alpha_i x_i, T^{i+1}(p))$$

$$\leq r_i + d((1-\alpha_i)f(x_i) \oplus \alpha_i x_i, T^i(p))$$

$$\leq r_i + (1-\alpha_i)d(f(x_i), T^i(p)) + \alpha_i d(x_i, T^i(p))$$

$$\leq r_i + (1-\alpha_i)d(f(x_i), f(T^i(p))) + (1-\alpha_i)d(f(T^i(p)), T^i(p)) + \alpha_i d(x_i, T^i(p))$$

$$\leq r_i + (1-\alpha_i)\rho d(x_i, T^i(p)) + \alpha_i d(x_i, T^i(p))$$

$$+ (1-\alpha_i)(d(f(T^i(p)), f(\theta)) + d(f(\theta), \theta) + d(\theta, T^i(p)))$$

$$\leq r_i + (1-(1-\alpha_i)(1-\rho))d(x_i, T^i(p)) + (1-\alpha_i)(\rho M_0 + d(f(\theta), \theta) + M_0)$$

$$\leq r_i + M + \sum\{r_j : j \in \{0, \dots, i\} \setminus \{i\}\}$$

$$- (1-\alpha_i)((1-\rho)M - d(f(\theta), \theta) - 2M_0) \leq M + \sum_{j=0}^{i} r_j.$$

Thus we showed that (2.40) holds for $i+1$ too. Therefore by induction, it holds for each integer $i \geq 0$. (A) and (2.40) imply that (2.41) for each integer $i \geq 0$. Lemma 2.21 is proved.

Proposition 2.22 *For each integer $n \geq 0$,*

$$d(x_{n+2}, x_{n+1}) \leq r_n + r_{n+1} + (1-(1-\rho)(1-\alpha_{n+1}))d(x_{n+1}, x_n)$$

$$+ |\alpha_{n+1} - \alpha_n|(2M + 2\sum\{r_j : j \in \{0, \dots, n\} \setminus \{n\}\}), \tag{2.42}$$

$$d(f(x_n), x_n) \leq 2M + \sum\{r_j : j \in \{0, \dots, n\} \setminus \{n\}\}, \tag{2.43}$$

$$d(x_n, T(x_n)) \le d(x_n, x_{n+1}) + r_n + (1 - \alpha_n)(2M + \sum_{j=0}^{\infty} r_j).$$

Proof Let $n \ge 0$ be an integer. Lemma 2.21 implies that for each integer $i \ge 0$,

$$d(x_i, T^i(p)) \le M + \sum \{r_j : j \in \{0, \dots, i\} \setminus \{i\}\}, \tag{2.44}$$

$$d(x_i, \theta) \le M + M_0 + \sum \{r_j : j \in \{0, \dots, i\} \setminus \{i\}\}. \tag{2.45}$$

Lemma 2.20 implies that

$$d(x_{n+2}, x_{n+1})$$

$$\le (1 - (1 - \rho)(1 - \alpha_{n+1}))d(x_{n+1}, x_n) + |\alpha_{n+1} - \alpha_n| d(f(x_n), x_n) + r_{n+1} + r_n. \tag{2.46}$$

Assumption (A) and Eqs. (2.22), (2.25), (2.44), and (2.45) imply that

$$d(f(x_n), x_n) \le d(f(x_n), f(T^n(p))) + d(f(T^n(p)), f(\theta)) + d(f(\theta), \theta) + d(\theta, x_n)$$

$$\le \rho d(x_n, T^n(p)) + \rho d(T^n(p), \theta) + d(f(\theta), \theta) + d(\theta, x_n)$$

$$\le \rho M + \rho \sum \{r_j : j \in \{0, \dots, n\} \setminus \{n\}\}$$

$$+ \rho M_0 + d(f(\theta), \theta) + M + M_0 + \sum \{r_j : j \in \{0, \dots, n\} \setminus \{n\}\}$$

$$\le 2M + 2 \sum \{r_j : j \in \{0, \dots, n\} \setminus \{n\}\}. \tag{2.47}$$

By (2.46) and (2.47),

$$d(x_{n+2}, x_{n+1}) \le r_n + r_{n+1} + (1 - (1 - \rho)(1 - \alpha_{n+1}))d(x_{n+1}, x_n)$$

$$+ |\alpha_{n+1} - \alpha_n|(2M + 2 \sum \{r_j : j \in \{0, \dots, n\} \setminus \{n\}\}). \tag{2.48}$$

Proposition 2.1 and (2.20), (2.37), and (2.47) imply that

$$d(x_n, T(x_n)) \le d(x_n, x_{n+1}) + d(x_{n+1}, T(x_n))$$

$$= d(x_n, x_{n+1}) + d(x_{n+1}, T((1 - \alpha_n) f(x_n) \oplus \alpha_n x_n))$$

$$+ d(T((1 - \alpha_n) f(x_n) \oplus \alpha_n x_n), T(x_n))$$

$$\leq d(x_n, x_{n+1}) + r_n + d((1-\alpha_n)f(x_n) \oplus \alpha_n x_n, x_n)$$

$$\leq d(x_n, x_{n+1}) + r_n + (1-\alpha_n)d(f(x_n), x_n)$$

$$\leq d(x_n, x_{n+1}) + r_n + (1-\alpha_n)(2M + 2\sum_{j=0}^{\infty} r_j).$$

Proposition 2.22 is proved.

Theorem 2.23 *Let $\gamma = 1-\rho$, $J = 2\lfloor(1-\rho)^{-1}\rfloor + 1$ and for each integer $n \geq 0$,*

$$\alpha_n = 1 - 2(1-\rho)^{-1}(n+J)^{-1}, \ r_n = 2(1-\rho)^{-1}(n+J+1)^{-2}.$$

Then for each integer $n \geq 0$,

$$d(x_n, x_{n+1}) \leq (2M+3)J(1-\rho)^{-1}(n+J)^{-1}$$

and

$$d(x_n, T(x_n)) \leq (2M+3)(J+2)(1-\rho)^{-1}(n+J)^{-1}.$$

Proof Recall that p is as guaranteed by (A). Assumption (A) and Lemma 2.21 imply that

$$d(x_0, x_1) \leq d(x_0, \theta) + d(\theta, T(p)) + d(T(p), x_1)$$

$$\leq d(x_0, \theta) + M_0 + M + \sum_{j=0}^{\infty} r_j \leq 2M + \sum_{j=0}^{\infty} r_j.$$

Set

$$a_n = 1 - \alpha_n, \ n = 0, 1, \ldots, \ N = 2. \tag{2.49}$$

Clearly,

$$\sum_{j=0}^{\infty} r_j \leq 2(1-\rho)^{-1}J^{-1}.$$

For each integer $n \geq 0$ set

$$s_n = d(x_n, x_{n+1}).$$

By (2.24) and (2.49),

$$s_{n+1} \le (1 - \gamma a_{n+1})s_n + |a_{n+1} - a_n|(2M + 2\sum_{j=0}^{\infty} r_j) + r_n + r_{n+1}. \tag{2.50}$$

Clearly for each integer $n \ge 0$,

$$|a_n - a_{n+1}| = 2(1-\rho)^{-1}(n+J)^{-1}(n+J+1)^{-1},$$

$$r_n + r_{n+1} = 2(1-\rho)^{-1}((n+J+1)^{-2} + (n+J+2)^{-2}) \le 2|a_n - a_{n+1}|. \tag{2.51}$$

In view of (2.50) and (2.51), for each integer $n \ge 0$,

$$s_{n+1} \le (1 - \gamma a_{n+1})s_n + |a_{n+1} - a_n|(2M + 2\sum_{j=0}^{\infty} r_j + 2).$$

For each integer $n \ge 0$ set

$$c_n = L = 2M + 2 + \sum_{j=0}^{\infty} r_j.$$

Lemma 2.2, Proposition 2.22, and the relations above imply that for each integer $n \ge 0$,

$$d(x_n, x_{n+1}) \le (2M+3)J(1-\rho)^{-1}(n+J)^{-1},$$

$$d(x_n, T(x_n)) \le (2M+3)(J+2)(1-\rho)^{-1}(n+J)^{-1}.$$

Theorem 2.23 is proved.

Note that our asymptotic regularity result for alternative iterative method with errors does not follow from the analogous result for the inexact Halpern-type iterations.

2.7 Set-Valued Mappings

We continue to use the notation and the definitions introduced in Sect. 2.2. In this section we assume that the set C is closed. Recall that for each set $A \subset X$, we denote by $\mathrm{cl}(A)$ its closure. Denote by $S(C)$ the set of all nonempty bounded closed subsets of $S(C)$. It is known [168] that $(S(C), H)$ is a complete metric space.

We assume that

$$\theta \in \mathrm{St}(C).$$

Denote by $\mathcal{M}$ the set of all mappings $T : C \to S(C)$ such that

$$H(T(x), T(y)) \leq d(x, y), \; x, y \in C. \tag{2.52}$$

The space $\mathcal{M}$ is equipped with the uniformity determined by the base

$$\mathcal{U}(m) = \{(T_1, T_2) \in \mathcal{M} :$$

$$H(T_1(x), T_2(x)) \leq m^{-1}, \; x \in B(\theta, m)\}, \; m = 1, 2, \dots . \tag{2.53}$$

The space $\mathcal{M}$ with this uniformity is metrizable and complete [168]. We show that a set-valued analog of assumption (A), used for a single-valued mapping, holds for a typical (generic) set-valued mapping.

Theorem 2.24 *There exists an open everywhere dense set $\mathcal{F} \subset \mathcal{M}$ such that for each $S \in \mathcal{F}$, there exist $M > 0$ and an open neighborhood V of S in $\mathcal{F}$ such that for each $T \in V$*

$$T(B(\theta, M) \cap C) \subset B(\theta, M) \cap C.$$

Proof Let $A \in \mathcal{M}$ and $\gamma \in (0, 1)$. Set

$$A_\gamma(x) = \mathrm{cl}(\{(1 - \gamma)\xi \oplus \gamma\theta : \; \xi \in A(x)\}), \; x \in C. \tag{2.54}$$

Let $x \in C$. Proposition 2.1 and (2.52) and (2.54) imply that

$$H(A(x), A_\gamma(x)) \leq \gamma \sup\{d(\xi, \theta) : \; \xi \in A(x)\}$$

$$\leq \gamma(\sup\{d(\xi, \theta) : \; \xi \in A(\theta)\} + H(A(x), A(\theta)))$$

$$\leq \gamma \sup\{d(\xi, \theta) : \; \xi \in A(\theta)\} + \gamma d(x, \theta). \tag{2.55}$$

This implies that

$$\{A_\gamma : \; A \in \mathcal{M}, \; \gamma \in (0, 1)\}$$

is an everywhere dense in $\mathcal{M}$.

Let $A \in \mathcal{M}$ and $\gamma \in (0, 1)$. Proposition 2.1, (2.54), and the fixed point result of Nadler [137] imply that there exists $x_{A,\gamma} \in C$ such that

$$x_{A,\gamma} \in A_\gamma(x_{A,\gamma}). \tag{2.56}$$

Choose a number

$$M > \gamma^{-2} \sup\{d(x_{A,\gamma}, \eta) : \; \eta \in A_\gamma(x_{A,\gamma})\} + 1. \tag{2.57}$$

There exists an open neighborhood $V(A,\gamma)$ of A_γ such that

$$H(T(x), A_\gamma(x)) \le 1,\ x \in B(x_{A,\gamma}, M) \cap C,\ T \in V(A,\gamma). \tag{2.58}$$

Let

$$T \in V(A,\gamma),\ x \in B(x_{A,\gamma}, M) \cap C. \tag{2.59}$$

By (2.58) and (2.59),

$$H(T(x), A_\gamma(x)) \le 1,$$

$$H(A_\gamma(x), A_\gamma(x_{A,\gamma})) \le (1-\gamma)M,$$

$$H(T(x), A_\gamma(x_{A,\gamma})) \le (1-\gamma)M + 1. \tag{2.60}$$

In view of (2.54),

$$\mathrm{cl}((1-\gamma)A(x_{A,\gamma}) \oplus \gamma\theta)$$

$$:= \mathrm{cl}(\{(1-\gamma)\xi \oplus \gamma\theta :\ \xi \in A(x_{A,\gamma})\}) = A_\gamma(x_{A,\gamma}). \tag{2.61}$$

Clearly, for each integer $i \ge 0$, the set $A_\gamma^i(x_{A,\gamma})$ is bounded. Let $i \ge 0$ be an integer. We show that

$$H(\mathrm{cl}(A_\gamma^{i+1}(x_{A,\gamma})), \mathrm{cl}(A_\gamma^{i+2}(x_{A,\gamma}))) \le (1-\gamma)H(\mathrm{cl}(A_\gamma^{i}(x_{A,\gamma})), \mathrm{cl}(A_\gamma^{i+1}(x_{A,\gamma}))). \tag{2.62}$$

In order to prove (2.62) it is sufficient to show that for each

$$y \in A_\gamma^{i+2}(x_{A,\gamma})) \tag{2.63}$$

we have

$$d(y, \mathrm{cl}(A_\gamma^{i+1}(x_{A,\gamma}))) \le (1-\gamma)H(\mathrm{cl}(A_\gamma^{i}(x_{A,\gamma})), \mathrm{cl}(A_\gamma^{i+1}(x_{A,\gamma}))).$$

Let (2.63) hold. Then

$$y = (1-\gamma)\xi \oplus \gamma\theta \text{ where } \xi \in A(A_\gamma^{i+1}(x_{A,\gamma})). \tag{2.64}$$

In order to prove (2.62), it is sufficient to show that

$$d((1-\gamma)\xi \oplus \gamma\theta, A_\gamma^{i+1}(x_{A,\gamma})) \le (1-\gamma)H(\mathrm{cl}(A_\gamma^{i}(x_{A,\gamma})), \mathrm{cl}(A_\gamma^{i+1}(x_{A,\gamma}))). \tag{2.65}$$

There exists

$$\xi_0 \in A_\gamma^{i+1}(x_{A,\gamma}) \tag{2.66}$$

such that

$$\xi \in A(\xi_0). \tag{2.67}$$

Let $\epsilon > 0$. In view of (2.66), there exists

$$\eta_0 \in A_\gamma^{i+1}(x_{A,\gamma}) \tag{2.68}$$

such that

$$d(\xi_0, \eta_0) \leq d(\xi_0, \mathrm{cl}(A_\gamma^i(x_{A,\gamma}))) + \epsilon \leq H(\mathrm{cl}(A_\gamma^i(x_{A,\gamma})), \mathrm{cl}(A_\gamma^{i+1}(x_{A,\gamma}))) + \epsilon. \tag{2.69}$$

By (2.52) and (2.67), there exists

$$\eta \in A(\eta_0) \tag{2.70}$$

such that

$$d(\xi, \eta) \leq d(\xi, A(\eta_0)) + \epsilon \leq H(\mathrm{cl}(A(\xi_0)), \mathrm{cl}(A(\eta_0))) + \epsilon \leq d(\xi_0, \eta_0) + \epsilon. \tag{2.71}$$

Eqs. (2.68) and (2.70) imply that

$$(1-\gamma)\eta \oplus \gamma\theta \in A_\gamma(\eta_0) \subset A_\gamma^{i+1}(x_{A,\gamma}). \tag{2.72}$$

Proposition 2.1 and (2.69) and (2.72) imply that

$$\begin{aligned} &d((1-\gamma)\xi \oplus \gamma\theta, A_\gamma^{i+1}(x_{A,\gamma})) \\ &\leq d((1-\gamma)\xi \oplus \gamma\theta, (1-\gamma)\eta \oplus \gamma\theta) \\ &\leq (1-\gamma)d(\xi, \eta) \leq (1-\gamma)d(\xi_0, \eta_0) + \epsilon \\ &\leq (1-\gamma)H(\mathrm{cl}(A_\gamma^i(x_{A,\gamma})), \mathrm{cl}(A_\gamma^{i+1}(x_{A,\gamma}))) + 2\epsilon. \end{aligned}$$

Since ϵ is an arbitrary positive number, we conclude that

$$d((1-\gamma)\xi \oplus \gamma\theta, A_\gamma^{i+1}(x_{A,\gamma})) \leq (1-\gamma)H(\mathrm{cl}(A_\gamma^i(x_{A,\gamma})), \mathrm{cl}(A_\gamma^{i+1}(x_{A,\gamma}))).$$

Thus (2.65) holds, and this implies that (2.62) holds for each integer $i \geq 0$. Equation (2.62) implies that for each integer $n \geq 0$,

$$H(\mathrm{cl}(A_\gamma^{n+1}(x_{A,\gamma})), \mathrm{cl}(A_\gamma^n(x_{A,\gamma}))) \leq (1-\gamma)^n H(\{x_{A,\gamma}\}, \mathrm{cl}(A_\gamma(x_{A,\gamma})))$$

and

$$H(\{x_{A,\gamma}\}, \mathrm{cl}(A_\gamma^n(x_{A,\gamma}))) \leq H(\{x_{A,\gamma}\}, \mathrm{cl}(A_\gamma(x_{A,\gamma}))) \sum_{i=0}^{\infty}(1-\gamma)^i$$

$$\leq H(\{x_{A,\gamma}\}, \mathrm{cl}(A_\gamma(x_{A,\gamma})))\gamma^{-1}.$$

Therefore

$$\cup_{n=0}^{\infty} A_\gamma^n(x_{A,\gamma}) \subset B(x_{A,\gamma}, H(\{x_{A,\gamma}\}, \mathrm{cl}(A_\gamma(x_{A,\gamma})))\gamma^{-1}).$$

Together with (2.57) and (2.61), this implies that

$$H(T(x), \{x_{A,\gamma}\}) \leq (1-\gamma)M + 1$$

$$+ H(\{x_{A,\gamma}\}, \mathrm{cl}(A_\gamma(x_{A,\gamma})))\gamma^{-1} \leq M.$$

Since the equation above holds for each $T \in V(A, \gamma)$ and each

$$x \in B(x_{A,\gamma}, M) \cap C$$

this completes the proof of Theorem 2.24.

2.8 Halpern-Type Iterations of Set-Valued Mappings

We assume that (X, d, W) is a W-hyperbolic space which is denoted by X for simplicity. We use the notation and definitions introduced in Sects. 2.1 and 2.2.

Recall that C is a nonempty star-shaped subset of X. Let $\rho \in [0, 1)$, $T : C \to 2^C \setminus \{\emptyset\}$, $T(z)$ be bounded for each $z \in C$,

$$H(T(x), T(y)) \leq d(x, y), \ x, y \in C \tag{2.73}$$

and that $f : C \to \mathrm{St}(C)$ satisfy

$$d(f(x), f(y)) \leq \rho d(x, y), \ x, y \in C. \tag{2.74}$$

We consider Halpern-type iterations of T which are defined below.

Assume that $\{\alpha_n\}_{n=0}^{\infty} \subset (0, 1)$, $\{r_n\}_{n=0}^{\infty} \subset (0, 1]$,

$$x_0 = x \in C, \ \ y_0 \in T(x_0)$$

and that for each integer $n \geq 0$,

$$y_n \in T(x_n), \tag{2.75}$$

$$x_{n+1} = (1 - \alpha_n) f(x_n) \oplus \alpha_n y_n, \tag{2.76}$$

$$d(y_n, y_{n+1}) \leq H(T(x_n), T(x_{n+1})) + r_n \leq d(x_n, x_{n+1}) + r_n. \tag{2.77}$$

Let $\theta \in C$,

$$M_0 > H(\{\theta\}, T(\theta)\}$$

and

$$M > (1-\rho)^{-1}(2M_0 + d(x, \theta) + d(f(\theta), \theta) + 2\sup\{d(\theta), \eta) : \ \eta \in T(\theta)\}). \tag{2.78}$$

We assume that the following assumption holds.
(A) There exists $p \in C \cap B(\theta, M_0)$ such that for each integer $n \geq 0$,

$$T^n(p) \subset B(\theta, M_0).$$

We will prove the following result which shows that our iterative process generates approximate fixed points of the set-valued mapping T.

Theorem 2.25 *Let $J = 2\lfloor(1-\rho)^{-1}\rfloor + 1$ and for each integer $n \geq 0$,*

$$\alpha_n = 1 - 2(1-\rho)^{-1}(n+J)^{-1}.$$

Then for each integer $n \geq 0$,

$$d(x_n, x_{n+1}) \leq (2M+1)J(1-\rho)^{-1}(n+J)^{-1}$$

and

$$d(x_n, T(x_n)) \leq (2M+1)(J+2)(1-\rho)^{-1}(n+J)^{-1}.$$

We begin with auxiliary results.

Lemma 2.26 *Let $n \geq 0$ be an integer. Then*

$$d(x_{n+2}, x_{n+1})$$

$$\leq (1-(1-\rho)(1-\alpha_{n+1}))d(x_{n+1},x_n)+\alpha_{n+1}r_n+|\alpha_{n+1}-\alpha_n|d(f(x_n),y_n), \tag{2.79}$$

$$d(x_n,y_n)\leq d(x_n,x_{n+1})+(1-\alpha_n)d(f(x_n),y_n).$$

Proof Proposition 2.1 and (2.74), (2.76), and (2.77) imply that

$$d(x_{n+2},x_{n+1})$$

$$=d((1-\alpha_{n+1})f(x_{n+1})\oplus\alpha_{n+1}y_{n+1}),(1-\alpha_n)f(x_n)\oplus\alpha_n y_n)$$

$$\leq (1-\alpha_{n+1})d(f(x_{n+1}),f(x_n))+\alpha_{n+1}d(y_n,y_{n+1})$$

$$+|\alpha_{n+1}-\alpha_n|d(f(x_n),y_n)$$

$$\leq \rho(1-\alpha_{n+1})d(x_{n+1},x_n)+\alpha_{n+1}d(x_{n+1},x_n)+\alpha_{n+1}r_n$$

$$+|\alpha_{n+1}-\alpha_n|d(f(x_n),y_n)$$

and (2.79) holds.

Proposition 2.1 and (2.76) imply that

$$d(x_n,y_n)\leq d(x_n,x_{n+1})+d(x_{n+1},y_n)$$

$$\leq d(x_n,x_{n+1})+(1-\alpha_n)d(f(x_n),y_n).$$

Lemma 2.26 is proved.

Lemma 2.27 *Let*

$$p\in B(\theta,M_0)$$

be as guaranteed by assumption (A). Then for each integer $n\geq 0$,

$$d(x_n,T^n(p))\leq M, \tag{2.80}$$

$$d(x_n,\theta)\leq M+M_0.$$

Proof We prove (2.80) by induction. By (A) and (2.78),

$$d(x_0,p)\leq M_0+d(\theta,x_0)<M.$$

Assume that (2.80) holds for a nonnegative integer n. Let $\epsilon>0$. In view of (2.80), there exists

$$\xi\in T^n(p) \tag{2.81}$$

such that

$$d(x_n, \xi) \le M + \epsilon. \tag{2.82}$$

By (2.73), (2.75), (2.81), and (2.82), there exists

$$\eta \in T(\xi) \tag{2.83}$$

such that

$$d(y_n, \eta) \le H(T(x_n), T(\xi)) + \epsilon \le d(x_n, \xi) + \epsilon \le M + 2\epsilon. \tag{2.84}$$

By Proposition 2.1, assumption (A) and (2.74), (2.76), and (2.81)–(2.84),

$$\begin{aligned}
d(x_{n+1}, T^{n+1}(p)) &\le d(x_{n+1}, \eta)\\
&\le d((1-\alpha_n)f(x_n) \oplus \alpha_n y_n, \eta)\\
&\le (1-\alpha_n)d(f(x_n), \eta) + \alpha_n d(y_n, \eta)\\
&\le (1-\alpha_n)d(f(x_n), f(\xi)) + (1-\alpha_n)d(f(\xi), \eta) + \alpha_n d(x_n, \xi) + \alpha_n \epsilon\\
&\le (1-\alpha_n)\rho d(x_n, \xi) + \alpha_n d(x_n, \xi) + \alpha_n \epsilon\\
&\quad + (1-\alpha_n)(d(f(\xi), f(\theta)) + d(f(\theta), \theta) + d(\theta, \eta))\\
&\le (1-(1-\alpha_n)(1-\rho))d(x_n, \xi) + \alpha_n \epsilon\\
&\quad + (1-\alpha_n)(\rho d(\xi, \theta) + d(f(\theta), \theta) + d(\theta, \eta))\\
&\le (1-(1-\alpha_n)(1-\rho))(M+2\epsilon) + \alpha_n \epsilon + (1-\alpha_n)(2M_0 + d(f(\theta), \theta)).
\end{aligned}$$

Since ϵ is any positive number together with (2.78), the relation above implies that

$$\begin{aligned}
&d(x_{n+1}, T^{n+1}(p))\\
&\le (1-(1-\alpha_n)(1-\rho))M + (1-\alpha_n)(2M_0 + d(f(\theta), \theta))\\
&\le M - (1-\alpha_n)((1-\rho)M - 2M_0 - d(f(\theta), \theta)) \le M.
\end{aligned}$$

Thus we showed that (2.80) holds for $n+1$ too. Therefore by induction, it holds for each integer $n \ge 0$. Let $n \ge 0$ be an integer. In view of (2.80) and the assumptions of the lemma,

$$d(x_n, \theta) \le d(x_n, T^n(p)) + \sup\{d(\theta, \eta) : \eta \in T^n(p)\} \le M + M_0.$$

Lemma 2.27 is proved.

Proposition 2.28 *For each integer $n \ge 0$,*

$$d(f(x_n), y_n) \le 2M,$$

$$d(x_{n+2}, x_{n+1}) \le (1 - (1-\rho)(1-\alpha_{n+1}))d(x_{n+1}, x_n) + 2M|\alpha_{n+1} - \alpha_n| + \alpha_{n+1}r_n,$$

$$d(x_n, T(x_n)) \le d(x_n, x_{n+1}) + 2M(1-\alpha_n).$$

Proof Let $n \ge 0$ be an integer. Lemma 2.26 implies that

$$d(x_{n+2}, x_{n+1}) \le (1 - (1-\rho)(1-\alpha_{n+1}))d(x_{n+1}, x_n)$$
$$+ \alpha_{n+1}r_n + |\alpha_{n+1} - \alpha_n| d(f(x_n), y_n). \quad (2.85)$$

Lemma 2.27, (2.73)–(2.75), and (2.78) imply that

$$d(f(x_n), y_n)$$
$$\le d(f(x_n), f(\theta)) + d(f(\theta), \theta) + \sup\{d(\theta, \eta) : \eta \in T(\theta)\} + d(y_n, T(\theta))$$
$$\le \rho d(x_n, \theta) + d(f(\theta), \theta) + \sup\{d(\theta, \eta) : \eta \in T(\theta)\} + H(T(x_n), T(\theta))$$
$$\le \rho d(x_n, \theta) + d(f(\theta), \theta) + \sup\{d(\theta, \eta) : \eta \in T(\theta)\} + d(x_n, \theta)$$
$$\le (1+\rho)d(x_n, \theta) + d(f(\theta), \theta) + \sup\{d(\theta, \eta) : \eta \in T(\theta)\}$$
$$\le (1+\rho)(M + M_0) + d(f(\theta), \theta) + \sup\{d(\theta, \eta) : \eta \in T(\theta)\} \le 2M. \quad (2.86)$$

In view of (2.85),

$$d(x_{n+2}, x_{n+1}) \le (1 - (1-\rho)(1-\alpha_{n+1}))d(x_{n+1}, x_n) + \alpha_{n+1}r_n + 2M|\alpha_{n+1} - \alpha_n|.$$

Proposition 2.1 and (2.75), (2.76), and (2.86) imply that for each integer $n \ge 0$,

$$d(x_n, T(x_n)) \le d(x_n, y_n) \le d(x_n, x_{n+1}) + d(x_{n+1}, y_n)$$
$$\le d(x_n, x_{n+1}) + (1-\alpha_n)d(f(x_n), y_n)$$
$$\le d(x_n, x_{n+1}) + 2M(1-\alpha_n).$$

Proposition 2.28 is proved.

Proof of Theorem 2.25 Set

$$\gamma = 1 - \rho, \ L = 2M + 1, \ N = 2,$$

$$c_n = L = 2M + 1, \ n = 0, 1, \ldots, \ r_n = (n + J)^{-1}, \ n = 0, 1, \ldots,$$

$$s_n = d(x_n, x_{n+1}), \ a_n = 1 - \alpha_n, \ n = 0, 1, \ldots.$$

Lemma 2.27 and (2.78) imply that

$$d(x_0, x_1) \le d(x_0, \theta) + d(\theta, x_1) \le d(x_0, \theta) + M + M_0 \le 2M.$$

Proposition 2.28 and the relation above imply that for each integer $n \ge 0$,

$$s_{n+1} \le (1 - \gamma a_{n+1})s_n + 2|a_{n+1} - a_n|M + \alpha_{n+1}r_n$$

$$\le (1 - \gamma a_{n+1})s_n$$

$$+ 2M(1-\rho)^{-1}((n+J)^{-1} - (n+J+1)^{-1}) + 2(1-\rho)^{-1}(n+J)^{-1}(n+J+1)^{-1}$$

$$\le (1 - \gamma a_{n+1})s_n + |a_{n+1} - a_n|(2M + 1).$$

Lemma 2.2 and the relation above imply that for each integer $n \ge 0$,

$$d(x_n, x_{n+1}) \le (2M + 1)J(n + J)^{-1}(1 - \rho)^{-1}.$$

Together with Proposition 2.28, this implies that for each integer $n \ge 0$,

$$d(x_n, T(x_n)) \le d(x_n, x_{n+1})$$

$$+ 4M(1 - \rho)^{-1}(n + J)^{-1} \le (2M + 1)(J + 2)(1 - \rho)^{-1}(n + J)^{-1}.$$

Theorem 2.25 is proved.

2.9 Alternative Iterative Method for Set-Valued Maps

We use the notation and definitions introduced in Sects. 2.1, 2.2, and 2.8. In particular we assume that assumption (A) from Sect. 2.8 holds.

We consider the alternative iterative method for the set-valued mapping T which is described below.

Assume that $\{\alpha_n\}_{n=0}^{\infty} \subset (0, 1)$, $\{r_n\}_{n=0}^{\infty} \subset (0, 1]$,

$$x_0 = x \in C$$

and that for each integer $n \ge 0$,

$$x_{n+1} \in T((1-\alpha_n)f(x_n) \oplus \alpha_n x_n), \tag{2.87}$$

$$d(x_n, x_{n+1}) \le d(x_n, T((1-\alpha_n)f(x_n) \oplus \alpha_n x_n)) + r_n. \tag{2.88}$$

We will prove the following result which shows that our iterative process generates approximate fixed points of the set-valued mapping T.

Theorem 2.29 *Let $J = 2\lfloor(1-\rho)^{-1}\rfloor + 1$ and for each integer $n \ge 0$,*

$$\alpha_n = 1 - 2(1-\rho)^{-1}(n+J)^{-1}.$$

Then for each integer $n \ge 0$,

$$d(x_n, x_{n+1}) \le (2M+1)J(1-\rho)^{-1}(n+J)^{-1}$$

and

$$d(x_n, T(x_n)) \le (2M+1)(J+2)(1-\rho)^{-1}(n+J)^{-1}.$$

We begin with auxiliary results.

Lemma 2.30 *Let $n \ge 0$ be an integer. Then*

$$d(x_{n+2}, x_{n+1}) \le (1-(1-\rho)(1-\alpha_{n+1}))d(x_{n+1}, x_n)$$

$$+ |\alpha_{n+1} - \alpha_n| d(f(x_n), x_n) + r_{n+1},$$

$$d(x_n, T(x_n)) \le d(x_n, x_{n+1}) + (1-\alpha_n)d(f(x_n), x_n).$$

Proof Proposition 2.1 and (2.73), (2.74), (2.87), and (2.88) imply that

$$d(x_{n+2}, x_{n+1})$$

$$\le d(x_{n+1}, T((1-\alpha_{n+1})f(x_{n+1}) \oplus \alpha_{n+1}x_{n+1})) + r_{n+1}$$

$$\le H(T((1-\alpha_n)f(x_n) \oplus \alpha_n x_n), T((1-\alpha_{n+1})f(x_{n+1}) \oplus \alpha_{n+1}x_{n+1})) + r_{n+1}$$

$$\le d((1-\alpha_n)f(x_n) \oplus \alpha_n x_n, (1-\alpha_{n+1})f(x_{n+1}) \oplus \alpha_{n+1}x_{n+1}) + r_{n+1}$$

$$\le (1-\alpha_{n+1})d(f(x_{n+1}), f(x_n)) + \alpha_{n+1}d(x_n, x_{n+1})$$

$$+ |\alpha_{n+1} - \alpha_n| d(f(x_n), x_n) + r_{n+1}$$

$$\le \rho(1-\alpha_{n+1})d(x_{n+1}, x_n) + \alpha_{n+1}d(x_{n+1}, x_n)$$

$$+ r_{n+1} + |\alpha_{n+1} - \alpha_n| d(f(x_n), x_n) + r_{n+1}$$

$$\leq (1 - (1 - \rho)(1 - \alpha_{n+1})) d(x_{n+1}, x_n) + |\alpha_{n+1} - \alpha_n| d(f(x_n), x_n) + r_{n+1}.$$

Proposition 2.1 and (2.73) and (2.87) imply that

$$d(x_n, T(x_n)) \leq d(x_n, x_{n+1}) + d(x_{n+1}, T(x_n))$$

$$\leq d(x_n, x_{n+1}) + H(T((1 - \alpha_n) f(x_n) \oplus \alpha_n x_n), T(x_n))$$

$$\leq d(x_n, x_{n+1}) + d((1 - \alpha_n) f(x_n) \oplus \alpha_n x_n, x_n)$$

$$\leq d(x_n, x_{n+1}) + (1 - \alpha_n) d(f(x_n), x_n).$$

Lemma 2.30 is proved.

Lemma 2.31 *For each integer $n \geq 0$,*

$$d(x_n, T^n(p)) \leq M, \tag{2.89}$$

$$d(x_n, \theta) \leq M + M_0.$$

Proof By (A) and (2.78),

$$d(x_0, p) \leq M_0 + d(\theta, x_0) < M.$$

Assume that (2.89) holds for a nonnegative integer n. Let $\epsilon > 0$. In view of (2.89), there exists

$$\xi \in T^n(p) \tag{2.90}$$

such that

$$d(x_n, \xi) \leq M + \epsilon. \tag{2.91}$$

There exists

$$\eta \in T(\xi) \tag{2.92}$$

such that

$$d(x_{n+1}, \eta) \leq d(x_{n+1}, T(\xi)) + \epsilon. \tag{2.93}$$

By Proposition 2.1, assumption (A) and (2.73), (2.74), (2.87), and (2.90)–(2.93),

$$d(x_{n+1}, T^{n+1}(p)) \le d(x_{n+1}, \eta) \le d(x_{n+1}, T(\xi)) + \epsilon$$

$$\le H(T((1-\alpha_n)f(x_n) \oplus \alpha_n x_n), T(\xi)) + \epsilon$$

$$\le d((1-\alpha_n)f(x_n) \oplus \alpha_n x_n, \xi) + \epsilon$$

$$\le (1-\alpha_n)d(f(x_n), \xi) + \alpha_n d(x_n, \xi) + \epsilon$$

$$\le (1-\alpha_n)d(f(x_n), f(\xi)) + (1-\alpha_n)d(f(\xi), \xi) + \alpha_n d(x_n, \xi) + \epsilon$$

$$\le (1-\alpha_n)\rho d(x_n, \xi) + \alpha_n d(x_n, \xi)$$

$$+ (1-\alpha_n)(d(f(\xi), f(\theta)) + d(f(\theta), \theta) + d(\theta, \xi)) + \epsilon$$

$$\le (1-(1-\alpha_n)(1-\rho))d(x_n, \xi) + (1-\alpha_n)(\rho M_0 + d(f(\theta), \theta) + M_0) + \epsilon$$

$$\le (1-(1-\alpha_n)(1-\rho))(M+\epsilon) + (1-\alpha_n)(\rho M_0 + d(f(\theta), \theta) + M_0).$$

Since ϵ is any positive number together with (2.78), the relation above implies that

$$d(x_{n+1}, T^{n+1}(p))$$

$$\le (1-(1-\alpha_n)(1-\rho))M + (1-\alpha_n)(\rho M_0 + d(f(\theta), \theta) + M_0)$$

$$\le M - (1-\alpha_n)((1-\rho)M - 2M_0 - d(f(\theta), \theta) - M_0) \le M.$$

Thus we showed that (2.89) holds for $n+1$ too. Therefore by induction, it holds for each integer $n \ge 0$. Let $n \ge 0$ be an integer. In view of (2.89) and the choice of M_0,

$$d(x_n, \theta) \le d(x_n, T^n(p)) + \sup\{d(\theta, \eta) : \eta \in T^n(p)\} \le M + M_0.$$

Lemma 2.31 is proved.

Proposition 2.32 *For each integer* $n \ge 0$,

$$d(f(x_n), y_n) \le 2M,$$

$$d(x_{n+2}, x_{n+1}) \le (1-(1-\rho)(1-\alpha_{n+1}))d(x_{n+1}, x_n) + 2M|\alpha_{n+1} - \alpha_n| + r_{n+1},$$

$$d(x_n, T(x_n)) \le d(x_n, x_{n+1}) + 2M(1-\alpha_n).$$

Proof Let $n \ge 0$ be an integer. Lemma 2.30 implies that

$$d(x_{n+2}, x_{n+1}) \le (1-(1-\rho)(1-\alpha_{n+1}))d(x_{n+1}, x_n)$$

$$+ |\alpha_{n+1} - \alpha_n| d(f(x_n), x_n) + r_{n+1}. \tag{2.94}$$

Lemma 2.31 and (2.74) and (2.79) imply that

$$d(f(x_n), x_n)$$

$$\leq d(f(x_n), f(\theta)) + d(f(\theta), \theta) + \sup\{d(\theta, \eta) : \eta \in T(\theta)\} + d(x_n, T(\theta))$$

$$\leq \rho d(x_n, \theta) + d(f(\theta), \theta) + \sup\{d(\theta, \eta) : \eta \in T(\theta)\} + d(x_n, \theta) + d(\theta, T(\theta))$$

$$\leq (1 + \rho) d(x_n, \theta) + d(f(\theta), \theta) + d(\theta, T(\theta)) + \sup\{d(\theta, \eta) : \eta \in T(\theta)\}$$

$$\leq (1 + \rho) M + (1 + \rho) M_0 + d(f(\theta), \theta) + 2 \sup\{d(\theta, \eta) : \eta \in T(\theta)\} \leq 2M. \tag{2.95}$$

In view of (2.94) and (2.95),

$$d(x_{n+2}, x_{n+1}) \leq (1 - (1 - \rho)(1 - \alpha_{n+1})) d(x_{n+1}, x_n) + 2M |\alpha_{n+1} - \alpha_n| + r_{n+1}.$$

Lemma 2.30 and (2.95) imply that

$$d(x_n, T(x_n)) \leq d(x_n, x_{n+1}) + 2M(1 - \alpha_n).$$

Proposition 2.32 is proved.

Theorem 2.29 is proved analogously to Theorem 2.25 using Proposition 2.32 and Lemmas 2.2 and 2.31.

2.10 Inexact Halpern-Type Iterations of Set-Valued Mappings

We assume that (X, d, W) is a W-hyperbolic space which is denoted by X for simplicity. We use the notation and definitions introduced in Sects. 2.1, 2.2, and 2.8.

Recall that $\rho \in [0, 1)$, $T : C \to 2^C \setminus \{\emptyset\}$, $T(z)$ is bounded for each $z \in C$,

$$H(T(x), T(y)) \leq d(x, y), \ x, y \in C \tag{2.96}$$

and that $f : C \to \mathrm{St}(C)$ satisfies

$$d(f(x), f(y)) \leq \rho d(x, y), \ x, y \in C. \tag{2.97}$$

Assume that $\{\alpha_n\}_{n=0}^{\infty} \subset (0, 1)$, $\{r_n\}_{n=0}^{\infty}$, $\{\Delta_n\}_{n=0}^{\infty} \subset (0, 1]$,

$$x_0 = x \in C, \{x_n\}_{n=1}^{\infty}, \ \{y_n\}_{n=0}^{\infty} \subset C \tag{2.98}$$

and that for each integer $n \geq 0$,

$$B(y_n, r_n) \cap T(x_n) \neq \emptyset, \tag{2.99}$$

$$x_{n+1} \in B((1-\alpha_n)f(x_n) \oplus \alpha_n y_n, \Delta_n) \tag{2.100}$$

$$B(y_{n+1}, r_{n+1}) \cap \{y \in T(x_{n+1}) :$$

$$d(y_n, y) \leq d(y_n, T(x_{n+1})) + r_{n+1}\} \neq \emptyset. \tag{2.101}$$

Let $\theta \in C$, $M_0 > 0$,

$$M > (1-\rho)^{-1}(2M_0 + d(x_0, \theta) + d(f(\theta), \theta) + H(\{\theta\}, T(\theta)) + 1)$$

$$+ 2\sup\{d(\theta), \eta) : \ \eta \in T(\theta)\}. \tag{2.102}$$

We assume that assumption (A) introduced in Sect. 2.8 holds. We prove the following result which shows that our iterative process generates approximate fixed points of the set-valued mapping T under the presence of computational errors.

Theorem 2.33 *Let $J = 2\lfloor(1-\rho)^{-1}\rfloor + 1$ and for each integer $n \geq 0$,*

$$\alpha_n = 1 - 2(1-\rho)^{-1}(n+J)^{-1}, \ r_n = \Delta_n = (n+J+1)^{-2}.$$

Then for each integer $n \geq 0$,

$$d(x_n, x_{n+1}) \leq (2M+6)J(1-\rho)^{-1}(n+J)^{-1}$$

and

$$d(x_n, T(x_n)) \leq (2M+6)(J+3)(1-\rho)^{-1}(n+J)^{-1}.$$

We begin with auxiliary results.

Lemma 2.34 *Let $n \geq 0$ be an integer. Then*

$$d(x_{n+2}, x_{n+1}) \leq (1-(1-\rho)(1-\alpha_{n+1}))d(x_{n+1}, x_n) + \Delta_n + \Delta_{n+1}$$

$$+ \alpha_{n+1}(2r_{n+1} + r_n) + |\alpha_{n+1} - \alpha_n| d(f(x_n), y_n).$$

Proof Proposition 2.1 and (2.97) and (2.100) imply that

$$d(x_{n+2}, x_{n+1})$$

$$\le d(x_{n+2}, (1-\alpha_{n+1})f(x_{n+1}) \oplus \alpha_{n+1} y_{n+1})$$

$$+ d((1-\alpha_{n+1})f(x_{n+1}) \oplus \alpha_{n+1} y_{n+1}, (1-\alpha_n)f(x_n) \oplus \alpha_n y_n)$$

$$+ d((1-\alpha_n)f(x_n) \oplus \alpha_n y_n, x_{n+1})$$

$$\le \Delta_n + \Delta_{n+1} + (1-\alpha_{n+1})d(f(x_{n+1}), f(x_n)) + \alpha_{n+1} d(y_n, y_{n+1})$$

$$+ |\alpha_{n+1} - \alpha_n| d(f(x_n), y_n)$$

$$\le \Delta_n + \Delta_{n+1} + \rho(1-\alpha_{n+1})d(x_{n+1}, x_n) + \alpha_{n+1} d(y_{n+1}, y_n) + \alpha_{n+1} d(f(x_n), y_n). \tag{2.103}$$

By (2.101), there exists

$$\xi \in T(x_{n+1}) \cap B(y_{n+1}, r_{n+1}) \tag{2.104}$$

such that

$$d(\xi, y_n) \le d(\eta, y_n) + r_{n+1}, \ \eta \in T(x_{n+1}). \tag{2.105}$$

By (2.96), (2.99), (2.104), and (2.105),

$$d(y_n, y_{n+1}) \le d(y_{n+1}, \xi) + d(\xi, y_n)$$

$$\le d(y_n, T(x_{n+1})) + 2r_{n+1}$$

$$\le 2r_{n+1} + d(y_n, T(x_n)) + H(T(x_n), T(y_n))$$

$$\le 2r_{n+1} + r_n + d(x_n, x_{n+1}). \tag{2.106}$$

By (2.103) and (2.106),

$$d(x_{n+2}, x_{n+1}) \le \Delta_n + \Delta_{n+1} + \rho(1-\alpha_{n+1})d(x_{n+1}, x_n)$$

$$+ \alpha_{n+1}(2r_{n+1} + r_n) + \alpha_{n+1} d(x_{n+1}, x_n) + |\alpha_{n+1} - \alpha_n| d(f(x_n), y_n).$$

Lemma 2.34 is proved.

Lemma 2.35 *For each integer $n \ge 0$,*

$$d(x_n, T^n(p)) \le M + \sum\{\alpha_i r_i + \Delta_i : i \in \{0, \dots, n\} \setminus \{n\}\}, \tag{2.107}$$

$$d(x_n, \theta) \le M + M_0 + \sum\{\alpha_i r_i + \Delta_i : i \in \{0, \dots, n\} \setminus \{n\}\}.$$

Proof By (A) and (2.78),

$$d(x_0, p) \le M_0 + d(\theta, x_0) < M.$$

Assume that (2.107) holds for a nonnegative integer n. Let $\epsilon > 0$. In view of (2.107), there exists

$$\xi \in T^n(p) \tag{2.108}$$

such that

$$d(x_n, \xi) \le M + \epsilon + \sum\{\alpha_i r_i + \Delta_i : i \in \{0, \dots, n\} \setminus \{n\}\}. \tag{2.109}$$

In view of (2.99), there exists

$$\widehat{y} \in T(x_n) \tag{2.110}$$

such that

$$d(y_n, \widehat{y}_n) \le r_n. \tag{2.111}$$

By (2.96), (2.109), and (2.110), there exists

$$\eta \in T(\xi) \tag{2.112}$$

such that

$$d(\widehat{y}_n, \eta) \le H(T(x_n), T(\xi)) + \epsilon \le d(x_n, \xi) + \epsilon$$

$$\le M + 2\epsilon + \sum\{\alpha_i r_i + \Delta_i : i \in \{0, \dots, n\} \setminus \{n\}\}. \tag{2.113}$$

By Proposition 2.1, assumption (A), and (2.97), (2.100), (2.108), (2.109), and (2.111)–(2.113),

$$d(x_{n+1}, T^{n+1}(p)) \le d(x_{n+1}, \eta)$$

$$\le d(x_{n+1}, (1 - \alpha_n) f(x_n) \oplus \alpha_n y_n) + d((1 - \alpha_n) f(x_n) \oplus \alpha_n y_n, \eta)$$

$$\le \Delta_n + (1 - \alpha_n) d(f(x_n), \eta) + \alpha_n d(y_n, \eta)$$

$$\leq \Delta_n + (1-\alpha_n)d(f(x_n), f(\xi)) + (1-\alpha_n)d(f(\xi), \eta) + \alpha_n(d(y_n, \widehat{y}) + d(\widehat{y}, \eta))$$

$$\leq \Delta_n + (1-\alpha_n)\rho d(x_n, \xi) + (1-\alpha_n)(d(f(\xi), f(\theta)) + d(f(\theta), \theta) + d(\theta, \eta))$$

$$+ \alpha_n(r_n + d(x_n, \xi) + \epsilon)$$

$$\leq (1-(1-\alpha_n)(1-\rho))d(x_n, \xi) + \alpha_n r_n + \alpha_n \epsilon$$

$$+ (1-\alpha_n)(\rho d(\xi, \theta) + d(f(\theta), \theta) + d(\theta, \eta)) + \Delta_n$$

$$\leq (1-(1-\alpha_n)(1-\rho))(M + \epsilon + \sum\{\alpha_i r_i + \Delta_i : i \in \{0, \dots, n\} \setminus \{n\}\})$$

$$+ \Delta_n + \alpha_n r_n + \alpha_n \epsilon + (1-\alpha_n)(\rho M_0 + d(f(\theta), \theta) + M_0). \tag{2.114}$$

Since ϵ is any positive number, Eqs. (2.78) and (2.114) imply that

$$d(x_{n+1}, T^{n+1}(p))$$

$$\leq (1-(1-\alpha_n)(1-\rho))(M + \sum\{\alpha_i r_i + \Delta_i : i \in \{0, \dots, n\} \setminus \{n\}\})$$

$$+ \Delta_n + \alpha_n r_n + (1-\alpha_n)(\rho M_0 + M_0 + d(f(\theta), \theta))$$

$$\leq M + \sum_{i=0}^{n}(\alpha_i r_i + \Delta_i) - (1-\alpha_n)((1-\rho)M - (1+\rho)M_0 - d(f(\theta), \theta))$$

$$\leq M + \sum_{i=0}^{n}(\alpha_i r_i + \Delta_i).$$

Thus we showed that (2.107) holds for $n+1$ too. Therefore by induction (2.107) holds for each integer $n \geq 0$. Let $n \geq 0$ be an integer. In view of (2.107),

$$d(x_n, \theta) \leq d(x_n, T^n(p)) + H(T^n(p), \{\theta\})$$

$$\leq M + \sum\{\alpha_i r_i + \Delta_i : i \in \{0, \dots, n\} \setminus \{n\}\} + M_0.$$

Lemma 2.35 is proved.

Proposition 2.36 *For each integer $n \geq 0$,*

$$d(f(x_n), y_n) \leq 2M + (1+\rho)\sum_{i=0}^{n}(\alpha_i r_i + \Delta_i),$$

$$d(x_{n+2}, x_{n+1}) \leq (1-(1-\rho)(1-\alpha_{n+1}))d(x_{n+1}, x_n) + 2M|\alpha_{n+1} - \alpha_n|$$

$$+ \Delta_{n+1} + \Delta_n + \alpha_{n+1}(2r_{n+1} + r_n) + (1+\rho)|\alpha_{n+1} - \alpha_n| \sum_{i=0}^{n} (\alpha_i r_i + \Delta_i),$$

$$d(x_n, T(x_n)) \le d(x_n, x_{n+1}) + 2M(1 - \alpha_n)$$

$$+ (1 - \alpha_n)(1+\rho) \sum_{i=0}^{n} (\alpha_i r_i + \Delta_i) + \Delta_n + r_n.$$

Proof Let $n \ge 0$ be an integer. Lemma 2.35 and (2.96), (2.97), (2.99), and (2.102) imply that

$$d(f(x_n), y_n) \le d(f(x_n), f(\theta)) + d(f(\theta), \theta)$$

$$+ \sup\{d(\theta, \eta) : \eta \in T(\theta)\} + H(T(\theta), T(x_n)) + d(y_n, T(x_n))$$

$$\le \rho d(x_n, \theta) + d(f(\theta), \theta) + \sup\{d(\theta, \eta) : \eta \in T(\theta)\} + d(\theta, x_n) + r_n$$

$$\le (1+\rho) d(x_n, \theta) + d(f(\theta), \theta) + \sup\{d(\theta, \eta) : \eta \in T(\theta)\} + r_n$$

$$\le (1+\rho)M + (1+\rho)M_0 + (1+\rho) \sum \{\alpha_i r_i + \Delta_i : i \in \{0, \dots, n\} \setminus \{n\}\},$$

$$+ r_n + d(f(\theta), \theta) + H(\{\theta\}, T(\theta))$$

$$\le 2M + (1+\rho) \sum_{i=0}^{n} (\alpha_i r_i + \Delta_i). \tag{2.115}$$

Lemmas 2.34 and (2.115) imply that

$$d(x_{n+2}, x_{n+1}) \le (1 - (1-\rho)(1 - \alpha_{n+1})) d(x_{n+1}, x_n) + \Delta_{n+1} + \Delta_n$$

$$+ \alpha_{n+1}(r_n + 2r_{n+1}) + |\alpha_{n+1} - \alpha_n| d(f(x_n), y_n)$$

$$\le (1 - (1-\rho)(1 - \alpha_{n+1})) d(x_{n+1}, x_n) + \Delta_{n+1} + \Delta_n$$

$$+ \alpha_{n+1}(r_n + 2r_{n+1}) + |\alpha_{n+1} - \alpha_n| 2M + (1+\rho)|\alpha_{n+1} - \alpha_n| \sum_{i=0}^{n} (\alpha_i r_i + \Delta_i).$$

Proposition 2.1 and (2.99), (2.100), and (2.115) imply that

$$d(x_n, T(x_n)) \le d(x_n, y_n) + d(y_n, T(x_n)) \le d(x_n, y_n) + r_n$$

$$\le d(x_n, x_{n+1}) + d(x_{n+1}, y_n) + r_n$$

$$\leq d(x_n, x_{n+1}) + r_n + d(x_n, (1-\alpha_n) f(x_n) \oplus \alpha_n y_n) + d((1-\alpha_n) f(x_n) \oplus \alpha_n y_n), y_n)$$

$$\leq d(x_n, x_{n+1}) + r_n + \Delta_n + (1-\alpha_n) d(f(x_n), y_n)$$

$$\leq d(x_n, x_{n+1}) + 2M(1-\alpha_n) + (1-\alpha_n)(1+\rho) \sum_{i=0}^{n} (\alpha_i r_i + \Delta_i) + \Delta_n + r_n.$$

Proposition 2.36 is proved.

Proof of Theorem 2.33 Set

$$\gamma = 1-\rho, \; L = 2M + 6, \; N = 2,$$

$$c_n = L = 2M+6, \; n = 0, 1, \ldots, \; a_n = 1-\alpha_n = 2(1-\rho)^{-1}(n+J)^{-1}, \; n = 0, 1, \ldots,$$

$$\Delta_n = r_n = (n+J+1)^{-2}, \; n = 0, 1, \ldots,$$

$$s_n = d(x_n, x_{n+1}), , \; n = 0, 1, \ldots.$$

Lemma 2.35 implies that

$$d(x_0, x_1) \leq d(x_0, \theta) + d(\theta, x_1) \leq d(x_0, \theta) + M + M_0 + 4 \leq 2M + 4.$$

Proposition 2.36 implies that for each integer $n \geq 0$,

$$s_{n+1} \leq s_n(1 - \gamma a_{n+1})$$

$$+ 4M(1-\rho)^{-1}((n+J)^{-1} - (n+J+1)^{-1}) + 2((n+J+1)^{-2} + (n+J+2)^{-2})$$

$$+ 2(1-\rho)^{-1}(1+\rho)((n+J+1)^{-1} - (n+J+2)^{-1}) + 2\sum_{i=0}^{\infty}(n+J+i)^2$$

$$\leq s_n(1 - \gamma a_{n+1}) + 2(1-\rho)^{-1}(n+J)^{-1}(n+J+1)^{-1}(2M+6)$$

$$\leq (1 - \gamma a_{n+1}) s_n + (a_n - a_{n+1})(2M+6).$$

Lemma 2.2, Proposition 2.36, and the relation above imply that for each integer $n \geq 0$,

$$d(x_n, x_{n+1}) \leq (2M+6) J (1-\rho)^{-1}(n+J)^{-1},$$

$$d(x_n, T(x_n)) \leq (2M+6) J (1-\rho)^{-1}(n+J)^{-1}$$

$$+ 4M(1-\rho)^{-1}(n+J)^{-1} + 16(1-\rho)^{-1}(n+J)^{-1}$$

$$\leq (2M+6)J(1-\rho)^{-1}(n+J)^{-1} + (1-\rho)^{-1}(n+J)^{-1}(4M+17)$$

$$\leq (2M+6)(J+3)(1-\rho)^{-1}(n+J)^{-1}.$$

Theorem 2.33 is proved.

2.11 Inexact Alternative Iterative Method for Set-Valued Maps

We assume that (X, d, W) is a W-hyperbolic space which is denoted by X for simplicity. We use the notation and definitions introduced in Sect. 2.8. In particular, we assume that (A) holds.

Recall that $\rho \in [0, 1)$, $T : C \to 2^C \setminus \{\emptyset\}$, $T(z)$ is bounded for each $z \in C$,

$$H(T(x), T(y)) \leq d(x, y), \; x, y \in C \tag{2.116}$$

and that $f : C \to \mathrm{St}(C)$ satisfies

$$d(f(x), f(y)) \leq \rho d(x, y), \; x, y \in C. \tag{2.117}$$

Assume that $\{\alpha_n\}_{n=0}^{\infty} \subset (0, 1)$, $\{r_n\}_{n=0}^{\infty} \subset (0, 1]$,

$$x_0 = x \in C, \{x_n\}_{n=1}^{\infty} \subset C$$

and that for each integer $n \geq 0$,

$$B(x_{n+1}, r_n) \cap \{\xi \in T((1-\alpha_n)f(x_n) \oplus \alpha_n x_n) :$$

$$d(x_n, \xi) \leq d(x_n, T((1-\alpha_n)f(x_n) \oplus \alpha_n x_n)) + r_n\} \neq \emptyset. \tag{2.118}$$

Let $\theta \in C$, $M_0 > 0$,

$$M > (1-\rho)^{-1}(2M_0 + d(f(\theta), \theta) + d(\theta, T(\theta)) + H(\{\theta\}, T(\theta))).$$

We prove the following result which shows that our iterative process generates approximate fixed points of the set-valued mapping T under the presence of computational errors.

Theorem 2.37 *Let $J = 2\lfloor (1-\rho)^{-1} \rfloor + 1$ and for each integer $n \geq 0$,*

$$\alpha_n = 1 - 2(1-\rho)^{-1}(n+J)^{-1}, \; r_n = (n+J+1)^{-2}.$$

Then for each integer $n \geq 0$,

$$d(x_n, x_{n+1}) \leq (2M+6)J(1-\rho)^{-1}(n+J)^{-1}$$

and

$$d(x_n, T(x_n)) \leq (2M+6)(J+3)(1-\rho)^{-1}(n+J)^{-1}.$$

By (2.118), for each integer $n \geq 0$, there exists

$$y_n \in B(x_{n+1}, r_n) \cap T((1-\alpha_n)f(x_n) \oplus \alpha_n x_n) \tag{2.119}$$

such that

$$d(x_n, y_n) \leq d(x_n, T((1-\alpha_n)f(x_n) \oplus \alpha_n x_n)) + r_n. \tag{2.120}$$

We begin with auxiliary results.

Lemma 2.38 *Let $n \geq 0$ be an integer. Then*

$$d(x_{n+2}, x_{n+1}) \leq (1-(1-\rho)(1-\alpha_{n+1}))d(x_{n+1}, x_n)$$

$$+ 2r_{n+1} + r_n + |\alpha_{n+1} - \alpha_n| d(f(x_n), x_n).$$

Proof Proposition 2.1 and (2.116), (2.119), and (2.120) imply that

$$d(x_{n+2}, x_{n+1})$$

$$\leq d(x_{n+2}, y_{n+1}) + d(y_{n+1}, x_{n+1})$$

$$\leq r_{n+1} + d(x_{n+1}, T((1-\alpha_{n+1})f(x_{n+1}) \oplus \alpha_{n+1}x_{n+1})) + r_{n+1}$$

$$\leq 2r_{n+1} + d(x_{n+1}, y_n) + d(y_n, T((1-\alpha_{n+1})f(x_{n+1}) \oplus \alpha_{n+1}x_{n+1}))$$

$$\leq 2r_{n+1} + r_n + H(T((1-\alpha_n)f(x_n) \oplus \alpha_n x_n), T((1-\alpha_{n+1})f(x_{n+1}) \oplus \alpha_{n+1}x_{n+1}))$$

$$\leq 2r_{n+1} + r_n + d((1-\alpha_n)f(x_n) \oplus \alpha_n x_n, (1-\alpha_{n+1})f(x_{n+1}) \oplus \alpha_{n+1}x_{n+1})$$

$$\leq 2r_{n+1} + r_n + (1-\alpha_{n+1})d(f(x_n, f(x_{n+1}))$$

$$+ \alpha_{n+1}d(x_n, x_{n+1})) + |\alpha_{n+1} - \alpha_n| d(f(x_n), x_n)$$

$$\leq r_n + 2r_{n+1} + \rho(1-\alpha_{n+1})d(x_{n+1}, x_n)$$

$$+ \alpha_{n+1}d(x_{n+1}, x_n) + |\alpha_{n+1} - \alpha_n| d(x_{n+1}, x_n).$$

Lemma 2.38 is proved.

Lemma 2.39 *For each integer* $n \geq 0$,

$$d(x_n, T^n(p)) \leq M + \sum\{r_i : i \in \{0, \dots, n\} \setminus \{n\}\}, \tag{2.121}$$

$$d(x_n, \theta) \leq M + M_0 + \sum\{(\alpha_i + 1)r_i : i \in \{0, \dots, n\} \setminus \{n\}\}.$$

Proof By (A) and (2.119),

$$d(x_0, p) \leq M_0 + d(\theta, x_0) < M.$$

Assume that (2.121) holds for a nonnegative integer n. Let $\epsilon > 0$. In view of (2.121), there exists

$$\xi \in T^n(p) \tag{2.122}$$

such that

$$d(x_n, \xi) \leq M + \epsilon + \sum\{r_i : i \in \{0, \dots, n\} \setminus \{n\}\}. \tag{2.123}$$

There exists

$$\eta \in T(\xi) \tag{2.124}$$

such that

$$d(x_{n+1}, \eta) \leq d(x_{n+1}, T(\xi)) + \epsilon. \tag{2.125}$$

By Proposition 2.1, assumption (A), and (2.116), (2.118), (2.122), (2.124), and (2.125),

$$\begin{aligned}
d(x_{n+1}, T^{n+1}(p)) &\leq d(x_{n+1}, \eta) \leq d(x_{n+1}, T(\xi)) + \epsilon \\
&\leq d(x_{n+1}, T((1 - \alpha_n) f(x_n) \oplus \alpha_n x_n)) \\
&+ H(T((1 - \alpha_n) f(x_n) \oplus \alpha_n x_n), T(\xi)) + \epsilon \\
&\leq r_n + d((1 - \alpha_n) f(x_n) \oplus \alpha_n x_n), \xi) + \epsilon \\
&\leq (1 - \alpha_n) d(f(x_n), \xi) + \alpha_n d(x_n, \xi) + \epsilon + r_n \\
&\leq (1 - \alpha_n) d(f(x_n), f(\xi)) + (1 - \alpha_n) d(f(\xi), \xi) + \alpha_n d(x_n, \xi) + \epsilon + r_n \\
&\leq (1 - \alpha_n) \rho d(x_n, \xi) + \alpha_n d(x_n, \xi) \\
&+ (1 - \alpha_n)(d(f(\xi), f(\theta)) + d(f(\theta), \theta) + d(\theta, \xi)) + \epsilon + r_n
\end{aligned}$$

$$\leq (1 - (1 - \alpha_n)(1 - \rho))d(x_n, \xi) +$$

$$+ (1 - \alpha_n)(\rho M_0 + d(f(\theta), \theta) + M_0) + \epsilon + r_n.$$

Since ϵ is any positive number together with (2.119) and (2.123), the relation above implies that

$$d(x_{n+1}, T^{n+1}(p))$$

$$\leq (1 - (1 - \alpha_n)(1 - \rho))(M + \sum\{r_i : i \in \{0, \dots, n\} \setminus \{n\}\})$$

$$+ (1 - \alpha_n)(\rho M_0 + M_0 + d(f(\theta), \theta)) + r_n$$

$$\leq M + \sum\{r_i : i \in \{0, \dots, n\} \setminus \{n\}\}$$

$$- (1 - \alpha_n)((1 - \rho)M - (1 + \rho)M_0 - d(f(\theta), \theta)) + r_n$$

$$\leq M + \sum_{i=0}^{n} r_i.$$

Thus we showed that (2.121) holds for $n+1$ too. Therefore, we showed by induction that (2.121) holds for each integer $n \geq 0$. Let $n \geq 0$ be an integer. In view of (2.121),

$$d(x_n, \theta) \leq d(x_n, T^n(p)) + H(T^n(p), \{\theta\})$$

$$\leq M + M_0 + \sum\{r_i : i \in \{0, \dots, n\} \setminus \{n\}\}.$$

Lemma 2.39 is proved.

Proposition 2.40 *Let $n \geq 0$ be an integer. Then*

$$d(f(x_n), x_n) \leq 2M + (1 + \rho)\sum\{r_i : i \in \{0, \dots, n\} \setminus \{n\}\},$$

$$d(x_{n+2}, x_{n+1}) \leq (1 - (1 - \rho)(1 - \alpha_{n+1}))d(x_{n+1}, x_n)$$

$$+ |\alpha_{n+1} - \alpha_n|(2M + \sum_{j=0}^{\infty} r_j) + 2r_{n+1} + r_n,$$

$$d(x_n, T(x_n)) \leq d(x_n, x_{n+1}) + r_n + (1 - \alpha_n)(2M + \sum_{j=0}^{\infty} r_j).$$

Proof Lemma 2.39 and (2.117) and (2.119) imply that

$$d(f(x_n), x_n) \le d(f(x_n), f(\theta)) + d(f(\theta), \theta)$$

$$+ H(T(\theta), \{\theta\}) + d(x_n, T(\theta))$$

$$\le \rho d(x_n, \theta) + d(f(\theta), \theta) + H(\{\theta\}, T(\theta)) + d(x_n, \theta) + d(\theta, T(\theta))$$

$$\le (1+\rho) d(x_n, \theta) + d(f(\theta), \theta) + H(\{\theta\}, T(\theta)) + d(\theta, T(\theta))$$

$$\le (1+\rho)(M + M_0) + (1+\rho) \sum \{r_i : i \in \{0, \dots, n\} \setminus \{n\}\}$$

$$+ d(f(\theta), \theta) + H(\{\theta\}, T(\theta) + d(\theta, T(\theta))$$

$$\le 2M + (1+\rho) \sum_{i=0}^{n} r_i. \tag{2.126}$$

Lemma 2.38 and (2.126) imply that

$$d(x_{n+2}, x_{n+1}) \le (1 - (1-\rho)(1-\alpha_{n+1})) d(x_{n+1}, x_n)$$

$$+ |\alpha_{n+1} - \alpha_n| d(f(x_n), x_n) + 2r_{n+1} + r_n$$

$$\le (1 - (1-\rho)(1-\alpha_{n+1})) d(x_{n+1}, x_n)$$

$$+ |\alpha_{n+1} - \alpha_n| (1+\rho)(2M + \sum_{j=0}^{\infty} r_j) + 2r_{n+1} + r_n.$$

Proposition 2.1 and (2.116), (2.118), and (2.126) imply that

$$d(x_n, T(x_n)) \le d(x_n, x_{n+1}) + d(x_{n+1}, T(x_n))$$

$$\le d(x_n, x_{n+1}) + d(x_{n+1}, T((1-\alpha_n) f(x_n) \oplus \alpha_n x_n))$$

$$+ H(T((1-\alpha_n) f(x_n) \oplus \alpha_n x_n), T(x_n))$$

$$\le d(x_n, x_{n+1}) + r_n + d((1-\alpha_n) f(x_n) \oplus \alpha_n x_n, x_n)$$

$$\le d(x_n, x_{n+1}) + r_n + (1-\alpha_n) d(f(x_n), x_n)$$

$$\le d(x_n, x_{n+1}) + r_n + (1-\alpha_n)(2M + \sum_{j=0}^{\infty} r_j)(1+\rho).$$

Proposition 2.40 is proved.

Theorem 2.37 is proved analogously to Theorem 2.33 using Lemma 2.33 and Proposition 2.40.

Chapter 3
Asymptotic Regularity for Iterations of Monotone Nonexpansive Mappings

In this chapter, we study asymptotic regularity for Halpern-type iterations and for iterations of alternative iterative methods generated by a monotone nonexpansive self-mapping of a star-shape set in a metric space with a hyperbolic structure and an order. Linear rates of this asymptotic regularity are computed under assumption that there exists a bounded orbit of the monotone nonexpansive mapping. We study asymptotic regularity for single-valued and set-valued mappings. All the results of this chapter are new.

3.1 Halpern-Type Iterations of a Monotone Nonexpansive Mapping

Assume that $X = (X, d, W)$ is a W-hyperbolic space, C is a nonempty star-shaped subset of X, C is equipped with an order $\leq$, and $T : C \to C$ satisfies

$$d(T(x), T(y)) \leq d(x, y), \ \ T(x) \leq T(y) \tag{3.1}$$

for each $x, y \in C$ satisfying $x \leq y$. Assume that $\{\alpha_n\}_{n=0}^{\infty} \subset (0, 1)$,

$$\alpha_n \leq \alpha_{n+1} \text{ for each integer } n \geq 0, \tag{3.2}$$

$$x_0 = x \in \mathrm{St}(C), \tag{3.3}$$

$$x_0 \leq T(x_0), \tag{3.4}$$

$$M_0 > 0, \ \ M > 4M_0$$

and that for each integer $n \geq 0$,

A. J. Zaslavski, *Approximate Fixed Points of Nonexpansive Mappings*,
Developments in Mathematics 80, https://doi.org/10.1007/978-3-031-70710-0_3

$$x_{n+1} = (1 - \alpha_n)x_0 \oplus \alpha_n T(x_n). \tag{3.5}$$

We study the asymptotic regularity of the sequence $\{x_n\}_{n=0}^{\infty}$ and assume that the following assumptions hold.

(A1) For each integer $n \geq 0$.

$$d(T^i(x_0), x_0) \leq M_0.$$

(A2) If $y \in C$ satisfies $x_0 \leq y$ and if $0 \leq \gamma_1 < \gamma_2 \leq 1$, then

$$(1 - \gamma_1)x_0 \oplus \gamma_1 y \leq (1 - \gamma_2)x_0 \oplus \gamma_2 y.$$

(A3) If $y_1, y_2 \in C$ satisfy $x_0 \leq y_1 \leq y_2$ and if $0 < \gamma < 1$, then

$$(1 - \gamma)x_0 \oplus \gamma y_1 \leq (1 - \gamma_2)x_0 \oplus \gamma_2 y.$$

(Note that (A2) and (A3) hold if X is a normed space ordered by a convex cone.)

In this chapter, our goal is to show that analogs of the results of Chap. 2 hold for monotone nonexpansive operators. For these operators, relation (3.1) is true only if $x \leq y$. Therefore in order to obtain such analogs, we should guarantee that for our sequences of iterates the inequality $x_n \leq x_{n+1}$ holds for any integer $n \geq 0$. In order to meet this goal, we need assumptions (A1)–(A3) on the connection between the hyperbolic structure and the order.

Lemma 3.1 *For each integer $n \geq 0$,*

$$x_n \leq x_{n+1} \leq T(x_n).$$

Proof Assumption (A2) and (3.4) and (3.5) imply that

$$x_0 \leq x_1 \leq T(x_0).$$

Assume that $n \geq 1$ is an integer and that for each integer $i \in \{0, \ldots, n-1\}$,

$$x_i \leq x_{i+1}. \tag{3.6}$$

By (3.4) and (3.6),

$$T(x_i) \leq T(x_{i+1}),\ i = 0, \ldots, n-1, \tag{3.7}$$

$$x_0 \leq T(x_0) \leq T(x_i),\ i = 0, \ldots, n. \tag{3.8}$$

In view of (3.5),

$$x_n = (1 - \alpha_{n-1})x_0 \oplus \alpha_{n-1}T(x_{n-1}), \tag{3.9}$$

$$x_{n+1} = (1 - \alpha_n)x_0 \oplus \alpha_n T(x_n). \tag{3.10}$$

Assumptions (A2) and (A3) and Eqs. (3.2) and (3.7)–(3.10) imply that

$$x_n \le (1 - \alpha_n)x_0 \oplus \alpha_n T(x_{n-1})$$

$$\le (1 - \alpha_n)x_0 \oplus \alpha_n T(x_n) = x_{n+1} \le T(x_n).$$

Thus,

$$x_n \le x_{n+1} \le T(x_n)$$

and Lemma 3.1 is proved by induction.

Corollary 3.2 *For each integer $n \ge 0$, $x_n \le T^n(x_0)$.*

In this section, we prove the following result which shows that our iterative process generates approximate fixed points of T.

Theorem 3.3 *Let for each integer $n \ge 0$,*

$$\alpha_n = 1 - 2(n+3)^{-1}.$$

Then for each integer $n \ge 0$,

$$d(x_n, x_{n+1}) \le 3M(n+3)^{-1}$$

and

$$d(x_n, T(x_n)) \le 7M(n+3)^{-1}.$$

We begin with auxiliary results.

Lemma 3.4 *Let $n \ge 0$ be an integer. Then*

$$d(x_{n+2}, x_{n+1}) \le \alpha_{n+1}d(x_{n+1}, x_n) + |\alpha_{n+1} - \alpha_n| d(x_0, T(x_n)),$$

$$d(x_n, T(x_n)) \le d(x_n, x_{n+1}) + (1 - \alpha_n)d(x_0, T(x_n)).$$

Proof Proposition 2.1, Lemma 3.1, and (3.1) and (3.5) imply that

$$d(x_{n+2}, x_{n+1})$$

$$= d((1 - \alpha_{n+1})x_0 \oplus \alpha_{n+1}T(x_{n+1}), (1 - \alpha_n)x_0 \oplus \alpha_n T(x_n))$$

$$\leq \alpha_{n+1}d(T(x_{n+1}), T(x_n)) + |\alpha_{n+1} - \alpha_n|d(x_0, T(x_n))$$

$$\leq \alpha_{n+1}d(x_{n+1}, x_n) + |\alpha_{n+1} - \alpha_n|d(x_0, T(x_n)).$$

Proposition 2.1 and (3.5) imply that

$$d(x_n, T(x_n)) \leq d(x_n, x_{n+1}) + d((1-\alpha_n)x_0 \oplus \alpha_n T(x_n), T(x_n))$$

$$\leq d(x_n, x_{n+1}) + (1-\alpha_n)d(x_0, T(x_n)).$$

Lemma 3.4 is proved.

Lemma 3.5 *For each integer $n \geq 0$,*

$$d(x_n, T^n(x_0)) \leq M. \tag{3.11}$$

Proof We prove (3.11) by induction. Clearly, (3.11) holds for $n = 0$. Assume that (3.11) holds for an integer $n \geq 0$. By Proposition 2.1, (A1), Corollary 3.2, and (3.1) and (3.11),

$$d(x_{n+1}, T^{n+1}(x_0)) \leq d((1-\alpha_n)x_0 \oplus \alpha_n T(x_n), T^{n+1}(x_0))$$

$$\leq (1-\alpha_n)d(x_0, T^{n+1}(x_0)) + \alpha_n d(T(x_n), T^{n+1}(x_0))$$

$$\leq 2M_0(1-\alpha_n) + \alpha_n d(x_n, T^n(x_0))$$

$$\leq M - (1-\alpha_n)(M - 2M_0) \leq M.$$

Thus we showed that (3.11) holds for $n+1$ too. Therefore by induction, it holds for each integer $n \geq 0$. Lemma 3.5 is proved.

Proposition 3.6 *Let $n \geq 0$ be an integer. Then*

$$d(x_{n+2}, x_{n+1}) \leq \alpha_{n+1}d(x_{n+1}, x_n) + M|\alpha_{n+1} - \alpha_n|, \tag{3.12}$$

$$d(x_n, T(x_n)) \leq d(x_n, x_{n+1}) + 2M(1-\alpha_n).$$

Proof Equation (3.12) follows from Lemmas 3.4 and 3.5. Lemmas 3.1 and 3.5, (A1), and Eq. (3.1) imply that

$$d(x_0, T(x_n)) \leq d(x_0, T(x_0)) + d(T(x_0), T(x_n))$$

$$\leq d(x_0, T(x_0)) + d(x_0, x_n)$$

$$\leq d(x_0, T(x_0)) + d(x_0, T^n(x_0)) + d(T^n(x_0), x_n)$$

$$\leq 4M_0 + M \leq 2M,$$

$$d(x_n, T(x_n)) \leq d(x_n, x_{n+1}) + 2M(1 - \alpha_n).$$

Proposition 3.6 is proved.

Proof of Theorem 3.3
Set

$$J = 3, \ N = 2,$$

and for each integer $n \geq 0$,

$$a_n = 1 - \alpha_n, \ s_n = d(x_n, x_{n+1}), \ c_n = L = M, \ \gamma = 1.$$

By (A1) and (3.5),

$$d(x_0, x_1) \leq d(x_0, T(x_0)) \leq 2M_0 \leq M.$$

Proposition 3.6 implies that for each integer $n \geq 0$,

$$s_{n+1} \leq (1 - a_n) d(x_n, x_{n+1}) + M|a_n - a_{n+1}|.$$

By Lemmas 2.2 and Proposition 3.3, for each integer $n \geq 0$,

$$d(x_n, x_{n+1}) \leq 3M(n+3)^{-1},$$

$$d(x_n, T(x_n)) \leq 7M(n+3)^{-1}.$$

Theorem 3.3 is proved.

3.2 Alternative Iterative Methods for a Monotone Mapping

We use the notation and definitions and assume the assumptions introduced in Sect. 3.1. In particular, we assume that assumptions (A1)–(A3) hold. Assume that

$$x_0 = x \in \mathrm{St}(C) \tag{3.13}$$

and that for each integer $n \geq 0$,

$$x_{n+1} = T((1 - \alpha_n)x_0 \oplus \alpha_n x_n). \tag{3.14}$$

By (A1), for each integer $i \geq 0$.

$$d(T^i(x_0), x_0) \le M_0.$$

We will prove the following result which shows that our iterative process generates approximate fixed points of T.

Theorem 3.7 *Let for each integer $n \ge 0$,*

$$\alpha_n = 1 - 2(n+3)^{-1}.$$

Then for each integer $n \ge 0$,

$$d(x_n, x_{n+1}) \le 6M(n+3)^{-1}$$

and

$$d(x_n, T(x_n)) \le 14M(n+3)^{-1}.$$

We begin with auxiliary results.

Lemma 3.8 *For each integer $i \ge 0$,*

$$x_i \le x_{i+1} \le T(x_i). \tag{3.15}$$

Proof Equations (3.4) and (3.14) imply that

$$x_1 = T(x_0) \ge x_0.$$

By (A2) and (3.1) and (3.14),

$$T(x_1) \ge x_2 = T((1-\alpha_1)x_0 \oplus \alpha_1 x_1) \ge T(x_0) = x_1.$$

Thus (3.15) holds for $i = 0, 1$.

Assume that $i \ge 1$ is an integer and that (3.15) holds. By (3.15),

$$(1-\alpha_{i+1})x_{i+1} \oplus \alpha_{i+1}x_{i+1} \ge x_0. \tag{3.16}$$

In view of (3.4), (3.14), and (3.16),

$$x_{i+2} = T((1-\alpha_{i+1})x_0 \oplus \alpha_{i+1}x_{i+1}) \ge T(x_0) \ge x_0. \tag{3.17}$$

Assumptions (A2) and (A3) and (3.14) and (3.15) imply that

$$x_{i+1} = T((1-\alpha_i)x_0 \oplus \alpha_i x_i) \le T((1-\alpha_i)x_0 \oplus \alpha_i x_{i+1})$$

$$\le T((1-\alpha_{i+1})x_0 \oplus \alpha_{i+1}x_{i+1}) = x_{i+2}$$

and

$$x_{i+2} \le T(x_{i+1}).$$

Thus (3.15) holds $i + 1$ too. and Lemma 3.8 is proved by induction.

Corollary 3.9 *For each integer $i \ge 0$, $x_i \le T^i(x_0)$ and*

$$(1 - \alpha_i)x_0 \oplus \alpha_i x_i \le (1 - \alpha_{i+1})x_0 \oplus \alpha_{i+1}x_{i+1}.$$

Lemma 3.10 *Let $n \ge 0$ be an integer. Then*

$$d(x_{n+2}, x_{n+1}) \le \alpha_{n+1}d(x_{n+1}, x_n) + |\alpha_{n+1} - \alpha_n|d(x_0, x_n),$$

$$d(x_n, T(x_n)) \le d(x_n, x_{n+1}) + (1 - \alpha_n)d(x_0, x_n).$$

Proof Proposition 2.1, Corollary 3.10, and (3.1) and (3.14) imply that

$$d(x_{n+2}, x_{n+1})$$

$$= d(T((1 - \alpha_{n+1})x_0 \oplus \alpha_{n+1}x_{n+1}), T((1 - \alpha_n)x_0 \oplus \alpha_n x_n))$$

$$\le d((1 - \alpha_{n+1})x_0 \oplus \alpha_{n+1}x_{n+1}, (1 - \alpha_n)x_0 \oplus \alpha_n x_n)$$

$$\le \alpha_{n+1}d(x_{n+1}, x_n) + |\alpha_{n+1} - \alpha_n|d(x_0, x_n).$$

Proposition 2.1, (A2), (A3), and (3.14) and (3.15) imply that

$$d(x_n, T(x_n)) \le d(x_n, x_{n+1}) + d(T((1 - \alpha_n)x_0 \oplus \alpha_n x_n), T(x_n))$$

$$\le d(x_n, x_{n+1}) + d((1 - \alpha_n)x_0 \oplus \alpha_n x_n, x_n)$$

$$\le d(x_n, x_{n+1}) + (1 - \alpha_n)d(x_0, x_n).$$

Lemma 3.10 is proved.

Lemma 3.11 *For each integer $n \ge 0$,*

$$d(x_n, T^n(x_0)) \le M, \tag{3.18}$$

$$d(x_n, x_0) \le M + 2M_0.$$

Proof We prove (3.18) by induction. Clearly,

$$d(x_0, T^0(x_0)) = 0.$$

Assume that (3.18) holds for an integer $n \geq 0$. By Proposition 2.1, (A2), (A3), (3.1), and (3.18),

$$\begin{aligned} d(x_{n+1}, T^{n+1}(x_0)) &= d(T((1-\alpha_n)x_0 \oplus \alpha_n x_n), T^{n+1}(x_0)) \\ &\leq d((1-\alpha_n)x_0 \oplus \alpha_n x_n, T^n(x_0)) \\ &\leq (1-\alpha_n)d(x_0, T^n(x_0)) + \alpha_n d(x_n, T^n(x_0)) \\ &\leq \alpha_n M + (1-\alpha_n)d(x_n, T^n(x_0)) \\ &\leq M - (1-\alpha_n)(M - d(x_0, T^n(x_0))) \leq M. \end{aligned}$$

Thus we showed that (3.18) holds for $n+1$ too. Therefore by induction, it holds for each integer $n \geq 0$. By (A1) and (3.18), for each integer $n \geq 0$,

$$d(x_n, x_0) \leq d(x_n, T^n(x_0)) + d(T^n(x_0), x_0) \leq M + 2M_0.$$

Lemma 3.11 is proved.

Lemmas 3.10 and 3.11 imply the following result.

Proposition 3.12 *Let $n \geq 0$ be an integer. Then*

$$d(x_{n+2}, x_{n+1}) \leq \alpha_{n+1} d(x_{n+1}, x_n) + 2M|\alpha_{n+1} - \alpha_n|,$$

$$d(x_n, T(x_n)) \leq d(x_n, x_{n+1}) + 2M(1-\alpha_n).$$

Theorem 3.7 is proved analogously to Theorem 3.3 using Proposition 3.12 and Lemma 2.2.

3.3 Halpern-Type Iterations of a Monotone Nonexpansive Set-Valued Map

We continue to use the notation and the definitions and assume the assumptions introduced in Sect. 3.1. In particular we assume that X is equipped with an order $\leq$ and C is star-shaped. Recall that for each $x \in X$ and each set $E \subset X$,

$$d(x, E) = \inf\{d(x, y) : y \in E\}$$

and that

$$T : C \to 2^C \setminus \{\emptyset\}.$$

We continue to assume that (A1)–(A3) hold and and that the following assumption holds.

(A4) For each $x, y \in C$ satisfying $x \le y$ and each $\xi \in T(x)$,

$$d(\xi, \{u \in T(y) : \xi \le u\}) \le d(x, y).$$

Note that in Sect. 3.1 in (A1), T was single-valued map, while here it is a set-valued mapping. Namely, here we assume the following modification of that assumption.

(A1) For each integer $i \ge 0$,

$$T^i(x_0) \subset B(x_0, M_0).$$

Assume that $\{\alpha_n\}_{n=0}^{\infty}, \{r_n\}_{n=0}^{\infty} \subset (0, 1)$,

$$\alpha_n \le \alpha_{n+1} \text{ for each integer } n \ge 0, \tag{3.19}$$

$$x_0 = x \in \mathrm{St}(C), \tag{3.20}$$

$$y_0 \in T(x_0),\ x_0 \le y_0 \tag{3.21}$$

and that for each integer $n \ge 0$,

$$y_n \in T(x_n), \tag{3.22}$$

$$x_{n+1} = (1 - \alpha_n)x_0 \oplus \alpha_n y_n, . \tag{3.23}$$

$$y_{n+1} \in T(x_{n+1}),\ y_n \le y_{n+1}, \tag{3.24}$$

$$d(y_n, y_{n+1}) \le d(y_n, \{u \in T(x_{n+1}) :\ y_n \le y\}) + r_n. \tag{3.25}$$

First we show that the sequences $\{x_i\}_{i=0}^{\infty}, \{y_i\}_{i=0}^{\infty}$ are well-defined.

Assume that $p \ge 0$, x_i, y_i, $i = 0, \dots, p$ are well-defined, (3.22) holds for all $n = 0, \dots, p$, and (3.23)–(3.25) hold for all nonnegative integers $n < p$. Then x_{p+1} is defined by (3.23) with $n = p$. If $p = 0$, then by (3.21) and (3.23),

$$x_{p+1} = x_1 \ge x_0 = x_p.$$

If $p > 0$, then by (3.23) and (3.24),

$$x_{p+1} \ge x_p.$$

By (3.22),

$$y_p \in T(x_p).$$

Assumption (A4) and the relation above imply that there exists $y_{p+1} \in C$ satisfying (3.24) and (3.25) with $n = p$. Thus our assumption holds for $p + 1$ too and the sequences $\{x_i\}_{i=0}^{\infty}$, $\{y_i\}_{i=0}^{\infty}$ are well-defined. By (3.23) and (3.24), for each integer $n \geq 0$,

$$x_n \leq x_{n+1}. \tag{3.26}$$

We will prove the following result which shows that our iterative process generates approximate fixed points of the set-valued mapping T.

Theorem 3.13 *Let for each integer $n \geq 0$,*

$$\alpha_n = 1 - 2(n+3)^{-1}, \; r_n = (n+4)^{-1}.$$

Then for each integer $n \geq 0$,

$$d(x_n, x_{n+1}) \leq 3(2M+1)(n+3)^{-1}$$

and

$$d(x_n, T(x_n)) \leq (10M+3)(n+3)^{-1}.$$

We begin with auxiliary results.

Lemma 3.14 *Let $n \geq 0$ be an integer. Then*

$$d(x_{n+2}, x_{n+1}) \leq \alpha_{n+1} d(x_{n+1}, x_n) + \alpha_{n+1} r_n + |\alpha_{n+1} - \alpha_n| d(x_0, y_n),$$

$$d(x_n, T(x_n)) \leq d(x_n, x_{n+1}) + (1 - \alpha_n) d(x_0, y_n).$$

Proof Proposition 2.1, (A4), (3.22)–(3.26) imply that

$$\begin{aligned} & d(x_{n+2}, x_{n+1}) \\ & = d((1-\alpha_{n+1})x_0 \oplus \alpha_{n+1} y_{n+1}, (1-\alpha_n)x_0 \oplus \alpha_n y_n) \\ & \leq \alpha_{n+1} d(y_n, y_{n+1}) + |\alpha_{n+1} - \alpha_n| d(x_0, y_n) \\ & \leq \alpha_{n+1} d(y_n, \{u \in T(x_{n+1}) : \; y_n \leq u\}) + \alpha_{n+1} r_n \\ & \quad + |\alpha_{n+1} - \alpha_n| d(x_0, y_n) \\ & \leq \alpha_{n+1} d(x_{n+1}, x_n) + \alpha_{n+1} r_n + |\alpha_{n+1} - \alpha_n| d(x_0, y_n). \end{aligned}$$

Proposition 2.1 and (3.22) and (3.23) imply that

$$d(x_n, T(x_n)) \le d(x_n, y_n)$$

$$\le d(x_n, x_{n+1}) + d(x_{n+1}, y_n)$$

$$= d(x_n, x_{n+1}) + (1 - \alpha_n)d(x_0, y_n) + \alpha_{n+1}r_{n+1}.$$

Lemma 3.14 is proved.

Lemma 3.15 *For each integer $n \ge 0$,*

$$d(x_n, \{y \in T^n(x_0) : x_n \le y\}) \le M, \tag{3.27}$$

$$d(y_n, \{y \in T^{n+1}(x_0) : y_n \le y\}) \le M. \tag{3.28}$$

Proof Clearly, (3.27) holds for $n = 0$. Assume that (3.27) holds for a nonnegative integer n. We show that (3.28) is true. Let $\epsilon > 0$. In view of (3.27), there exists

$$\xi \in T^n(x_0) \tag{3.29}$$

such that

$$x_n \le \xi, \; d(x_n, \xi) \le M + \epsilon. \tag{3.30}$$

By (A4), (3.22), (3.29), and (3.30), there exists

$$\eta \in \{z \in T(\xi) : y_n \le z\} \subset T^{n+1}(x_0) \tag{3.31}$$

such that

$$d(y_n, \eta) \le d(x_n, \xi) + \epsilon. \tag{3.32}$$

Equations (3.30)–(3.32) imply that

$$d(y_n, \{u \in T^{n+1}(x_0) : y_n \le u\}) \le d(y_n, \eta) \le d(x_n, \xi) + \epsilon \le M + \epsilon.$$

Since ϵ is any positive number, Eq. (3.28) holds.

By (3.21), (3.23), (3.24), and (3.31),

$$x_{n+1} = (1 - \alpha_n)x_0 \oplus \alpha_n y_n \le y_n \le \eta. \tag{3.33}$$

Proposition 2.1, assumption (A1), and (3.30)–(3.33) imply that

$$d(x_{n+1}, \eta) = d((1 - \alpha_n)x_0 \oplus \alpha_n y_n, \eta)$$

$$\le (1 - \alpha_n)d(x_0, \eta) + \alpha_n d(y_n, \eta)$$

$$\leq 2M_0(1-\alpha_n) + \alpha_n d(x_n, \xi) + \alpha_n \epsilon$$

$$\leq 2M_0(1-\alpha_n) + M\alpha_n + 2\alpha_n \epsilon.$$

Together with (3.31) and (3.33), this implies that

$$d(x_{n+1}, \{y \in T^{n+1}(x_0) : \ y \leq x_{n+1}\}) \leq d(x_{n+1}, \eta) \leq 2M + 2\alpha_n \epsilon.$$

Since ϵ is any positive number, we conclude that

$$d(x_{n+1}, \{y \in T^{n+1}(x_0) : \ y \leq x_{n+1}\}) \leq 2M$$

and (3.27) holds for $n+1$ too. Lemma 3.15 is proved by induction.

Lemmas 3.14 and 3.15 and (A1) imply the following result.

Proposition 3.16 *For each integer $n \geq 0$,*

$$d(x_{n+2}, x_{n+1}) \leq \alpha_{n+1} d(x_{n+1}, x_n) + 2M|\alpha_{n+1} - \alpha_n| + \alpha_{n+1} r_n,$$

$$d(x_n, T(x_n)) \leq d(x_n, x_{n+1}) + 2M(1-\alpha_n).$$

Proof of Theorem 3.13 We use Proposition 3.16 and Lemma 2.2 with $J = 3$, $N = 2$, $\gamma = 1$, $L = 2M + 1$,

$$a_n = 1 - \alpha_n = 2(n+3)^{-1}, c_n = L, \ s_n = d(x_n, x_{n+1}), \ n = 0, 1, \ldots$$

and obtain that for each integer $n \geq 0$,

$$s_{n+1} \leq (1 - a_{n+1}) s_n + |a_{n+1} - a_n|(2M + 1).$$

Lemma 2.2, Proposition 3.16, and the relation above imply that for each integer $n \geq 0$,

$$d(x_n, x_{n+1}) \leq 3(2M + 2)(n+3)^{-1},$$

$$d(x_n, T(x_n)) \leq (10M + 3)(n+3)^{-1}.$$

Theorem 3.13 is proved.

3.4 Alternative Iterative Methods for a Monotone Set-Valued Map

We continue to use the notation and the definitions and assume the assumptions introduced in Sect. 3.3.

Assume that

$$x_0 = x \in \mathrm{St}(C),$$

$$x_1 \in T(x_0), \ x_0 \le x_1$$

and that for each integer $n \ge 0$,

$$x_{n+1} \in T((1-\alpha_n)x_0 \oplus \alpha_n x_n), \tag{3.34}$$

$$x_n \le x_{n+1}, \tag{3.35}$$

$$d(x_n, x_{n+1}) \le d(x_n, \{y \in T((1-\alpha_n)x_0 \oplus \alpha_n x_n), \ x_n \le y\}) + r_n. \tag{3.36}$$

It is not difficult to show by induction that the sequence $\{x_i\}_{i=0}^{\infty}$ is well-defined and that for each integer $n \ge 0$,

$$(1-\alpha_n)x_0 \oplus \alpha_n x_n \le x_n, \tag{3.37}$$

$$(1-\alpha_n)x_0 \oplus \alpha_n x_n \le (1-\alpha_{n+1})x_0 \oplus \alpha_{n+1}x_{n+1}. \tag{3.38}$$

We will prove the following result.

Theorem 3.17 *Let for each integer $n \ge 0$,*

$$\alpha_n = 1 - 2(n+3)^{-1}, \ r_n = (n+4)^{-2}.$$

Then for each integer $n \ge 0$,

$$d(x_n, x_{n+1}) \le 3(2M+1)(n+3)^{-1}$$

and

$$d(x_n, T(x_n)) \le (10M+3)(n+3)^{-1}.$$

We begin with auxiliary results.

Lemma 3.18 *Let $n \ge 0$ be an integer. Then*

$$d(x_{n+2}, x_{n+1}) \le \alpha_{n+1} d(x_{n+1}, x_n) + r_{n+1} + |\alpha_{n+1} - \alpha_n| d(x_0, x_n),$$

$$d(x_n, T(x_n)) \le d(x_n, x_{n+1}) + (1 - \alpha_n)d(x_0, x_n).$$

Proof Proposition 2.1, (A4), and (3.34), (3.36), and (3.38) imply that

$$d(x_{n+2}, x_{n+1})$$

$$\le d(x_{n+1}, \{y \in T((1 - \alpha_{n+1})x_0 \oplus \alpha_{n+1}x_{n+1}) : x_{n+1} \le y\}) + r_{n+1}$$

$$\le d((1 - \alpha_n)x_0 \oplus \alpha_n x_n, (1 - \alpha_{n+1})x_0 \oplus \alpha_{n+1}x_{n+1}) + r_{n+1}$$

$$\le \alpha_{n+1}d(x_{n+1}, x_n) + r_{n+1} + |\alpha_{n+1} - \alpha_n|d(x_0, x_n).$$

By (A4) and (3.34) and (3.35),

$$d(x_n, T(x_n)) \le d(x_n, x_{n+1}) + d(x_{n+1}, T(x_n))$$

$$\le d(x_n, x_{n+1}) + d(x_{n+1}, \{v \in T(x_n) : x_{n+1} \le v\})$$

$$\le d(x_n, x_{n+1}) + d((1 - \alpha_n)x_0 \oplus \alpha_n x_n, x_n)$$

$$\le d(x_n, x_{n+1}) + (1 - \alpha_n)d(x_0, x_n).$$

Lemma 3.18 is proved.

Lemma 3.19 *For each integer $n \ge 0$,*

$$\{y \in T^n(x_0) : x_n \le y\} \ne \emptyset$$

and

$$d(x_n, \{y \in T^n(x_0) : x_n \le y\}) \le M. \tag{3.39}$$

Proof Clearly, (3.39) holds for $n = 0$. Assume that (3.39) holds for a nonnegative integer n. Let $\epsilon > 0$. In view of (3.39), there exists

$$\xi \in T^n(x_0) \tag{3.40}$$

such that

$$x_n \le \xi,\ d(x_n, \xi) \le M + \epsilon. \tag{3.41}$$

We have

$$x_{n+1} \in T((1 - \alpha_n)x_0 \oplus \alpha_n x_n). \tag{3.42}$$

In view of (3.35) and (3.41),

$$(1-\alpha_n)x_0 \oplus \alpha_n x_n \le x_n \le \xi. \tag{3.43}$$

Proposition 2.1, assumptions (A1) and (A4), and (3.40)–(3.43) imply that

$$d(x_{n+1}, \{y \in T(\xi) : x_{n+1} \le y\})$$

$$\le d((1-\alpha_n)x_0 \oplus \alpha_n x_n, \xi)$$

$$\le (1-\alpha_n)d(x_0, \xi) + \alpha_n d(x_n, \xi) \le 2M_0(1-\alpha_n) + \alpha_n(M+\epsilon).$$

Since ϵ is any positive number, we conclude that

$$d(x_{n+1}, \{y \in T(\xi) : x_{n+1} \le y\}) \le 2M_0(1-\alpha_n) + \alpha_n M$$

$$\le M - \alpha_n(M - 2M_0) \le M.$$

Thus (3.39) holds for $n+1$ too, and Lemma 3.19 is proved by induction.

Lemma 3.18 and (3.19) and (A1) imply the following result.

Proposition 3.20 *For each integer $n \ge 0$,*

$$d(x_{n+2}, x_{n+1}) \le \alpha_{n+1} d(x_{n+1}, x_n) + 2M|\alpha_{n+1} - \alpha_n| + r_{n+1},$$

$$d(x_n, T(x_n)) \le d(x_n, x_{n+1}) + 2M(1-\alpha_n).$$

Theorem 13.17 is proved analogously to Theorem 3.13 using Proposition 3.20.

Chapter 4
Asymptotic Regularity for Iterations of Uniformly Locally Nonexpansive Mappings

In this chapter, we study asymptotic regularity for Halpern-type iterations and for iterations of alternative iterative methods generated by a uniformly locally nonexpansive self-mapping of a star-shape set in a metric space with a hyperbolic structure. Linear rates of this asymptotic regularity are computed under the assumption that there exists a bounded orbit of the mapping. We study asymptotic regularity for single-valued and set-valued mapping and also under the presence of computational errors. All the results of this chapter are new.

4.1 Halpern-Type Iterations

In this chapter we assume that (X, d, W) is a W-hyperbolic space which is denoted by X for simplicity. Recall that nonempty set $C \subset X$ is called star-shaped and $x_* \in C$ is its star if for each $x \in C$ and each $\lambda \in [0, 1]$

$$(1 - \lambda)x \oplus \lambda x_* \in C.$$

If C is star-shaped we denote by $\mathrm{St}(C)$ the sets of all stars of C.

Assume that C is a nonempty star-shaped subset of X, $\Delta > 0$, $T : C \to C$ satisfies

$$d(T(x), T(y)) \le d(x, y) \text{ for each } x, y \in C \text{ satisfying } d(x, y) \le \Delta. \tag{4.1}$$

Such mapping is called a uniformly locally nonexpansive mapping. Assume that $\{\alpha_n\}_{n=0}^{\infty} \subset (0, 1)$,

$$x_0 = x \in C \tag{4.2}$$

A. J. Zaslavski, *Approximate Fixed Points of Nonexpansive Mappings*,
Developments in Mathematics 80, https://doi.org/10.1007/978-3-031-70710-0_4

$$\theta \in \mathrm{St}(C) \tag{4.3}$$

and that for each integer $n \geq 0$,

$$x_{n+1} = (1 - \alpha_n)\theta \oplus \alpha_n T(x_n). \tag{4.4}$$

In this chapter, our goal is to show that analogs of the results of Chap. 2 hold for uniformly locally nonexpansive operators. For these operators relation (4.1) is true only if $d(x, y) \leq \Delta$. This makes our study more difficult and less understood then in the case of nonexpansive operators. To overcome this difficulty, we introduce the quasi-metric d_1 (see (4.5)). It turns out that the estimations which are main ingredients in our proofs for nonexpansive mappings have corresponding analogs for uniformly locally nonexpansive operators when the metric d is replaced by quasi-metric d_1.

For each $x, y \in C$, define

$$d_1(x, y) := \inf\{\sum_{i=0}^{q-1} d(y_i, y_{i+1}) : \ q \geq 1 \text{ is an integer},$$

$$y_i \in C,\ i = 0, \ldots, q,\ y_0 = x,\ y_q = y,\ d(y_i, y_{i+1}) \leq \Delta,\ i \in \{0, \ldots, q\} \setminus \{q\}\}. \tag{4.5}$$

It follows from (4.3) and (4.5) that for each $x, y, z \in C$,

$$d(x, y) \leq d_1(x, y) < \infty,$$

$$d_1(x, z) \leq d_1(x, y) + d_1(y, z)$$

and

$$\text{if } d(x, y) \leq \Delta, \text{ then } d_1(x, y) = d(x, y), \tag{4.6}$$

Clearly, d_1 is a metric on X.

Proposition 4.1 *Let $y \in C$ and $0 \leq \gamma_1 < \gamma_2 \leq 1$. Then*

$$d_1((1 - \gamma_1)\theta \oplus \gamma_1 y, (1 - \gamma_2)\theta \oplus \gamma_2 y) \leq (\gamma_2 - \gamma_1)d(\theta, y)$$

and, in particular, $d_1(\theta, y) = d(\theta, y)$.

Proof Proposition 2.1 implies that for each $\lambda_1, \lambda_2 \in [0, 1]$ satisfying $\lambda < \lambda_2$,

$$d_1((1 - \lambda_1)\theta \oplus \lambda_1 y, (1 - \lambda_2)\theta \oplus \lambda_2 y) \leq |\lambda_2 - \lambda_1| d(\theta, y). \tag{4.7}$$

Clearly, there exist an integer $q \geq 1$ and $\lambda_i \in [0, 1]$, $i = 0, 1, \dots, q$ such that

$$\lambda_i < \lambda_{i+1},\ i = 0, \dots, q-1,$$

$$\lambda_0 = \gamma_1,\ \lambda_q = \gamma_2,, \tag{4.8}$$

$$(\lambda_{i+1} - \lambda_i)d(\theta, y) \leq \Delta,\ i = 0, \dots, q-1. \tag{4.9}$$

Proposition 2.1 and (4.5), (4.8), and (4.9) imply that

$$d_1((1-\gamma_1)\theta \oplus \gamma_1 y, (1-\gamma_2)\theta \oplus \gamma_2 y)$$

$$\leq \sum_{i=0}^{q-1} d_1((1-\lambda_i)\theta \oplus \lambda_i y, (1-\lambda_{i+1})\theta \oplus \lambda_{i+1} y)$$

$$\leq \sum_{i=0}^{q-1} (\lambda_{i+1} - \lambda_i)d(\theta, y) \leq (\gamma_2 - \gamma_1)d(\theta, y).$$

Proposition 4.1 is proved.

Let $M > 0$. We assume the following assumption.

(A) $d(\theta, T^i(\theta)) \leq M$ for each integer $i \geq 0$ and $d_1(x_0, \theta) \leq M$.

We will prove the following result which shows that our iterative process generates approximate fixed points of the mapping T.

Theorem 4.2 *Let for each integer $n \geq 0$,*

$$\alpha_n = 1 - 2(n+3)^{-1}.$$

Then for each integer $n \geq 0$,

$$d_1(x_n, x_{n+1}) \leq 9M(n+3)^{-1}$$

and

$$d(x_n, T(x_n)) \leq 13M(n+3)^{-1}.$$

We begin with auxiliary results.

Proposition 4.3 *Let $y_i \in C$, $i = 1, 2$ and $\gamma \in (0, 1)$. Then*

$$d_1((1-\gamma)\theta \oplus \gamma y_1, (1-\gamma)\theta \oplus \gamma y_2) \leq \gamma d_1(y_1, y_2).$$

Proof Let $\epsilon > 0$. By (4.5), there exist an integer $q \geq 1$ and $z_i \in C$, $i = 0, \dots, q$ such that

$$z_0 = y_1, \; z_q = y_2, \tag{4.10}$$

$$d(z_i, z_{i+1}) \le \Delta, \; i \in \{0, \dots, q-1\}, \tag{4.11}$$

$$\sum_{i=0}^{q-1} d(z_i, z_{i+1}) \le d_1(y_1, y_2) + \epsilon. \tag{4.12}$$

Proposition 2.1 implies that for each $i \in \{0, \dots, q-1\}$,

$$d_1((1-\gamma)\theta \oplus \gamma z_i, (1-\gamma)\theta \oplus \gamma z_{i+1}) \le \gamma d(z_i, z_{i+1}). \tag{4.13}$$

By (4.5), (4.10), and (4.11)–(4.13),

$$d_1((1-\gamma)\theta \oplus \gamma y_1, (1-\gamma)\theta \oplus \gamma y_2)$$

$$\le \sum_{i=0}^{q-1} d((1-\gamma)\theta \oplus \gamma z_i, (1-\gamma)\theta \oplus \gamma z_{i+1})$$

$$\le \gamma \sum_{i=0}^{q-1} d(z_i, z_{i+1}) \le \gamma d_1(y_1, y_2) + \gamma\epsilon.$$

Since ϵ is any positive number, we conclude that

$$d_1((1-\gamma)\theta \oplus \gamma y_1, (1-\gamma)\theta \oplus \gamma y_2) \le \gamma d_1(y_1, y_2).$$

Proposition 4.3 is proved.

By (4.1) and (4.5), for each $x, y \in C$,

$$d_1(T(x), T(y)) \le d_1(x, y). \tag{4.14}$$

Lemma 4.4 *Let $n \ge 0$ be an integer. Then*

$$d_1(x_{n+2}, x_{n+1}) \le \alpha_{n+1} d_1(x_{n+1}, x_n) + |\alpha_{n+1} - \alpha_n| d(\theta, T(x_n)),$$

$$d_1(x_n, T(x_n)) \le d_1(x_n, x_{n+1}) + (1-\alpha_n) d(\theta, T(x_n)).$$

Proof Propositions 4.1 and 4.3 and (4.1) imply that

$$d_1(x_{n+2}, x_{n+1})$$

$$= d_1((1-\alpha_{n+1})\theta \oplus \alpha_{n+1} T(x_{n+1}), (1-\alpha_n)\theta \oplus \alpha_n T(x_n))$$

$$\leq d_1((1-\alpha_{n+1})\theta \oplus \alpha_{n+1}T(x_{n+1}), (1-\alpha_{n+1})\theta \oplus \alpha_{n+1}T(x_n))$$

$$+ d_1((1-\alpha_{n+1})\theta \oplus \alpha_{n+1}T(x_n), (1-\alpha_n)\theta \oplus \alpha_n T(x_n))$$

$$\leq \alpha_{n+1}d_1(T(x_{n+1}), T(x_n)) + |\alpha_{n+1} - \alpha_n| d(\theta, T(x_n))$$

$$\leq \alpha_{n+1}d_1(x_{n+1}, x_n) + |\alpha_{n+1} - \alpha_n| d(\theta, T(x_n)).$$

Proposition 4.1 and (4.4) imply that

$$d_1(x_n, T(x_n)) \leq d_1(x_n, x_{n+1}) + d_1(x_{n+1}, T(x_n))$$

$$= d_1(x_n, x_{n+1}) + d_1((1-\alpha_n)\theta \oplus \alpha_n T(x_n), T(x_n))$$

$$\leq d_1(x_n, x_{n+1}) + (1-\alpha_n)d(f(\theta, T(x_n))$$

and Lemma 4.4 is proved.

Lemma 4.5 *For each integer $n \geq 0$,*

$$d_1(x_n, T^n(\theta)) \leq M, \tag{4.15}$$

$$d_1(x_n, \theta) \leq 2M.$$

Proof By (A), (4.15) holds for $n = 0$. Assume that (4.15) holds for an integer $n \geq 0$. By (A), Propositions 4.1 and 4.3, and (4.4), (4.14), and (4.15),

$$d_1(x_{n+1}, T^{n+1}(\theta)) \leq d_1((1-\alpha_n)\theta \oplus \alpha_n T(x_n), T^{n+1}(\theta))$$

$$\leq d_1((1-\alpha_n)\theta \oplus \alpha_n T(x_n), (1-\alpha_n)\theta \oplus \alpha_n T^{n+1}(\theta))$$

$$+ d_1((1-\alpha_n)\theta \oplus \alpha_n T^{n+1}(\theta), T^{n+1}(\theta))$$

$$\leq \alpha_n d_1(T(x_n), T^{n+1}(\theta)) + (1-\alpha_n)d(\theta, T^{n+1}(\theta))$$

$$\leq \alpha_n d_1(x_n, T^n(\theta)) + (1-\alpha_n)d(\theta, T^{n+1}(\theta))$$

$$\leq \alpha_n M + (1-\alpha_n)M = M.$$

Thus we showed that (4.15) holds for $n+1$ too. Therefore by induction, it holds for each integer $n \geq 0$.

(A) and (4.15) imply that for each integer $n \geq 0$,

$$d_1(x_n, \theta) \leq d_1(x_n, T^n(\theta)) + d_1(T^n(\theta), \theta) \leq M + d(T^n(\theta), \theta) \leq 2M.$$

Lemma 4.5 is proved.

Proposition 4.6 *For each integer $n \geq 0$,*

$$d(T(x_n), \theta) = d_1(T(x_n), \theta) \leq 2M,$$

$$d_1(x_{n+2}, x_{n+1}) \leq \alpha_{n+1} d_1(x_{n+1}, x_n) + 2M|\alpha_{n+1} - \alpha_n|,$$

$$d_1(x_n, T(x_n)) \leq d_1(x_n, x_{n+1}) + 2M(1 - \alpha_n).$$

Proof Let $n \geq 0$ be an integer. Lemma 4.5, Proposition 4.1, (A), and (4.14) imply that

$$\begin{aligned} d(T(x_n), \theta) = d_1(T(x_n), \theta) &\leq d_1(T(x_n), T^{n+1}(\theta)) + d_1(T^{n+1}(\theta), \theta) \\ &\leq d_1(x_n, T^n(\theta)) + d(T^{n+1}(\theta), \theta) \leq 2M. \end{aligned} \tag{4.16}$$

Lemma 4.4 and (4.16) imply that

$$d_1(x_{n+2}, x_{n+1}) \leq \alpha_{n+1} d_1(x_{n+1}, x_n) + 2M|\alpha_{n+1} - \alpha_n|$$

and that

$$\begin{aligned} d_1(x_n, T(x_n)) &\leq d_1(x_n, x_{n+1}) + (1 - \alpha_n) d(T(x_n), \theta) \\ &\leq d_1(x_n, x_{n+1}) + 2M(1 - \alpha_n). \end{aligned}$$

Proposition 4.6 is proved.

Proof of Theorem 4.2 We apply Lemma 2.2 with $J = 3$, $N = 2$ and assume that for each $n \geq 0$,

$$a_n = 1 - \alpha_n, \ s_n = d_1(x_n, x_{n+1}), \ c_n = L = 3M.$$

By Proposition 4.6, for each integer $n \geq 0$,

$$s_{n+1} \leq s_n(1 - a_n) + 2M(a_n - a_{n+1}).$$

Lemma 2.2, Proposition 4.6, and the equation above imply that for each integer $n \geq 0$,

$$d_1(x_n, x_{n+1}) \leq 9M(n+3)^{-1}$$

and

$$d(x_n, T(x_n)) \leq 13M(n+3)^{-1}.$$

Theorem 4.2 is proved.

4.2 The Alternative Iterative Method

We use the notation and definitions and assume the assumptions introduced in Sects. 4.1. In particular, (A) holds. Assume that

$$x_0 = x \in C$$

and for each integer $n \geq 0$,

$$x_{n+1} = T((1 - \alpha_n)\theta \oplus \alpha_n x_n). \tag{4.17}$$

We will prove the following result which shows that our iterative process generates approximate fixed points of the mapping T.

Theorem 4.7 *Let for each integer $n \geq 0$,*

$$\alpha_n = 1 - 2(n + 3)^{-1}.$$

Then for each integer $n \geq 0$,

$$d_1(x_n, x_{n+1}) \leq 9M(n + 3)^{-1}$$

and

$$d_1(x_n, T(x_n)) \leq 13M(n + 3)^{-1}.$$

We begin with auxiliary results.

Lemma 4.8 *Let $n \geq 0$ be an integer. Then*

$$d_1(x_{n+2}, x_{n+1}) \leq \alpha_{n+1} d_1(x_{n+1}, x_n) + |\alpha_{n+1} - \alpha_n| d(\theta, x_n),$$

$$d_1(x_n, T(x_n)) \leq d_1(x_n, x_{n+1}) + (1 - \alpha_n) d(\theta, x_n).$$

Proof Propositions 4.1 and 4.3 and (4.14) and (4.17) imply that

$$d_1(x_{n+2}, x_{n+1})$$

$$= d_1(T((1 - \alpha_{n+1})\theta \oplus \alpha_{n+1} x_{n+1}), T((1 - \alpha_n)\theta \oplus \alpha_n x_n))$$

$$\leq d_1((1 - \alpha_{n+1})\theta \oplus \alpha_{n+1} x_{n+1}, (1 - \alpha_n)\theta \oplus \alpha_n x_n)$$

$$\leq d_1((1-\alpha_{n+1})\theta \oplus \alpha_{n+1}x_{n+1}, (1-\alpha_{n+1})\theta \oplus \alpha_{n+1}x_n)$$

$$+ d_1((1-\alpha_{n+1})\theta \oplus \alpha_{n+1}x_n, (1-\alpha_n)\theta \oplus \alpha_n x_n)$$

$$\leq \alpha_{n+1}d_1(x_{n+1}, x_n) + |\alpha_{n+1} - \alpha_n| d(\theta, x_n).$$

By (4.14) and (4.17) and Proposition 4.1,

$$d_1(x_n, T(x_n)) \leq d_1(x_n, x_{n+1}) + d_1(x_{n+1}, T(x_n))$$

$$\leq d_1(x_n, x_{n+1}) + d_1(T((1-\alpha_n)\theta \oplus \alpha_n x_n), T(x_n))$$

$$\leq d_1(x_n, x_{n+1}) + d_1((1-\alpha_n)\theta \oplus \alpha_n x_n, x_n)$$

$$\leq d_1(x_n, x_{n+1}) + (1-\alpha_n)d(\theta, x_n).$$

Lemma 4.8 is proved.

Lemma 4.9 *For each integer $n \geq 0$,*

$$d_1(x_n, T^n(\theta)) \leq M, \tag{4.18}$$

$$d(x_n, \theta) = d_1(x_n, \theta) \leq 2M.$$

Proof In view of (A), (4.18) holds for $n = 0$. Assume that $n \geq 0$ is an integer and (4.18) holds. By (A), Propositions 4.1 and 4.3, and (4.14), (4.17), and (4.18),

$$d_1(x_{n+1}, T^{n+1}(\theta)) \leq d_1(T((1-\alpha_n)\theta \oplus \alpha_n x_n), T^{n+1}(\theta))$$

$$\leq d_1((1-\alpha_n)\theta \oplus \alpha_n x_n, T^n(\theta))$$

$$\leq d_1((1-\alpha_n)\theta \oplus \alpha_n x_n, (1-\alpha_n)\theta \oplus \alpha_n T^n(\theta))$$

$$+ d_1((1-\alpha_n)\theta \oplus \alpha_n T^n(\theta), T^n(\theta))$$

$$\leq \alpha_n d_1(x_n, T^n(\theta)) + (1-\alpha_n)d(\theta, T^n(\theta))$$

$$\leq \alpha_n M + (1-\alpha_n)M \leq M.$$

Thus by induction we showed that (4.18) holds for all integers $n \geq 0$. By (A) and (4.18), for each integer $n \geq 0$,

$$d_1(x_n, \theta) \leq d_1(x_n, T^n(\theta)) + d_1(T^n(\theta), \theta) \leq 2M.$$

Lemma 4.9 is proved.

Lemmas 4.8 and 4.9 imply the following result.

Proposition 4.10 *For each integer $n \geq 0$,*

$$d_1(x_{n+2}, x_{n+1}) \leq \alpha_{n+1} d_1(x_{n+1}, x_n) + 2M|\alpha_{n+1} - \alpha_n|,$$

$$d_1(x_n, T(x_n)) \leq d_1(x_n, x_{n+1}) + 2M(1 - \alpha_n).$$

We can prove now Theorem 4.7 exactly as Theorem 4.2 using Lemma 4.2 and Proposition 4.5.

4.3 Inexact Halpern-Type Iterations

We use the notation and definitions and assume the assumptions introduced in Sect. 4.1. In particular, assumption (A) holds.

Assume that

$$\{r_n\}_{n=0}^{\infty} \subset [0, \Delta], \tag{4.19}$$

$$x_0 = x \in C \tag{4.20}$$

and that for each integer $n \geq 0$,

$$x_{n+1} \in B((1 - \alpha_n)\theta \oplus \alpha_n T(x_n), r_n) \cap C. \tag{4.21}$$

We study the asymptotic regularity of the sequence $\{x_n\}_{n=0}^{\infty}$ and prove the following result which shows that our iterative process generates approximate fixed points of the mapping T under the presence of computational errors.

Theorem 4.11 *Let for each integer $n \geq 0$,*

$$\alpha_n = 1 - 2(n + 3)^{-1}, \; r_n = (n + 4)^{-2}.$$

Then for each integer $n \geq 0$,

$$d_1(x_n, x_{n+1}) \leq (9M + 15)(n + 3)^{-1}$$

and

$$d_1(x_n, T(x_n)) \leq (13M + 25)(n + 3)^{-1}.$$

We begin with the following lemma.

Lemma 4.12 *Let $n \geq 0$ be an integer. Then*

$$d_1(x_{n+2}, x_{n+1}) \le \alpha_{n+1} d_1(x_{n+1}, x_n) + |\alpha_{n+1} - \alpha_n| d(\theta, T(x_n)) + r_{n+1} + r_n,$$

$$d_1(x_n, T(x_n)) \le d_1(x_n, x_{n+1}) + r_n + (1 - \alpha_n) d(\theta, T(x_n)).$$

Proof By (4.19) and (4.21),

$$d(x_{n+1}, (1 - \alpha_n)\theta \oplus \alpha_n T(x_n)) \le r_n \le \Delta, \tag{4.22}$$

$$d(x_{n+2}, (1 - \alpha_{n+1})\theta \oplus \alpha_{n+1} T(x_{n+1})) \le r_{n+1} \le \Delta. \tag{4.23}$$

Propositions 4.1 and 4.3 and (4.14), (4.22), and (4.23) imply that

$$\begin{aligned}
& d_1(x_{n+2}, x_{n+1}) \\
& \le d_1(x_{n+2}, (1 - \alpha_{n+1})\theta \oplus \alpha_{n+1} T(x_{n+1})) \\
& + d_1((1 - \alpha_{n+1})\theta \oplus \alpha_{n+1} T(x_{n+1}), (1 - \alpha_n)\theta \oplus \alpha_n T(x_n)) \\
& + d_1((1 - \alpha_n)\theta \oplus \alpha_n T(x_n), x_{n+1}) \\
& \le r_{n+1} + r_n + d_1((1 - \alpha_{n+1})\theta \oplus \alpha_{n+1} T(x_{n+1}), (1 - \alpha_n)\theta \oplus \alpha_n T(x_n)) \\
& \le r_{n+1} + r_n + d_1((1 - \alpha_{n+1})\theta \oplus \alpha_{n+1} T(x_{n+1}), (1 - \alpha_{n+1})\theta \oplus \alpha_{n+1} T(x_n)) \\
& + d_1((1 - \alpha_{n+1})\theta \oplus \alpha_{n+1} T(x_n), (1 - \alpha_n)\theta \oplus \alpha_n T(x_n)) \\
& \le r_n + r_{n+1} + \alpha_{n+1} d_1(T(x_{n+1}), T(x_n)) + |\alpha_{n+1} - \alpha_n| d(\theta, T(x_n)) \\
& \le r_n + r_{n+1} + \alpha_{n+1} d_1(x_{n+1}, x_n) + |\alpha_{n+1} - \alpha_n| d(\theta, T(x_n)).
\end{aligned}$$

Proposition 4.1 and (4.22) imply that

$$\begin{aligned}
& d_1(x_n, T(x_n)) \le d_1(x_n, x_{n+1}) + d_1(x_{n+1}, T(x_n)) \\
& \le d_1(x_n, x_{n+1}) + d_1(x_{n+1}, (1 - \alpha_n)\theta \oplus \alpha_n T(x_n)) \\
& + d_1((1 - \alpha_n)\theta \oplus \alpha_n T(x_n), T(x_n)) \\
& \le d_1(x_n, x_{n+1}) + r_n + (1 - \alpha_n) d(\theta, T(x_n)).
\end{aligned}$$

Lemma 4.12 is proved.

Define

$$r_{-1} = 0.$$

Lemma 4.13 *For each integer $n \geq 0$,*

$$d_1(x_n, T^n(\theta)) \leq M + \sum\{\alpha_i r_i : i \in \{0, \dots, n\} \setminus \{n-1, n\}\} + r_{n-1}, \tag{4.24}$$

$$d_1(x_n, \theta) \leq 2M + \sum\{\alpha_i r_i : i \in \{0, \dots, n\} \setminus \{n-1, n\}\} + r_{n-1}.$$

Proof Clearly (4.24) holds for $n = 0$ in view of (A). Assume that (4.24) holds for an integer $n \geq 0$. By (A), Propositions 4.1 and 4.3, and (4.14), (4.21), and (4.24),

$$d_1(x_{n+1}, T^{n+1}(\theta))$$

$$\leq d_1(x_{n+1}, (1-\alpha_n)\theta \oplus \alpha_n T(x_n))$$

$$+ d_1((1-\alpha_n)\theta \oplus \alpha_n T(x_n), T^{n+1}(\theta))$$

$$\leq r_n + d_1((1-\alpha_n)\theta \oplus \alpha_n T(x_n), (1-\alpha_n)\theta \oplus \alpha_n T^{n+1}(\theta))$$

$$+ d_1((1-\alpha_n)\theta \oplus \alpha_n T^n(\theta), T^{n+1}(\theta))$$

$$\leq r_n + \alpha_n d_1(T(x_n), T^{n+1}(\theta)) + (1-\alpha_n) d_1(\theta, T^{n+1}(\theta))$$

$$\leq r_n + \alpha_n d_1(x_n, T^n(\theta)) + M(1-\alpha_n)$$

$$\leq r_n + M + \sum\{\alpha_i r_i : i \in \{0, \dots, n\} \setminus \{n\}\}.$$

Thus we showed that (4.24) holds for $n+1$ too. Therefore by induction, it holds for each integer $n \geq 0$. (A) and (4.24) imply that for each integer $n \geq 0$,

$$d_1(x_n, \theta) \leq d_1(x_n, T^n(\theta)) + d_1(T^n(\theta), \theta)$$

$$\leq 2M + \sum\{\alpha_i r_i : i \in \{0, \dots, n\} \setminus \{n-1, n\}\} + r_{n-1}.$$

Lemma 4.13 is proved.

Proposition 4.14 *For each integer $n \geq 0$,*

$$d_1(T(x_n), \theta) = d(T(x_n), \theta) \leq 2M + \sum_{i=0}^{\infty} \alpha_i r_i + r_{n-1},$$

$$d_1(x_{n+2}, x_{n+1})$$

$$\leq \alpha_{n+1} d_1(x_{n+1}, x_n) + 2M|\alpha_{n+1} - \alpha_n| + r_n + r_{n+1}$$

$$+|\alpha_{n+1}-\alpha_n|(\sum_{i=0}^{\infty}\alpha_i r_i + r_{n-1}),$$

$$d_1(x_n, T(x_n)) \le d_1(x_n, x_{n+1}) + r_n + (1-\alpha_n)(2M + 2\sum_{i=0}^{\infty}\alpha_i r_i + r_{n-1}).$$

Proof Let $n \ge 0$ be an integer. Lemma 4.13, Proposition 4.1, assumption (A), and (4.14) imply that

$$d(T(x_n), \theta) = d_1(T(x_n), \theta) \le d_1(T(x_n), T^{n+1}(\theta)) + d_1(T^{n+1}(\theta), \theta)$$

$$\le d_1(x_n, T^n(\theta)) + d(T^{n+1}(\theta), \theta)$$

$$\le 2M + \sum\{\alpha_i r_i : i \in \{0, \ldots, n\} \setminus \{n-1, n\}\} + r_{n-1}. \tag{4.25}$$

Lemma 4.12 implies that

$$d_1(x_{n+2}, x_{n+1})$$

$$\le \alpha_{n+1} d_1(x_{n+1}, x_n) + r_n + r_{n+1} + |\alpha_{n+1} - \alpha_n|(2M + \sum_{i=0}^{\infty}\alpha_i r_i + r_{n-1}),$$

$$d_1(x_n, T(x_n)) \le d_1(x_n, x_{n+1}) + r_n + (1-\alpha_n) d(\theta, T(x_n))$$

$$\le d_1(x_n, x_{n+1}) + r_n + (1-\alpha_n)(2M + \sum_{i=0}^{\infty}\alpha_i r_i + r_{n-1}).$$

Proposition 4.14 is proved.

Proof of Theorem 4.11 Let $\gamma = 1$, $L = 3M + 5$, $N = 2$, $c_n = L$, $n = 0, 1, \ldots$, $J = 3$, $a_n = 1 - \alpha_n$, $n = 0, 1, \ldots$. For each integer $n \ge 0$ set

$$s_n = d(x_n, x_{n+1}).$$

Assumption (A) and Lemma 4.13 imply that

$$d_1(x_0, x_1) \le d_1(x_0, \theta) + d_1(T(\theta), x_1) + d_1(\theta, T(\theta)) \le 3M + 2 \le L.$$

Proposition 4.14 implies that for each integer $n \ge 0$,

$$s_{n+1} \le (1 - a_{n+1}) s_n + |a_{n+1} - a_n|(2M + 2) + r_n + r_{n+1}$$

$$\le (1 - a_{n+1}) s_n + |a_{n+1} - a_n|(3M + 5).$$

Lemma 2.2 and the relation above imply that for each integer $n \geq 0$

$$d_1(x_n, x_{n+1}) \leq (9M + 15)(n + 3)^{-1}.$$

Together with Proposition 4.14, this implies that for each integer $n \geq 0$,

$$d_1(x_n, T(x_n)) \leq (9M + 15)(n + 3)^{-1} + 2(2M + 4)(n + 3)^{-1} + (n + 4)^{-2}$$

$$\leq (13M + 25(n + 3)^{-1}.$$

Theorem 4.11 is proved.

4.4 The Alternative Iterative Method with Errors

We use the notation and definitions and assume the assumptions introduced in Sects. 4.1 and 4.3. In particular, we assume that (A) holds.

Let

$$\{r_n\}_{n=0}^{\infty} \subset [0, \Delta], \tag{4.26}$$

$$x_0 = x \in C \tag{4.27}$$

and that for each integer $n \geq 0$,

$$x_{n+1} \in B(T(1 - \alpha_n)\theta \oplus \alpha_n x_n), r_n). \tag{4.28}$$

We study the asymptotic regularity of the sequence $\{x_n\}_{n=0}^{\infty}$ and prove the following result which shows that our iterative process generates approximate fixed points of the mapping T under the presence of computational errors.

Theorem 4.15 *Let for each integer $n \geq 0$,*

$$\alpha_n = 1 - 2(n + 3)^{-1}, \; r_n = (n + 4)^{-2}.$$

Then for each integer $n \geq 0$,

$$d_1(x_n, x_{n+1}) \leq (9M + 15)(n + 3)^{-1}$$

and

$$d_1(x_n, T(x_n)) \leq (13M + 25)(n + 3)^{-1}.$$

We begin with the following auxiliary results.

Lemma 4.16 *Let $n \geq 0$ be an integer. Then*

$$d_1(x_{n+2}, x_{n+1})$$

$$\leq \alpha_{n+1} d_1(x_{n+1}, x_n) + |\alpha_{n+1} - \alpha_n| d_1(\theta, x_n) + r_n + r_{n+1},$$

$$d_1(x_n, T(x_n)) \leq d_1(x_n, x_{n+1}) + (1 - \alpha_n) d_1(\theta, x_n) + r_n.$$

Proof Propositions 4.1 and 4.3 and (4.14) and (4.28) imply that

$$d_1(x_{n+2}, x_{n+1})$$

$$\leq d_1(x_{n+2}, T((1 - \alpha_{n+1})\theta \oplus \alpha_{n+1} x_{n+1}))$$

$$+ d_1(T((1 - \alpha_{n+1})\theta \oplus \alpha_{n+1} x_{n+1}), T((1 - \alpha_n)\theta \oplus \alpha_n x_n))$$

$$+ d_1(T((1 - \alpha_n)\theta \oplus \alpha_n x_n), x_{n+1})$$

$$\leq r_{n+1} + r_n + d_1((1 - \alpha_{n+1})\theta \oplus \alpha_{n+1} x_{n+1}, (1 - \alpha_n)\theta \oplus \alpha_n x_n)$$

$$\leq r_{n+1} + r_n + d_1((1 - \alpha_{n+1})\theta \oplus \alpha_{n+1} x_{n+1}, (1 - \alpha_{n+1})\theta \oplus \alpha_{n+1} x_n)$$

$$+ d_1((1 - \alpha_{n+1})\theta \oplus \alpha_{n+1} x_n, (1 - \alpha_n)\theta \oplus \alpha_n x_n)$$

$$\leq r_{n+1} + r_n + \alpha_{n+1} d_1(x_{n+1}, x_n) + |\alpha_{n+1} - \alpha_n| d(\theta, x_n).$$

Proposition 4.1 and (4.14) and (4.28) imply that

$$d_1(x_n, T(x_n)) \leq d_1(x_n, x_{n+1}) + d_1(x_{n+1}, T(x_n))$$

$$\leq d_1(x_n, x_{n+1}) + d_1(x_{n+1}, T((1 - \alpha_n)\theta \oplus \alpha_n x_n))$$

$$+ d_1(T((1 - \alpha_n)\theta \oplus \alpha_n x_n), T(x_n))$$

$$\leq d_1(x_n, x_{n+1}) + r_n + d_1((1 - \alpha_n) x_n \oplus \alpha_n x_n, x_n)$$

$$\leq d_1(x_n, x_{n+1}) + (1 - \alpha_n) d_1(\theta, x_n) + r_n.$$

Lemma 4.16 is proved.

Set $r_i = 0$ for each integer $i < 0$.

Lemma 4.17 *For each integer $n \geq 0$,*

$$d_1(x_n, T^n(\theta)) \leq M + \sum \{\alpha_{i+1} r_i : i \in \{0, \ldots, n\} \setminus \{n - 1, n\}\} + r_{n-1}, \tag{4.29}$$

$$d_1(x_n, \theta) \le 2M + \sum\{\alpha_{i+1} r_i : i \in \{0, \dots, n\} \setminus \{n-1, n\}\} + r_{n-1}.$$

Proof We prove the lemma by induction. By (A) and Proposition 4.1,

$$d_1(x_0, \theta) = d(x_0, \theta) \le M$$

and (4.29) holds for $n = 0$. Assume that (4.29) holds for an integer $n \ge 0$. Propositions 4.1 and 4.3 and (4.14), (4.28), and (4.29) imply that

$$d_1(x_{n+1}, T^{n+1}(\theta))$$

$$\le d_1(x_{n+1}, T((1-\alpha_n)\theta \oplus \alpha_n x_n))$$

$$+ d_1(T((1-\alpha_n)\theta \oplus \alpha_n x_n), T^{n+1}(\theta))$$

$$\le r_n + d_1((1-\alpha_n)\theta \oplus \alpha_n x_n, T^n(\theta))$$

$$\le r_n + d_1((1-\alpha_n)\theta \oplus \alpha_n x_n, (1-\alpha_n)\theta \oplus \alpha_n T^n(\theta))$$

$$+ d_1((1-\alpha_n)\theta \oplus \alpha_n T^n(\theta), T^n(\theta))$$

$$\le r_n + \alpha_n d_1(x_n, T^n(\theta)) + (1-\alpha_n) d_1(\theta, T^n(\theta))$$

$$\le \alpha_n M + \sum\{\alpha_{i+1} r_i : i \in \{0, \dots, n\} \setminus \{n-1, n\}\}$$

$$+ \alpha_n r_{n-1} + r_n + (1-\alpha_n)M$$

$$\le M + \sum\{\alpha_{i+1} r_i : i \in \{0, \dots, n\} \setminus \{n\}\} + r_n$$

and (4.29) holds for $n+1$ too. Therefore by induction, it holds for each integer $n \ge 0$. (A) and (4.29) imply that for each integer $n \ge 0$,

$$d_1(x_n, \theta) \le d_1(x_n, T^n(\theta)) + d_1(\theta, T(\theta))$$

$$\le 2M + \sum\{\alpha_{i+1} r_i : i \in \{0, \dots, n\} \setminus \{n\}\} + r_{n-1}.$$

Lemma 4.17 is proved.

Lemmas 4.10 and 4.17 imply the following result.

Proposition 4.18 *For each integer $n \ge 0$,*

$$d_1(x_{n+2}, x_{n+1})$$

$$\leq r_n + r_{n+1} + \alpha_{n+1} d_1(x_{n+1}, x_n) + |\alpha_{n+1} - \alpha_n|(2M + 2\sum_{i=0}^{\infty} r_i),$$

$$d_1(x_n, T(x_n)) \leq d_1(x_n, x_{n+1}) + r_n + (1 - \alpha_n)(2M + \sum_{i=0}^{\infty} r_i).$$

Theorem 4.15 is proved analogously to Theorem 4.11 using Proposition 4.18 and Lemma 4.17.

4.5 Halpern-Type Iterations for Set-Valued Maps

We assume that (X, d, W) is a W-hyperbolic space which is denoted by X for simplicity and continue to use the notation and the definitions introduced in Sect. 4.1.

Recall that C is a nonempty star-shaped set in X. Assume that $\Delta > 0$, $T : C \to 2^C \setminus \{\emptyset\}$ satisfies

$$H(T(x), T(y)) \leq d(x, y) \tag{4.30}$$

for each $x, y \in C$ satisfying $d(x, y) \leq \Delta$. For each $\xi \in C$, $T^0(\xi) = \xi$, $T^1(\xi) = T(\xi)$ and $T^{i+1}(\xi) = T(T^i(\xi))$, $i = 0, 1, \ldots$.

Assume that

$$\{r_n\}_{n=0}^{\infty} \subset (0, \min\{\Delta, 1\}, \tag{4.31}$$

$$\theta \in \mathrm{St}(C),$$

$M > 0$ and assume that the following assumption holds.

(A)

$$T^i(\theta) \subset B(\theta, M), \ i = 0, 1, \ldots.$$

Assume that

$$x^{(0)} \in C, \ d(x_0, \theta) < M, \tag{4.32}$$

$$y^{(0} \in T(x^{(0)}), \ d_1(x^{(0}, y^{(0)}) \leq M, \tag{4.33}$$

$$x^{(1)} = (1 - \alpha_0)\theta \oplus \alpha_0 y^{(0)}. \tag{4.34}$$

Proposition 4.1 and (4.33) and (4.34) imply that

$$d_1(x^{(1)}, \theta) = d_1((1 - \alpha_0)\theta \oplus \alpha_0 y^{(0}, \theta) \le d_1(\theta, y^{(0)}) < 2M. \tag{4.35}$$

For each $y \in C$ and each set $A \subset C$, put

$$d_1(y, E) = \inf\{d_1(y, a) : \ a \in A\}$$

and for every pair of sets $A, B \subset C$,

$$H_1(A, B) = \max\{\sup\{d_1(b, A) : \ b \in B\}, \ \sup\{d_1(a, B) : \ a \in A\}\}. \tag{4.36}$$

Proposition 4.19 *Let $x, y \in C$. Then $H_1(T(x), T(y)) \le d_1(x, y)$.*

Proof We may assume that $x \neq y$. Let $\epsilon > 0$. By (4.5), there exist a natural number q and $\{\xi_i\}_{i=0}^q \subset C$ such that

$$d(\xi_i, \xi_{i+1}) \le \Delta, \ i \in \{0, \dots, q\} \setminus \{q\}, \ \xi_0 = x, \ \xi_q = y, \tag{4.37}$$

$$\sum_{i=0}^{q-1} d(\xi_i, \xi_{i+1}) \le d_1(x, y) + \epsilon. \tag{4.38}$$

Let

$$u_0 \in T(x) = T(\xi_0). \tag{4.39}$$

By (4.30), (4.37), and (4.39), there exist $u_i \in C$, $i = 0, \dots, q$ such that

$$u_i \in T(\xi_i), \ i = 0, \dots, q, \tag{4.40}$$

$$d(u_i, u_{i+1}) \le d(\xi_i, \xi_{i+1}) + \epsilon/q, \ i = 0, \dots, q-1. \tag{4.41}$$

In view of (4.37), (4.38), and (4.40),

$$u_q \in T(\xi_q) = T(y).$$

It follows from the relation above and (4.41) that

$$d_1(u_0, T(y)) \le d_1(u_0, u_q) \le \sum_{i=0}^{q-1} d(u_i, u_{i+1})$$

$$\le \sum_{i=0}^{q-1} d(\xi_i, \xi_{i+1}) + \epsilon \le d_1(x, y) + 2\epsilon.$$

Since ϵ is any positive number, we conclude that

$$d_1(u_0, T(y)) \le d_1(x, y)$$

for each $u_0 \in T(x)$ and that

$$\sup\{d_1(u, T(y)) : u \in T(x)\} \le d_1(x, y).$$

This implies that

$$H_1(T(x), T(y)) \le d_1(x, y).$$

Proposition 4.19 is proved.

Proposition 4.20 *Let* $\{\alpha_i\}_{i=0}^{\infty} \subset (0, \infty)$,

$$x_0 \in C,\ d(x_0, \theta) < M, \tag{4.42}$$

for each integer $i \ge 0$,

$$y_i \in T(x_i), \tag{4.43}$$

$$x_{i+1} = (1 - \alpha_i)\theta \oplus y_i. \tag{4.44}$$

Then for each integer $n \ge 0$,

$$d_1(x_n, T^n(\theta)) < M, \tag{4.45}$$

$$d_1(x_n, \theta)) < 2M,$$

$$d_1(y_n, T^{n+1}(\theta)) < M,$$

$$d_1(y_n, \theta)) < 2M,$$

Proof We prove (4.45) by induction. By (4.42), (4.45) holds for $n = 0$. Assume that (4.45) holds for a nonnegative integer n. By (4.45), there exists

$$\xi \in T^n(\theta) \tag{4.46}$$

such that

$$d_1(x_n, \xi) < M. \tag{4.47}$$

In view of (4.47), choose a positive number

$$\epsilon < M - d_1(x_n, \xi). \tag{4.48}$$

By (4.44),

$$x_{n+1} = (1-\alpha_n)\theta \oplus \alpha_n y_n. \tag{4.49}$$

Proposition 4.19 and (4.43), (4.47), and (4.48) imply that there exists

$$\eta \in T(\xi) \tag{4.50}$$

such that

$$d_1(y_n, \eta) \le d_1(x_n, \xi) + \epsilon < M. \tag{4.51}$$

Propositions 4.1 and 4.3, assumption (A), and (4.46), (4.49), (4.50), and (4.51) imply that

$$d_1(x_{n+1}, T^{n+1}(\theta)) = d_1((1-\alpha_n)\theta \oplus \alpha_n y_n, T^{n+1}(\theta))$$

$$\le d_1((1-\alpha_n)\theta \oplus \alpha_n y_n, (1-\alpha_n)\theta \oplus \alpha_n \eta)$$

$$+ d_1((1-\alpha_n)\theta \oplus \alpha_n \eta, \eta)$$

$$\le \alpha_n d_1(y_n, \eta) + (1-\alpha_n) d_1(y_n, \eta) < M$$

and (4.45) holds for $n+1$ too. Therefore by induction, it holds for each integer $n \ge 0$. In view of Proposition 4.19, (4.43), and (4.45), for each integer $n \ge 0$,

$$d_1(y_n, T^{n+1}(\theta)) \le d_1(x_n, T^n(\theta)) < M.$$

Together with (A), this completes the proof of Proposition 4.20.

Lemma 4.21 *Let* $\epsilon > 0$, $\beta_1, \beta_2 \in (0, 1)$,

$$\epsilon < (1-\beta_1)\Delta\beta_1^{-1}/4, \tag{4.52}$$

$$|\beta_2 - \beta_1| < (3/4)(2M)^{-1}\Delta(1-\beta_1), \tag{4.53}$$

$$\xi_1, \xi_2 \in C,\ d(\xi_1, \xi_2) \le \Delta, \tag{4.54}$$

$$\eta_i \in T(\xi_i) \cap B(\theta, 2M),\ i = 1, 2, \tag{4.55}$$

$$d(\eta_1, \eta_2) \le H(T(\xi_1), T(\xi_2)) + \epsilon. \tag{4.56}$$

Then

$$d((1-\beta_1)\theta \oplus \beta_1\eta_1, (1-\beta_2)\theta \oplus \beta_2\eta_2) \le \Delta.$$

Proof Proposition 2.1 and (4.30) and (4.52)–(4.56) imply that

$$d((1-\beta_1)\theta \oplus \beta_1\eta_1, (1-\beta_2)\theta \oplus \beta_2\eta_2)$$

$$\leq \beta_1 d(\eta_1, \eta_2) + |\beta_2 - \beta_1| d(\theta, \eta_2)$$

$$\leq \beta_1 d(\xi_1, \xi_2) + \beta_1\epsilon + 2M|\beta_2 - \beta_1|$$

$$< \beta_1\Delta + \beta_1\epsilon + (3/4)\Delta(1-\beta_1) \leq \Delta.$$

Lemma 4.21 is proved.

Let for each integer $i \geq 0$,

$$\alpha_i \in (0, 1),\ \alpha_i < \alpha_{i+1}.$$

Now we construct our Halperin-type iterative process induced by the mapping T. In view of (4.32) and (4.35),

$$d_1(x^{(0)}, x^{(1)}) < 2M. \tag{4.57}$$

By (4.57), there exist a natural number q_0 and $x_i \in C$, $i = 0, \ldots, q_0$ such that

$$d(x_i, x_{i+1}) \leq \Delta,\ i \in \{0, \ldots, q_0 - 1\}, \tag{4.58}$$

$$x_0 = x^{(0)},\ x_{q_0} = x^{(1)}. \tag{4.59}$$

Set

$$y_0 = y^{(0)} \in T(x_0). \tag{4.60}$$

By (4.58),

$$H(T(x_i), T(x_{i+1})) \leq d(x_i, x_{i+1}),\ i = 0, \ldots, q-1. \tag{4.61}$$

Equations (4.60) and (4.61) imply that for each $i \in \{0, \ldots, q_0\}$, there exists

$$y_i \in T(x_i) \tag{4.62}$$

such that for each $i \in \{0, \ldots, q_0 - 1\}$,

$$d(y_i, y_{i+1}) \leq d(y_i.T(x_{i+1})) + r_0$$

$$\leq H(T(x_i), T(x_{i+1})) + r_0 \leq d(x_i, x_{i+1}) + r_0. \tag{4.63}$$

Set

$$x_{i+q_0} = (1-\alpha_0)\theta \oplus \alpha_0 y_i, \ i = 0, \ldots, q_0 - 1. \tag{4.64}$$

In view of (4.59) and (4.62),

$$y_{q_0} \in T(x_{q_0}) = T(x^{(1)}). \tag{4.65}$$

Set

$$x^{(2)} = (1-\alpha_1)\theta \oplus \alpha_1 y_q. \tag{4.66}$$

If

$$d(x^{(2)}, x_{2q_0-1}) \leq \Delta, \tag{4.67}$$

then set

$$q_1 = 2q_0, \ x_{2q_0} = x^{(2)}.$$

Assume that (4.67) does not hold. Set

$$x_{2q_0} = (1-\alpha_0)\theta \oplus \alpha_0 y_{q_0}. \tag{4.68}$$

By Propositions 2.1, 4.19, and 4.20 and Eqs. (4.65), (4.66), and (4.68),

$$d(x^{(2)}, x_{2q_0}) \leq |\alpha_0 - \alpha_1| d(y_{q_0}, \theta) \leq |\alpha_0 - \alpha_1| H_1(\{\theta\}, T(x^{(1)})$$

$$\leq |\alpha_0 - \alpha_1|(H_1(\{\theta\}, T^2(\theta)) + H_1(T^2(\theta), T(x^{(1)}))) \leq 2M|\alpha_0 - \alpha_1|.$$

If

$$2M|\alpha_0 - \alpha_1| \leq \Delta, \tag{4.69}$$

then $p_0 = 1$ and $q_1 = 2q_0 + 1$, otherwise

$$p_0 = \lfloor 2M|\alpha_0 - \alpha_1|\Delta^{-1} \rfloor + 1, \ q_1 = 2q_0 + p_0.$$

Assume that (4.69) does not hold. Set

$$\beta = (\alpha_1 - \alpha_0) p_0^{-1},$$

for $i = 0, \ldots, p_0$,

$$\beta_i = \alpha_0 + \beta i$$

and

$$x_{2q_0+i} = (1 - \beta_i)\theta \oplus \beta_i y_{q_0}.$$

It is not difficult to see that for $i = 0, \dots, p_0 - 1$,

$$d(x_{2q_0+i}, x_{2q_0+i+1}) \le \beta d(y_{q_0}, \theta) \le 2\beta M \le 2M|\alpha_1 - \alpha_0| p_0^{-1} \le \Delta.$$

Now we describe the induction step. Assume that $m \ge 0$ is an integer, we already defined $x^{(i)} \in C, i = 0, \dots, m+1$,

$$y^{(i)} \in T(x^{(i)}),\ i = 0, \dots, m, \tag{4.70}$$

such that

$$x^{(i+1)} = (1 - \alpha_i)\theta \oplus \alpha_i y^{(i)},\ i = 0, \dots, m \tag{4.71}$$

and defined natural numbers $q_i, i = 0, \dots, m+1$ such that

$$q_k < q_{k+1},\ k \in \{0, \dots, m\} \tag{4.72}$$

and

$$x_{q_m+i} \in C,\ i = 0, \dots, q_{m+1} - q_m$$

such that

$$d(x_i, x_{i+1}) \le \Delta,\ i = q_m, \dots, q_{m+1} - 1, \tag{4.73}$$

$$x_{q_m} = x^{(m)}, \tag{4.74}$$

$$x_{q_{m+1}} = x^{(m+1)}, \tag{4.75}$$

and assume that for each $k = 0, \dots, m$,

$$y_{q_k} = y^{(k)} \in T(x^{(k)}) \tag{4.76}$$

are defined.

We assume that

$$r_m < (1 - \alpha_m)4^{-1}\Delta,$$

$$\alpha_m r_m \le (\alpha_{m+1} - \alpha_m)(q_{m+1} - q_m + 1)^{-1}. \tag{4.77}$$

By (4.73), for each $i \in \{q_m, \dots, q_{m+1} - 1\}$, there exists

$$y_{i+1} \in T(x_{i+1}) \tag{4.78}$$

such that

$$d(y_i, y_{i+1}) \le d(y_i, T(x_{i+1})) + r_m \le d(x_i, x_{i+1}) + r_m. \tag{4.79}$$

For each $i \in \{0, \ldots, q_{m+1} - q_m\} \setminus \{q_{m+1} - q_m\}$ set

$$x_{q_{m+1}+i} = (1 - \alpha_m)\theta \oplus \alpha_m y_{q_m+i}. \tag{4.80}$$

Proposition 2.1 implies that for each $i \in \{q_m, \ldots, q_{m+1} - q_m\} \setminus \{q_{m+1} - q_m - 1, q_{m+1} - q_m\}$,

$$\begin{aligned} d(x_{q_{m+1}+i}, x_{q_{m+1}+i+1}) &\le \alpha_m d(y_{q_m+i}, y_{q_m+i+1}) \\ &\le \alpha_m d(x_{q_m+i}, x_{q_m+i+1}) + \alpha_m r_m. \end{aligned} \tag{4.81}$$

Set

$$x^{(m+2)} = (1 - \alpha_{m+1})\theta \oplus \alpha_{m+1} y_{q_{m+1}}, \; y^{(m+1)} = y_{q_{m+1}}. \tag{4.82}$$

Assume that

$$|\alpha_{m+1} - \alpha_m| < (3/2)\Delta M^{-1}(1 - \alpha_m). \tag{4.83}$$

(Note that (4.83) holds for large enough n is $\lim_{n\to\infty} \alpha_n = 1$.) Set

$$q_{m+2} = 2q_{m+1} - q_m, \; x_{q_{m+2}} = x^{(m+2)}. \tag{4.84}$$

Propositions 2.1 and 4.20 and Eqs. (4.77)–(4.80) imply that

$$\begin{aligned} &d(x_{q_{m+2}-1}, x_{q_{m+2}}) \\ &= d((1 - \alpha_m)\theta \oplus \alpha_m y_{q_{m+1}-1}, (1 - \alpha_{m+1})\theta \oplus \alpha_{m+1} y_{q_{m+1}}) \\ &\le \alpha_m d(y_{q_{m+1}-1}, y_{q_{m+1}}) + |\alpha_m - \alpha_{m+1}| d(\theta, y_{q_{m+1}}) \\ &\le \alpha_m d(x_{q_{m+1}-1}, x_{q_{m+1}}) + \alpha_m r_m + 2M|\alpha_m - \alpha_{m+1}|. \end{aligned} \tag{4.85}$$

It follows from (4.73), (4.77), (4.81), (4.83), and (4.85) that for each $i \in \{0, \ldots, q_{m+1} - q_m\} \setminus \{q_{m+1} - q_m\}$,

$$\begin{aligned} d(x_{q_{m+1}+i}, x_{q_{m+1}+i+1}) &\le \alpha_m \Delta + \alpha_m r_m + 2M|\alpha_m - \alpha_{m+1}| \\ &< \alpha_m \Delta + \alpha_m r_m + (3/4)\Delta(1 - \alpha_m) < \Delta. \end{aligned} \tag{4.86}$$

and our assumption made for m holds for $m+1$ too.

Assume that (4.83) does not hold. Define

$$x_{2q_{m+1}-q_m} = (1-\alpha_m)\theta \oplus \alpha_m y_{q_{m+1}}. \tag{4.87}$$

Proposition 2.1 and Eqs. (4.80) and (4.87) imply that

$$d(x_{2q_{m+1}-q_m-1}, x_{2q_{m+1}-q_m})$$

$$\leq \alpha_m d(y_{q_{m+1}-1}, y_{q_{m+1}}) \leq \alpha_m d(x_{q_{m+1}-1}, x_{q_{m+1}}) + \alpha_m r_m.)$$

Assume that

$$2m|\alpha_m - \alpha_{m+1}| \leq \Delta. \tag{4.88}$$

Set

$$q_{m+2} = 2q_{m+1} - q_m + 1, \ \ x_{q_{m+2}} = x^{(m+2)}.$$

Propositions 2.1 and 4.20 and Eqs. (4.78), (4.82), and (4.87) imply that

$$d(x_{2q_{m+1}-q_m}, x_{2q_{m+1}-q_m+1})$$

$$= d((1-\alpha_m)\theta \oplus \alpha_m y_{q_{m+1}}, (1-\alpha_{m+1})\theta \oplus \alpha_{m+1} y_{q_{m+1}})$$

$$\leq |\alpha_m - \alpha_{m+1}| d(\theta, y_{q_{m+1}}) \leq 2M|\alpha_m - \alpha_{m+1}|. \tag{4.89}$$

In view of (4.89),

$$d(x_{2q_{m+1}-q_m}, x_{2q_{m+1}-q_m+1}) \leq \Delta.$$

Together with (4.73), (4.77), and (4.81), this implies that $i \in \{0, \dots, \dots, q_{m+1} - q_m\}$,

$$d(x_{q_{m+1}+i}, x_{q_{m+1}+i+1}) \leq \max\{\alpha_m \Delta + \alpha_m r_m, \Delta\} \leq \Delta$$

and our assumption made for m holds for $m+1$ too.

Assume now that

$$2M|\alpha_m - \alpha_{m+1}| > \Delta. \tag{4.90}$$

Let

$$p_m = \lfloor 2M|\alpha_{m+1} - \alpha_m|\Delta^{-1} \rfloor + 1, \tag{4.91}$$

Set

$$\beta = (\alpha_{m+1} - \alpha_m) p_m^{-1}, \tag{4.92}$$

$$q_{m+2} = 2q_{m+1} - q_m + p, \tag{4.93}$$

for all $i = 1, \dots, p_m$,

$$x_{2q_{m+1}-q_m+i} = (1 - \alpha_m - \beta i)\theta \oplus (\alpha_m + \beta_i) y_{q_{m+1}}. \tag{4.94}$$

By (4.82) and (4.92)–(4.94),

$$x_{q_{m+2}} = x_{2q_{m+1}-q_m+p_m} = (1 - \alpha_{m+1})\theta \oplus \alpha_{m+1} y_{q_{m+1}} = x^{(m+1)}.$$

Propositions 2.1 and 4.20 and Eqs. (4.78), (4.87), (4.91), (4.92), and (4.94) imply that

$$d(x_{2q_{m+1}-q_m+i}, x_{2q_{m+1}-q_m+1+i})$$

$$\leq \beta d(\theta, y_{q_{m+1}}) \leq 2M\beta \leq 2M(\alpha_{m+1} - \alpha_m) p_m^{-1} \leq \Delta. \tag{4.95}$$

Together with (4.73), (4.77), and (4.81), this implies that for each

$$i \in \{q_{m+1}, \dots, q_{m+2} - q_1\},$$

we have

$$d(x_i, x_{i+1}) \leq \max\{\alpha_m \Delta + \alpha_m r_m, \Delta\} \leq \Delta$$

and our assumption made for m holds for $m + 1$ too. Thus by induction, we have constructed $x_i, x^{(i)}$ for each integer $i \geq 0$.

Proposition 4.22 *Let $m \geq 0$ be an integer and $\alpha_m < \alpha_{m+1}$. Then*

$$\sum_{i=q_{m+1}}^{q_{m+2}-1} d(x_i, x_{i+1}) \leq \alpha_m \sum_{i=q_m}^{q_{m+1}-1} d(x_i, x_{i+1}) + (2M + 1)|\alpha_{m+1} - \alpha_m|,$$

$$d_1(x^{(m)}, T(x^{(m)})) \leq d_1(x^{(m)}, x^{(m+1)}) + 2M(1 - \alpha_m).$$

Proof In all the three cases of our construction of $\{x_i\}_{i=0}^{\infty}$, it follows from (4.81) that

$$\sum \{d(x_{q_{m+1}+i}, x_{q_{m+1}+1+i}) :$$

$$i \in \{0, \dots, q_{m+1} - q_m\} \setminus \{q_{m+1} - q_m, q_{m+1} - q_m - 1\}\}$$

$$\le \alpha_m \sum \{d(x_{q_m+i}, x_{q_m+1+i}) :$$

$$i \in \{0, \dots, q_{m+1}-q_m\}\setminus\{q_{m+1}-q_m, q_{m+1}-q_m-1\}\}+\alpha_m r_m(q_{m+1}-q_m). \tag{4.96}$$

Now we consider the three cases separately. First we consider the case when (4.83) holds. By (4.84), (4.85), and (4.96),

$$\sum_{i=q_{m+1}}^{q_{m+2}-1} d(x_i, x_{i+1}) \le \alpha_m \sum_{i=q_m}^{q_{m+1}-1} d(x_i, x_{i+1})$$

$$+ \alpha_m r_m(q_{m+1} - q_m + 1) + 2M|\alpha_m - \alpha_{m+1}|. \tag{4.97}$$

Now assume that (4.83) does not hold and (4.88) is true. It follows from (4.5), (4.89), and (4.96) that

$$\sum_{i=q_{m+1}}^{q_{m+2}-1} d(x_i, x_{i+1}) \le \alpha_m \sum_{i=q_m}^{q_{m+1}-1} d(x_i, x_{i+1}) + \alpha_m r_m(q_{m+1} - q_m + 1)$$

$$+ 2M|\alpha_m - \alpha_{m+1}|$$

and (4.97) holds.

Assume that (4.90) holds. Equations (4.5), (4.92), (4.93), (4.95), and (4.96) imply that (4.97) holds. Thus (4.97) holds in all the cases. In view of (4.77) and (4.97),

$$\sum_{i=q_{m+1}}^{q_{m+2}-1} d(x_i, x_{i+1}) \le \alpha_m \sum_{i=q_m}^{q_{m+1}-1} d(x_i, x_{i+1}) + (2M + 1)|\alpha_m - \alpha_{m+1}|. \tag{4.98}$$

Proposition 4.3 and 4.20 and Eqs. (4.70), (4.71), (4.73), (4.74), and (4.98) imply that

$$d_1(x^{(m)}, T(x^{(m)})) \le d_1(x^{(m)}, x^{(m+1)}) + d_1(x^{(m+1)}, T(x^{(m)}))$$

$$\le d_1(x^{(m)}, x^{(m+1)}) + d_1((1 - \alpha_m)\theta \oplus y^{(m)}, T(x^{(m)}))$$

$$\le d_1(x^{(m)}, x^{(m+1)}) + d_1((1 - \alpha_m)\theta \oplus y^{(m)}, y^{(m)})$$

$$\le d_1(x^{(m)}, x^{(m+1)}) + (1 - \alpha_m)d(\theta, y^{(m)}) \le d_1(x^{(m)}, x^{(m+1)}) + 2M(1 - \alpha_m).$$

Proposition 4.22 is proved.

We prove the following result which shows that our iterative process generates approximate fixed points of the set-valued mapping T.

Theorem 4.23 *Let*

$$\alpha_n = 1 - 2(n+3)^{-1},\ n = 0, 1 \ldots.$$

Then for each integer $n \geq 0$,

$$d(x^{(n)}, x^{(n+1)}) \leq d_1(x^{(n)}, x^{(n+1)}) \leq (9M+3)(n+3)^{-1},$$

and

$$d_1(x^{(n)}, T(x^{(n)})) \leq (14M+5)(n+3)^{-1}.$$

Proof Set

$$N = 2,\ J = 3,\ \gamma = 1,$$

$$a_n = 1 - \alpha_n = (2n+3)^{-1},\ c_n = L = 3M+1,\ n = 0, 1, \ldots,$$

for each integer $n \geq 0$,

$$s_n = \sum_{i=q_n}^{q_{n+1}-1} d(x_i, x_{i+1}).$$

By Proposition 4.22 and the relations above, for each integer $n \geq 0$,

$$s_{n+1} \leq (1 - a_n)s_n + 2|a_{n+1} - a_n|L.$$

In view of (4.33) and (4.35),

$$d_1(x_0, x_1) \leq 3M.$$

By Lemma 2.2 and the relations above, for each integer $n \geq 0$,

$$d(x^{(n)}, x^{(n+1)}) \leq d_1(x^{(n)}, x^{(n+1)})$$

$$\leq \sum_{i=q_n}^{q_{n+1}-1} d(x_i, x_{i+1}) = s_n \leq (9M+3)(n+3)^{-1}.$$

Together with Proposition 4.22, this implies that for each integer $n \geq 0$,

$$d_1(x^{(n)}, T(x^{(n)})) \leq (14M+5)(n+3)^{-1}.$$

Theorem 4.23 is proved.

4.6 The Alternative Iterative Method for Set-Valued Mappings

We use the notation and definitions and assume the assumptions introduced in Sects. 4.1 and 4.5. Recall that

$$\theta \in \mathrm{St}(C),$$

$$\{r_n\}_{n=0}^{\infty} \subset (0, \min\{\Delta, 1\}$$

(see (4.31)), $T : C \to 2^C \setminus \{\emptyset\}$ satisfies (4.30) for each $x, y \in C$ satisfying $d(x, y) \leq \Delta$ and assumption (A) holds. Let

$$x^{(0)} \in C \cap B(\theta, M). \tag{4.99}$$

and

$$x^{(1} \in T((1 - \alpha_0)\theta \oplus \alpha_0 x^{(0)}). \tag{4.100}$$

Propositions 2.1, 4.1, and 4.14 and (4.99) imply that

$$H_1(T((1 - \alpha_0)\theta \oplus \alpha_0 x^{(0)}), T(\theta))$$

$$\leq d_1((1 - \alpha_0)\theta \oplus \alpha_0 x^{(0)}, \theta) \leq \alpha_0 d(\theta, x^{(0)}) \leq \alpha_0 M. \tag{4.101}$$

By (A) and (4.100) and (4.101),

$$d(x^{(1}, T(\theta)) \leq \alpha_0 M, \ d(x^{(1)}, \theta) \leq 2M. \tag{4.102}$$

Proposition 4.24 *Assume that $n \geq 0$ is an integer, $z \in C$ satisfies*

$$d_1(z, T^n(\theta)) < M, \tag{4.103}$$

$\alpha \in (0, 1)$,

$$\xi \in T((1 - \alpha)\theta \oplus \alpha z), \tag{4.104}$$

$$\eta \in T(\xi). \tag{4.105}$$

Then

$$d_1(\xi, T^{n+1}(\theta)) < M, \ d_1(\eta, T^{n+2}(\theta)) < M,$$

$$d_1(\xi, \theta) < 2M, \ d_1(\eta, \theta) < 2M.$$

Proof In view of (4.103), there exists

$$u \in T^n(\theta) \text{ such that } d_1(z, u) < M. \tag{4.106}$$

Proposition 4.3 and (4.106) imply that

$$d_1((1-\alpha)\theta \oplus \alpha z, (1-\alpha)\theta \oplus \alpha u) \le \alpha d_1(u, z) < \alpha M. \tag{4.107}$$

Proposition 4.1, (A), and (4.106) imply that

$$d_1((1-\alpha)\theta \oplus \alpha u, u) \le (1-\alpha)d_1(\theta, u) < (1-\alpha)M. \tag{4.108}$$

It follows from (4.107) and (4.108) that

$$d_1((1-\alpha)\theta \oplus \alpha z, u) < M. \tag{4.109}$$

Proposition 4.19 and (4.104), (4.106), and (4.109) imply that

$$\begin{aligned} d_1(\xi, T^{n+1}(\theta)) &\le d_1(\xi, T(u)) \le H_1(T((1-\alpha)\theta \oplus \alpha z), T(u)) \\ &\le d_1((1-\alpha)\theta \oplus \alpha z, u) < M. \end{aligned} \tag{4.110}$$

In view of (4.110), there exists

$$v \in T^{n+1}(\theta) \tag{4.111}$$

such that

$$d_1(\xi, v) < M. \tag{4.112}$$

Proposition 4.19, (A), and (4.105), (4.110), (4.111), and (4.113) imply that

$$d_1(\eta, T^{n+2}(\theta)) \le d_1(\eta, T(v)) \le H_1(T(\xi), T(v)) \le d_1(\xi, v) < M.$$

Proposition 4.24 is proved.

In view of (4.99) and (4.102),

$$d_1(x^{(0)}, x^{(1)}) \le d_1(x^{(0)}, \theta) + d_1(\theta, x^{(1)}) = d(x^{(0)}, \theta) + d(\theta, x^{(1)}) \le 3M. \tag{4.113}$$

Assume that q_0 is a natural number, $x_i \in C, i = 0, \dots, q_0$ and that

$$x_0 = x^{(0)}, \ x_{q_0} = x^{(1)}, \tag{4.114}$$

$$d(x_i, x_{i+1}) \le \Delta,\ i \in \{0, \dots, q_0 - 1\}. \tag{4.115}$$

Now we describe the induction step. Assume that $k \ge 0$ is an integer; we already defined natural numbers $q_i, i = 0, \dots, k$ such that

$$q_i < q_{i+1},\ i \in \{0, \dots, k\} \setminus \{k\},$$

$$x_{q_k} = x^{(k)} \in C,\ x_{q_{k+1}} = x^{(k+1)} \in C, \tag{4.116}$$

$$x_i \in C,\ i = 0, \dots, q_{k+1},$$

$$d(x_i, x_{i+1}) \le \Delta,\ i = q_k, \dots, q_{k+1} - 1, \tag{4.117}$$

$$x_{q_{i+1}} \in T((1 - \alpha_i)\theta \oplus \alpha_i x_{q_i}),\ i = 0, \dots, k. \tag{4.118}$$

Assume that a positive number r_k satisfies

$$r_k < 8^{-1}(1 - \alpha_k)\Delta,\ r_k(q_{k+1} - q_k + 1) < |\alpha_k - \alpha_{k+1}|. \tag{4.119}$$

Set

$$\tilde{x}_{q_{k+1}} = x_{q_{k+1}}.$$

By (4.30) and (4.116)–(4.118) and Proposition 4.3, for each $i \in \{0, \dots, q_{k+1} - q_k - 1$, there exists

$$\tilde{x}_{q_{k+1}+i+1} \in T((1 - \alpha_k)\theta \oplus \alpha_k x_{q_k+i+1}) \tag{4.120}$$

such that

$$d(\tilde{x}_{q_{k+1}+i}, \tilde{x}_{q_{k+1}+i+1}) \le d(\tilde{x}_{q_{k+1}+i}, T((1 - \alpha_k)\theta \oplus \alpha_k x_{q_k+i+1})) + r_k$$

$$\le H(T((1 - \alpha_k)\theta \oplus \alpha_k x_{q_k+i}), T((1 - \alpha_k)\theta \oplus \alpha_k x_{q_k+i+1})) + r_k$$

$$\le d((1 - \alpha_k)\theta \oplus \alpha_k x_{q_k+i}, (1 - \alpha_k)\theta \oplus \alpha_k x_{q_k+i+1}) + r_k$$

$$\le \alpha_k d(x_{q_k+i}, x_{q_k+i+1}) + r_k \tag{4.121}$$

$$\le \alpha_k \Delta + r_k < \Delta. \tag{4.122}$$

Set

$$x_i = \tilde{x}_i,\ i \in \{q_{k+1} + 1, \dots, 2q_{k+1} - q_k\} \setminus \{2q_{k+1} - q_k\}. \tag{4.123}$$

Now we define an integer

$$q_{k+2} \geq 2q_{k+1} - q_k$$

and set

$$x_{q_{k+2}} = x^{(k+2)}.$$

Clearly,

$$x_{2q_{k+1}-q_k-1} \in T((1-\alpha_k)\theta \oplus \alpha_k x_{q_{k+1}-1}). \tag{4.124}$$

Proposition 4.24 and (4.99) and (4.118) imply that

$$d_1(\theta, x_{q_i}) \leq 2M, \ i = 0, \ldots, k+1. \tag{4.125}$$

Proposition 2.1 and (4.117) and (4.125) imply that

$$\begin{aligned} &d((1-\alpha_k)\theta \oplus \alpha_k x_{q_{k+1}-1}, (1-\alpha_{k+1})\theta \oplus \alpha_{k+1} x_{q_{k+1}}) \\ &\leq \alpha_k d(x_{q_{k+1}-1}, x_{q_{k+1}}) + |\alpha_k - \alpha_{k+1}| d(\theta, x_{q_{k+1}}) \\ &\leq \alpha_k d(x_{q_{k+1}-1}, x_{q_{k+1}}) + 2M|\alpha_k - \alpha_{k+1}| \\ &\leq \alpha_k \Delta + 2M|\alpha_k - \alpha_{k+1}|. \end{aligned} \tag{4.126}$$

Assume that

$$2M|\alpha_{k+1} - \alpha_k| \leq \Delta(1-\alpha_k). \tag{4.127}$$

(Note that (4.127) holds for large enough n is $\lim_{n\to\infty} \alpha_n = 1$.) By (4.126) and (4.127),

$$d((1-\alpha_k)\theta \oplus \alpha_k x_{q_{k+1}-1}, (1-\alpha_{k+1})\theta \oplus \alpha_{k+1} x_{q_{k+1}}) \leq \Delta. \tag{4.128}$$

Set

$$q_{k+2} = 2q_{k+1} - q_k. \tag{4.129}$$

Propositions 2.1 and (4.25), (4.30), (4.120), (4.123), and (4.128) imply that there exists

$$x_{2q_{k+1}-q_k} \in T((1-\alpha_{k+1})\theta \oplus \alpha_{k+1} x_{q_{k+1}}) \tag{4.130}$$

such that

$$d(x_{2q_{k+1}-q_k-1}, x_{2q_{k+1}-q_k})$$

$$\leq d(x_{2q_{k+1}-q_k-1}, T((1-\alpha_{k+1})\theta \oplus \alpha_{k+1}x_{q_{k+1}})) + r_k$$

$$\leq H(T((1-\alpha_k)\theta \oplus \alpha_k x_{q_{k+1}-1}), T((1-\alpha_{k+1})\theta \oplus \alpha_{k+1}x_{q_{k+1}})) + r_k$$

$$\leq d((1-\alpha_k)\theta \oplus \alpha_k x_{q_{k+1}-1}, (1-\alpha_{k+1})\theta \oplus \alpha_{k+1}x_{q_{k+1}}) + r_k$$

$$\leq \alpha_k d(x_{q_{k+1}-1}, x_{q_{k+1}}) + 2M|\alpha_k - \alpha_{k+1}| + r_k. \tag{4.131}$$

Note that x_i, $i = q_{k+1}, \dots, 2q_{k+1} - q_k$ are defined. Set

$$x^{(k+2)} = x_{2q_{k+1}-q_k}. \tag{4.132}$$

By (4.117), (4.119), and (4.131),

$$d(x_{2q_{k+1}-q_k-1}, x_{2q_{k+1}-q_k}) \leq \alpha_k \Delta + r_k + 2M|\alpha_k - \alpha_{k+1}| \leq \Delta_k. \tag{4.133}$$

In view of (4.122),(4.132), (4.133), (4.139), and (4.142), our assumption holds for $k+1$ too. It follows from (4.119), (4.121), and (4.123) that

$$\sum_{i=q_{k+1}}^{q_{k+2}-1} d(x_i, x_{i+1}) \leq \alpha_k \sum_{i=q_k}^{q_{k+1}-1} d(x_i, x_{i+1}) + (q_{k+1} + 1 - q_k)r_k + 2M|\alpha_k - \alpha_{k+1}|$$

$$\leq \alpha_k \sum_{i=q_k}^{q_{k+1}-1} d(x_i, x_{i+1}) + (2M+1)|\alpha_k - \alpha_{k+1}|. \tag{4.134}$$

Assume that (4.127) does not hold. Let

$$\Delta > 2M|\alpha_{k+1} - \alpha_k|. \tag{4.135}$$

Set

$$x_{2q_{k+1}-q_k} = \tilde{x}_{2q_{k+1}-q_k}. \tag{4.136}$$

Propositions 2.1 and 4.24 and (4.99), (4.118), and (4.135) imply that

$$d((1-\alpha_k)\theta \oplus \alpha_k x_{q_{k+1}}, (1-\alpha_{k+1})\theta \oplus \alpha_{k+1}x_{q_{k+1}})$$

$$\leq |\alpha_k - \alpha_{k+1}| d(\theta, x_{q+1}) \leq 2M|\alpha_k - \alpha_{k+1}| \leq \Delta. \tag{4.137}$$

Fix a positive number

$$\tilde{r}_k < \min\{r_k,\ \Delta - 2M|\alpha_k - \alpha_{k+1}|\}. \tag{4.138}$$

Set

$$q_{k+2} = 2q_{k+1} - q_k + 1. \tag{4.139}$$

Propositions 2.1 and 4.24 and Eqs. (4.30), (4.118), (4.120), (4.123), and (4.136) imply that there exists

$$x_{2q_{k+1}-q_k+1} \in T((1-\alpha_{k+1})\theta \oplus \alpha_{k+1}x_{q_{k+1}}) \tag{4.140}$$

such that

$$\begin{aligned}
d(x_{2q_{k+1}-q_k}, x_{2q_{k+1}-q_k+1}) &\le d(x_{2q_{k+1}-q_k}, T((1-\alpha_{k+1})\theta \oplus \alpha_{k+1}x_{q_{k+1}})) + \tilde{r}_k \\
&\le H(T((1-\alpha_k)\theta \oplus \alpha_k x_{q_{k+1}}), T((1-\alpha_{k+1})\theta \oplus \alpha_{k+1}x_{q_{k+1}})) + \tilde{r}_k \\
&\le d((1-\alpha_k)\theta \oplus \alpha_k x_{q_{k+1}}, (1-\alpha_{k+1})\theta \oplus \alpha_{k+1}x_{q_{k+1}}) + \tilde{r}_k \\
&\le |\alpha_k - \alpha_{k+1}| d(\theta, x_{q_{k+1}}) + \tilde{r}_k \\
&\le 2M|\alpha_k - \alpha_{k+1}| + \min\{r_k,\ \Delta - 2M|\alpha_k - \alpha_{k+1}|\}.
\end{aligned} \tag{4.141}$$

By (4.135) and (4.141),

$$d(x_{2q_{k+1}-q_k}, x_{2q_{k+1}-q_k+1}) \le \Delta. \tag{4.142}$$

Thus the assumption made for k also holds for $k+1$. It follows from (4.119), (4.121), (4.123), (4.136), and (4.141) that

$$\begin{aligned}
\sum_{i=q_{k+1}}^{q_{k+2}-1} d(x_i, x_{i+1}) &\le \alpha_k \sum_{i=q_k}^{q_{k+1}-1} d(x_i, x_{i+1}) + (q_{k+1} + 1 - q_k)r_k + 2M|\alpha_k - \alpha_{k+1}| \\
&\le \alpha_k \sum_{i=q_k}^{q_{k+1}-1} d(x_i, x_{i+1}) + (2M+1)|\alpha_k - \alpha_{k+1}|.
\end{aligned} \tag{4.143}$$

Assume that (4.135) does not hold. Then

$$2M|\alpha_{k+1} - \alpha_k| \ge \Delta. \tag{4.144}$$

Set

$$p_k = \lfloor 4M|\alpha_{k+1} - \alpha_k|\Delta^{-1} \rfloor,\ \beta = (\alpha_{k+1} - \alpha_k)p_k^{-1} \tag{4.145}$$

and for each $i \in \{0, \dots, p_k\}$,

$$x_{2q_{k+1}-q_k+i} = (1 - \alpha_k - \beta i)\theta \oplus (\alpha_k + i\beta)x_{q_{k+1}}. \tag{4.146}$$

In view of (4.145) and (4.146),

$$x_{2q_{k+1}-q_k+p_k} = (1 - \alpha_{k+1})\theta \oplus \alpha_{k+1}x_{q_{k+1}}.$$

Propositions 2.1 and 4.24 and Eqs. (4.115), (4.118), and (4.146) imply that for each $i \in \{0, \dots, p_k - 1\}$,

$$d(x_{2q_{k+1}-q_k+i}, x_{2q_{k+1}-q_k+i+1}) \le \beta d(\theta, x_{q_{k+1}}) \le 2M\beta \le \Delta/2. \tag{4.147}$$

Thus the assumption made for k also holds for $k + 1$. It follows from (4.145) and (4.147) and Proposition 4.24 that

$$\sum_{i=2q_{k+1}-q_k}^{q_{k+2}-1} d(x_i, x_{i+1}) \le p_k\beta d(\theta, x_{q_{k+1}}) \le 2M|\alpha_k - \alpha_{k+1}|.$$

Together with (4.119), (4.121), (4.123), and (4.136), this implies that

$$\sum_{i=q_{k+1}}^{q_{k+2}-1} d(x_i, x_{i+1}) = \sum_{i=q_{k+1}}^{2q_{k+1}-q_k-1} d(x_i, x_{i+1}) + \sum_{i=2q_{k+1}-q_k}^{q_{k+2}-1} d(x_i, x_{i+1})$$

$$\le \alpha_k \sum_{i=q_k}^{q_{k+1}-1} d(x_i, x_{i+1}) + r_k(q_{k+1} - q_k) + 2M|\alpha_k - \alpha_{k+1}|$$

$$\le \alpha_k \sum_{i=q_k}^{q_{k+1}-1} d(x_i, x_{i+1}) + (2M + 1)|\alpha_k - \alpha_{k+1}|.$$

Thus in all the cases

$$\sum_{i=q_{k+1}}^{q_{k+2}-1} d(x_i, x_{i+1}) \le \alpha_k \sum_{i=q_k}^{q_{k+1}-1} d(x_i, x_{i+1}) + (2M + 1)|\alpha_k - \alpha_{k+1}|. \tag{4.148}$$

We prove the following result which shows that our iterative process generates approximate fixed points of the set-valued mapping T.

Theorem 4.25 *Let*

$$\alpha_n = 1 - 2(n + 3)^{-1}, \ n = 0, 1 \dots.$$

Then for each integer $n \ge 0$,

$$d(x^{(n)}, x^{(n+1)}) \le d_1(x^{(n)}, x^{(n+1)}) \le (9M + 3)(n + 3)^{-1}, \tag{4.149}$$

and

$$d_1(x^{(n)}, T(x^{(n)}))) \le (13M + 14)(n + 3)^{-1}.$$

Proof This theorem is proved analogously to Theorem 4.23. Using (4.148), we obtain that (4.149) holds for each integer $n \ge 0$. By (4.118) and (4.149) and Propositions 4.1, 4.19, and 4.24, for each integer $n \ge 0$,

$$d_1(x^{(n)}, T(x^{(n)}) \le d_1(x^{(n)}, x^{(n+1)}) + d_1(x^{(n+1)}, T(x^{(n)})$$

$$\le (9M + 3)(n + 3)^{-1} + d_1(x^{(n+1)}, T(x^{(n)})$$

$$\le (9M + 3)(n + 3)^{-1} + H_1(T((1 - \alpha_n)\theta \oplus \alpha_n x^{(n)}), T(x^{(n)}))$$

$$\le (9M + 3)(n + 3)^{-1} + d_1((1 - \alpha_n)\theta \oplus \alpha_n x^{(n)}, x^{(n)})$$

$$\le (9M + 3)(n + 3)^{-1} + \alpha_n d(\theta, x^{(n)})$$

$$\le (9M + 3)(n + 3)^{-1} + 4(n + 3)^{-1} M.$$

Theorem 4.25 is proved.

Chapter 5
Asymptotic Regularity Property in Spaces with Graphs

In this chapter we study asymptotic regularity for Halpern-type iterations and for iterations of alternative iterative methods generated by a nonexpansive self-mapping of a star-shaped set in a metric space with a hyperbolic structure and a graph. Linear rates of this asymptotic regularity are computed. We study asymptotic regularity for single-valued and set-valued mappings. All the results of this chapter are new.

5.1 Halpern-Type Iterations in a Metric Space with a Graph

Assume that $X = (X, d, W)$ is a W-hyperbolic space, C is a nonempty star-shaped subset of X, C is equipped with a graph G, $V(G) \subset C$ is the set of its vertices, $E(G) \subset C \times C$ is the set of its edges, and $T : C \to C$ satisfies for each $(x, y) \in E(G)$,

$$(T(x), T(y)) \in E(G) \text{ and } d(T(x), T(y)) \le d(x, y). \tag{5.1}$$

We assume that $(z, z) \in E(G)$ for each $z \in C$. Let

$$x_0 = x \in \mathrm{St}(C), \tag{5.2}$$

$$(x_0, T(x_0)) \in E(G). \tag{5.3}$$

Set

$$C_x = \{z \in C : \ (x, z) \in E(G)\} \tag{5.4}$$

and assume that

A. J. Zaslavski, *Approximate Fixed Points of Nonexpansive Mappings*,
Developments in Mathematics 80, https://doi.org/10.1007/978-3-031-70710-0_5

$$T(C_x) \subset C_x. \tag{5.5}$$

Assume that $\{\alpha_n\}_{n=0}^{\infty} \subset (0, 1)$,

$$\alpha_n \le \alpha_{n+1} \text{ for each integer } n \ge 0 \tag{5.6}$$

and that the following assumptions hold.

(A1) For each pair of numbers β_1, β_2 satisfying $0 \le \beta_1 < \beta_2 \le 1$ and each $z_1, z_2 \in C_x$ satisfying $(z_1, z_2) \in E(G)$ the relations

$$(1 - \beta_i)x_0 \oplus \beta_i z \in C_x, \ i = 1, 2$$

and

$$((1 - \beta_1)x_0 \oplus \beta_1 z_1, (1 - \beta_2)x_0 \oplus \beta_2 z_2) \in E(G)$$

are true.

(A2) There exists $M > 0$ such that

$$\rho(T^i(x_0), x_0) \le M, \ i = 0, 1, \ldots.$$

In this chapter, our goal is to show that analogs of the results of Chap. 2 hold for nonexpansive operators in metric spaces with graphs. For these operators, relation (5.1) is true only if $(x, y) \in E(G)$. Therefore in order to obtain such analogs, we should guarantee that for our sequences of iterates, the inclusion $(x_n, x_{n+1}) \in E(G)$ holds for any integer $n \ge 0$. In order to meet this goal, we need assumption (A1) on the connection between the hyperbolic structure and the graph.

Assume that for each integer $n \ge 0$,

$$x_{n+1} = (1 - \alpha_n)x_0 \oplus \alpha_n T(x_n). \tag{5.7}$$

We prove the following result which shows that our iterative process generates approximate fixed points of the set-valued mapping T.

Theorem 5.1 *Let for each integer $n \ge 0$,*

$$\alpha_n = 1 - 2(n + 3)^{-1}.$$

Then for each integer $n \ge 0$,

$$d(x_n, x_{n+1}) \le 6M(n + 3)^{-1}$$

and

$$d(x_n, T(x_n)) \le 10M(n + 3)^{-1}.$$

We begin with the following auxiliary results.

Lemma 5.2 *For each integer $n \geq 0$,*

$$(x_0, x_n) \in E(G),\ (x_n, x_{n+1}) \in E(G),\ (x_{n+1}, T(x_n)) \in E(G). \tag{5.8}$$

Proof Assumption (A1) and (5.3) imply that

$$(x_0, T(x_0)) \in E(G),\ (x_0, x_1) \in E(G),\ (x_1, T(x_0)) \in E(G).$$

Thus (5.8) holds for $n = 0$. Assume that $n \geq 0$ is an integer and that (5.8) holds. Assumption (A1) and (5.5), (5.7), and (5.8) imply that

$$(x_0, T(x_n)) \in E(G),\ (x_0, x_{n+1}) \in E(G),\ (x_0, T(x_{n+1})) \in E(G),$$

$$(x_0, x_{n+2}) \in E(G),\ (x_{n+2}, T(x_{n+1})) \in E(G).$$

By (5.7),

$$x_{n+1} = (1 - \alpha_n)x_0 \oplus \alpha_n T(x_n),$$

$$x_{n+2} = (1 - \alpha_{n+1})x_0 \oplus \alpha_{n+1} T(x_{n+1}).$$

Together with (A1), (5.1), (5.6), and (5.8), this implies that

$$(x_{n+1}, x_{n+2}) \in E(G).$$

Thus our assumption holds for $n + 1$ and Lemma 5.2 is proved by induction.

Lemma 5.3 *Let $n \geq 0$ be an integer. Then*

$$d(x_{n+2}, x_{n+1}) \leq \alpha_{n+1} d(x_{n+1}, x_n) + |\alpha_{n+1} - \alpha_n| d(x_0, T(x_n)),$$

$$d(x_n, T(x_n)) \leq d(x_n, x_{n+1}) + (1 - \alpha_n) d(x_0, T(x_n)).$$

Proof By (5.1), (5.6), and (5.8),

$$\alpha_n \leq \alpha_{n+1},\ (T(x_n), T(x_{n+1})) \in E(G).$$

By the relations above, (5.1) and (5.7), Proposition 2.1, and Lemma 5.2,

$$d(x_{n+2}, x_{n+1})$$

$$= d((1 - \alpha_{n+1})x_0 \oplus \alpha_{n+1} T(x_{n+1}), (1 - \alpha_n)x_0 \oplus \alpha_n T(x_n))$$

$$\leq \alpha_{n+1} d(T(x_{n+1}), T(x_n)) + |\alpha_{n+1} - \alpha_n| d(x_0, T(x_n))$$

$$\leq \alpha_{n+1}d(x_{n+1}, x_n) + |\alpha_{n+1} - \alpha_n|d(x_0, T(x_n)).$$

Proposition 2.1 and (5.7) imply that

$$d(x_n, T(x_n)) \leq d(x_n, x_{n+1}) + d(x_{n+1}, T(x_n))$$

$$\leq d(x_n, x_{n+1}) + d((1 - \alpha_n)x_0 \oplus \alpha_n T(x_n), T(x_n))$$

$$\leq d(x_n, x_{n+1}) + (1 - \alpha_n)d(x_0, T(x_n)).$$

Lemma 5.3 is proved.

Lemma 5.4 *For each integer $n \geq 0$,*

$$(x_n, T^n(x_0)) \in E(G), \tag{5.9}$$

$$d(x_n, T^n(x_0)) \leq M. \tag{5.10}$$

Proof We prove the lemma by induction. Clearly, (5.9) and (5.10) hold for $n = 0$. Assume that (5.9) and (5.10) hold for an integer $n \geq 0$. In view of (5.1), (5.9), and (5.10),

$$d(T(x_n), T^{n+1}(x_0)) \leq d(x_n, T^n(x_0)) \leq M, \tag{5.11}$$

$$(T(x_n), T^{n+1}(x_0)) \in E(G).$$

It follows from (5.3), (5.5), and (5.7) that

$$(x_{n+1}, T^{n+1}(x_0)) \in E(G).$$

Proposition 2.1, (A2), and (5.7), (5.9) and (5.11) imply that

$$d(x_{n+1}, T^{n+1}(x_0)) \leq d((1 - \alpha_n)x_0 \oplus \alpha_n T(x_n), T^{n+1}(x_0))$$

$$\leq (1 - \alpha_n)d(x_0, T^{n+1}(x_0)) + \alpha_n d(T(x_n), T^{n+1}(x_0))$$

$$\leq 2M(1 - \alpha_n) + \alpha_n d(x_n, T^n(x_0))$$

$$\leq 2M(1 - \alpha_n) + \alpha_n M \leq M.$$

Thus our assumption holds for $n + 1$ too. Therefore by induction (5.9) and (5.10) hold for each integer $n \geq 0$. Lemma 5.4 is proved.

Corollary 5.5 *Let $n \geq 0$ be an integer. Then $d(x_0, T(x_n)) \leq 2M$.*

Proof Lemma 5.4, (A2), and (5.1) imply that

$$d(x_0, T(x_n)) \leq d(x_0, T^{n+1}(x_0)) + d(T(x_n), T^{n+1}(x_0))$$

$$\leq M + d(x_n, T^n(x_0)) \leq 2M.$$

Corollary 5.5 is proved.

Lemma 5.3 implies the following result.

Proposition 5.6 *Let $n \geq 0$ be an integer. Then*

$$d(x_{n+2}, x_{n+1}) \leq \alpha_{n+1} d(x_{n+1}, x_n) + 2M|\alpha_{n+1} - \alpha_n|,$$

$$d(x_n, T(x_n)) \leq d(x_n, x_{n+1}) + 2M(1 - \alpha_n).$$

Proof of Theorem 5.1 Set

$$J = 3, \; N = 2, \; \gamma = 1.$$

and for each integer $n \geq 0$,

$$a_n = 1 - \alpha_n, \; s_n = d(x_n, x_{n+1}), \; c_n = L = 2M.$$

By (A2), Proposition 2.1, and (5.7),

$$d(x_0, x_1) \leq M.$$

Proposition 5.6 implies that for each integer $n \geq 0$,

$$s_{n+1} \leq (1 - a_n)s_n + c_n|a_n - a_{n+1}|.$$

By Lemma 2.2 and the relations above, for each integer $n \geq 0$,

$$d(x_n, x_{n+1}) \leq 6M(n+3)^{-1}.$$

Together with Proposition 5.6, this implies that for each integer $n \geq 0$,

$$d(x_n, T(x_n)) \leq 10M(n+3)^{-1}.$$

Theorem 5.1 is proved.

5.2 Alternative Iterative Method in a Space with a Graph

We use the notation, definitions, and the assumptions introduced in Sect. 5.1. In particular, we assume that assumptions (A1) and (A2) hold. Assume that

$$x_0 = x \in \mathrm{St}(C), \tag{5.12}$$

$$(x_0, T(x_0)) \in E(G) \tag{5.13}$$

and that for each integer $n \geq 0$,

$$x_{n+1} = T((1 - \alpha_n)x_0 \oplus \alpha_n x_n). \tag{5.14}$$

We will prove the following result which shows that our iterative process generates approximate fixed points of the mapping T.

Theorem 5.7 *Let for each integer $n \geq 0$,*

$$\alpha_n = 1 - 2(n+3)^{-1}.$$

Then for each integer $n \geq 0$,

$$d(x_n, x_{n+1}) \leq 6M(n+3)^{-1}$$

and

$$d(x_n, T(x_n)) \leq 10M(n+3)^{-1}.$$

We begin with auxiliary results.

Lemma 5.8 *For each integer $n \geq 0$,*

$$(x_0, x_n) \in E(G),\ (x_n, x_{n+1}) \in E(G),\ (x_{n+1}, T(x_n)) \in E(G). \tag{5.15}$$

Proof In view of (5.13) and (A1),

$$x_1 = T((1 - \alpha_0)x_0 \oplus \alpha_0 x_0) = T(x_0),$$

$$(x_0, x_1) \in E(G),\ (T(x_0, T(x_0)) = (x_1, T(x_0)) \in E(G)$$

and (5.15) holds for $n = 0$. Assume that $n \geq 0$ is an integer and that (5.15) holds. Assumptions (A1) and (A2) and (5.1), (5.5), and (5.14) imply that

$$((1 - \alpha_{n+1})x_0 \oplus \alpha_{n+1}x_{n+1}) \in E(G),\ (x_{n+2}, T(x_{n+1}))$$

$$\in E(G),\ (x_0, x_{n+1}) \in E(G).$$

By (A1), (5.1), (5.14), and (5.15),

$$(x_{n+1}, x_{n+2}) = (T((1-\alpha_n)x_0 \oplus \alpha_n x_n), T((1-\alpha_{n+1})x_0 \oplus \alpha_{n+1}x_{n+1})) \in E(G).$$

Lemma 5.8 is proved by induction.

Lemma 5.9 *Let $n \geq 0$ be an integer. Then*

$$d(x_{n+2}, x_{n+1}) \leq \alpha_{n+1}d(x_{n+1}, x_n) + |\alpha_{n+1} - \alpha_n| d(x_0, x_n),$$

$$d(x_n, T(x_n)) \leq d(x_n, x_{n+1}) + (1-\alpha_n)d(x_0, x_n).$$

Proof By (A1), Proposition 2.1, and (5.1), (5.8), and (5.14),

$$d(x_{n+2}, x_{n+1})$$

$$= d(T((1-\alpha_{n+1})x_0 \oplus \alpha_{n+1}x_{n+1}), T((1-\alpha_n)x_0 \oplus \alpha_n x_n))$$

$$\leq d((1-\alpha_{n+1})x_0 \oplus \alpha_{n+1}x_{n+1}, (1-\alpha_n)x_0 \oplus \alpha_n x_n)$$

$$\leq \alpha_{n+1}d(x_{n+1}, x_n) + |\alpha_{n+1} - \alpha_n| d(x_0, x_n).$$

Proposition 2.1, Lemma 5.8, (A1), and (5.1) and (5.14) imply that

$$d(x_n, T(x_n)) \leq d(x_n, x_{n+1}) + d(x_{n+1}, T(x_n))$$

$$\leq d(x_n, x_{n+1}) + d(T((1-\alpha_n)x_0 \oplus \alpha_n x_n), T(x_n))$$

$$\leq d(x_n, x_{n+1}) + d((1-\alpha_n)x_0 \oplus \alpha_n x_n, x_n)$$

$$\leq d(x_n, x_{n+1}) + (1-\alpha_n)d(x_0, x_n).$$

Lemma 5.9 is proved.

Lemma 5.10 *For each integer $n \geq 0$,*

$$(x_n, T^n(x_0)) \in E(G),\ d(x_n, T^n(x_0)) \leq M,\ d(x_0, x_n) \leq 2M. \tag{5.16}$$

Proof We prove the lemma by induction. Lemma 5.8 implies that

$$(x_0, x_1) \in E(G),\ (x_1, T(x_0)) \in E(G).$$

Clearly, (5.16) holds for $n = 0$. Assume that (5.16) holds for an integer $n \geq 0$. In view of (5.14),

$$x_{n+1} = T((1 - \alpha_n)x_0 \oplus \alpha_n x_n). \tag{5.17}$$

By (A1), (5.1), (5.16), and (5.17),

$$((1 - \alpha_n)x_0 \oplus \alpha_n x_n, T^n(x_0)) \in E(G), \ (x_{n+1}, T^{n+1}(x_0)) \in E(G). \tag{5.18}$$

Proposition 2.1, (A2), and (5.1), (5.14), (5.16), and (5.16) imply that

$$\begin{aligned} d(x_{n+1}, T^{n+1}(x_0)) &\le d(T((1 - \alpha_n)x_0 \oplus \alpha_n x_n), T(T^n(x_0))) \\ &\le d((1 - \alpha_n)x_0 \oplus \alpha_n x_n, T^n(x_0)) \\ &\le (1 - \alpha_n)d(x_0, T^n(x_0)) + \alpha_n d(x_n, T^n(x_0)) \\ &\le M(1 - \alpha_n) + \alpha_n M. \end{aligned}$$

Together with (A2) this implies that

$$d(x_0, x_{n+1}) \le d(x_{n+1}, T^{n+1}(x_0)) + d(T^{n+1}(x_0), x_0) \le 2M.$$

Lemma 5.10 is proved.

Lemmas 5.9 and 5.10 imply the following result.

Proposition 5.11 *Let $n \ge 0$ be an integer. Then*

$$d(x_{n+2}, x_{n+1}) \le \alpha_{n+1} d(x_{n+1}, x_n) + 2M|\alpha_{n+1} - \alpha_n|,$$

$$d(x_n, T(x_n)) \le d(x_n, x_{n+1}) + 2M(1 - \alpha_n).$$

Theorem 5.7 is proved analogously to Theorem 5.1 using Proposition 5.11.

5.3 Halpern-Type Iterations for Set-Valued Maps

We continue to use the notation and the definitions introduced in Sect. 5.1. Namely, assume that $X = (X, d, W)$ is a W-hyperbolic space, C is a nonempty star-shaped subset of X, and C is equipped with a graph G, $V(G) \subset C$ is the set of its vertices, and $E(G) \subset C \times C$ is the set of its edges and $T : C \to 2^C \setminus \{\emptyset\}$. Assume that for each $(x, y) \in E(G)$ and each $\xi \in T(x)$,

$$d(\xi, \{u \in T(y) : \ (\xi, u) \in E(G)\}) \le d(x, y). \tag{5.19}$$

Assume that $\{\alpha_n\}_{n=0}^{\infty}, \{r_n\}_{n=0}^{\infty} \subset (0, 1)$,

$$\alpha_n \le \alpha_{n+1} \text{ for each integer } n. \tag{5.20}$$

$$x_0 = x \in \mathrm{St}(C), \tag{5.21}$$

$$y_0 \in T(x_0),\ (x_0, y_0) \in E(G), \tag{5.22}$$

$$T^i(x_0) \subset B(x, M),\ i = 0, 1, \ldots. \tag{5.23}$$

Set

$$C_x = \{z \in C :\ (x, z) \in E(G)\} \tag{5.24}$$

and assume that

$$T(C_x) \subset C_x, \tag{5.25}$$

$$(x, x) \in E(G),\ (z, z) \in E(G),\ z \in C_x$$

and that the following assumptions hold.

(A1) For each pair of numbers β_1, β_2 satisfying $0 \le \beta_1 < \beta_2 \le 1$ and each $z_1, z_2 \in C_x$ satisfying $(z_1, z_2) \in E(G)$ the relations

$$(1 - \beta_i)x_0 \oplus \beta_i z \in C_x,\ 1, 2$$

and

$$((1 - \beta_1)x_0 \oplus \beta_1 z_1, (1 - \beta_2)x_0 \oplus \beta_2 z_2) \in E(G)$$

hold.

We define by induction $y_i, x_i \in C, i = 1, 2, \ldots$ such that for each integer $n \ge 0$,

$$y_n \in T(x_n), \tag{5.26}$$

$$x_{n+1} = (1 - \alpha_n)x_0 \oplus \alpha_n y_n, \tag{5.27}$$

$$(x_0, x_n) \in E(G),\ (x_n, x_{n+1}) \in E(G), \tag{5.28}$$

$$(y_n, y_{n+1}) \in E(G), \tag{5.29}$$

$$d(y_n, y_{n+1}) \le d(y_n, \{u \in T(x_{n+1}) :\ (y_n, u) \in E(G)\}) + r_n. \tag{5.30}$$

It is not difficult to see that the sequences $\{x_i\}_{i=0}^{\infty}, \{y_i\}_{i=0}^{\infty}$ are well-defined.

Lemma 5.12 *For each integer* $n \ge 0$,

$$(x_0, x_n) \in E(G),\ (x_0, y_n) \in E(G),\ (x_n, x_{n+1}) \in E(G),\ (x_{n+1}, y_n) \in E(G). \tag{5.31}$$

Proof Assumption (A1) and (5.22), (5.25), and (5.27) imply that

$$(x_0, y_0) \in E(G),\ (x_0, x_1) \in E(G),\ (x_1, y_0) \in E(G)$$

and (5.31) holds for $n = 0$. Assume that $n \geq 0$ is an integer and that (5.31) holds. Assumption (A1) and (5.24)–(5.27) and (5.31) imply that

$$(x_0, y_{n+1}) \in E(G),\ (x_0, x_{n+2}) \in E(G),\ (x_{n+2}, y_{n+1}) \in E(G),$$

$$(y_n, y_{n+1}),\ (x_{n+1}, x_{n+2}) \in E(G)$$

and our assumption holds for $n + 1$ and Lemma 5.12 is proved by induction.

Lemma 5.13 *Let $n \geq 0$ be an integer. Then*

$$d(x_{n+2}, x_{n+1}) \leq \alpha_{n+1} d(x_{n+1}, x_n) + \alpha_{n+1} r_n + |\alpha_{n+1} - \alpha_n| d(x_0, y_n),$$

$$d(x_n, T(x_n)) \leq d(x_n, y_n) \leq d(x_n, x_{n+1}) + (1 - \alpha_n) d(x_0, y_n).$$

Proof Proposition 2.1 and (5.19), (5.26), (5.27), (5.30), and (5.31) imply that

$$\begin{aligned} &d(x_{n+2}, x_{n+1}) \\ &= d((1 - \alpha_{n+1})x_0 \oplus \alpha_{n+1} y_{n+1}, (1 - \alpha_n)x_0 \oplus \alpha_n y_n) \\ &\leq \alpha_{n+1} d(y_n, y_{n+1}) + |\alpha_{n+1} - \alpha_n| d(x_0, y_n) \\ &\leq \alpha_{n+1} d(y_n, \{u \in T(x_{n+1}) :\ (y_n, u) \in E(G)\}) + \alpha_{n+1} r_n \\ &\quad + |\alpha_{n+1} - \alpha_n| d(x_0, y_n) \\ &\leq \alpha_{n+1} d(x_{n+1}, x_n) + \alpha_{n+1} r_n + |\alpha_{n+1} - \alpha_n| d(x_0, y_n). \end{aligned}$$

Proposition 2.1 and (5.26) and (5.27) imply that

$$\begin{aligned} d(x_n, T(x_n)) &\leq d(x_n, y_n) \\ &\leq d(x_n, x_{n+1}) + d(x_{n+1}, y_n) \\ &= d(x_n, x_{n+1}) + (1 - \alpha_n d(x_0, y_n). \end{aligned}$$

Lemma 5.13 is proved.

Lemma 5.14 *For each integer $n \geq 0$,*

$$d(x_n, \{y \in T^n(x_0) : (x_n, y) \in E(G)\}) \leq M, \tag{5.32}$$

$$d(y_n, \{y \in T^{n+1}(x_0) : (y_n, y) \in E(G)\}) \leq M. \tag{5.33}$$

Proof Clearly, (5.32) holds for $n = 0$. Assume that (5.32) holds for a nonnegative integer n. Let $\epsilon > 0$. In view of (5.32), there exists

$$\xi \in T^n(x_0) \tag{5.34}$$

such that

$$(x_n, \xi) \in E(G), \tag{5.35}$$

and

$$d(x_n, \xi) \leq M + \epsilon. \tag{5.36}$$

Since $y_n \in T(x_n)$ it follows from (5.19) and (5.34)–(5.36) that there exists

$$\eta \in \{z \in T(\xi) : (y_n, z) \in E(G)\} \subset T^{n+1}(x_0) \tag{5.37}$$

such that

$$d(y_n, \eta) \leq d(x_n, \xi) + \epsilon \leq M + 2\epsilon \tag{5.38}$$

and

$$d(y_n, \{u \in T^{n+1}(x_0) : (y_n, u) \in E(G)\}) \leq M + 2\epsilon.$$

Since ϵ is any positive number, we conclude that

$$d(y_n, \{u \in T^{n+1}(x_0) : (y_n, u) \in E(G)\}) \leq M. \tag{5.39}$$

By (5.27),

$$x_{n+1} = (1 - \alpha_n)x_0 \oplus \alpha_n y_n. \tag{5.40}$$

Lemma 5.12 implies that

$$(x_0, y_n) \in E(G).$$

In view of (A1) and (5.25), (5.35), and (5.37),

$$\eta \in T^{n+1}(x_0) \subset C_x. \tag{5.41}$$

(A1) and (5.39)–(5.41) imply that

$$(x_{n+1}, \eta) \in E(G). \tag{5.42}$$

Proposition 2.1 and (5.23), (5.38), (5.39), and (5.41) imply that

$$\begin{aligned} d(x_{n+1}, \eta) &= d((1 - \alpha_n)x_0 \oplus \alpha_n y_n, \eta) \\ &\le (1 - \alpha_n)d(x_0, \eta) + \alpha_n d(y_n, \eta) \\ &\le 2M(1 - \alpha_n) + \alpha_n(2M + \epsilon). \end{aligned}$$

Since ϵ is any positive number, we conclude in view of (5.42) that

$$d(x_{n+1}, \{u \in T^{n+1}(x_0) : (x_{n+1}, u) \in E(G)\}) \le M.$$

Lemma 5.14 is proved by induction.

Lemmas 5.14 and 5.13 and (5.23) imply the following result.

Proposition 5.15 *For each integer $n \ge 0$,*

$$d(x_{n+2}, x_{n+1}) \le \alpha_{n+1} d(x_{n+1}, x_n) + 2M|\alpha_{n+1} - \alpha_n| + \alpha_{n+1} r_n,$$

$$d(x_n, T(x_n)) \le d(x_n, y_n) \le d(x_n, x_{n+1}) + 2M(1 - \alpha_n).$$

We will prove the following result which shows that our iterative process generates approximate fixed points of the mapping T.

Theorem 5.16 *Let for each integer $n \ge 0$,*

$$\alpha_n = 1 - 2(n + 3)^{-1}, \ r_n = (n + 4)^{-1}.$$

Then for each integer $n \ge 0$,

$$d(x_n, x_{n+1}) \le 3(2M + 1)(n + 3)^{-1}$$

and

$$d(x_n, T(x_n)) \le (10M + 3)(n + 3)^{-1}.$$

Proof Set $J = 3$, $N = 2$, $\gamma = 1$, $L = 2M + 1$,

$$a_n = 1 - \alpha_n = 2(n + 3)^{-1}, c_n = L, \ s_n = d(x_n, x_{n+1}), \ n = 0, 1, \ldots.$$

By (5.23) and (5.27),

$$d(x_0, x_1) \le M.$$

Proposition 5.15 implies that for each integer $n \ge 0$,

$$s_{n+1} \le (1 - a_{n+1})s_n + |a_{n+1} - a_n|(2M + 1).$$

Lemma 2.2 implies that for each integer $n \ge 0$,

$$d(x_n, x_{n+1}) \le 3(2M + 1)(n + 3))^{-1}.$$

Together with Proposition 5.15, this implies that for each integer $n \ge 0$,

$$d(x_n, T(x_n)) \le (10M + 3)(n + 3)^{-1}.$$

Theorem 5.16 is proved.

5.4 Alternative Iterative Method for a Set-Valued Mapping

We continue to use the notation and the definitions and assume the assumptions introduced in Sect. 5.3. Recall that

$$x_0 = x \in \mathrm{St}(C), \tag{5.43}$$

$$x_1 \in T(x_0),\ \ (x_0, x_1) \in E(G). \tag{5.44}$$

Choose for each integer $n \ge 0$,

$$x_{n+1} \in T((1 - \alpha_n)x_0 \oplus \alpha_n x_n) \tag{5.45}$$

which satisfies

$$(x_n, x_{n+1}) \in E(G), \tag{5.46}$$

$$d(x_n, x_{n+1}) \le d(x_n, \{y \in T((1-\alpha_n)x_0 \oplus \alpha_n x_n) :\ (x_n, y) \in E(G)\}) + r_n. \tag{5.47}$$

It is not difficult to see that the sequence $\{x_i\}_{i=0}^{\infty}$ is well-defined be the equations above.

We prove the following result which shows that our iterative process generates approximate fixed points of the set-valued mapping T.

Theorem 5.17 *Let for each integer $n \ge 0$,*

$$\alpha_n = 1 - 2(n+3)^{-1}, \ r_n = (n+4)^{-1}.$$

Then for each integer $n \ge 0$,

$$d(x_n, x_{n+1}) \le 3(2M+1)(n+3)^{-1}$$

and

$$d(x_n, T(x_n)) \le (10M+3)(n+3)^{-1}.$$

We begin with auxiliary results. Using (5.25) we can easily show by induction that the following auxiliary result is true.

Lemma 5.18 *For each integer $n \ge 0$,*

$$(x_0, x_n) \in E(G), \ (x_n, x_{n+1}) \in E(G).$$

Lemma 5.19 *Let $n \ge 0$ be an integer. Then*

$$d(x_{n+2}, x_{n+1}) \le \alpha_{n+1} d(x_{n+1}, x_n) + r_{n+1} + |\alpha_{n+1} - \alpha_n| d(x_0, x_n), \tag{5.48}$$

$$d(x_n, \{u \in T(x_n) : \ (x_{n+1}, u) \in E(G)\}) \le d(x_n, x_{n+1}) + (1-\alpha_n) d(x_0, x_n).$$

Proof In view of (5.45),

$$x_{n+1} \in T((1-\alpha_n)x_0 \oplus \alpha_n x_n)$$

Proposition 2.1, Lemma 5.18, and (5.19) and (5.47) and the inclusion above imply that

$$d(x_{n+2}, x_{n+1})$$

$$\le d(x_{n+1}, \{y \in T((1-\alpha_{n+1})x_0 \oplus \alpha_{n+1}x_{n+1}) : \ (x_{n+1}, y) \in E(G)\}) + r_{n+1}$$

$$\le d((1-\alpha_{n+1})x_0 \oplus \alpha_{n+1}x_{n+1}, (1-\alpha_n)x_0 \oplus \alpha_n x_n) + r_{n+1}$$

$$\le \alpha_{n+1} d(x_n, x_{n+1}) + |\alpha_{n+1} - \alpha_n| d(x_0, x_n) + r_{n+1}.$$

Proposition 2.1, Lemma 5.18, and (5.19), (5.45), and (5.46) imply that

$$d(x_n, \{v \in T(x_n) : \ (x_{n+1}, v) \in E(G)\})$$

$$\le d(x_n, x_{n+1}) + d(x_{n+1}, \{v \in T(x_n) : \ (x_{n+1}, v) \in E(G)\})$$

$$\leq d(x_n, x_{n+1}) + d((1 - \alpha_n)x_0 \oplus \alpha_n x_n, x_n)$$

$$\leq d(x_n, x_{n+1}) + (1 - \alpha_n)d(x_0, x_n).$$

Lemma 5.19 is proved.

Lemma 5.20 *For each integer $n \geq 0$,*

$$\{y \in T^n(x_0) : (x_n, y) \in E(G)\} \neq \emptyset,$$

$$d(x_n, \{y \in T^n(x_0) : (x_n, y) \in E(G)\}) \leq M. \tag{5.49}$$

Proof Clearly, (5.49) holds for $n = 0$. Assume that (5.49) holds for a nonnegative integer n. Let $\epsilon > 0$. In view of (5.49), there exists

$$\xi \in T^n(x_0) \tag{5.50}$$

such that

$$(x_n, \xi) \in E(G), \tag{5.51}$$

and

$$d(x_n, \xi) \leq M + \epsilon. \tag{5.52}$$

By (A1), Lemma 5.18, and (5.24), (5.45), and (5.50),

$$((1 - \alpha_n)x_0 \oplus \alpha_n x_n), \xi) \in E(G), \tag{5.53}$$

$$x_{n+1} \in T((1 - \alpha_n)x_0 \oplus \alpha_n x_n). \tag{5.54}$$

Proposition 2.1 and (5.19), (5.23), (5.50), and (5.52)–(5.54) imply that

$$d(x_{n+1}, \{y \in T(\xi) : (x_{n+1}, y) \in E(G)\})$$

$$\leq d((1 - \alpha_n)x_0 \oplus \alpha_n x_n, \xi)$$

$$\leq (1 - \alpha_n)d(x_0, \xi) + \alpha_n d(x_n, \xi)$$

$$\leq M(1 - \alpha_n) + \alpha_n(M + \epsilon). \tag{5.55}$$

Since ϵ is any positive number, we conclude that in view of (5.50) and (5.55),

$$T(\xi) \subset T^{n+1}(x_0),$$

$$\{y \in T^{n+1}(x_0) : (x_{n+1}, y) \in E(G)\} \neq \emptyset,$$

$$d(x_{n+1}, \{y \in T^{n+1}(x_0) : (x_{n+1}, y) \in E(G)\}) \leq M.$$

Lemma 5.20 is proved by induction.

Lemmas 5.19 and 5.20 and (5.23) imply the following result.

Proposition 5.21 *For each integer* $n \geq 0$,

$$d(x_{n+2}, x_{n+1}) \leq \alpha_{n+1} d(x_{n+1}, x_n) + 2M|\alpha_{n+1} - \alpha_n| + r_{n+1},$$

$$d(x_n, \{v \in T(x_n) : (x_{n+1}, v) \in E(G)\}) \leq d(x_n, x_{n+1}) + 2M(1 - \alpha_n).$$

Theorem 5.17 is proved analogously to Theorem 5.16 using Lemma 2.2 and Proposition 5.21.

5.5 An Extension of Theorem 5.1

Assume that $X = (X, d, W)$ is a W-hyperbolic space, C is a nonempty star-shaped subset of X, and C is equipped with a graph G, $V(G) \subset C$ is the set of its vertices, $E(G) \subset C \times C$ is the set of its edges, and $T : C \to C$ satisfies for each $(x, y) \in E(G)$

$$(T(x), T(y)) \in E(G) \text{ and } d(T(x), T(y)) \leq d(x, y), \tag{5.56}$$

$$x_0 = x \in \mathrm{St}(C), \tag{5.57}$$

$$(x_0, T(x_0) \in E(G). \tag{5.58}$$

Set

$$C_x = \{z \in C : (x, z) \in E(G)\}$$

and assume that

$$(y, y) \in E(G), \ y \in C, \ T(C_x) \subset C_x. \tag{5.59}$$

Assume that $\{\alpha_n\}_{n=0}^{\infty} \subset (0, 1)$,

$$\alpha_n \leq \alpha_{n+1} \text{ for each integer } n \geq 0, \tag{5.60}$$

and that the following assumptions hold.

(A2) For each number $\beta \in [0, 1]$ and each $z_1, z_2 \in C_x$ satisfying $(z_1, z_2) \in E(G)$,

$$((1-\beta)x_0 \oplus \beta z_1, (1-\beta)x_0 \oplus \beta z_2) \in E(G).$$

(A2) For each pair of numbers β_1, β_2 satisfying $0 \le \beta_1 < \beta_2 \le 1$ and each $z \in C_x$,

$$((1-\beta_1)x_0 \oplus \beta_1 z, (1-\beta_2)x_0 \oplus \beta_2 z) \in E(G).$$

Assume that there exists $M > 0$ such that

$$\rho(T^i(x_0), x_0) \le M, \ i = 0, 1, \dots. \tag{5.61}$$

Define

$$y_0 = x_0, \tag{5.62}$$

and for each integer $n \ge 0$,

$$x_{n+1} = (1-\alpha_n)x_0 \oplus \alpha_n T(x_n), \tag{5.63}$$

$$y_{n+1} = (1-\alpha_{n+1})x_0 \oplus \alpha_{n+1} T(x_n). \tag{5.64}$$

Lemma 5.22 *Let $n \ge 0$ be an integer. Then*

$$(x_0, x_n) \in E(G), \ (x_n, y_n) \in E(G), \ (x_{n+1}, T(x_n))$$

$$\in E(G), \ (y_{n+1}, T(x_n)) \in E(G).$$

Proof In view of (5.59),

$$(x_0, x_0) \in E(G).$$

Assumption (A3) and (5.58), (5.60), (5.63), and (5.64) imply that

$$(x_1, T(x_0)) \in E(G), \ (y_1, T(x_0)) \in E(G), \ (x_1, y_1) \in E(G), \ (x_0, x_1) \in E(G).$$

Thus our lemma holds for $n = 0$. Assume that $n \ge 0$ is an integer and that the assertion of the lemma holds. Then Assumption (A3) and (5.59) and (5.63) imply that

$$(x_0, T(x_n)) \in E(G),$$

$$(x_0, x_{n+1}) = (x_0, (1-\alpha_n)x_0 \oplus \alpha_n T(x_n)) \in E(G),$$

$$(x_0, T(x_{n+1})) \in E(G), \ (x_{n+2}, T(x_{n+1})) \in E(G), \ (y_{n+2}, T(x_{n+1})) \in E(G),$$

$$(x_0, y_{n+1}) = (x_0, (1 - \alpha_{n+1})x_0 \oplus \alpha_{n+1}T(x_n)) \in E(G),$$

$$(x_{n+1}, y_{n+1}) \in E(G).$$

Thus our assumption holds for $n + 1$ and Lemma 5.22 is proved by induction.

Lemma 5.22 implies that

$$(x_0, x_1) \in E(G).$$

In view of (5.63) and Proposition 2.1,

$$d(x_1, x_2) = d((1 - \alpha_1)x_0 \oplus \alpha_1 T(x_1), (1 - \alpha_0)x_0 \oplus \alpha_0 T(x_0))$$

$$\leq \alpha_1 d(T(x_0), T(x_1)) + |\alpha_1 - \alpha_0| d(x_0, T(x_0)).$$

Lemma 5.22 implies that

$$(x_1, y_1) \in E(G).$$

It follows from (A3) and (5.63) and (5.64) that

$$x_1 = (1 - \alpha_0)x_0 \oplus \alpha_0 T(x_0), \ (x_0, x_1) \in E(G),$$

$$x_2 = (1 - \alpha_1)x_0 \oplus \alpha_1 T(x_1),$$

$$y_1 = (1 - \alpha_1)x_0 \oplus \alpha_1 T(x_0).$$

Set

$$y_0^{(0)} = x_0, \ y_1^{(0)} = x_1.$$

Clearly,

$$(y_0^{(0)}, y_1^{(0)}) \in E(G).$$

Set

$$y_0^{(1)} = x_1, \ y_1^{(1)} = (1 - \alpha_0)x_0 \oplus \alpha_0 T(x_1), \ y_2^{(2)} = x_2. \tag{5.65}$$

Assumptions (A2) and (A3), the relation $(x_0, x_1) \in E(G)$, and (5.63), (5.64), and (5.65) imply that

$$(y_0^{(1)}, y_1^{(1)}) \in E(G),\ (y_1^{(1)}, y_2^{(1)}) \in E(G),$$

$$y_i^{(1)} \in C_x,\ i = 0, 1, 2.$$

Assume that $n \geq 1$ is an integer and we defined

$$y_i^{(n)} \in C_x,\ i = 0, \dots, n+1 \tag{5.66}$$

such that

$$y_0^{(n)} = x_n,\ (x_n, y_1^{(n)}) \in E(G),\ (y_i^{(n)}, y_{i+1}^{(n)}) \in E(G),\ i = 0, \dots, n,\ y_{n+1}^{(n)} = x_{n+1}. \tag{5.67}$$

(Note that our assumption holds for $n = 1$.) By (5.63),

$$x_{n+2} = (1 - \alpha_{n+1})x_0 \oplus \alpha_{n+1} T(x_{n+1}). \tag{5.68}$$

Set

$$y_0^{(n+1)} = x_{n+1} = (1 - \alpha_n)x_0 \oplus \alpha_n T(y_0^{(n)}), \tag{5.69}$$

$$y_i^{(n+1)} = (1 - \alpha_n)x_0 \oplus \alpha_n T(y_i^{(n)}),\ i = 1, \dots, n+1, \tag{5.70}$$

$$y_{n+2}^{(n+1)} = (1 - \alpha_{n+1})x_0 \oplus \alpha_{n+1} T(y_{n+1}^{(n)}) = (1 - \alpha_{n+1})x_0 \oplus \alpha_{n+1} T(x_{n+1}) = x_{n+2}. \tag{5.71}$$

(A2) and (A3) and (5.56), (5.59), (5.66), (5.67), and (5.69)–(5.71) imply that

$$y_i^{(n+1)} \in C_x,\ i = 0, \dots, n+2,,\ T(y_i^{(n)}) \in C_x,\ i = 0, \dots, n+1,$$

$$(y_i^{(n+1)}, y_{i+1}^{(n+1)}) \in E(G),\ i = 0, \dots, n+1. \tag{5.72}$$

Therefore our assumption holds for $n + 1$ too, and by induction, we constructed for each integer $n \geq 0$, the finite sequence $\{y_i^{(n)}\}_{i=0}^{n+1} \subset C$ such that for each integer $n \geq 1$, (5.66), (5.67), and (5.69)–(5.72) hold.

We show that for each integer $n \geq 0$, and all $i = 1, \dots, n+1$,

$$(x_0, y_i^{(n)}) \in E(G), \tag{5.73}$$

$$d(T^n(x_0), y_i^{(n)}) \leq M. \tag{5.74}$$

It is easy to see that in view of (5.66), (5.73) holds for each integer $n \geq 0$ and each $i \in \{0, \ldots, n+1\}$. We show that for each integer $n \geq 0$ and each integer $i \in \{0, \ldots, n+1\}$,

$$(y_i^{(n)}, T^n(x_0)) \in E(G),\ d(T^n(x_0), y_i^{(n)}) \leq M. \tag{5.75}$$

Clearly, (5.75) holds for $n = 0$. Assume that $n \geq 0$ is an integer and (5.75) holds for $i = 0, \ldots, n+1$. By (5.56)–(5.58), (5.65), and (5.75), for $i = 0, \ldots .n+1$,

$$(T(y_i^{(n)}), T^{n+1}(x_0)) \in E(G),\ d(T(y_i^{(n)}), T^{n+1}(x_0)) \leq d(y_i^{(n)}, T^n(x_0)). \tag{5.76}$$

It follows from Proposition 2.1 and (5.56), (5.60), (5.75), and (5.76) that for each $i \in \{0, \ldots, n+2\}$,

$$(y_i^{(n+1)}, T^{n+1}(x_0)) \in E(G))$$

and for each $i \in \{0, \ldots, n+1\}$,

$$d(y_i^{(n+1)}, T^{n+1}(x_0)) = d(T^{n+1}(x_0), (1-\alpha_n)x_0 \oplus \alpha_n T(y_i^{(n)}))$$

$$\leq (1-\alpha_n)d(x_0, T^{n+1}(x_0)) + \alpha_n d(T(y_i^{(n)}), T^{n+1}(x_0)) \leq M,$$

$$d(y_{n+2}^{(n+1)}, T^{n+1}(x_0)) = d(T^{n+1}(x_0), (1-\alpha_{n+1})x_0 \oplus \alpha_{n+1} T(y_n^{(n+1)}))$$

$$\leq (1-\alpha_{n+1})d(x_0, T^{n+1}(x_0)) + \alpha_{n+1} d(T(y_{n+1}^{(n)}), T^{n+1}(x_0)) \leq M,$$

our assumption holds for $n+1$ too, and we showed by induction that it holds for each integer $n \geq 0$.

Let $n \geq 0$ be an integer. Proposition 2.1 and (5.56), (5.67), (5.69)–(5.71), and (5.75) imply that

$$\sum_{i=0}^{n+1} d(y_i^{(n+1)}, y_{i+1}^{(n+1)})$$

$$= \sum_{i=0}^{n+1} d((1-\alpha_n)x_0 \oplus \alpha_n T(y_i^{(n)}), (1-\alpha_n)x_0 \oplus \alpha_n T(y_{i+1}^{(n)}))$$

$$+ d((1-\alpha_n)x_0 \oplus \alpha_n T(y_{n+1}^{(n)}), (1-\alpha_{n+1})x_0 \oplus \alpha_{n+1} T(y_{n+1}^{(n)}))$$

$$\leq \sum_{i=0}^{n} \alpha_n d(T(y_i^{(n)}), T(y_{i+1}^{(n)})) + |\alpha_{n+1} - \alpha_n| d(x_0, T(y_{n+1}^{(n)}))$$

$$\leq \alpha_n \sum_{i=0}^{n} d(y_i^{(n)}), y_{i+1}^{(n)}) + 2M|\alpha_{n+1} - \alpha_n|. \tag{5.77}$$

By (5.56), (5.63), and (5.75),

$$d(x_n, T(x_n)) \leq d(x_n, x_{n+1}) + d(x_{n+1}, T(x_n))$$

$$\leq d(x_n, x_{n+1}) + (1 - \alpha_n) d(x_0, T(x_n))$$

$$\leq d(x_n, x_{n+1}) + (1 - \alpha_n)(d(x_0, T^{n+1}(x_0)) + d(T^{n+1}(x_0), T(y_{n+1}^{(n)})))$$

$$\leq d(x_n, x_{n+1}) + 2M(1 - \alpha_n). \tag{5.78}$$

Thus (5.77) and (5.78) hold for each integer $n \geq 0$.

Theorem 5.23 *Let for each integer $n \geq 0$,*

$$\alpha_n = 1 - 2(n+3)^{-1}.$$

Then for each integer $n \geq 0$,

$$d(x_n, x_{n+1}) \leq 6M(n+3)^{-1}$$

and

$$d(x_n, T(x_n)) \leq 10M(n+3)^{-1}.$$

Set

$$J = 3, \ N = 2, \ \gamma = 1.$$

and for each integer $n \geq 0$,

$$a_n = 1 - \alpha_n, \ s_n = \sum_{i=0}^{n} d(y_i^{(n)}, y_{i+1}^{(n)}), \ c_n = L = 2M.$$

Clearly,

$$s_0 = d(x_0, x_1) \leq M.$$

By (5.77), for each integer $n \geq 0$,

$$s_{n+1} \leq (1 - a_n)s_n + c_n|a_n - a_{n+1}|.$$

By Lemmas 2.2 (5.78), for each integer $n \geq 0$,

$$d(x_n, x_{n+1}) \leq 6M(n+3)^{-1},$$

$$d(x_n, T(x_n)) \leq 10M(n+3)^{-1}.$$

Theorem 5.23 is proved.

5.6 The Extension of Theorem 5.7

We use the notation and definitions and assume the assumptions introduced in Sect. 5.5.

Recall that

$$(x_0, T(x_0)) \in E(G). \tag{5.79}$$

Define for each integer $n \geq 0$,

$$x_{n+1} = T((1 - \alpha_n)x_0 \oplus \alpha_n x_n), \tag{5.80}$$

$$y_{n+1} = T((1 - \alpha_{n+1})x_0 \oplus \alpha_{n+1} x_n). \tag{5.81}$$

Clearly,

$$x_1 = T(x_0) = y_1. \tag{5.82}$$

In view of (5.80) and (5.81),

$$x_2 = T((1 - \alpha_1)x_0 \oplus \alpha_1 x_1), \tag{5.83}$$

$$y_1 = T((1 - \alpha_1)x_0 \oplus \alpha_1 x_0) = T(x_0). \tag{5.84}$$

(A2) and (A3) and (5.44), (5.79), (5.83), and (5.84) imply that

$$(x_1, y_1) \in E(G), \ (y_1, x_2) \in E(G). \tag{5.85}$$

Set

$$y_0^{(0)} = x_0, \ y_1^{(0)} = x_1,$$

$$y_0^{(1)} = x_1,\ y_1^{(1)} = y_1,\ y_2^{(1)} = x_2.$$

Assume $n \geq 1$ is an integer and that

$$y_i^{(n)} \in C_x,\ ,i = 0, \ldots, n+1, \tag{5.86}$$

$$y_0^{(n)} = x_n,\ y_{n+1}^{(n)} = x_{n+1}, \tag{5.87}$$

$$(y_i^{(n)}, y_{i+1}^{(n)}) \in E(G),\ i = 0, \ldots, n, \tag{5.88}$$

$$(x_0, y_i^{(n)}) \in E(G),\ i = 0, \ldots, n+1, \tag{5.89}$$

$$(y_i^{(n)}, T^n(x_0)) \in E(G),\ i = 0, \ldots, n+1, \tag{5.90}$$

$$d(T^n(x_0), y_i^{(n)}) \leq M,\ i = 0, \ldots, n+‘1. \tag{5.91}$$

(Clearly, our assumption holds f or $n = 0$.) For $i = 0, \ldots, n+1$, define

$$y_i^{(n+1)} = T((1-\alpha_n)x_0 \oplus \alpha_n y_i^{(n)}), \tag{5.92}$$

$$y_{n+2}^{(n+1)} = T((1-\alpha_{n+1})x_0 \oplus \alpha_{n+1} y_{n+1}^{(n+1)}). \tag{5.93}$$

It follows from (A2) and (A3) and (5.44), (5.74), (5.87), (5.89), (5.92), and (5.93) that

$$y_0^{(n+1)} = x_{n+1},\ y_{n+2}^{(n+1)} = x_{n+2},$$

$$(x_0, y_i^{(n+1)}) \in E(G),\ i = 0, \ldots, n+2.$$

By (A2) and (A3) and (5.43), (5.44), and (5.90),

$$((1-\alpha_n)x_0 \oplus \alpha_n y_i^{(n)}, T^n(x_0)) \in E(G),\ i = 0, \ldots, n+1, \tag{5.94}$$

$$((1-\alpha_{n+1})x_0 \oplus \alpha_{n+1} y_{n+1}^{(n)}, T^n(x_0)) \in E(G). \tag{5.95}$$

In view of the relations above and (5.42), (5.92), and (5.93),

$$(y_i^{(n+1)}, T^{n+1}(x_0)) \in E(G),\ i = 0, \ldots, n+2.$$

(A2), (A3) and (5.42), (5.48), (5.88) and (5.93) imply that for each $i \in \{0, \ldots, n+1\}$,

$$(y_i^{(n+1)}, y_{i+1}^{(n+1)}) \in E(G).$$

It follows from Proposition 2.1, (5.42), (5.46), (5.90), (5.92), (5.94) and (5.95) that for each $i \in \{0, \dots, n+1\}$,

$$d(y_i^{(n+1)}, T^{n+1}(x_0)) \le d(T^n(x_0), (1-\alpha_n)x_0 \oplus \alpha_n y_i^{(n)})$$

$$\le (1-\alpha_n)d(x_0, T^n(x_0)) + \alpha_n d(y_i^{(n)}, T^n(x_0)) \le M,$$

$$d(y_{n+2}^{(n+1)}, T^{n+1}(x_0)) \le d((1-\alpha_{n+1})x_0 \oplus \alpha_{n+1} y_n^{(n+1)}, T^n(x_0))$$

$$\le (1-\alpha_{n+1})d(x_0, T^n(x_0)) + \alpha_n d(y_{n+1}^{(n)}, T^n(x_0) \le M$$

our assumption holds for $n+1$ too. Therefore by induction, we constructed for each integer $n \ge 0$, $y_i^{(n)}$, $i = 0, \dots, n+1$ satisfying for each integer $n \ge 0$ (5.85)–(5.93).

Let $n \ge 0$ be an integer. Proposition 2.1, (A2) and (A3), and (5.42), (5.46), (5.88), (5.92), and (5.93) imply that

$$\sum_{i=0}^{n+1} d(y_i^{(n+1)}, y_{i+1}^{(n+1)})$$

$$= \sum_{i=0}^{n} d(T((1-\alpha_n)x_0 \oplus \alpha_n y_i^{(n)}), T((1-\alpha_n)x_0 \oplus \alpha_n y_{i+1}^{(n)}))$$

$$+ d(T((1-\alpha_n)x_0 \oplus \alpha_n y_{n+1}^{(n)}), T((1-\alpha_{n+1})x_0 \oplus \alpha_{n+1} y_{n+1}^{(n)}))$$

$$\le \sum_{i=0}^{n} d((1-\alpha_n)x_0 \oplus \alpha_n y_i^{(n)}, (1-\alpha_n)x_0 \oplus \alpha_n y_{i+1}^{(n)})$$

$$+ d((1-\alpha_n)x_0 \oplus \alpha_n y_{n+1}^{(n)}, (1-\alpha_{n+1})x_0 \oplus \alpha_{n+1} y_{n+1}^{(n)})$$

$$\le \alpha_n \sum_{i=0}^{n} d(y_i^{(n)}), y_{i+1}^{(n)}) + |\alpha_{n+1} - \alpha_n| d(x_0, y_{n+1}^{(n)})$$

$$\le \sum_{i=0}^{n} \alpha_n d(y_i^{(n)}), y_{i+1}^{(n)}) + 2M|\alpha_{n+1} - \alpha_n|. \tag{5.96}$$

By (A2) and (A3) and (5.42), (5.80), and (5.91),

$$d(x_n, T(x_n)) \le d(x_n, x_{n+1}) + d(x_{n+1}, T(x_n))$$

$$\le d(x_n, x_{n+1}) + d((1-\alpha_n)x_0 \oplus x_n, x_n)$$

$$\leq d(x_n, x_{n+1}) + (1 - \alpha_n)d(x_0, x_n)$$

$$\leq d(x_n, x_{n+1}) + 2M(1 - \alpha_n). \tag{5.97}$$

Theorem 5.24 *Let for each integer* $n \geq 0$,

$$\alpha_n = 1 - 2(n+3)^{-1}.$$

Then for each integer $n \geq 0$,

$$d(x_n, x_{n+1}) \leq 6M(n+3)^{-1}$$

and

$$d(x_n, T(x_n)) \leq 10M(n+3)^{-1}.$$

Theorem 5.24 is proved analogously to Theorem 5.23 by using (5.96) and (5.97).

5.7 An Extension of Theorem 5.16

We continue to use the notation and definitions introduced in Sect. 5.5. Assume that $X = (X, d, W)$ is a W-hyperbolic space, C is a nonempty star-shaped subset of X, and C is equipped with a graph G, $V(G) \subset C$ is the set of its vertices, $E(G) \subset C \times C$ is the set of its edges (A2) and (A3) hold, $\{r_n\}_{n=0}^{\infty} \subset (0, 1)$, and $T : C \to 2^C \setminus \{\emptyset\}$.

Assume that the following assumption holds.

(A4) For each $x, y \in C$ satisfying $(x, y) \in E(G)$ and each $\xi \in T(x)$,

$$d(\xi, \{u \in T(y) : (\xi, u) \in E(G)\}) \leq d(x, y).$$

(Recall that $d(x, \emptyset) = \infty$, $x \in C$.) Assume that

$$0 < \alpha_n \leq \alpha_{n+1} < 1 \text{ for each integer } n \geq 0, \tag{5.98}$$

$$x_0 = x \in \mathrm{St}(C), \tag{5.99}$$

$$(x_0, x_0) \in E(G), \ y_0 \in T(x_0), \ (x_0, y_0) \in E(G). \tag{5.100}$$

$$T^i(x_0) \subset B(x_0, M), \ i = 0, 1, \ldots, \tag{5.101}$$

Set

$$C_x = \{z \in C : (x, z) \in E(G)\}$$

and assume that

$$T(C_x) \subset C_x. \tag{5.102}$$

We construct our Halpern-type iterations. Set

$$x_1 = (1 - \alpha_0)x_0 \oplus \alpha_0 y_0, \tag{5.103}$$

$$x_0^{(0)} = x_0, \; y_0^{(0)} = y_0, \; x_1^{(0)} = x_1. \tag{5.104}$$

By (A2) and (A3) and (5.100) and (5.103),

$$(x_0, y_0 \in E(G), \; (x_1, y_0) \in E(G), \; (x_0, x_1) \in E(G). \tag{5.105}$$

Set

$$x_0^{(1)} = x_1.$$

By (A4) and (5.100) and (5.105), there exists

$$y_0^{(1)} \in T(x_0^{(1)}) \tag{5.106}$$

such that

$$(y_0^{(0)}, y_0^{(1)}) \in E(G), \tag{5.107}$$

$$d(y_0^{(0)}, y_0^{(1)}) \le d(y_0^{(0)}, \{u \in T(x_0^{(1)}) : \; (y_0^{(0)}, u) \in E(G)\}) + r_0. \tag{5.108}$$

Set

$$x_1^{(1)} = (1 - \alpha_0)x_0 \oplus \alpha_0 y_0^{(1)}, \; x_2^{(1)} = (1 - \alpha_1)x_0 \oplus \alpha_1 y_0^{(1)}. \tag{5.109}$$

(A2) and (A3) and (5.103), (5.107), and (5.109) imply that

$$(x_1^{(0)}, x_1^{(1)}) = (x_1, x_1^{(1)}) = ((1 - \alpha_0)x_0 \oplus \alpha_0 y_0, (1 - \alpha_0)x_0 \oplus \alpha_0 y_0^{(1)}) \in E(G),$$

$$(x_1^{(1)}, x_1^{(2)}) \in E(G).$$

Assume that $n \ge 0$ is an integer and that

$$x_i^{(n)} \in C_x, \; i = 0, \dots, n + 1, \tag{5.110}$$

$$(x_i^{(n)}, x_{i+1}^{(n)}) \in E(G), \; i = 0, \dots, n, \tag{5.111}$$

$$x_{n+1}^{(n)} \in (1-\alpha_n)x_0 \oplus \alpha_n T(x_0^{(n)}). \tag{5.112}$$

(Note that our assumption holds for $n = 0$.) Set

$$x_0^{(n+1)} = x_{n+1}^{(n)}. \tag{5.113}$$

By (5.112), there exists

$$y_0^{(n+1)} \in T(x_0^{(n)}) \tag{5.114}$$

such that

$$x_{n+1}^{(n)} = (1-\alpha_n)x_0 \oplus \alpha_n y_0^{(n)}. \tag{5.115}$$

By induction, (A4), and (5.109) and (5.115), for each $i \in \{0, \dots, n+1\}$, there exists

$$y_i^{(n+1)} \in T(x_i^{(n)}) \tag{5.116}$$

such that for each $i \in \{0, \dots, n\}$,

$$(y_i^{(n+1)}, y_{i+1}^{(n+1)}) \in E(G), \tag{5.117}$$

$$d(y_i^{(n+1)}, y_{i+1}^{(n+1)}) \le d(y_i^{(n+1)}, \{u \in T(x_{i+1}^{(n)}) : (y_i^{(n+1)}, u) \in E(G)\}) + r_{n+1}, \tag{5.118}$$

$$x_i^{(n+1)} = (1-\alpha_n)x_0 \oplus \alpha_n y_i^{(n+1)}, \ i = 1, \dots, n+1 \tag{5.119}$$

and set

$$x_{n+2}^{(n+1)} = (1-\alpha_{n+1})x_0 \oplus \alpha_{n+1} y_{n+1}^{(n+1)}. \tag{5.120}$$

By (5.108), (5.113), (5.116), and (5.120),

$$x_{n+2}^{(n+1)} \in (1-\alpha_{n+1})x_0 \oplus \alpha_{n+1} T(x_{n+1}^{(n)}) = (1-\alpha_{n+1})x_0 \oplus \alpha_{n+1} T(x_0^{(n+1)}). \tag{5.121}$$

(A2) and (A3) and (5.108), (5.117), and (5.119)–(5.121) imply that

$$x_i^{(n+1)} \in C_x, \ i = 0, \dots, n+1,$$

$$(x_i^{(n+1)}, x_{i+1}^{(n+1)}) \in E(G), \ i = 0, \dots, n+1.$$

By (5.113), (5.116), and (5.120),

$$y_{n+1}^{(n+1)} \in T(x_{n+1}^{(n)}) = T(x_0^{(n+1)}),$$

$$x_{n+2}^{(n+1)} \in (1-\alpha_{n+1})x_0 \oplus \alpha_{n+2}T(x_0^{(n+1)}).$$

Thus by induction we constructed for each integer $n \geq 0$, $x_i^{(n)} \in C_x$, $y_i^{(n+1)} \in C$, $i = 0, \ldots, n+1$ such that (5.109)–(5.121) hold.

Let $n \geq 0$ be an integer. Proposition 2.1, (A4), and (5.117)–(5.120) imply that

$$\sum_{i=0}^{n+1} d(x_i^{(n+1)}, x_{i+1}^{(n+1)})$$

$$= \sum_{i=0}^{n} d((1-\alpha_n)x_0 \oplus \alpha_n y_i^{(n+1)}, (1-\alpha_n)x_0 \oplus \alpha_n y_{i+1}^{(n+1)})$$

$$+ d((1-\alpha_n)x_0 \oplus \alpha_n y_{n+1}^{(n+1)}, (1-\alpha_{n+1})x_0 \oplus \alpha_{n+1} y_{n+1}^{(n+1)}))$$

$$\leq \sum_{i=0}^{n} \alpha_n d(y_i^{(n+1)}, y_{i+1}^{(n+1)}) + |\alpha_{n+1} - \alpha_n| d(x_0, y_{n+1}^{(n+1)})$$

$$\leq \sum_{i=0}^{n} \alpha_n d(x_i^{(n)}, x_{i+1}^{(n)}) + |\alpha_{n+1} - \alpha_n| d(x_0, y_{n+1}^{(n+1)}) + \alpha_n r_n(n+1). \tag{5.122}$$

We show by induction that for each integer $n \geq 0$ and each $i \in \{0, \ldots, n+1\}$,

$$(x_0, y_i^{(n)}) \in E(G), \tag{5.123}$$

$$E_i^{(n)} := \{u \in T^n(x_0) :\ (y_i^{(n)}, u) \in E(G)\} \neq \emptyset, \tag{5.124}$$

$$d(y_i^{(n)}, E_i^{(n)}) \leq M. \tag{5.125}$$

(Clearly, (5.123)–(5.125) hold for $n = 0$.) Assume that $n \geq 0$ is an integer and (5.123)–(5.125) hold for each $i \in \{0, \ldots, n+1\}$. Let $\epsilon > 0$ and $i \in \{0, \ldots, n+1\}$. In view of (5.125), there exists

$$\xi_i \in T^n(x_0) \tag{5.126}$$

such that

$$(y_i^{(n)}.\xi_i) \in E(G), \tag{5.127}$$

$$d(y_i^{(n)}, \xi_i) \le M + \epsilon. \tag{5.128}$$

Proposition 2.1, (A2) and (A3), and (5.119), (5.120), and (5.128) imply that

$$(x_i^{(n)}, \xi_i) \in E(G), \tag{5.129}$$

$$d(x_i^{(n)}, \xi_i) \le \max\{d(x_0, \xi_i),\ d(y_i^{(n)}, \xi_i)\} \le M + \epsilon, \tag{5.130}$$

$$(x_{n+1}^{(n)}, \xi_n) \in E(G), \tag{5.131}$$

$$d(x_{n+1}^{(n)}, \xi_n) \le \max\{d(x_0, \xi_n),\ d(y_{n+1}^{(n)}, \xi_n)\} \le M + \epsilon.$$

By (A4), (5.116) and (5.129)–(5.131), and the relation above, for each $i \in \{0, \dots, n\}$,

$$d(y_i^{(n+1)}, \{u \in T(\xi_i) :\ (y_i^{(n+1)}, u) \in E(G)\}) \le d(x_i^{(n)}, \xi_i) \le M + \epsilon.$$

$$d(y_{n+1}^{(n+1)}, \{u \in T(\xi_n) :\ (y_{n+1}^{(n+1)}, u) \in E(G)\}) \le d(x_n^{(n)}, \xi_i) \le M + \epsilon.$$

Since ϵ is any positive number, we obtain that the assumption made for n also holds for $n + 1$. Thus we showed by induction that (5.123)–(5.125) hold for each integer $n \ge 0$. Combined with (5.101), (5.113), (5.116), (5.120), and (5.122), this implies that for each integer $n \ge 0$,

$$\sum_{i=0}^{n+1} d(x_i^{(n+1)}, x_{i+1}^{(n+1)}) \le \sum_{i=0}^{n} \alpha_n d(x_i^{(n)}), x_{i+1}^{(n)}) + 2M|\alpha_{n+1} - \alpha_n| + \alpha_n r_{n+1}(n+1), \tag{5.132}$$

$$d(x_0^{(n)}, x_0^{(n+1)}) \le \sum_{i=0}^{n} d(x_i^{(n)}, x_{i+1}^{(n)}), \tag{5.133}$$

$$\begin{aligned}
d(x_0^{(n+1)}, T(x_0^{(n+1)})) &\le d(x_0^{(n+1)}, x_{n+2}^{(n+1)}) + d(x_{n+2}^{(n+1)}, T(x_0^{(n+1)})) \\
&\le d(x_0^{(n+1)}, x_{n+2}^{(n+1)}) + d((1 - \alpha_{n+1})x_0 \oplus \alpha_{n+1} y_{n+1}^{(n+1)}, T(x_0^{(n+1)})) \\
&\le d(x_0^{(n+1)}, x_{n+2}^{(n+1)}) + d((1 - \alpha_{n+1})x_0 \oplus \alpha_{n+1} y_{n+1}^{(n+1)}, y_{n+1}^{(n+1)}) \\
&\le d(x_0^{(n+1)}, x_{n+2}^{(n+1)}) + (1 - \alpha_{n+1}) d(x_0, y_{n+1}^{(n+1)}) \\
&\le d(x_0^{(n+1)}, x_{n+2}^{(n+1)}) + 2M(1 - \alpha_{n+1}).
\end{aligned} \tag{5.134}$$

Theorem 5.25 *Let for each integer $n \ge 0$,*

$$\alpha_n = 1 - 2(n+3)^{-1},\ r_n = (n+4)^{-2}.$$

Then for each integer $n \geq 0$,

$$d(x_0^{(n)}, x_0^{(n+1)}) \leq (6M+3)(n+3)^{-1}$$

and

$$d(x_0^{(n+1)}, T(x_0^{(n+1)})) \leq (10M+3)(n+3)^{-1}.$$

Proof Set

$$J = 3,\ N = 2,\ \gamma = 1.$$

and for each integer $n \geq 0$,

$$a_n = 1 - \alpha_n,\ s_n = \sum_{i=0}^{n} d(x_i^{(n)}, x_{i+1}^{(n)}),\ c_n = L = 2M+1.$$

By (5.132), for each integer $n \geq 0$,

$$s_{n+1} \leq (1-a_n)s_n + (2M+1)|a_n - a_{n+1}|.$$

By Lemmas 2.2 and (5.133), for each integer $n \geq 0$,

$$d(x_0^{(n)}, x_0^{n+1}) \leq s_n \leq (6M+3)(n+3)^{-1}.$$

Together with (5.134), this implies that for each integer $n \geq 0$,

$$d(x_0^{n+1)}, T(x_0^{(n+1)})) \leq (10M+3)(n+3)^{-1}.$$

Theorem 5.25 is proved.

5.8 An Extension of Theorem 5.17

We continue to use all the notation and definitions of Sect. 5.7 and assume that all the assumptions introduced there hold.

Let

$$x_0^{(0)} = x_0 = x \in \mathrm{St}(C),$$

$$x_1^{(0)} \in T(x_0^{(0)}),\ (x_0^{(0)}, x_1^{(0)}) \in E(G).$$

We construct iterates by induction. Assume that $n \geq 0$ is an integer and that

$$x_i^{(n)} \in C_x,\ i = 0, \ldots, n+1, \tag{5.135}$$

$$(x_i^{(n)}, x_{i+1}^{(n)}) \in E(G),\ i = 0, \ldots, n, \tag{5.136}$$

$$x_{n+1}^{(n)} \in T((1-\alpha_n)x_0 \oplus \alpha_n x_0^{(n)}). \tag{5.137}$$

(Note that our assumption holds for $n = 0$.) Set

$$x_0^{(n+1)} = x_{n+1}^{(n)}. \tag{5.138}$$

(A2)-(A4) and (5.136) imply that for each $i \in \{0, \ldots, n+1\}$,

$$(1-\alpha_n)x_0 \oplus \alpha_n x_i^{(n)} \in C_x,$$

$$((1-\alpha_n)x_0 \oplus \alpha_n x_i^{(n)}, (1-\alpha_n)x_0 \oplus \alpha_n x_{i+1}^{(n)}) \in E(G). \tag{5.139}$$

By induction and (5.136)–(5.139), there exist $x_i^{(n+1)} \in C,\ i = 0, \ldots, n+1$ such that for each $i \in \{0, \ldots, n\}$,

$$(x_i^{(n+1)}, x_{i+1}^{(n+1)}) \in E(G), \tag{5.140}$$

$$x_{i+1}^{(n+1)} \in T((1-\alpha_n)x_0 \oplus \alpha_n x_{i+1}^{(n)}), \tag{5.141}$$

$$d(x_i^{(n+1)}, x_{i+1}^{(n+1)})$$

$$\leq d(x_i^{(n+1)}, \{u \in T((1-\alpha_n)x_0 \oplus \alpha_n x_{i+1}^{(n)}) :\ (x_{i+1}^{(n)}, u) \in E(G)\}) + r_n$$

and

$$x_{n+2}^{(n+1)} \in T((1-\alpha_{n+1})x_0 \oplus \alpha_{n+1} x_{n+1}^{(n)})$$

$$= T((1-\alpha_n)x_0 \oplus \alpha_{n+1} x_0^{(n+1)})$$

such that

$$(x_{n+1}^{(n+1)}, x_{n+2}^{(n+1)}) \in E(G),$$

$$d(x_{n+1}^{(n+1)}, x_{n+2}^{(n+1)})$$

$$\leq d(x_{n+1}^{(n+1)}, \{u \in T((1-\alpha_{n+1})x_0 \oplus \alpha_{n+1}x_{n+1}^{(n)}) : (x_{n+1}^{(n)}, u) \in E(G)\}) + r_n. \tag{5.142}$$

Clearly, our assumption holds for $n+1$ too. Thus by induction, it holds for each integer $n \geq 0$.

By the relations above, (A2)–(A4), and (5.139), (5.141)–(5.142), for each integer $n \geq 0$,

$$\sum_{i=0}^{n+1} d(x_i^{(n+1)}, x_{i+1}^{(n+1)})$$

$$= \sum_{i=0}^{n} d(x_i^{(n+1)}, x_{i+1}^{(n+1)}) + d(x_{n+1}^{(n+1)}, x_{n+2}^{(n+1)})$$

$$\leq \sum_{i=0}^{n} (d(x_i^{(n+1)}, \{u \in T((1-\alpha_n)x_0 \oplus \alpha_n x_i^{(n+1)}) : (x_{i+1}^{(n)}, u) \in E(G)\}) + r_n)$$

$$+ d(x_{n+1}^{(n+1)}, \{u \in T((1-\alpha_{n+1})x_0 \oplus \alpha_{n+1}x_{n+1}^{(n+1)}) : (x_{n+1}^{(n+1)}, u) \in E(G)\}) + r_n$$

$$\leq \sum_{i=0}^{n} d((1-\alpha_n)x_0 \oplus \alpha_n x_i^{(n)}, (1-\alpha_n)x_0 \oplus \alpha_n x_i^{(n)})$$

$$+ d((1-\alpha_n)x_0 \oplus \alpha_n x_{n+1}^{(n)}, (1-\alpha_{n+1})x_0 \oplus \alpha_{n+1}x_{n+1}^{(n)}) + (n+2)r_n$$

$$\leq \sum_{i=0}^{n} \alpha_n d(x_i^{(n)}, x_{i+1}^{(n)}) + |\alpha_{n+1} - \alpha_n| d(x_0, x_{n+1}^{(n)}) + (n+2)r_n. \tag{5.143}$$

We show by induction that for each integer $n \geq 0$, and each $i \in \{0, \ldots, n+1\}$,

$$d(x_i^{(n)}, \{u \in T^n(x_0) : (x_i^{(n)}, u) \in E(G)\}) \leq M. \tag{5.144}$$

Clearly, (5.144) holds for $n = 0$. Assume that $n \geq 0$ is an integer and (5.144) holds for each $i \in \{0, \ldots, n+1\}$. Let $\epsilon > 0$. By (5.144), for each $i \in \{0, \ldots, n+1\}$, there exists

$$\xi_i \in T^n(x_0) \tag{5.145}$$

such that

$$(x_i^{(n)}, \xi_i) \in E(G), \tag{5.146}$$

$$d(x_i^{(n)}, \xi_i) \leq M + \epsilon. \tag{5.147}$$

Let $i \in \{0, \dots, n+1\}$. In view of (5.102), (5.145), and (5.146),

$$((1-\alpha_n)x_0 \oplus \alpha_n x_i^{(n)}, \xi_i) \in E(G),$$

$$((1-\alpha_{n+1})x_0 \oplus \alpha_{n+1} x_{n+1}^{(n)}, \xi_{n+1}) \in E(G). \tag{5.148}$$

Proposition 2.1 and (5.101), (5.145), and (5.147) imply that

$$d((1-\alpha_n)x_0 \oplus \alpha_n x_i^{(n)}, \xi_i) \le (1-\alpha_n)d(x_0, \xi_i) + \alpha_n d(x_i^{(n)}, \xi_i)$$

$$\le M(1-\alpha_n) + \alpha_n M + \epsilon \le M + \epsilon,$$

$$d((1-\alpha_{n+1})x_0 \oplus \alpha_{n+1} x_{n+1}^{(n)}, \xi_{n+1}) \le M + \epsilon.$$

By the relation above, (5.141), and (A4), there exists

$$\eta_i \in T(\xi_i) \tag{5.149}$$

such that

$$(x_i^{(n+1)}, \eta_i) \in E(G),\ d(x_i^{(n+1)}, \eta_i) \le d((1-\alpha_n)x_0 \oplus \alpha_n x_i^{(n)}, \xi_i) \le M + \epsilon \tag{5.150}$$

and

$$\eta_{n+2} \in T(\xi_{n+1}) \tag{5.151}$$

such that

$$(x_{n+2}^{(n+1)}, \eta_{n+2}) \in E(G),\ d(x_{n+2}^{(n+1)}, \eta_{n+2})$$

$$\le d((1-\alpha_{n+1})x_0 \oplus \alpha_{n+1} x_{n+1}^{(n)}, \xi_{n+1}) \le M + \epsilon. \tag{5.152}$$

Since ϵ is any positive number, it follows from (5.145), (5.149), and (5.152) that for each $i \in \{0, \dots, n+2\}$,

$$d(x_i^{(n+1)}, \{u \in T^{n+1}(x_0) : (x_i^{(n+1)}, u) \in E(G)\}) \le M.$$

Thus our assumption holds for $n+1$ too. Therefore we have showed by induction that (5.144) holds for each integer $n \ge 0$, and each $i \in \{0, \dots, n+1\}$. Together with (5.143), this implies that for each integer $n \ge 0$,

$$\sum_{i=0}^{n+1} d(x_i^{(n+1)}, x_{i+1}^{(n+1)}) \le \sum_{i=0}^{n} \alpha_n d(x_i^{(n)}, x_{i+1}^{(n)}) + 2M|\alpha_{n+1} - \alpha_n| + (n+2)r_n. \tag{5.153}$$

By (A4) and (5.135), (5.137), and (5.144),

$$d(x_0^{(n)}, T(x_0^{(n)}) \leq \sum_{i=0}^{n} d(x_i^{(n)}, x_{i+1}^{(n)}) + d(x_n^{(n+1)}, T(x_0^{(n)})$$

$$\leq \sum_{i=0}^{n} d(x_i^{(n)}, x_{i+1}^{(n)}) + d((1-\alpha_n)x_0 \oplus \alpha_n x_0^{(n)}, x_0^{(n)})$$

$$\leq \sum_{i=0}^{n} d(x_i^{(n)}, x_{i+1}^{(n)}) + 2M(1-\alpha_n). \tag{5.154}$$

The following theorem is proved analogously to Theorem 5.25 using (5.153) and (5.154).

Theorem 5.26 *Let for each integer $n \geq 0$,*

$$\alpha_n = 1 - 2(n+3)^{-1},\ r_n = (n+4)^{-2}.$$

Then for each integer $n \geq 0$,

$$d(x_0^{(n)}, x_0^{n+1}) \leq (6M+3)(n+3)^{-1}$$

and

$$d(x_0^{n+1)}, T(x_0^{(n+1)})) \leq (10M+3)(n+3)^{-1}.$$

Chapter 6
Inexact Viscosity Approximation Methods in Hilbert Spaces

In this chapter, we study the viscosity approximation method for solving fixed point problems for nonexpansive mappings in a Hilbert space and for finding a common element of the set of solutions of an equilibrium problem and the set of fixed points of a nonexpansive mapping in a Hilbert space. We show the uniform convergence of inexact iterates generated by the viscosity approximation method.

6.1 A Fixed Point Problem

In [133] it was introduced the viscosity approximation method for solving of fixed point problems for nonexpansive mappings in a Hilbert space. In this chapter, we show the uniform convergence of inexact iterates generated by this viscosity approximation method. As a matter of fact, our goal is to show that the iterative processes generated by the viscosity approximation methods possess convergence properties which are analogous to the convergence properties of other iterative processes arising in the fixed point theory which were studied in our previous research.

Let $(X, \langle \cdot, \cdot \rangle)$ be a Hilbert space equipped with an inner product $\langle \cdot, \cdot \rangle$ which induces the norm $\| \cdot \|$. Let C be a nonempty, closed, and convex set in the space X and let a mapping $P : C \to C$ satisfy

$$\|P(x) - P(y)\| \le \|x - y\| \text{ for all } x, y \in C. \tag{6.1}$$

In other words, the mapping P is nonexpansive. Set

$$S = \{x \in C : \ P(x) = x\}. \tag{6.2}$$

Suppose that

A. J. Zaslavski, *Approximate Fixed Points of Nonexpansive Mappings*,
Developments in Mathematics 80, https://doi.org/10.1007/978-3-031-70710-0_6

$$S \neq \emptyset. \tag{6.3}$$

Let $\{\epsilon_k\}_{k=1}^{\infty}$ be a sequence of positive numbers such that

$$\lim_{k\to\infty} \epsilon_k = 0 \tag{6.4}$$

$\theta \in [0, 1)$ and let a mapping $\pi : X \to C$ satisfy

$$\|\pi(x) - \pi(y)\| \leq \theta\|x - y\|, \ x, y \in X. \tag{6.5}$$

It is clear that for each integer $k \geq 1$, there exists a unique point $x_k \in C$ such that

$$x_k = (1 + \epsilon_k)^{-1}P(x_k) + \epsilon_k(1 + \epsilon_k)^{-1}\pi(x_k). \tag{6.6}$$

Denote by $I : X \to X$ the identity operator: $I(x) = x$ for all $x \in X$.

The following result was proved in [133].

Theorem 6.1 *The sequence $\{x_k\}_{k=1}^{\infty}$ converges in the norm topology to the unique solution of the variational inequality*

$$\text{find } \tilde{x} \in S \text{ such that } \langle (I - \pi)(\tilde{x}), \tilde{x} - x\rangle \leq 0 \text{ for all } x \in S,$$

in other words, the unique fixed point $\tilde{x}$ of the mapping $proj_S \circ \pi$, where $proj_S$ is the projection on S.

Suppose that

$$\sum_{k=1}^{\infty} \epsilon_k = \infty \tag{6.7}$$

and

$$\lim_{k\to\infty} |\epsilon_k^{-1} - \epsilon_{k-1}^{-1}| = 0. \tag{6.8}$$

The following result, which was also obtained in [133], establishes the convergence of the viscosity approximation method to the fixed point $\tilde{x}$ of the mapping P.

Theorem 6.2 *Let $z_0 \in C$ and let for all integers $k \geq 1$,*

$$z_k = (1 + \epsilon_k)^{-1}P(z_{k-1}) + \epsilon_k(1 + \epsilon_k)^{-1}\pi(z_{k-1}).$$

Then

$$\lim_{k\to\infty} \|z_k - \tilde{x}\| = 0.$$

In [133], in the proof of Theorem 6.2, it was shown that there exists a constant $c_0 > 0$ such that

$$\|x_k - x_{k-1}\| \le c_0 |1 - \epsilon_k(\epsilon_{k-1})^{-1}| \text{ for all integers } k \ge 1. \tag{6.9}$$

This inequality is an important ingredient in the proof of Theorem 6.2.

In this chapter we prove the following three results which show the uniform convergence of (inexact) iterates generated by the viscosity approximation method to the fixed point $\tilde{x}$. Our first result is on the convergence of exact iterates, while the second and third results cover the case with inexact iterates.

Theorem 6.3 *Let $\epsilon, M > 0$. Then there exists a natural number n_0 such that for each sequence $\{z_i\}_{i=0}^{\infty} \subset C$ satisfying*

$$\|z_0\| \le M,$$

$$z_i = (1 + \epsilon_i)^{-1} P(z_{i-1}) + \epsilon_i (1 + \epsilon_i)^{-1} \pi(z_{i-1}), \ i = 1, 2, \ldots$$

the inequality $\|z_i - \tilde{x}\| \le \epsilon$ holds for all integers $i \ge n_0$.

Theorem 6.4 *Let $\gamma, M > 0$, let a natural number n_0 be as guaranteed by Theorem 6.3 with $\epsilon = \gamma/2$ and let*

$$\delta = 2^{-1} \gamma n_0^{-1}. \tag{6.10}$$

Assume that a finite sequence $\{z_i\}_{i=0}^{n_0} \subset C$ satisfies

$$\|z_0\| \le M \tag{6.11}$$

and

$$\|z_i - (1 + \epsilon_i)^{-1} P(z_{i-1}) - \epsilon_i (1 + \epsilon_i)^{-1} \pi(z_{i-1})\| \le \delta \tag{6.12}$$

for integers $i = 1, \ldots, n_0$. Then the inequality $\|z_{n_0} - \tilde{x}\| \le \gamma$ holds.

Theorem 6.5 *Let $\gamma, M > 0$ and $\epsilon_i \le 1$ for all natural numbers i. Then there exist $\delta > 0$ and a natural number n_1 such that for each sequence $\{z_i\}_{i=0}^{\infty} \subset C$ satisfying*

$$\|z_0\| \le M$$

and

$$\|z_i - (1 + \epsilon_i)^{-1} P(z_{i-1}) - \epsilon_i (1 + \epsilon_i)^{-1} \pi(z_{i-1})\| \le \delta \epsilon_i$$

for all natural numbers i, the inequality $\|z_n - \tilde{x}\| \le \gamma$ holds for all integers $n \ge n_1$.

Clearly, our final result is a generalization of Theorem 6.3 which is used in the proof of Theorem 6.5. Theorems 6.3–6.5 were obtained in [215].

6.2 Proof of Theorem 6.3

By Theorem 6.1,

$$\lim_{i\to\infty} \|x_i - \tilde{x}\| = 0. \tag{6.13}$$

We may assume without loss of generality that

$$\epsilon < 2^{-1},\ c_0 > 1,\ M - 1 > \|x_i\| \text{ for all integers } i \geq 1, \tag{6.14}$$

$$c_0 > |(\epsilon_{i-1} - \epsilon_i)\epsilon_{i-1}^{-1}|,\ i = 1, 2, \ldots. \tag{6.15}$$

Choose a number $\gamma_0 \in (0, 1]$ such that

$$\gamma_0 \leq 16^{-1}\epsilon(1-\theta)c_0^{-1}. \tag{6.16}$$

In view of (6.4), (6.8), and (6.13), there exists an integer $p_0 \geq 1$ such that for each integer $i \geq p_0$,

$$\epsilon_{i-1} \leq 1,\ |\epsilon_{i-1}^{-1} - \epsilon_i^{-1}| \leq \gamma_0,\ \|x_i - \tilde{x}\| \leq \epsilon/2. \tag{6.17}$$

By (6.7), there exists a natural number $n_0 > 2p_0 + 4$ such that

$$\sum_{i=p_0}^{n_0-1} \epsilon_{i+1} > 16(2M + p_0c_0^2)\epsilon^{-1}(1-\theta)^{-1}. \tag{6.18}$$

Assume that $\{z_i\}_{i=0}^{\infty} \subset C$ satisfies

$$\|z_0\| \leq M, \tag{6.19}$$

$$z_i = (1+\epsilon_i)^{-1}P(z_{i-1}) + \epsilon_i(1+\epsilon_i)^{-1}\pi(z_{i-1}),\ i = 1, 2, \ldots. \tag{6.20}$$

Assume that an integer $i \geq 0$. By (6.1), (6.5), (6.6), (6.9), and (6.20),

$$\begin{aligned}
&\|z_i - x_i\| \\
&= \|(1+\epsilon_i)^{-1}P(z_{i-1}) + \epsilon_i(1+\epsilon_i)^{-1}\pi(z_{i-1}) \\
&\quad - (1+\epsilon_i)^{-1}P(x_i) + \epsilon_i(1+\epsilon_i)^{-1}\pi(x_i)\|
\end{aligned}$$

$$\leq (1+\epsilon_i)^{-1}\|P(z_{i-1}) - P(x_i)\| + \epsilon_i(1+\epsilon_i)^{-1}\|\pi(z_{i-1}) - \pi(x_i)\|$$

$$\leq (1+\epsilon_i)^{-1}\|z_{i-1} - x_i\| + \theta\epsilon_i(1+\epsilon_i)^{-1}\|z_{i-1} - x_i\|$$

$$\leq (1+\epsilon_i)^{-1}(1+\theta\epsilon_i)\|z_{i-1} - x_i\|$$

$$\leq (1+\epsilon_i)^{-1}(1+\theta\epsilon_i)(\|z_{i-1} - x_{i-1}\| + \|x_{i-1} - x_i\|)$$

$$\leq (1+\epsilon_i)^{-1}(1+\theta\epsilon_i)\|z_{i-1} - x_{i-1}\|$$

$$+ (1+\epsilon_i)^{-1}(1+\theta\epsilon_i)c_0|1 - \epsilon_i\epsilon_{i-1}^{-1}|. \tag{6.21}$$

It follows (6.15) and (6.21) that for all natural numbers i,

$$\|z_i - x_i\| \leq \|z_{i-1} - x_{i-1}\| + c_0^2, \tag{6.22}$$

$$\|z_i - x_i\| \leq \|z_0 - x_0\| + c_0^2 i. \tag{6.23}$$

By (6.14), (6.19), and (6.23), for all integers $i \in \{0, \dots, p_0\}$,

$$\|z_i - x_i\| \leq \|z_0 - x_0\| + p_0 c_0^2 \leq 2M + p_0 c_0^2. \tag{6.24}$$

We show that there exists $j \in \{p_0, \dots, n_0\}$ such that

$$\|x_j - z_j\| \leq \epsilon/4. \tag{6.25}$$

Assume the contrary. Then for each

$$i \in \{p_0, \dots, n_0\}, \tag{6.26}$$

we have

$$\|x_i - z_i\| > \epsilon/4. \tag{6.27}$$

It follows from (6.16), (6.17), (6.21), and (6.27) that for each $i \in \{p_0, \dots, n_0\}$,

$$\|z_i - x_i\| - \|z_{i+1} - x_{i+1}\|$$

$$\geq \|z_i - x_i\|(1+\epsilon_{i+1})^{-1}\epsilon_{i+1}(1-\theta)$$

$$- c_0(1+\epsilon_{i+1})^{-1}(1+\theta\epsilon_{i+1})|\epsilon_i - \epsilon_{i+1}|\epsilon_i^{-1}$$

$$\geq 4^{-1}\epsilon(1+\epsilon_{i+1})^{-1}\epsilon_{i+1}(1-\theta) - c_0(1+\epsilon_{i+1})^{-1}(1+\theta\epsilon_{i+1})\epsilon_{i+1}\gamma_0$$

$$\geq \epsilon_{i+1}(1+\epsilon_{i+1})^{-1}(4^{-1}\epsilon(1-\theta) - c_0(1+\theta\epsilon_{i+1})\gamma_0)$$

$$\geq \epsilon_{i+1}(1+\epsilon_{i+1})^{-1}(4^{-1}\epsilon(1-\theta) - 2c_0\gamma_0)$$

$$\geq 8^{-1}\epsilon(1-\theta)\epsilon_{i+1}(1+\epsilon_{i+1})^{-1}$$

$$\geq 16^{-1}\epsilon(1-\theta)\epsilon_{i+1}. \tag{6.28}$$

By (6.24) and (6.28),

$$2M + p_0c_0^2 \geq \|z_{p_0} - x_{p_0}\| \geq \|z_{p_0} - x_{p_0}\| - \|z_{n_0} - x_{n_0}\|$$

$$= \sum_{i=p_0}^{n_0-1}(\|z_i - x_i\| - \|z_{i+1} - x_{i+1}\|)$$

$$\geq 16^{-1}\epsilon(1-\theta)\sum_{i=p_0}^{n_0-1}\epsilon_{i+1}.$$

This contradicts (6.18). The contradiction we have reached proves that there is $j \in \{p_0, \ldots, n_0\}$ such that

$$\|x_j - z_j\| \leq \epsilon/4. \tag{6.29}$$

We show that for all integers $i \geq j$,

$$\|x_i - z_i\| \leq \epsilon/2. \tag{6.30}$$

In view of (6.29), (6.30) holds for $i = j$.

Assume that an integer $i \geq j$ and (6.30) holds. There are two cases:

$$\|x_i - z_i\| \leq \epsilon/4; \tag{6.31}$$

$$\|x_i - z_i\| > \epsilon/4. \tag{6.32}$$

Assume that (6.31) holds. By (6.16), (6.17), (6.21), and (6.31) and the inequality $i \geq j \geq p_0$,

$$\|z_{i+1} - x_{i+1}\|$$

$$\leq \|z_i - x_i\|(1+\epsilon_{i+1})^{-1}(1+\theta\epsilon_{i+1})$$

$$+ (1+\epsilon_{i+1})^{-1}(1+\theta\epsilon_{i+1})c_0|\epsilon_i^{-1} - \epsilon_{i+1}^{-1}|\epsilon_{i+1}$$

$$\leq \epsilon/4 + c_0\epsilon_{i+1}\gamma_0 \leq \epsilon/2.$$

Assume that (6.32) holds. In view of (6.16), (6.17), (6.21), and (6.32), relation (6.28) holds. Together with (6.30), this implies that

$$\|z_{i+1} - x_{i+1}\| \leq \|z_i - x_i\| \leq \epsilon/2.$$

This implies that

$$\|z_{i+1} - x_{i+1}\| \leq \epsilon/2$$

in both cases. Therefore we have shown by induction that for all integers $i \geq j$,

$$\|z_i - x_i\| \leq \epsilon/2.$$

Together with (6.17) and the inequality $n_0 \geq j \geq p_0$, this implies that for all integers $i \geq n_0$,

$$\|z_i - \tilde{x}\| \leq \epsilon.$$

Theorem 6.3 is proved.

6.3 Proof of Theorem 6.4

Set

$$\xi_0 = z_0, \tag{6.33}$$

$$\xi_k = (1 + \epsilon_k)^{-1} P(\xi_{k-1}) + \epsilon_k (1 + \epsilon_k)^{-1} \pi(\xi_{k-1}) \tag{6.34}$$

for all natural numbers k. By (6.11), (6.33), and (6.34), the choice of n_0 and Theorem 6.3,

$$\|\xi_{n_0} - \tilde{x}\| \leq \gamma/2. \tag{6.35}$$

We show that for all integers $i = 0, \ldots, n_0$,

$$\|z_i - \xi_i\| \leq i\delta. \tag{6.36}$$

In view of (6.35), (6.36) holds for $i = 0$. Assume that $i \in \{0, \ldots, n_0 - 1\}$ and (6.36) holds. It follows from (6.1), (6.5), (6.10), (6.12), (6.34), and (6.36) that

$$\|z_{i+1} - \xi_{i+1}\|$$

$$\leq \|z_{i+1} - (1+\epsilon_{i+1})^{-1}P(z_i) - \epsilon_{i+1}(1+\epsilon_{i+1})^{-1}\pi(z_i)\|$$

$$+ \|(1+\epsilon_{i+1})^{-1}P(z_i) + \epsilon_{i+1}(1+\epsilon_{i+1})^{-1}\pi(z_i)$$

$$- (1+\epsilon_{i+1})^{-1}P(\xi_i) - \epsilon_{i+1}(1+\epsilon_{i+1})^{-1}\pi(\xi_i)\|$$

$$\leq \delta + (1+\epsilon_{i+1})^{-1}\|P(z_i) - P(\xi_i)\|$$

$$+ \epsilon_{i+1}(1+\epsilon_{i+1})^{-1}\|\pi(z_i) - \pi(\xi_i)\|$$

$$\leq \delta + (1+\epsilon_{i+1})^{-1}i\delta + \epsilon_{i+1}(1+\epsilon_{i+1})^{-1}i\delta \leq (i+1)\delta.$$

Thus (6.34) holds for all integers $i = 0, \dots, n_0$ and

$$\|\xi_{n_0} - z_{n_0}\| \leq n_0\delta \leq \gamma/2.$$

Together with (6.35) this implies that

$$\|z_{n_0} - \tilde{x}\| \leq \|z_{n_0} - \xi_{n_0}\| + \|\xi_{n_0} - \tilde{x}\| \leq \gamma.$$

Theorem 6.4 is proved.

6.4 Proof of Theorem 6.5

Let a natural number n_0 be as guaranteed by Theorem 6.3 with $\epsilon = \gamma/4$. In view of Theorem 6.1, we may assume without loss of generality that

$$\|x_i - \tilde{x}\| \leq \gamma/8 \text{ for all integers } i \geq n_0. \tag{6.37}$$

We may assume that $c_0 \geq 1$ (see (6.9)). By (6.8), there exists an integer $n_1 > n_0$ such that for each integer $i \geq n_1$,

$$c_0|\epsilon_i^{-1} - \epsilon_{i+1}^{-1}| \leq 16^{-1}\gamma(1-\theta). \tag{6.38}$$

Choose a positive number δ such that

$$\delta \leq 16^{-1}\gamma(1-\theta),\ \delta n_1 \leq 4^{-1}\gamma. \tag{6.39}$$

Assume that a sequence $\{z_i\}_{i=0}^{\infty} \subset C$ satisfies

$$\|z_0\| \leq M \tag{6.40}$$

and

$$\|z_i - (1+\epsilon_i)^{-1}P(z_{i-1}) - \epsilon_i(1+\epsilon_i)^{-1}\pi(z_{i-1})\| \le \delta\epsilon_i \tag{6.41}$$

for all natural numbers i. Set

$$\xi_0 = z_0, \tag{6.42}$$

$$\xi_i = (1+\epsilon_i)^{-1}P(\xi_{i-1}) + \epsilon_i(1+\epsilon_i)^{-1}\pi(\xi_{i-1}) \tag{6.43}$$

for all natural numbers i. By (6.40), (6.42), and(6.43), the choice of n_0 and Theorem 6.3,

$$\|\xi_i - \tilde{x}\| \le \gamma/4 \text{ for all integers } i \ge n_0. \tag{6.44}$$

We show that for all integers $i \ge 0$,

$$\|z_i - \xi_i\| \le i\delta. \tag{6.45}$$

In view of (6.42), (6.45) holds for $i = 0$. Assume that $i \ge 0$ is an integer and (6.45) holds. It follows from (6.1), (6.5), (6.41), (6.43), and (6.45) that

$$\|z_{i+1} - \xi_{i+1}\|$$

$$\le \|z_{i+1} - (1+\epsilon_{i+1})^{-1}P(z_i) - \epsilon_{i+1}(1+\epsilon_{i+1})^{-1}\pi(z_i)\|$$

$$+ \|(1+\epsilon_{i+1})^{-1}P(z_i) + \epsilon_{i+1}(1+\epsilon_{i+1})^{-1}\pi(z_i)$$

$$- (1+\epsilon_{i+1})^{-1}P(\xi_i) - \epsilon_{i+1}(1+\epsilon_{i+1})^{-1}\pi(\xi_i)\|$$

$$\le \delta + (1+\epsilon_{i+1})^{-1}\|P(z_i) - P(\xi_i)\|$$

$$+ \epsilon_{i+1}(1+\epsilon_{i+1})^{-1}\|\pi(z_i) - \pi(\xi_i)\|$$

$$\le \delta + \|z_i - \xi_i\| \le (i+1)\delta.$$

Thus we have shown that (6.45) holds for all integers $i \ge 0$. By (6.37), (6.39), (6.44), and (6.45), for all integers $i \in [n_0, n_1]$,

$$\|z_i - \tilde{x}\| \le \|z_i - \xi_i\| + \|\xi_i - \tilde{x}\|$$

$$\le \delta n_1 + \gamma/4 \le \gamma/2,$$

$$\|z_i - x_i\| \le \gamma/2 + \gamma/8. \tag{6.46}$$

We show that (6.46) holds for all integers $i \geq n_1$.

Assume that an integer $i \geq n_1$ and (6.46) holds. By (6.1), (6.5), (6.6), (6.9), and (6.41),

$$\|z_{i+1} - x_{i+1}\|$$

$$\leq \|z_{i+1} - (1+\epsilon_{i+1})^{-1}P(z_i) + \epsilon_{i+1}(1+\epsilon_{i+1})^{-1}\pi(z_i)\|$$

$$+ \|(1+\epsilon_{i+1})^{-1}P(z_i) + \epsilon_{i+1}(1+\epsilon_{i+1})^{-1}\pi(z_i)$$

$$- (1+\epsilon_{i+1})^{-1}P(x_{i+1}) + \epsilon_{i+1}(1+\epsilon_{i+1})^{-1}\pi(x_{i+1})\|$$

$$\leq \delta\epsilon_{i+1} + (1+\epsilon_{i+1})^{-1}\|P(z_i) - P(x_{i+1})\|$$

$$+ \epsilon_{i+1}(1+\epsilon_{i+1})^{-1}\|\pi(z_i) - \pi(x_{i+1})\|$$

$$\leq \delta\epsilon_{i+1} + (1+\epsilon_{i+1})^{-1}\|z_i - x_{i+1}\|$$

$$+ \theta\epsilon_{i+1}(1+\epsilon_{i+1})^{-1}\|z_i - x_{i+1}\|$$

$$\leq \delta\epsilon_{i+1} + (1+\epsilon_{i+1})^{-1}(1+\theta\epsilon_{i+1})\|z_i - x_{i+1}\|$$

$$\leq \delta\epsilon_{i+1} + (1+\epsilon_{i+1})^{-1}(1+\theta\epsilon_{i+1})\|z_i - x_i\|$$

$$+ (1+\epsilon_{i+1})^{-1}(1+\theta\epsilon_{i+1})\|x_{i+1} - x_i\|$$

$$\leq \delta\epsilon_{i+1} + (1+\epsilon_{i+1})^{-1}(1+\theta\epsilon_{i+1})\|z_i - x_i\|$$

$$+ (1+\epsilon_{i+1})^{-1}(1+\theta\epsilon_{i+1})c_0|1 - \epsilon_{i+1}\epsilon_i^{-1}|$$

$$\leq \delta\epsilon_{i+1} + (1+\epsilon_{i+1})^{-1}(1+\theta\epsilon_{i+1})\|z_i - x_i\|$$

$$+ (1+\epsilon_{i+1})^{-1}(1+\theta\epsilon_{i+1})c_0|\epsilon_{i+1}^{-1} - \epsilon_i^{-1}|\epsilon_{i+1}. \tag{6.47}$$

There are two cases:

$$\|x_i - z_i\| \leq \gamma/2; \tag{6.48}$$

$$\|x_i - z_i\| > \gamma/2. \tag{6.49}$$

Assume that (6.48) holds. By (6.38), (6.39), (6.46), and (6.47),

$$\|z_{i+1} - x_{i+1}\|$$

$$\leq \delta\epsilon_{i+1} + \|z_i - x\| + c_0|\epsilon_i^{-1} - \epsilon_{i+1}^{-1}|\epsilon_{i+1}$$

$$\leq \delta + \gamma/2 + c_0|\epsilon_i^{-1} - \epsilon_{i+1}^{-1}|$$

$$\leq \gamma/16 + \gamma/2 + \gamma/16$$

and

$$\|z_{i+1} - x_{i+1}\| \leq \gamma/2 + \gamma/8.$$

Assume that (6.49) holds. In view of (6.38), (6.39), (6.47), and (6.49),

$$\|z_i - x_i\| - \|z_{i+1} - x_{i+1}\|$$

$$\geq \|z_i - x_i\|(1 + \epsilon_{i+1})^{-1}\epsilon_{i+1}(1 - \theta)$$

$$- \delta\epsilon_{i+1} - c_0|\epsilon_i^{-1} - \epsilon_{i+1}^{-1}|\epsilon_{i+1}$$

$$\geq \epsilon_{i+1}[4^{-1}\gamma(1 - \theta) - \delta - c_0|\epsilon_i^{-1} - \epsilon_{i+1}^{-1}|] \geq 0. \tag{6.50}$$

By (6.46) and (6.50),

$$\|z_{i+1} - x_{i+1}\| \leq \|z_i - x_i\| \leq \gamma/2 + \gamma/8.$$

Thus this inequality holds in both cases. Therefore we have shown that (6.46) holds for all integers $i \geq n_1$. By (6.37) and (6.46), for all integers $i \geq n_1$,

$$\|z_i - \tilde{x}\| \leq \|z_i - x_i\| + \|x_i - \tilde{x}\|$$

$$\leq \gamma/8 + \gamma/2 + \gamma/8 \leq \gamma.$$

Theorem 6.5 is proved.

6.5 Equilibrium Problems

Let $(H, \langle \cdot, \cdot \rangle)$ be a Hilbert space equipped with an inner product $\langle \cdot, \cdot \rangle$ which induces the norm $\| \cdot \|$. Let C be a nonempty, closed, and convex set in the space H and let a function $F : C \times C \to R^1$. The equilibrium problem for F is to find $x \in C$ such that

$$F(x, y) \geq 0 \text{ for all } y \in C. \tag{6.51}$$

Denote by $EP(F)$ the set of solutions of (6.51). A mapping $S : C \to H$ is called nonexpansive if

$$\|S(x) - S(y)\| \leq \|x - y\| \text{ for all } x, y \in C.$$

Set

$$F(S) = \{x \in C : S(x) = x\}.$$

Denote by $I : X \to X$ the identity operator: $I(x) = x$ for all $x \in X$.

Let D be a nonempty closed convex subset of H. Then, for every $x \in H$, there exists a unique nearest point in D, denoted by $P_D(x)$, such that

$$\|x - P_D(x)\| \leq \|x - y\| \text{ for all } y \in D,$$

$$\|P_(x) - P_D(y)\| \leq \|x - y\| \text{ for all } x, y \in D.$$

Suppose that the function F satisfies the following conditions:

(A1) $F(x, x) = 0$ for all $x \in C$;
(A2) $F(x, y) + F(y, x) \leq 0$ for all $x, y \in C$;
(A3) for each $x, y, z \in C$,

$$\lim_{t \to 0^+} F(tz + (1-t)x, y) \leq F(x, y);$$

(A4) for each $x \in C$, the function $F(x, \cdot) : C \to R^1$ is convex and lower semicontinuous.

The following lemma appears implicitly in [22].

Lemma 6.6 *Let $r > 0$ and $x \in H$. Then there exists $z \in C$ such that*

$$F(z, y) + r^{-1}\langle y - z, z - x\rangle \geq 0 \textit{ for all } y \in C.$$

For every $r > 0$ define a mapping $T_r : H \to C$ as follows:

$$T_r(x) = \{z \in C : F(z, y) + r^{-1}\langle y - z, z - x\rangle \geq 0 \text{ for all } y \in C\} \tag{6.52}$$

for all $x \in H$.

The following lemma was given in [60].

Lemma 6.7 *For every $r > 0$ the following properties hold:*

(1) the mapping T_r is single-valued;
(2) for each $x, y \in H$,

$$\|T_r(x) - T_r(y)\|^2 \leq \langle T_r(x) - T_r(y), x - y\rangle;$$

(3) $F(T_r) = EP(F)$;
(4) $EP(F)$ *is a closed convex set.*

Lemma 6.7 implies the following auxiliary results.

Lemma 6.8 *Let* $r > 0$. *Then for each* $x, y \in H$,

$$\|T_r(x) - T_r(y)\| \le \|x - y\|.$$

Proof Let $x, y \in H$ satisfy $x \neq y$. By Lemma 6.7,

$$\|T_r(x) - T_r(y)\| \cdot \|x - y\|$$

$$\ge \langle T_r(x) - T_r(y), x - y\rangle$$

$$\ge \|T_r(x) - T_r(y)\|^2$$

and

$$\|T_r(x) - T_r(y)\| \le \|x - y\|.$$

Lemma 6.8 is proved.

Lemma 6.9 *Let* $r > 0$, $x \in H$ *and* $z \in EP(F)$. *Then*

$$\|x - T_r(x)\|^2 \le \|x - z\|^2 - \|T_r(x) - z\|^2.$$

Proof By Lemma 6.7,

$$\|T_r(x) - z\|^2 = \|T_r(x) - T_r(z)\|^2$$

$$\le \langle T_r(x) - T_r(z), x - z\rangle$$

$$= \langle T_r(x) - z, x - z\rangle.$$

In view of the relation above,

$$\|x - T_r(x)\|^2 = \|x - z + z - T_r(x)\|^2$$

$$= \|x - z\|^2 + \|T_r(x) - z\|^2 + 2\langle x - z, z - T_r(x)\rangle$$

$$\le \|x - z\|^2 + \|T_r(x) - z\|^2 - 2\|T_r(x) - z\|^2$$

$$= \|x - z\|^2 - \|T_r(x) - z\|^2.$$

Lemma 6.9 is proved.

Assume that $S : C \to H$ satisfies

$$\|S(x) - S(y)\| \le \|x - y\|, \ x, y \in C, \tag{6.53}$$

$$F(S) \cap EP(F) \ne \emptyset, \tag{6.54}$$

$\theta \in [0, 1)$, $f : H \to H$ satisfies

$$\|f(x) - f(y)\| \le \theta \|x - y\| \text{ for all } x, y \in H \tag{6.55}$$

and that

$$\{\alpha_n\}_{n=1}^{\infty} \subset [0, 1], \ \{r_n\}_{n=1}^{\infty} \subset (0, \infty), \tag{6.56}$$

$$\lim_{n\to\infty} \alpha_n = 0, \ \sum_{n=1}^{\infty} \alpha_n = \infty, \ \sum_{n=1}^{\infty} |\alpha_n - \alpha_{n+1}| < \infty, \tag{6.57}$$

$$\liminf_{n\to\infty} r_n > 0, \ \sum_{n=1}^{\infty} |r_n - r_{n+1}| < \infty. \tag{6.58}$$

Clearly, there exists a unique point

$$\tilde{z} \in F(S) \cap EP(F) \tag{6.59}$$

such that

$$\tilde{z} = P_{F(S)\cap EP(F)} f(\tilde{z}). \tag{6.60}$$

The following result obtained in [186] establishes the convergence of viscosity approximation methods for finding a common element of the set of solutions of equilibrium problem (6.51) and the set of fixed points of the nonexpansive mapping S. Note that viscosity approximation methods for finding a fixed point of a nonexpansive mapping in a Hilbert space were introduced and studied in [133].

Theorem 6.10 *Assume that* $\{x_n\}_{n=1}^{\infty}, \{u_n\}_{n=1}^{\infty} \subset H$ *and that for all integers* $n \ge 1$,

$$F(u_n, y) + r_n^{-1}\langle y - u_n, u_n - x_n\rangle \ge 0 \textit{ for all } y \in C,$$

$$x_{n+1} = \alpha_n f(x_n) + (1 - \alpha_n)S(u_n).$$

Then

$$\lim_{n\to\infty} u_n = \lim_{n\to\infty} x_n = \tilde{z}.$$

In this chapter, we prove three results which show the uniform convergence of (inexact) iterates generated by the viscosity approximation method to the point $\tilde{z}$. Our first result is on the convergence of exact iterates, while the second and third results cover the case with inexact iterates. These results were obtained in [216].

6.6 The First Result

Theorem 6.11 *Let $\epsilon, M > 0$. Then there exist a natural number n_0 such that for each pair of sequences $\{x_n\}_{n=1}^{\infty}, \{u_n\}_{n=1}^{\infty} \subset H$ which satisfy*

$$\|x_1\| \leq M \tag{6.61}$$

and

$$u_n = T_{r_n}(x_n), \tag{6.62}$$

$$x_{n+1} = \alpha_n f(x_n) + (1 - \alpha_n)S(u_n) \tag{6.63}$$

for every integer $n \geq 1$, the inequalities

$$\|x_n - \tilde{z}\|, \ \|u_n - \tilde{z}\| \leq \epsilon$$

hold for all integers $n \geq n_0$.

Proof Set

$$\xi_1 = 0 \tag{6.64}$$

and for all integers $n \geq 1$ set

$$\eta_n = T_{r_n}(\xi_n), \tag{6.65}$$

$$\xi_{n+1} = \alpha_n f(\xi_n) + (1 - \alpha_n)S(\eta_n). \tag{6.66}$$

By (6.52) and (6.64)–(6.66) and Theorem 6.10,

$$\lim_{n\to\infty} \|\xi_n - \tilde{z}\| = 0, \ \lim_{n\to\infty} \|\eta_n - \tilde{z}\| = 0. \tag{6.67}$$

We may assume without loss of generality that

$$\epsilon < 2^{-1}, \ M > 1 + \|\xi_n\| + \|\eta_n\| \text{ for all integers } n \geq 1. \tag{6.68}$$

By (6.57) and (6.67), there exists a natural number $n_0 > 2$ such that

$$\sum_{i=1}^{n_0-1} \alpha_i > 4M(1-\theta)^{-1}\epsilon^{-1},$$

$$\|\xi_n - \tilde{z}\| \le \epsilon/4, \ \|\eta_n - \tilde{z}\| \le \epsilon/4 \text{ for all integers } n \ge n_0. \tag{6.69}$$

Assume that a pair of sequences $\{x_n\}_{n=1}^{\infty}, \{u_n\}_{n=1}^{\infty} \subset H$ satisfies (6.61) and (6.62) and (6.63) for every integer $n \ge 1$.

Let an integer $n \ge 1$. By (6.62), (6.65) and Lemma 6.8,

$$\|u_n - \eta_n\| = \|T_{r_n}(x_n) - T_{r_n}(\xi_n)\| \le \|x_n - \xi_n\|. \tag{6.70}$$

It follows from (6.53), (6.55), (6.56), (6.63), (6.66), and (6.70) that

$$\begin{aligned}
&\|x_{n+1} - \xi_{n+1}\| \\
&= \|\alpha_n f(x_n) + (1-\alpha_n)S(u_n) - \alpha_n f(\xi_n) - (1-\alpha_n)S(\eta_n)\| \\
&\le \alpha_n \|f(x_n) - f(\xi_n)\| + (1-\alpha_n)\|S(u_n) - S(\eta_n)\| \\
&\le \alpha_n\theta\|x_n - \xi_n\| + (1-\alpha_n)\|u_n - \eta_n\| \\
&\le \alpha_n\theta\|x_n - \xi_n\| + (1-\alpha_n)\|x_n - \xi_n\| \\
&\le \|x_n - \xi_n\|(1 - \alpha_n(1-\theta)).
\end{aligned} \tag{6.71}$$

By (6.56), (6.61), (6.64), (6.68), (6.70), and (6.71),

$$\|x_n - \xi_n\| \le \|x_1 - \xi_1\| \le M, \ \|x_n\| \le 2M, \tag{6.72}$$

$$\|u_n - \eta_n\| \le M, \ \|u_n\| \le 2M. \tag{6.73}$$

In view of (6.71), for every natural number n,

$$\begin{aligned}
&\|x_n - \xi_n\| - \|x_{n+1} - \xi_{n+1}\| \\
&\ge \|x_n - \xi_n\| - \|x_n - \xi_n\|(1 - \alpha_n(1-\theta)) \\
&= \|x_n - \xi_n\|\alpha_n(1-\theta).
\end{aligned} \tag{6.74}$$

We show that there exists $j \in \{1, \dots, n_0\}$ such that

$$\|x_j - \xi_j\| \le \epsilon/4.$$

Assume the contrary. Then

$$\|x_i - \xi_i\| > \epsilon/4, \ i = 1, \dots, n_0. \tag{6.75}$$

By (6.74) and (6.75), for all $n \in \{1, \dots, n_0\}$,

$$\|x_n - \xi_n\| - \|x_{n+1} - \xi_{n+1}\| \geq 4^{-1}\epsilon\alpha_n(1-\theta). \tag{6.76}$$

In view of (6.72) and (6.76),

$$M \geq \|x_1 - \xi_1\| \geq \|x_1 - \xi_1\| - \|x_{n_0} - \xi_{n_0}\|$$

$$= \sum_{i=1}^{n_0-1}(\|x_i - \xi_i\| - \|x_{i+1} - \xi_{i+1}\|)$$

$$\geq 4^{-1}\epsilon(1-\theta)\sum_{i=1}^{n_0-1}\alpha_i. \tag{6.77}$$

Equation (6.77) contradicts (6.69). The contradiction we have reached proves that there exists $j \in \{1, \dots, n_0\}$ such that

$$\|x_j - \xi_j\| \leq \epsilon/4. \tag{6.78}$$

It follows from (6.70), (6.71), and (6.78) that for all integers $n \geq j$,

$$\|x_n - \xi_n\| \leq \epsilon/4, \ \|u_n - \eta_n\| \leq \epsilon/4. \tag{6.79}$$

By (6.69) and (6.79), for all integers $n \geq n_0$,

$$\|x_n - \tilde{z}\| \leq \epsilon, \ \|u_n - \eta_n\| \leq \epsilon.$$

Theorem 6.11 is proved.

6.7 The Second Result

Theorem 6.12 *Let $\gamma, M > 0$, an integer $n_0 > 1$ be as guaranteed by Theorem 6.11 with $\epsilon = \gamma/2$ and let*

$$\delta = (4n_0)^{-1}\gamma. \tag{6.80}$$

Assume that finite sequences $\{x_n\}_{n=1}^{n_0}, \{u_n\}_{n=1}^{n_0-1} \subset H$ satisfy

$$\|x_1\| \le M \tag{6.81}$$

and that for every integer $n \in \{1, \ldots, n_0 - 1\}$

$$\|u_n - T_{r_n}(x_n)\| \le \delta, \tag{6.82}$$

$$\|x_{n+1} - \alpha_n f(x_n) - (1-\alpha_n)S(u_n)\| \le \delta. \tag{6.83}$$

Then

$$\|x_{n_0} - \tilde{z}\| \le \gamma.$$

Proof Set

$$\xi_1 = x_1 \tag{6.84}$$

and for all integers $n \ge 1$ set

$$\eta_n = T_{r_n}(\xi_n), \tag{6.85}$$

$$\xi_{n+1} = \alpha_n f(\xi_n) + (1-\alpha_n)S(\eta_n). \tag{6.86}$$

By (6.81), (6.84), and (6.86), Theorem 6.11, and the choice of n_0,

$$\|\xi_n - \tilde{z}\| \le \gamma/2 \text{ for all integers } n \ge n_0. \tag{6.87}$$

We show that for all integers $n = 1, \ldots, n_0$,

$$\|x_n - \xi_n\| \le 2(n-1)\delta. \tag{6.88}$$

Clearly, (6.88) holds for $n = 1$.

Assume that an integer $n \in \{1, \ldots, n_0 - 1\}$ and (6.88) holds. By (6.82) and (6.88) and Lemma 6.8,

$$\|u_n - \eta_n\| \le \|u_n - T_{r_n}(x_n)\| + \|T_{r_n}(x_n) - T_{r_n}(\xi_n)\|$$

$$\le \delta + \|x_n - \xi_n\| \le \delta + 2\delta(n-1). \tag{6.89}$$

By (6.53), (6.55), (6.83), (6.86), (6.88), and (6.89),

$$\|x_{n+1} - \xi_{n+1}\|$$

$$\le \|x_{n+1} - \alpha_n f(x_n) - (1-\alpha_n)S(u_n)\|$$

$$+ \|\alpha_n f(x_n) + (1-\alpha_n)S(u_n) - \alpha_n f(\xi_n) - (1-\alpha_n)S(\eta_n)\|$$

$$\leq \delta + \alpha_n \|f(x_n) - f(\xi_n)\| + (1 - \alpha_n)\|S(u_n) - S(\eta_n)\|$$

$$\leq \delta + \alpha_n \|x_n - \xi_n\| + (1 - \alpha_n)\|u_n - \eta_n\|$$

$$\leq \delta + \delta + 2\delta(n - 1) = 2\delta n.$$

Thus (6.87) holds for all $n = 1, \dots, n_0$. By (6.80), (6.87), and (6.88),

$$\|x_{n_0} - \tilde{z}\| \leq \|x_{n_0} - \xi_{n_0}\| + \|\xi_{n_0} - \tilde{z}\| \leq 2n_0\delta + \gamma/2 \leq \gamma.$$

Theorem 6.12 is proved.

6.8 The Third Result

Theorem 6.13 *Let $\gamma, M > 0$, and let $\alpha_i \leq 1$ for all natural numbers i. Then there exist $\delta > 0$ and a natural number n_0 such that for each pair of finite sequences $\{x_n\}_{n=1}^{\infty}, \{u_n\}_{n=1}^{\infty} \subset C$ which satisfy*

$$\|x_1\| \leq M \tag{6.90}$$

and for every integer $n \geq 1$,

$$\|u_n - T_{r_n}(x_n)\| \leq \delta\alpha_n, \tag{6.91}$$

$$\|x_{n+1} - \alpha_n f(x_n) - (1 - \alpha_n)S(u_n)\| \leq \delta\alpha_n. \tag{6.92}$$

the inequality

$$\|x_n - \tilde{z}\| \leq \gamma$$

holds for all integer $n \geq n_0$.

Proof Let a integer n_0 be as guaranteed by Theorem 6.11 with $\epsilon = \gamma/4$. Choose a positive number δ such that

$$\delta \leq (8n_0)^{-1}\gamma,\ \delta \leq 8^{-1}\gamma(1 - \theta). \tag{6.93}$$

Assume that finite sequences $\{x_n\}_{n=1}^{\infty}, \{u_n\}_{n=1}^{\infty} \subset C$ satisfy (6.90), (6.91), and (6.92) for all natural numbers n. Set

$$\xi_1 = x_1 \tag{6.94}$$

and for all integers $n \geq 1$ set

$$\eta_n = T_{r_n}(\xi_n), \tag{6.95}$$

$$\xi_{n+1} = \alpha_n f(\xi_n) + (1 - \alpha_n)S(\eta_n). \tag{6.96}$$

By (6.90) and (6.94)–(6.96), Theorem 6.11, and the choice of n_0,

$$\|\xi_n - \tilde{z}\| \le \gamma/4 \text{ for all integers } n \ge n_0. \tag{6.97}$$

We show that for all integers $n \ge 1$,

$$\|x_n - \xi_n\| \le 2(n-1)\delta. \tag{6.98}$$

Clearly, (6.98) holds for $n = 1$.

Assume that an integer $n \ge 1$ satisfies (6.98) holds. By (6.91), (6.95) and (6.98) and Lemma 6.8,

$$\|u_n - \eta_n\| \le \|u_n - T_{r_n}(x_n)\| + \|T_{r_n}(x_n) - T_{r_n}(\xi_n)\|$$

$$\le \delta + \|x_n - \xi_n\| \le \delta + 2\delta(n-1). \tag{6.99}$$

In view of (6.53), (6.55), (6.92), (6.96), (6.98), and (6.99),

$$\|x_{n+1} - \xi_{n+1}\|$$

$$\le \|x_{n+1} - \alpha_n f(x_n) - (1 - \alpha_n)S(u_n)\|$$

$$+ \|\alpha_n f(x_n) + (1 - \alpha_n)S(u_n) - \alpha_n f(\xi_n) - (1 - \alpha_n)S(\eta_n)\|$$

$$\le \delta + \alpha_n \|f(x_n) - f(\xi_n)\| + (1 - \alpha_n)\|S(u_n) - S(\eta_n)\|$$

$$\le \delta + \alpha_n \|x_n - \xi_n\| + (1 - \alpha_n)\|u_n - \eta_n\| \le 2\delta n.$$

Thus (6.98) holds for all integers $n \ge 1$.

By (6.97) and (6.98),

$$\|x_{n_0} - \tilde{z}\| \le \|x_{n_0} - \xi_{n_0}\| + \|\xi_{n_0} - \tilde{z}\|$$

$$\le 2n_0\delta + \gamma/4 \le \gamma/4 + \gamma/4.$$

$$\|x_{n_0} - \tilde{z}\| \le \gamma/2.$$

Assume that an integer $n \ge n_0$ and

$$\|x_n - \tilde{z}\| \le \gamma/2 + \gamma/4. \tag{6.100}$$

By (6.91) and Lemma 6.8,

$$\|u_n - \eta_n\| \le \|u_n - T_{r_n}(x_n)\| + \|T_{r_n}(x_n) - T_{r_n}(\xi_n)\|$$
$$\le \delta\alpha_n + \|x_n - \xi_n\|. \tag{6.101}$$

It follows from (6.53), (6.55), (6.92), (6.96), and (6.101) that

$$\|x_{n+1} - \xi_{n+1}\|$$
$$\le \|x_{n+1} - \alpha_n f(x_n) - (1-\alpha_n)S(u_n)\|$$
$$+ \|\alpha_n f(x_n) + (1-\alpha_n)S(u_n) - \alpha_n f(\xi_n) - (1-\alpha_n)S(\eta_n)\|$$
$$\le \delta\alpha_n + \alpha_n\|f(x_n) - f(\xi_n)\| + (1-\alpha_n)\|S(u_n) - S(\eta_n)\|$$
$$\le \delta\alpha_n + \alpha_n\theta\|x_n - \xi_n\| + (1-\alpha_n)\|u_n - \eta_n\|$$
$$\le \delta\alpha_n + \alpha_n\theta\|x_n - \xi_n\| + (1-\alpha_n)\delta\alpha_n + (1-\alpha_n)\|x_n - \xi_n\|$$
$$\le 2\delta\alpha_n + \|x_n - \xi_n\|(1 - \alpha_n(1-\theta)). \tag{6.102}$$

In view of (6.102),

$$\|x_n - \xi_n\| - \|x_{n+1} - \xi_{n+1}\|$$
$$\ge \|x_n - \xi_n\|\alpha_n(1-\theta) - 2\delta\alpha_n$$
$$= \alpha_n(\|x_n - \xi_n\|(1-\theta) - 2\delta). \tag{6.103}$$

There are two cases:

$$\|x_n - \xi_n\| \le \gamma/2; \tag{6.104}$$

$$\|x_n - \xi_n\| > \gamma/2. \tag{6.105}$$

Assume that (6.104) holds. In view of (6.93), (6.102), and (6.104),

$$\|x_{n+1} - \xi_{n+1}\| \le \|x_n - \xi_n\| + 2\delta\alpha_n \le \gamma/2 + \gamma/4.$$

Assume that (6.105) holds. By (6.93), (6.103), and (6.105),

$$\|x_n - \xi_n\| - \|x_{n+1} - \xi_{n+1}\|$$
$$\ge \alpha_n(2^{-1}\gamma(1-\theta) - 2\delta) \ge 0$$

and

$$\|x_{n+1} - \xi_{n+1}\| \le \gamma/2 + \gamma/4.$$

Thus for all integers $n \ge n_0$,

$$\|x_n - \xi_n\| \le \gamma/2 + \gamma/4.$$

Together with (6.97), this implies that for all integers $n \ge n_0$,

$$\|x_n - \tilde{z}\| \le \|x_n - \xi_n\| + \|\xi_n - \tilde{z}\| \le \gamma/4 + \gamma/2 + \gamma/4 = \gamma.$$

Theorem 6.13 is proved.

Chapter 7
Halpern-Mann Iterations and Krasnoselskii-Mann Iterations for a Common Fixed Point Problem

In this chapter we consider a common fixed point problem with a finite family of nonexpansive self-mappings of a convex subset of a metric space with the hyperbolic structure. For Halpern-Mann iterations, we prove a quantitative asymptotic regularity result for the common fixed point problem under assumption that this problem has approximate solutions. This result generalize recent results of [121] obtained for a family consisting only of two operators which have a common fixed point. We also obtain two asymptotic regularity results for Krasnoselskii-Mann iterations. All the results of Sects. 7.1–7.9 are new.

7.1 Rates of Convergence

Let $X = (X, d, W)$ be a W-hyperbolic space. We assume that the W-hyperbolic space X has a structure (X, η), where $\eta : (0, \infty) \times (0, 2] \to (0, 1]$ is a so-called modulus of uniform convexity such that the following assumptions hold.

(B1) For each $r > 0$, each $\epsilon \in (0, 2]$, and each $x, y, a \in X$, if

$$d(x, a),\ \ d(y, a) \leq r \text{ and } d(x, y) \geq \epsilon r,$$

then

$$d(2^{-1}x + 2^{-1}y, a) \leq (1 - \eta(r, \epsilon))r.$$

(B2) For each $\epsilon \in (0, 2]$ and each numbers s, r satisfying $0 < r \leq s$,

$$\eta(s, \epsilon) \leq \eta(r, s).$$

A. J. Zaslavski, *Approximate Fixed Points of Nonexpansive Mappings*,
Developments in Mathematics 80, https://doi.org/10.1007/978-3-031-70710-0_7

It turns out that this class of spaces is an appropriate setting for obtaining quantitative results on the asymptotic behavior of the Mann iteration for nonexpansive mappings (see [112, 118, 119]), as well as of the Picard iteration for firmly nonexpansive mappings [10]. It contains uniformly convex normed spaces and CAT(0) space.

In this chapter we consider a common fixed point problem with a finite family of nonexpansive self-mappings of a convex subset of a metric space with the hyperbolic structure. For Halpern-Mann iterations, we prove a quantitative asymptotic regularity result for the common fixed point problem under assumption that this problem has approximate solutions. This result generalize recent results of [121] obtained for a family consisting only of two operators which have a common fixed point. Reading the work [121], the author realized that the assumption on the existence of a common fixed point can be weakened if we assume only the existence of approximate solutions of the common fixed point problem. As usual the analysis of the iterative process is based on some estimations. If instead of a common fixed point one uses in these estimations ϵ-approximate common fixed points, then some additional terms containing ϵ appear. But since ϵ is an arbitrary positive number passing to the limit, we obtain desirable results.

In this chapter we use the following lemma. For its proof see Lemma 2.1 (iv) of [119].

Lemma 7.1 *Let* $r > 0$, $\epsilon \in (0, 2]$, $x, y, a \in X$,

$$d(x, a) \leq r,\ d(y, a) \leq r,\ d(x, y) \geq \epsilon r.$$

Then for each $\lambda \in [0, 1]$ *and each* $s \geq r$,

$$d((1-\lambda)x \oplus \lambda y, a) \leq (1 - 2\lambda(1-\lambda)\eta(s, \epsilon))r.$$

Let $\mathcal{N} = \{0, 1, \dots\}$ and (Y, ρ) be a metric space. Assume that $\{a_n\}_{n=0}^{\infty} \subset Y$, $a \in Y$ and $\phi : \mathcal{N} \to \mathcal{N}$. We say that:

(i) ϕ is a rate convergence of $\{a_n\}_{n=0}^{\infty}$ (toward a) if $\lim_{n\to\infty} a_n = a$ and for each integer $k \geq 0$ and each integer $n \geq \phi(k)$ we have $\rho(a_n, a) \leq (k+1)^{-1}$
and

(ii) ϕ is a Cauchy modulus of $\{a_n\}_{n=0}^{\infty}$ if it is a Cauchy sequence and for each integer $k \geq 0$, each integer $n \geq \phi(k)$ and each integer $p \geq 0$, we have $\rho(a_{n+p}, a_n) \leq (k+1)^{-1}$.

Consider the series $\sum_{n=0}^{\infty} b_n$, where $\{b_n\}_{n=0}^{\infty}$ is a sequence of nonnegative numbers. We say that it diverges with the rate of divergence $\theta : \mathcal{N} \to \mathcal{N}$ if $\sum_{i=0}^{\theta(n)} b_i \geq n$ for all integers $n \geq 0$.

Furthermore, if $\sum_{n=0}^{\infty} b_n$ converges, then a a Cauchy modulus of the series, is a Cauchy modulus of the sequence $\{\sum_{i=0}^{n} b_i\}_{n=0}^{\infty}$ of the partial sums.

A sequence $\{z_n\}_{n=0}^{\infty} \subset Y$ is asymptotical regular if $\lim_{n\to\infty} \rho(z_n, z_{n+1}) = 0$. A rate of asymptotic regularity of $\{z_n\}_{n=0}^{\infty}$ is the rate of convergence of $\{\rho(z_n, z_{n+1})\}_{n=0}^{\infty}$ toward 0.

Let $S : Y \to Y$ and $\{z_n\}_{n=0}^{\infty} \subset Y$. We say $\{z_n\}_{n=0}^{\infty} \subset Y$ is S-asymptotically regular if $\lim_{n\to\infty} \rho(z_n, S(z_n)) = 0$. A rate of S-asymptotic regularity of $\{z_n\}_{n=0}^{\infty}$ is the rate of convergence of $\{\rho(z_n, S(z_n))\}_{n=0}^{\infty}$ toward 0.

For the proof of the next result, see Lemma 2.9 of [63].

Proposition 7.2 *Let $\{s_n\}_{n=0}^{\infty}$, $\{c_n\}_{n=0}^{\infty} \subset [0, \infty)$, $\{a_n\}_{n=0}^{\infty} \subset [0, 1]$ satisfy for each integer $n \geq 0$*

$$s_{n+1} \leq (1 - a_n)s_n + c_n,$$

L is a natural number, $s_n \leq L$, $n = 0, 1, \dots$; $\sum_{n=0}^{\infty} a_n$ diverges with the rate of divergence θ; $\sum_{n=0}^{\infty} c_n$ converges with Cauchy modulus χ. Then $\lim_{n\to\infty} s_n = 0$ *with the rate of convergence Σ defined for each integer $n \geq 0$, by*

$$\Sigma(k) = \theta(\chi(2k + 1) + 1 + \lfloor \log (2L(k + 1)) \rfloor + 1).$$

Let $m \geq 2$ be an integer

$$\{\alpha_{i,n}\}_{n=0}^{\infty} \subset (0, 1),\ i = 1, \dots, m. \tag{7.1}$$

In the sequel we use the following assumptions.

(Q1) $\lim_{n\to\infty} \alpha_{1,n} = 0$ with the rate of convergence σ_1.
(Q2) $\sum_{n=0}^{\infty} \alpha_{1,n}$ diverges with the rate of divergence σ_2.
(Q3) $\sum_{n=0}^{\infty} |\alpha_{1,n+1} - \alpha_{1,n}|$ converges with a Cauchy modulus σ_3.
(Q4) $\sum_{n=0}^{\infty} |\alpha_{i,n+1} - \alpha_{i,n}|$ converges with a Cauchy modulus σ_4 for $i = 2, \dots, m$.
(Q5) $\Lambda \geq 2$ is an integer such that $\Lambda^{-1} \leq \alpha_{i,n} \leq 1 - \Lambda^{-1}$ for each $i \in \{1, \dots, m\}$ and each integer $n \geq 0$.

Assume that $C \subset X$ is a nonempty convex set. In other words, for each $x, y \in C$ and each $\lambda \in [0, 1]$,

$$\lambda x \oplus (1 - \lambda)y \in C.$$

Assume that $m \geq 2$ is a natural number, for $i = 1, \dots, m$, $T_i : C \to C$ is a nonexpansive mapping such that

$$d(T_i(x), T_i(y)) \leq d(x, y),\ x, y \in C,\ i = 1, \dots, m, \tag{7.2}$$

$$\alpha_{i,n} \in (0, 1),\ n = 0, 1, \dots,\ i \in \{1, \dots, m\}.$$

Fix

$$\theta \in C,\ M_0 > 0.$$

Assume that the following assumption holds.

(A1) For each $\epsilon \in (0, 1)$ there exists $p_\epsilon \in B(\theta, M_0) \cap C$ such that

$$d(p_\epsilon, T_i(p_\epsilon)) \le \epsilon,\ i = 1, \dots, m.$$

Now we describe our algorithm.
Initialization: Let $u, x_0 \in C$. Define

$$x_1 = (1 - \alpha_{1,0})T_1(x_0) \oplus \alpha_{1,0}u \tag{7.3}$$

and for each $i \in \{2, \dots, m\}$,

$$x_i = (1 - \alpha_{i,0})T_1(x_{i-1}) \oplus \alpha_{i,0}x_{i-1}. \tag{7.4}$$

Iterative step: Assume that $k \ge 1$ is an integer and $x_{km} \in C$ is defined, and define

$$x_{km+1} = (1 - \alpha_{1,k})T_1(x_{km}) \oplus \alpha_{1,k}u \tag{7.5}$$

and for each $i \in \{2, \dots, m\}$,

$$x_{km+i} = (1 - \alpha_{i,k})T_i(x_{km+i-1}) \oplus \alpha_{i,k}x_{km+i-1}. \tag{7.6}$$

Fix

$$M \ge M_0 + d(x_0, \theta) + 2 + d(u, \theta) + d(u, T_1(\theta)). \tag{7.7}$$

We will prove the following result which describes the asymptotic regularity of the sequence $\{x_i\}_{i=0}^\infty$.

Theorem 7.3 *Assume that (Q1)–(Q5) hold. Then for each $i \in \{1, \dots, m\}$, the sequence $\{x_{km+i}\}_{k=0}^\infty$ is asymptotically regular and the following assertions hold:*

1. *Let $\bar{K} > 2M + 1$, for each integer $k \ge 0$,*

$$\chi(k) = \max\{\sigma_3(2\bar{K}(k+1)m), \sigma_4(2(k+1)\bar{K}m)\},$$

$$\Gamma_1(k) = \sigma_2(\chi(2k+1) + 2 + \lfloor \log(4\bar{K}(k+1)) \rfloor),$$

$$\Gamma_2(k) = \max\{\Gamma_1(2k+2), \sigma_3(4(k+1)\bar{K}m)\}.$$

   Then Γ_1 is the rate of asymptotic regularity of $\{x_{km}\}_{k=0}^\infty$, and Γ_2 is the rate of asymptotic regularity of $\{x_{km+i}\}_{k=0}^\infty$ for each $i \in \{1, \dots, m-1\}$.
2. *Assume that $\sigma : \{0, 1, \dots\} \to \{0, 1, \dots\}$ is the rate of asymptotic regularity of $\{x_{km}\}_{k=0}^\infty$ and let for each integer $k \ge 0$,*

$$\Gamma(k) = \max\{\sigma_1(\lfloor M(4k+4)\Lambda^2\eta(M+1, (M+1)^{-1}(k+1)^{-1})^{-1} \rfloor + 1),$$

$$\sigma\lfloor(4k+4)\Lambda^2\eta(M+1,(M+1)^{-1}(k+1)^{-1})^{-1}\rfloor+2),$$

$$\sigma(\lfloor(2k+2)(M+1)\rfloor+1)\}.$$

Then for each $i \in \{1, \dots, m\}$ and each $j \in \{0, \dots, m-1\}$, the sequence $\{d(x_{nm+j}, T_i(x_{nm+j}))\}_{n=0}^{\infty}$ has the rate of convergence $\{\Gamma((k+1)(3m+1))\}_{k=0}^{\infty}$.

Example 7.4 Assume that $T : C \to C$ is nonexpansive map without fixed points and that for each $\epsilon \in (0, 1)$, there exists $p_\epsilon \in B(\theta, M_0) \cap C$ such that

$$d(p_\epsilon, T(p_\epsilon)) \le \epsilon.$$

For each $\lambda \in (0, 1)$ define

$$T_\lambda(x) = \lambda T(x) \oplus (1-\lambda)x,\ x \in X.$$

Then for each $\lambda \in (0, 1)$ and each $\epsilon > 0$,

$$d(p_\epsilon, T_\lambda(p_\epsilon)) \le \lambda d(T(p_\epsilon), p_\epsilon) \le \lambda\epsilon$$

and (A1) holds for each finite subset of $\{T_\lambda : \lambda \in (0, 1)\}$.

7.2 Auxiliary Results

Lemma 7.5 *For each integer $k \ge 0$ and for each $i \in \{2, \dots, m\}$,*

$$d(x_{km+1}, x_{(k+1)m+1})$$

$$\le (1-\alpha_{1,k+1})d(x_{km}, x_{(k+1)m}) + |\alpha_{1,k} - \alpha_{1,k+1}|d(T_1(x_{km}), u),$$

$$d(x_{km+i}, x_{(k+1)m+i})$$

$$\le d(x_{km+i-1}, x_{(k+1)m+i-1}) + |\alpha_{i,k} - \alpha_{i,k+1}|d(x_{km+i-1}, T_i(x_{km+i-1})).$$

Proof Let $k \ge 0$ be an integer. Proposition 2.1 and (7.2)–(7.6) imply that

$$d(x_{km+1}, x_{(k+1)m+1})$$

$$= d((1-\alpha_{1,k})T_1(x_{km}) \oplus \alpha_{1,k}u, (1-\alpha_{1,k+1})T_1(x_{(k+1)m}) \oplus \alpha_{1,k+1}u)$$

$$\le (1-\alpha_{1,k+1})d(T_1(x_{km}), T_1(x_{(k+1)m})) + |\alpha_{1,k} - \alpha_{1,k+1}|d(T_1(x_{km}), u)$$

$$\le (1-\alpha_{1,k+1})d(x_{km}, x_{(k+1)m}) + |\alpha_{1,k} - \alpha_{1,k+1}|d(T_1(x_{km}), u).$$

Proposition 2.1 and (7.2)–(7.6) imply that for each $i \in \{2, \dots, m\}$,

$$d(x_{km+i}, x_{(k+1)m+i})$$

$$= d((1-\alpha_{i,k})T_i(x_{km+i-1}) \oplus \alpha_{i,k}x_{km+i-1},$$

$$(1-\alpha_{i,k+1})T_i(x_{(k+1)m+i-1}) \oplus \alpha_{i,k+1}x_{(k+1)m+i-1})$$

$$\le (1-\alpha_{i,k+1})d(T_i(x_{km+i-1}), T_i(x_{(k+1)m+i-1}))$$

$$+ \alpha_{i,k+1}d(x_{km+i-1}, x_{(k+1)m+i-1})$$

$$+ |\alpha_{i,k} - \alpha_{i,k+1}|d(x_{km+i-1}, T_i(x_{km+i-1}))$$

$$\le (1-\alpha_{i,k+1})d(x_{km+i-1}, x_{(k+1)m+i-1}) + \alpha_{i,k+1}d(x_{km+i-1}, x_{(k+1)m+i-1})$$

$$+ |\alpha_{i,k} - \alpha_{i,k+1}|d(x_{km+i-1}, T_i(x_{km+i-1}))$$

$$\le d(x_{km+i-1}, x_{(k+1)m+i-1})$$

$$+ |\alpha_{i,k} - \alpha_{i,k+1}|d(x_{km+i-1}, T_i(x_{km+i-1})).$$

Lemma 7.5 is proved.

Lemma 7.6 *Let $k \ge 0$ be an integer, $\epsilon \in (0, 1)$,*

$$p_\epsilon \in C \cap B(\theta, M_0), \tag{7.8}$$

and

$$d(p_\epsilon, T_i(p_\epsilon)) \le \epsilon,\ i = 1, \dots, m. \tag{7.9}$$

Then

$$d(x_{km+1}, p_\epsilon) \le (1-\alpha_{1,k})d(x_{km}, p_\epsilon) + (1-\alpha_{1,k})\epsilon + \alpha_{1,k}(d(u, \theta) + M_0)$$

and for each $j \in \{2, \dots, m\}$,

$$d(x_{km+j}, p_\epsilon) \le d(x_{km+j-1}, p_\epsilon) + (1-\alpha_{j,k})\epsilon$$

and

$$d(x_{km+j}, p_\epsilon) \le (1-\alpha_{1,k})d(x_{km}, p_\epsilon) + \epsilon\sum_{i=1}^{j}(1-\alpha_{i,k}) + \alpha_{1,k}(d(u, \theta) + M_0).$$

Proof Proposition 2.1 and (7.2), (7.3), (7.5), (7.8), and (7.9) imply that

$$d(x_{km+1}, p_\epsilon) = d((1-\alpha_{1,k})T_1(x_{km}) \oplus \alpha_{1,k}u, p_\epsilon)$$

$$\leq (1-\alpha_{1,k})d(T_1(x_{km}), p_\epsilon) + \alpha_{1,k}d(u, p_\epsilon)$$

$$\leq (1-\alpha_{1,k})d(T_1(x_{km}), T_1(p_\epsilon)) + (1-\alpha_{1,k})d(T_1(p_\epsilon), p_\epsilon)$$

$$+ \alpha_{1,k}(d(u,\theta) + M_0)$$

$$\leq (1-\alpha_{1,k})d(x_{km}, p_\epsilon) + (1-\alpha_{1,k})\epsilon + \alpha_{1,k}(d(u,\theta) + M_0). \tag{7.10}$$

Proposition 2.1 and (7.2), (7.4), (7.6), and (7.9) imply that for each $i \in \{2, \dots, m\}$,

$$d(x_{km+i}, p_\epsilon) = d((1-\alpha_{i,k})T_i(x_{km+i-1}) \oplus \alpha_{i,k}x_{km+i-1}, p_\epsilon)$$

$$\leq (1-\alpha_{i,k})d(T_i(x_{km+i-1}), p_\epsilon) + \alpha_{i,k}d(x_{km+i-1}, p_\epsilon)$$

$$\leq (1-\alpha_{i,k})d(T_i(x_{km+i-1}), T_i(p_\epsilon)) + (1-\alpha_{ik})d(T_i(p_\epsilon), p_\epsilon)$$

$$+ \alpha_{i,k}d(x_{km+i-1}, p_\epsilon)$$

$$\leq (1-\alpha_{i,k})d(x_{km+i-1}, p_\epsilon) + (1-\alpha_{i,k})\epsilon + \alpha_{i,k}d(x_{km+i-1}, p_\epsilon)$$

$$\leq d(x_{km+i-1}, p_\epsilon) + (1-\alpha_{i,k})\epsilon.$$

Let $j \in \{2, \dots, m\}$. By (7.10),

$$d(x_{km+j}, p_\epsilon) - d(x_{km}, p_\epsilon)$$

$$= \sum_{i=1}^{j}(d(x_{km+i}, p_\epsilon) - d(x_{km+i-1}, p_\epsilon))$$

$$= d(x_{km+1}, p_\epsilon) - d(x_{km}, p_\epsilon) + \sum_{i=2}^{j}(d(x_{km+i}, p_\epsilon) - d(x_{km+i-1}, p_\epsilon))$$

$$\leq -\alpha_{1,k}d(x_{km}, p_\epsilon) + (1-\alpha_{1,k})\epsilon + \alpha_{1,k}(d(u,\theta) + M_0) + \epsilon\sum_{i=2}^{j}(1-\alpha_{i,k})$$

$$\leq -\alpha_{1,k}d(x_{km}, p_\epsilon) + +\epsilon\sum_{i=1}^{j}(1-\alpha_{i,k}) + \alpha_{1,k}(d(u,\theta) + M_0).$$

Lemma 7.6 is proved.

Lemma 7.7 *Let $s \geq 0$ be an integer, $\epsilon \in (0, 1)$,*

$$p_\epsilon \in C \cap B(\theta, M_0),$$

$$d(p_\epsilon, T_i(p_\epsilon)) \leq \epsilon, \ i = 1, \dots, m \tag{7.11}$$

and that for each $k \in \{0, \dots, s\}$,

$$\alpha_{1,k} > \epsilon \sum_{i=1}^{m} (1 - \alpha_{i,k}). \tag{7.12}$$

Then for each $i \in \{0, \dots, m\}$,

$$d(x_i, p_\epsilon) \leq M, \ d(x_i, \theta) \leq M + M_0. \tag{7.13}$$

Proof We show that (7.13) holds for each $i \in \{0, \dots, m\}$ by induction. In view of (7.7) and (7.11), $d(x_0, p_\epsilon) \leq M - 1$ and (7.13) holds for $i = 0$. Assume that $k \in \{0, \dots, s\}$ and (7.13) holds for $i = k$. Lemma 7.6 and (7.7) and (7.12) imply that

$$d(x_{km}, p_\epsilon) \leq M,$$

and for all $j = 1, \dots, m - 1$,

$$d(x_{km+j+1}, p_\epsilon) \leq (1 - \alpha_{1,k}) d(x_{km}, p_\epsilon) + \epsilon \sum_{i=1}^{m} (1 - \alpha_{i,k}) + \alpha_{1,k}(d(u, \theta) + M_0)$$

$$\leq (1 - \alpha_{1,k}) M + \epsilon \sum_{i=1}^{m} (1 - \alpha_{i,k}) + \alpha_{1,k}(d(u, \theta) + M_0)$$

$$\leq M - \alpha_{1,k}(M - d(u, \theta) - M_0) + \epsilon \sum_{i=1}^{m} (1 - \alpha_{i,k})$$

$$\leq M - \alpha_{1,k} + \epsilon \sum_{i=1}^{m} (1 - \alpha_{i,k}) \leq M.$$

This completes the proof of Lemma 7.7.

Proposition 7.8 *For each integer $k \geq 0$ and for each $i \in \{2, \dots, m\}$,*

$$d(x_{km+1}, x_{(k+1)m+1})$$

$$\leq (1-\alpha_{1,k+1})d(x_{km}, x_{(k+1)m}) + |\alpha_{1,k} - \alpha_{1,k+1}|(M + M_0 + 1 + d(\theta, u))$$

$$\leq (1-\alpha_{1,k+1})d(x_{km}, x_{(k+1)m}) + 2M|\alpha_{1,k} - \alpha_{1,k+1}|,$$

$$d(x_{km+i}, x_{(k+1)m+i}) \leq d(x_{km+i-1}, x_{(k+1)m+i-1}) + (2M+1)|\alpha_{i,k} - \alpha_{i,k+1}|$$

and

$$d(x_{(k+1)m}, x_{(k+2)m})$$

$$\leq (1-\alpha_{1,k+1})d(x_{km}, x_{(k+1)m}) + (2M+1)\sum_{i=1}^{m} |\alpha_{i,k} - \alpha_{i,k+1}|.$$

Proof Note that the third inequality in the statement of the proposition follows from the first and second inequalities. Let $k \geq 0$ be an integer. Lemma 7.5 implies that

$$d(x_{km+1}, x_{(k+1)m+1}) \leq (1-\alpha_{1,k+1})d(x_{km}, x_{(k+1)m})$$
$$+ |\alpha_{1,k} - \alpha_{1,k+1}|d(T_1(x_{km}), u) \tag{7.14}$$

and for each $i \in \{2, \ldots, m\}$,

$$d(x_{km+i}, x_{(k+1)m+i}) \leq d(x_{km+i-1}, x_{(k+1)m+i-1})$$
$$+ |\alpha_{i,k} - \alpha_{i,k+1}|d(x_{km+i-1}, T_i(x_{km+i-1})). \tag{7.15}$$

Assume that $\epsilon \in (0, 1)$, for each $q \in \{0, \ldots, k+1\}$,

$$\alpha_{1,q} > \epsilon \sum_{i=1}^{m} (1 - \alpha_{q,k})$$

and

$$p_\epsilon \in C \cap B(\theta, M_0)$$

is as guaranteed by (A1). Hence (7.11) is true. Lemma 7.7 implies that

$$d(x_i, p_\epsilon) \leq M, \; i = 1, \ldots, (k+1)m. \tag{7.16}$$

By (7.2), (7.7), (7.11), and (7.16),

$$d(T_1(x_{km}), u) \leq d(T_1(x_{km}), T_1(p_\epsilon)) + d(T_1(p_\epsilon), p_\epsilon) + d(p_\epsilon, \theta) + d(\theta, u)$$

$$\leq d(x_{km}, p_\epsilon) + 1 + M_0 + d(\theta, u)$$

$$\leq M + 1 + M_0 + d(\theta, u) \leq 2M. \tag{7.17}$$

In view of (7.2), (7.11), and (7.16), for each $i \in \{2, \ldots, m\}$,

$$d(x_{km+i-1}, T_i(x_{km+i-1}))$$

$$\leq d(x_{km+i-1}, p_\epsilon) + d(p_\epsilon, T_i(p_\epsilon))$$

$$+ d(T_i(p_\epsilon), T_i(x_{km+i-1}))$$

$$\leq 2d(x_{km+i-1}, p_\epsilon) + 1 \leq 2M + 1. \tag{7.18}$$

It follows from (7.14) and (7.17) that

$$d(x_{km+1}, x_{(k+1)m+1})$$

$$\leq (1 - \alpha_{1,k+1})d(x_{km}, x_{(k+1)m}) + |\alpha_{1,k} - \alpha_{1,k+1}|(M + 1 + M_0 + d(\theta, u))$$

$$\leq (1 - \alpha_{1,k+1})d(x_{km}, x_{(k+1)m}) + 2M|\alpha_{1,k} - \alpha_{1,k+1}|.$$

Equations (7.15) and (7.18) imply that for $i \in \{2, \ldots, m\}$,

$$d(x_{km+i}, x_{(k+1)m+i}) \leq d(x_{km+i-1}, x_{(k+1)m+i-1}) + (2M + 1)|\alpha_{i,k} - \alpha_{i,k+1}|.$$

This completes the proof of Proposition 7.8.

Let $\bar{K} \geq M + 1$ be an integer.

Proposition 7.9 *Assume that (Q2)–(Q4) hold, for each integer $k \geq 0$,*

$$\chi(k) = \max\{\sigma_3(2\bar{K}(k + 1)m), \sigma_4(2(k + 1)\bar{K}m)\}. \tag{7.19}$$

Then $\{x_{km}\}_{k=0}^{\infty}$ is asymptotically regular with the rate Γ_1 defined for each integer $k \geq 0$ by

$$\Gamma_1(k) = \sigma_2(\chi(2k + 1) + 2 + \lfloor \log(4\bar{K}(k + 1)) \rfloor) \tag{7.20}$$

and for each $i \in \{1, \ldots, m - 1\}$, $\{x_{km+i}\}_{k=0}^{\infty}$ is asymptotically regular with the rate Γ_2 defined for each integer $k \geq 0$ by

$$\Gamma_2(k) = \max\{\Gamma_1(2k + 2), \sigma_3(4(k + 1)\bar{K}m)\}. \tag{7.21}$$

Proof In order to apply Proposition 7.2, we set for each integer $n \geq 0$

$$s_n = d(x_{nm}, x_{(n+1)m}),\ a_n = \alpha_{1,n+1},\ c_n = 2\bar{K}\sum_{i=1}^{m} |\alpha_{i,n} - \alpha_{i,n+1}|.$$

In view of the choice of $\bar{K}$ and Lemma 7.7, for each integer $n \ge 0$,

$$s_n \le 2\bar{K}.$$

Proposition 7.8 implies that for each integer $n \ge 0$,

$$s_{n+1} \le (1 - a_n)s_n + c_n.$$

In view of (Q2), $\sum_{n=0}^{\infty} \alpha_{1,n}$ diverges with the rate of divergence σ_2. By (Q3) and (Q4), $\sum_{n=0}^{\infty} c_n$ converges. We define its Cauchy modulus. Let $k \ge 0$ be an integer. (Q3) and (Q4) and (7.19) imply that for each $i \in \{1, \dots, m\}$,

$$\sum_{n=\chi(k)}^{\infty} |\alpha_{i,n} - \alpha_{i,n+1}| \le (2\bar{K}m(k+1))^{-1}$$

and

$$\sum_{n=\chi(k)}^{\infty} c_n \le (k+1)^{-1}.$$

Thus $\sum_{n=0}^{\infty} c_n$ converges with the Cauchy modulus χ. Proposition 7.2 implies that

$$\lim_{n\to\infty} d(x_{mn}, x_{m(n+1)}) = 0$$

with the modulus of convergence

$$\Gamma_1(k) = \sigma_2(\chi(2k+1) + 2 + \lfloor \log(4\bar{K}(k+1)) \rfloor),\ k = 0, 1, \dots.$$

Proposition 7.8 implies that for each $i \in \{1, \dots, m-1\}$, $\{x_{km+i}\}_{k=0}^{\infty}$ is a asymptotically regular with the modulus of convergence

$$\Gamma_2(k) = \max\{\Gamma_1(2k+2), \sigma_3(4(k+1)\bar{K}m)\}.$$

Proposition 7.9 is proved.

Proposition 7.10 *Let*

$$\alpha_{1,n} = 2(n+2)^{-1},\ \alpha_{i,n} = \beta \in (0,1)$$

for each integer $n \geq 0$ and each $i \in \{2, \dots, m\}$. Then for each integer $k \geq 0$ and each $i \in \{2, \dots, m\}$,

$$d(x_{mk}, x_{m(k+1)}) \leq (4M + 2)(n + 2)^{-1},$$

$$d(x_{mk+i}, x_{m(k+1)+i}) \leq (4M + 2)(n + 2)^{-1} + M(n + 2)^{-2}.$$

Proof Proposition 7.8 implies that for each integer $k \geq 0$,

$$d(x_{(k+1)m}, x_{(k+2)m})$$

$$\leq (1 - 2(n + 3)^{-1})d(x_{km}, x_{(k+1)m}) + (2M + 1)(2(n + 2)^{-1} - 2(n + 3)^{-1}).$$

We apply Lemma 2.2 with

$$s_k = d(x_{km}, x_{(k+1)m}),\ k = 0, 1, \dots,\ L = 2M + 1,\ N = J = 2,\ \gamma = 1,$$

$$a_n = \alpha_{1,n} = 2(n + 2)^{-1},\ c_n = 2M + 1,\ n = 0, 1, \dots$$

and obtain that for each integer $k \geq 0$,

$$d(x_{km}, x_{(k+1)m}) \leq (4M + 2)(n + 2)^{-1}$$

and in view of Proposition 7.8 for each $i \in \{2, \dots, m\}$,

$$d(x_{mk+i}, x_{m(k+1)+i}) \leq (4M + 2)(k + 2)^{-1} + M(k + 2)^{-2}.$$

Proposition 7.10 is proved.

7.3 Proof of Theorem 7.3

The first assertion of the theorem follows from Proposition 7.9. Let us prove the second assertion. Let $k \geq 0$ be an integer and $n \geq \Gamma(k)$ be an integer. We show that

$$d(x_{nm+i-1}, T_i(x_{nm+i-1})) \leq (k + 1)^{-1},\ i = 2, \dots, m. \tag{7.22}$$

Assume the contrary. Then there exists $j \in \{2, \dots, m\}$ such that

$$d(x_{nm+j-1}, T_j(x_{nm+j-1})) > (k + 1)^{-1}. \tag{7.23}$$

Let $\delta \in (0, 1)$ satisfy

$$\delta < 4^{-1}(k+1)^{-1}, \tag{7.24}$$

$$\alpha_{1,q} > \delta \sum_{i=1}^{m} (1-\alpha_{i,q}),\ q = 0, \ldots, n+2. \tag{7.25}$$

By (A1), there exists

$$p_\delta \in C \cap B(\theta, M_0) \tag{7.26}$$

such that

$$d(p_\delta, T_i(p_\delta)) \le \delta,\ i = 1, \ldots, m. \tag{7.27}$$

Lemma 7.6 and (7.24) and (7.26) imply that

$$d(x_{nm+j-1}, p_\delta) \le (1-\alpha_{1,n})d(x_{nm}, p_\delta) + \delta \sum_{i=1}^{j-1}(1-\alpha_{i,n}) + \alpha_{1,n}(d(u,\theta) + M_0). \tag{7.28}$$

In view of (7.6),

$$d(x_{nm+j}, p_\delta) = d((1-\alpha_{j,n})T_j(x_{nm+j-1}) \oplus \alpha_{j,n}x_{nm+j-1}, p_\delta). \tag{7.29}$$

Lemma 7.7 and (7.2) and (7.25)–(7.27) imply that

$$d(x_{nm+j-1}, p_\delta) \le M,$$

$$d(T_j(x_{nm+j-1}), p_\delta) \le d(x_{nm+j-1}, p_\delta) + \delta \le M + 1. \tag{7.30}$$

By (7.1), (7.2), (7.23), and (7.24),

$$(k+1)^{-1} < d(x_{nm+j-1}, T_j(x_{nm+j-1}))$$

$$\le d(T_j(x_{nm+j-1}), T_j(p_\delta)) + d(p_\delta, x_{nm+j-1}) + d(T_j(p_\delta), p_\delta)$$

$$\le 2d(p_\delta, x_{nm+j-1}) + \delta$$

and

$$d(p_\delta, x_{nm+j-1}) \ge 4^{-1}(k+1)^{-1}. \tag{7.31}$$

Applying Lemma 7.1 with

$$x = T_j(x_{nm+j-1}),\ y = x_{nm+j-1},\ a = p_\delta,\ \epsilon = ((M+1)(k+1))^{-1},$$

$$r = d(x_{nm+j-1}, p_\delta) + \delta,\ s = M+1,\ \lambda = \alpha_{j,n}$$

and using (Q5), (7.23), (7.29), (7.30), and (7.31), we have

$$d(x_{nm+j}, p_\delta)$$

$$\leq (1 - 2\alpha_{j,n}(1-\alpha_{j,n})\eta(M+1, ((M+1)(k+1))^{-1}))(d(x_{nm+j-1}, p_\delta) + \delta)$$

$$\leq \delta(1 - 2\alpha_{j,n}(1-\alpha_{j,n})\eta(M+1, ((M+1)(k+1))^{-1}))$$

$$+ d(x_{nm+j-1}, p_\delta) - 2d(x_{nm+j-1}, p_\delta)\alpha_{j,n}(1-\alpha_{j,n})\eta(M+1, ((M+1)(k+1))^{-1})$$

$$\leq d(x_{nm+j-1}, p_\delta) + \delta(1 - 2\alpha_{j,n}(1-\alpha_{j,n})\eta(M+1, ((M+1)(k+1))^{-1})$$

$$- 2^{-1}(k+1)^{-1}\Lambda^{-2}\eta(M+1, ((M+1)(k+1))^{-1}). \tag{7.32}$$

By (7.28) and (7.32),

$$d(x_{nm+j}, p_\delta) \leq d(x_{nm}, p_\delta) + \alpha_{1,n}(d(u,\theta) + M_0) + \delta\sum_{i=1}^{j-1}(1-\alpha_{i,n})$$

$$+ \delta(1 - 2\alpha_{j,n}(1-\alpha_{j,n})\eta(M+1, ((M+1)(k+1))^{-1}))$$

$$- 2^{-1}(k+1)^{-1}\Lambda^{-2}\eta(M+1, ((M+1)(k+1))^{-1}). \tag{7.33}$$

Lemma 7.6 and (7.25)–(7.27) and (7.33) imply that

$$d(x_{(n+1)m}, p_\delta) \leq d(x_{nm+j}, p_\delta) + \delta\sum_{i=1}^{m}(1-\alpha_{i,n})$$

$$\leq d(x_{nm}, p_\delta) + 2\delta\sum_{i=1}^{m}(1-\alpha_{i,n}) + \alpha_{1,n}(d(u,\theta) + M_0)$$

$$+ \delta(1 - 2\alpha_{j,n}(1-\alpha_{j,n})\eta(M+1, ((M+1)(k+1))^{-1}))$$

$$- 2^{-1}(k+1)^{-1}\Lambda^{-2}\eta(M+1, ((M+1)(k+1))^{-1}). \tag{7.34}$$

By (7.34),

$$d(x_{(n+1)m}, x_{nm}) \geq d(x_{nm}, p_\delta) - d(x_{(n+1)m}, p_\delta)$$

$$\geq 2^{-1}(k+1)^{-1}\Lambda^{-2}\eta(M+1,((M+1)(k+1))^{-1}) - \alpha_{1,n}(d(u,\theta)+M_0)$$

$$- 2\delta\sum_{i=1}^{m}(1-\alpha_{i,n}) - \delta(1-2\alpha_{j,n}(1-\alpha_{j,n})\eta(M+1,((M+1)(k+1))^{-1})).$$

Since δ is any positive number satisfying (7.24) and (7.25), the relation above implies that

$$d(x_{(n+1)m}, x_{nm})$$

$$\geq 2^{-1}(k+1)^{-1}\Lambda^{-2}\eta(M+1,((M+1)(k+1))^{-1}) - \alpha_{1,n}(d(u,\theta)+M_0). \quad (7.35)$$

In view of the definition of Γ and the inequality $n \geq \Gamma(k)$,

$$\alpha_{1,n} \leq 4^{-1}(k+1)^{-1}\Lambda^{-2}\eta(M+1,((M+1)(k+1))^{-1})M^{-1} \quad (7.36)$$

and together with (7.35) this implies that

$$d(x_{(n+1)m}, x_{nm}) \geq 4^{-1}(k+1)^{-1}\Lambda^{-2}\eta(M+1,((M+1)(k+1))^{-1}). \quad (7.37)$$

Since $n \geq \Gamma(k)$ relation (7.37) contradicts the definition of Γ. The contradiction we have reached proves that (7.22) holds for each $i \in \{2, \ldots, m\}$ and each integer $n \geq \Gamma(k)$.

Let $n \geq \Gamma(k)$ be an integer. By Proposition 2.1 and (7.22), for each integer $i \in \{2, \ldots, m\}$,

$$d(x_{nm+i-1}, x_{nm+i})$$

$$\leq d(x_{nm+i-1}, (1-\alpha_{i,n})T_i(x_{nm+i-1}) \oplus \alpha_{i,n}x_{nm+i-1})$$

$$\leq (1-\alpha_{i,n})d(T_i(x_{nm+i-1}), x_{nm+i-1}) \leq (k+1)^{-1}. \quad (7.38)$$

This implies that for each $i, j \in \{1, \ldots, m\}$,

$$d(x_{nm+i}, x_{nm+j}) \leq (m-1)(k+1)^{-1}. \quad (7.39)$$

By the definition of Γ and the relation $n \geq \Gamma(k)$,

$$d(x_{nm}, x_{(n+1)m}) \leq (k+1)^{-1}. \quad (7.40)$$

Equations (7.8) and (7.39) imply that $i, j \in \{0, \ldots, m\}$,

$$d(x_{nm+i}, x_{nm+j}) \leq m(k+1)^{-1}. \quad (7.41)$$

By Proposition 2.1, Lemma 7.7, the definition of Γ, the relation $n \geq \Gamma(k)$ and (7.2), (7.7),

$$d(x_{nm+1}, T_1(x_{nm})) = d((1-\alpha_{1,n})T_1(x_{nm}) \oplus \alpha_{1,n}u, T_1(x_{nm}))$$

$$\leq \alpha_{1,n}d(u, T_1(x_{nm})) \leq \alpha_{1,n}d(T_1(x_{nm}), T_1(\theta)) + \alpha_{1,n}d(T_1(\theta), u)$$

$$\leq \alpha_{1,n}(M + M_0 + d(T_1(\theta), u)) \leq 2M\alpha_{1,n} \leq (k+1)^{-1}. \tag{7.42}$$

It follows from (7.41) and (7.42) that

$$d(x_{nm}, T_1(x_{nm})) \leq d(x_{nm}, x_{nm+1}) + d(x_{nm+1}, T_1(x_{nm})) \leq (m+1)(k+1)^{-1}. \tag{7.43}$$

In view of (7.22) and (7.43),

$$d(x_{nm+i-1}, T_i(x_{nm+i-1})) \leq (m+1)(k+1)^{-1},\ i = 1, \ldots, m. \tag{7.44}$$

By (7.2), (7.41), and (7.44), for each $j \in \{0, \ldots, m\}$ and each $i \in \{1, \ldots, m\}$,

$$d(x_{nm+j}, T_i(x_{nm+j}))$$

$$\leq d(x_{nm+j}, x_{nm+i-1}) + d(T_i(x_{nm+i-1}), x_{nm+i-1}) + d(T_i(x_{nm+i-1}), T_i(x_{nm+j}))$$

$$\leq m(k+1)^{-1} + (m+1)(k+1)^{-1} + m(k+1)^{-1} = (3m+1)(k+1)^{-1}.$$

Thus for each integer $k \geq 0$, each integer $n \geq \Gamma(k)$, each $j \in \{0, \ldots, m\}$ and each $i \in \{1, \ldots, m\}$,

$$d(x_{nm+j}, T_i(x_{nm+j})) \leq (3m+1)(k+1)^{-1}$$

and therefore for each integer $n \geq \Gamma((k+1)(3m+1))$, each $j \in \{0, \ldots, m\}$ and each $i \in \{1, \ldots, m\}$,

$$d(x_{nm+j}, T_i(x_{nm+j})) \leq (k+1)^{-1}.$$

Theorem 7.3 is proved.

7.4 The First Result on Krasnoselskii-Mann Iterations

We continue to use the notation and definitions and assume the assumptions introduced in Sect. 7.1. In particular, $X = (X, d, W)$ is a W-hyperbolic space equipped with the structure (X, η), where $\eta : (0, \infty) \times (0, 2] \to (0, 1]$ is the modulus of uniform convexity and that $C \subset X$ is a nonempty convex set. Assume that m is a natural number, for $i = 1, \dots, m$, $T_i : C \to C$ is a nonexpansive mapping satisfying (7.2) for each $x, y \in C$ and each $i \in \{1, \dots, m\}$. Fix

$$\theta \in C, \ M_0 > 0$$

and assume that assumption (A1) holds.

Assume that $\bar{N} \geq m$ is a natural number. Denote by $\mathcal{R}$ the set of all mappings $r : \{0, 1, \dots\} \to \{1, \dots, m\}$ such that for each integer $k \geq 0$,

$$\{1, \dots, m\} \subset \{r(i) : \ i \in \{k, \dots, k + \bar{N} - 1\}. \tag{7.45}$$

Now we describe our algorithm. Let

$$\Lambda \in (0, 2^{-1}).$$

Initialization: Choose

$$x \in C \cap B(\theta, M_0), \ r \in \mathcal{R} \tag{7.46}$$

and set

$$x_0 = x.$$

Iterative step: For each integer $i \geq 0$ choose

$$\alpha_i \in (\Lambda, 1 - \Lambda) \tag{7.47}$$

and set

$$x_{i+1} = (1 - \alpha_i)x_i \oplus \alpha_i T_{r(i)}(x_i). \tag{7.48}$$

Theorem 7.11 *Let q be a natural number. Then*

$$Card(\{n \in \{0, 1, \dots\} : \ \max\{d(x_n, T_s(x_n)) : \ s \in \{1, \dots, m\}\} > q^{-1}\})$$

$$\leq 4q(1-\Lambda)^{-1}\Lambda^{-1}(4M_0+1)\bar{N}(2\bar{N}+1)\eta(4M_0+1, q^{-1}(2\bar{N}+1)^{-1}(4M_0+1)^{-1})^{-1}.$$

7.5 Auxiliary Results

Proposition 7.12 *Assume that $r \in \mathcal{R}$, (4.46) holds and that for each integer $i \geq 0$ (7.47) and (7.48) hold. Then for each integer $k \geq 0$ and each $j \in \{1, \dots, m\}$,*

$$d(x_k, \theta),\ d(T_j(x_k), \theta) \le 3M_0.$$

Proof Fix $\epsilon \in (0, 1)$. By (A1), there exists

$$p_\epsilon \in C \cap B(\theta, M_0) \tag{7.49}$$

such that

$$d(p_\epsilon, T_i(p_\epsilon)) \le \epsilon,\ i = 1, \dots, m. \tag{7.50}$$

Let $k \ge 0$ be an integer. Proposition 2.1, (7.2), and (7.48) imply that

$$\begin{aligned} d(x_{k+1}, p_\epsilon) &= d((1-\alpha_k)x_k \oplus \alpha_k T_{r(k)}(x_k), p_\epsilon) \\ &\le (1-\alpha_k)d(x_k, p_\epsilon) + \alpha_k d(T_{r(k)}(x_k), p_\epsilon) \\ &\le (1-\alpha_k)d(x_k, p_\epsilon) + \alpha_k d(T_{r(k)}(x_k), T_{r(k)}(p_\epsilon)) + \alpha_k d(T_{r(k)}(p_\epsilon), p_\epsilon) \\ &\le (1-\alpha_k)d(x_k, p_\epsilon) + \alpha_k d(x_k, p_\epsilon) + \epsilon \\ &\le d(x_k, p_\epsilon) + \epsilon. \end{aligned} \tag{7.51}$$

In view of (7.2) and (7.50), for each $j \in \{1, \dots, m\}$,

$$\begin{aligned} d(T_j(x_k), p_\epsilon) &\le d(T_j(x_k), T_j(p_\epsilon)) + d(T_j(p_\epsilon), p_\epsilon) \\ &\le d(x_k, p_\epsilon) + \epsilon. \end{aligned} \tag{7.52}$$

It follows from (7.46), (7.49), (7.51), and (7.52) that

$$d(x_k, p_\epsilon) \le d(x_0, p_\epsilon) + k\epsilon \le 2M_0 + k\epsilon,$$

for each $j \in \{1, \dots, m\}$,

$$d(T_j(x_k), p_\epsilon) \le 2M_0 + k\epsilon + \epsilon, \tag{7.53}$$

$$d(x_k, \theta) \le d(x_k, p_\epsilon) + d(p_\epsilon, \theta) \le 2M_0 + k\epsilon + M_0, \tag{7.54}$$

for each $j \in \{1, \dots, m\}$,

$$d(T_j(x_k), \theta) \le d(T_j(x_k), p_\epsilon) + d(p_\epsilon, \theta) \le 3M_0 + (k+1)\epsilon. \tag{7.55}$$

Since ϵ is any element of $(0, 1)$, it follows from (7.54) and (7.55) that for each integer $k \ge 0$,

$$d(x_k, \theta) \le 3M_0$$

and

$$d(T_j(x_k), \theta) \le 3M_0, \ j = 1, \ldots .m.$$

Proposition 7.12 is proved.

Lemma 7.13 *Assume that k is a natural number, $n \ge 0$ is an integer,*

$$d(x_n, T_{r(n)}) \ge k^{-1}, \tag{7.56}$$

$$\delta \in (0, (4k)^{-1}\alpha_n(1-\alpha_n)\eta(4M_0+1, k^{-1}(4M_0+1)^{-1})), \tag{7.57}$$

$$p_\delta \in C \cap B(\theta, M_0), \tag{7.58}$$

$$d(p_\delta, T_j(p_\delta)) \le \delta, \ j = 1, \ldots, m. \tag{7.59}$$

Then

$$d(x_{n+1}, p_\delta) \le d(x_n, p_\delta) - (4k)^{-1}\eta(4M_0+1, k^{-1}(4M_0+1)^{-1})\alpha_n(1-\alpha_n).$$

Proof By (7.2), (7.56), (7.57), and (7.59),

$$k^{-1} \le d(x_n, T_{r(n)}(x_n))$$

$$\le d(x_n, p_\delta) + d(p_\delta, T_{r(n)}(p_\delta)) + d(T_{r(n)}(p_\delta), T_{r(n)}(x_n))$$

$$\le d(x_n, p_\delta) + \delta + d(x_n, p_\delta)$$

and

$$d(x_n, p_\delta) \ge 2^{-1}(k^{-1} - \delta) \ge (4k)^{-1}. \tag{7.60}$$

Proposition 7.12 and (7.2), (7.56), and (7.58) imply that

$$d(x_n, T_{r(n)}(x_n)) \ge k^{-1} \ge k^{-1}(4M_0+1)^{-1}(d(x_n, p_\delta) + \delta) \tag{7.61}$$

and

$$d(p_\delta, T_{r(n)}(x_n)) \le d(p_\delta, T_{r(n)}(p_\delta)) + d(T_{r(n)}(p_\delta), T_{r(n)}(x_n))$$

$$\le \delta + d(x_n, p_\delta).$$

Thus

$$d(x_n, p_\delta),\ d(p_\delta, T_{r(n)}(x_n)) \le \delta + d(x_n, p_\delta). \tag{7.62}$$

In view of (7.56) and (7.62), we apply Lemma 7.1 with

$$x = x_n,\ y = T_{r(n)}(x_n),\ a = p_\delta,$$

$$r = \delta + d(x_n, p_\delta),\ \epsilon = k^{-1}(4M_0 + 1)^{-1},\ s = 4M_0 + 1$$

and combining with (7.48), (7.57), and (7.60) obtain that

$$d(x_{n+1}, p_\delta) = d((1 - \alpha_n)x_n \oplus \alpha_n T_{r(n)}(x_n), p_\delta)$$

$$\le (1 - 2\alpha_n(1 - \alpha_n)\eta(4M_0 + 1, k^{-1}(4M_0 + 1)^{-1}))(\delta + d(x_n, p_\delta))$$

$$\le \delta + d(x_n, p_\delta) - 2d(x_n, p_\delta)\alpha_n(1 - \alpha_n)\eta(4M_0 + 1, k^{-1}(4M_0 + 1)^{-1})$$

$$\le \delta + d(x_n, p_\delta) - (2k)^{-1}\alpha_n(1 - \alpha_n)\eta(4M_0 + 1, k^{-1}(4M_0 + 1)^{-1})$$

$$\le d(x_n, p_\delta) - (4k)^{-1}\alpha_n(1 - \alpha_n)\eta(4M_0 + 1, k^{-1}(4M_0 + 1)^{-1}).$$

Lemma 7.13 is proved.

Lemma 7.14 *Let k, Q be natural numbers. Then*

$$Card(\{n \in \{0, \dots, Q\} : d(x_n, T_{r(n)})(x_n) \ge k^{-1}\})$$

$$\le 4k(4M_0 + 1)\eta(4M_0 + 1, k^{-1}(4M_0 + 1)^{-1})^{-1}\Lambda^{-1}(1 - \Lambda)^{-1}.$$

Proof Choose a positive number δ satisfying

$$\delta < Q^{-1},$$

$$\delta < (4k)^{-1}\eta(4M_0 + 1, k^{-1}(4M_0 + 1)^{-1}) \min\{\alpha_i(1 - \alpha_i) : i = 0, \dots, Q\}. \tag{7.63}$$

Set

$$\Delta = (4k)^{-1}\Lambda(1 - \Lambda)\eta(4M_0 + 1, k^{-1}(4M_0 + 1)^{-1}), \tag{7.64}$$

$$E = \{n \in \{0, \dots, Q\} : d(x_n, T_{r(n)}(x_n)) \ge k^{-1}\}. \tag{7.65}$$

Let p_δ satisfies (7.58) and (7.59). Lemma 7.12 and (7.47) and (7.63)–(7.65) imply that for each integer $n \in E$,

$$d(x_{n+1}, p_\delta) \le d(x_n, p_\delta) - \Delta. \tag{7.66}$$

Proposition 2.1 and (7.48) and (7.59) imply that for each $n \in \{0, \dots, Q\}$,

$$d(x_{n+1}, p_\delta) \le d((1-\alpha_n)x_n \oplus \alpha_n T_{r(n)}(x_n), p_\delta)$$

$$\le (1-\alpha_n)d(x_n, p_\delta) + \alpha_n d(T_{r(n)}(x_n), p_\delta)$$

$$\le (1-\alpha_n)d(x_n, p_\delta) + \alpha_n d(T_{r(n)}(x_n), T_{r(n)}(p_\delta)) + \delta$$

$$\le d(x_n, p_\delta) + \delta. \tag{7.67}$$

Proposition 7.1 and (7.58), (7.63), (7.66), and (7.67) imply that

$$4M_0 \ge d(x_0, p_\delta) \ge d(x_0, p_\delta) - d(x_{Q+1}, p_\delta)$$

$$= \sum_{n=0}^{Q} (d(x_n, p_\delta) - d(x_{n+1}, p_\delta))$$

$$= \sum_{n\in E} (d(x_n, p_\delta) - d(x_{n+1}, p_\delta))$$

$$+ \sum \{d(x_n, p_\delta) - d(x_{n+1}, p_\delta) : \ n \in \{0, \dots, Q\} \setminus E\}$$

$$\ge \Delta \mathrm{Card}(E) - \delta Q$$

and

$$\Delta \mathrm{Card}(E) \le 4M_0 + \delta Q \le 4M_0 + 1.$$

By the relation above and (7.64),

$$\mathrm{Card}(E) \le (4M_0 + 1)\Delta^{-1}$$

$$\le 4k(4M_0+1)\eta(4M_0+1, k^{-1}(4M_0+1)^{-1})^{-1}\Lambda^{-1}(1-\Lambda)^{-1}.$$

Lemma 7.14 is proved.

Corollary 7.15 *For each integer $k \ge 1$,*

$$Card(\{n \in \{0, 1, \dots\} : \ d(x_n, T_{r(n)}(x_n)) \ge k^{-1}\})$$

$$\le 4k(4M_0+1)\eta(4M_0+1, k^{-1}(4M_0+1)^{-1})^{-1}\Lambda^{-1}(1-\Lambda)^{-1}.$$

7.6 Proof of Theorem 7.11

Set

$$k = q(2\bar{N} + 1) \tag{7.68}$$

and

$$E_0 = \{n \in \{0, 1, \dots\} : d(x_n, T_{r(n)}(x_n)) \geq k^{-1}\}. \tag{7.69}$$

Corollary 7.15 implies that

$$\mathrm{Card}(E_0) \leq 4k(4M_0+1)\eta(4M_0+1, k^{-1}(4M_0+1)^{-1})^{-1}\Lambda^{-1}(1-\Lambda)^{-1}. \tag{7.70}$$

Set

$$E_1 = \{n \in \{0, 1, \dots\} : [n, n + \bar{N} - 1] \cap E_0 \neq \emptyset\}. \tag{7.71}$$

By (7.70) and (7.71),

$$\mathrm{Card}(E_1) \leq \bar{N}\mathrm{Card}(E_0)$$

$$\leq 4k\bar{N}(4M_0 + 1)\eta(4M_0 + 1, k^{-1}(4M_0 + 1)^{-1})^{-1}\Lambda^{-1}(1 - \Lambda)^{-1}.$$

Assume that

$$n \in \{0, 1, \dots\} \setminus E_1.$$

In view of (7.71),

$$[n, n + \bar{N} - 1] \cap E_0 = \emptyset. \tag{7.72}$$

By (7.69) and (7.72), for each $i \in \{n, \dots, n + \bar{N} - 1\}$,

$$d(x_i, T_{r(i)}(x_i)) \leq k^{-1} \tag{7.73}$$

and combining with Proposition 2.1 and (7.48) and (7.73), we obtain that

$$d(x_i, x_{i+1}) = d(x_i, (1-\alpha_i)x_i \oplus \alpha_i T_{r(i)}(x_i)) \leq \alpha_i d(x_i, T_{r(i)}(x_i)) \leq k^{-1}. \tag{7.74}$$

It follows from (7.74) that for each $i, j \in \{n, \dots, n + \bar{N}\}$,

$$d(x_i, x_j) \leq \bar{N}k^{-1}. \tag{7.75}$$

Let $s \in \{1, \dots, m\}$. Equations (7.45) and (7.75) imply that there exists

$$i \in \{n, \dots, n + \bar{N} - 1\} \tag{7.76}$$

such that

$$r(i) = s. \tag{7.77}$$

By (7.2), (7.73), (7.75), and (7.77),

$$d(x_n, T_s(x_n)) = d(x_n, T_{r(i)}(x_n))$$

$$\leq d(x_n, x_i) + d(x_i, T_{r(i)}(x_i)) + d(T_{r(i)}(x_i), T_{r(i)}(x_n))$$

$$\leq 2d(x_i, x_n) + k^{-1} + 2\bar{N}k^{-1} + k^{-1} \leq k^{-1}(2\bar{N} + 1).$$

Thus for each integer $n \in \{0, 1, \dots\} \setminus E_1$ and each $s \in \{1, \dots, m\}$,

$$d(x_n, T_s(x_n)) \leq k^{-1}(2\bar{N} + 1) = q^{-1}.$$

Theorem 7.11 is proved.

7.7 The Second Result on Krasnoselskii-Mann Iterations

We continue to assume that $X = (X, d, W)$ is a W-hyperbolic space equipped with the structure (X, η), where $\eta : (0, \infty) \times (0, 2] \to (0, 1]$ is the modulus of uniform convexity and that $C \subset X$ is a nonempty convex set. Let

$$\theta \in C, \ M_0 > 0.$$

Assume that m is a natural number, for $i = 1, \dots, m$, $T_i : C \to C$ and that the following assumption holds.

(A2) For each $\delta \in (0, 1)$ there exists $p_\delta \in B(\theta, M_0) \cap C$ such that for each $i \in \{1, \dots, m\}$ and each $x \in C$,

$$d(T_i(x), p_\delta) \leq d(x, p_\delta) + \delta.$$

Let $\delta \in (0, 1)$ and p_δ be as guaranteed by (A2). Clearly, for $i = 1, \dots, m$,

$$d(T_i(p_\delta), p_\delta) \leq \delta. \tag{7.78}$$

Assume that $\bar{N} \geq m$ is a natural number. Recall that $\mathcal{R}$ is the set of all mappings $r : \{0, 1, \dots\} \to \{1, \dots, m\}$ such that for each integer $k \geq 0$,

$$\{1, \dots, m\} \subset \{r(i) : i \in \{k, \dots, k + \bar{N} - 1\}\}. \tag{7.79}$$

Assume that

$$r \in \mathcal{R}, \ \{\alpha_i\}_{i=0}^{\infty} \subset (0, 1), \tag{7.80}$$

$$x_0 = x \in B(\theta, M_0) \cap C \tag{7.81}$$

and that for each integer $n \geq 0$,

$$x_{n+1} = (1 - \alpha_n)x_n \oplus \alpha_n T_{r(n)}(x_n). \tag{7.82}$$

We prove the following result which shows how many iterates should be done in order to obtain an approximate solution of our common fixed point problem.

Theorem 7.16 *Assume that*

$$\sum_{n=0}^{\infty} \min\{\alpha_i(1 - \alpha_i) : i \in \{n\bar{N}, \dots, (n+1)\bar{N} - 1\}\} = \infty.$$

Let $\epsilon \in (0, 1)$, $\gamma = \bar{N}^{-1}\epsilon$ and a natural number Q satisfy

$$\sum_{n=0}^{Q} \min\{\alpha_i(1 - \alpha_i) : i \in \{n\bar{N}, \dots, (n+1)\bar{N} - 1\}\}$$

$$> 16M_0\gamma^{-1}\eta(4M_0 + 1, \gamma(4M_0 + 1)^{-1})^{-1}.$$

Then there exists an integer $n \in \{0, \dots, Q\}$ such that for each

$$i \in \{n\bar{N}, \dots, (n+1)\bar{N} - 1\},$$

we have

$$d(x_i, T_{r(i)}(x_i)) \leq \epsilon\bar{N}^{-1},$$

for each $i, j \in \{n\bar{N}, \dots, (n+1)\bar{N}\}$,

$$d(x_i, x_j) \leq \epsilon$$

and for each $i \in \{n\bar{N}, \dots, (n+1)\bar{N}\}$ and each $s \in \{1, \dots, m\}$,

$$B(x_i, \epsilon) \cap \{z \in C : d(T_s(z), z) \leq \epsilon\bar{N}^{-1}\} \neq \emptyset.$$

7.8 Auxiliary Results

Proposition 7.17 *For each integer $k \ge 0$ and each $j \in \{1, \dots, m\}$,*

$$d(x_k, \theta),\ d(T_j(x_k), \theta) \le 3M_0.$$

Proof Fix $\delta \in (0, 1)$ and let p_δ be as guaranteed by (A2). Proposition 2.1, (A2), and (7.81) and (7.82) imply that for each integer $k \ge 0$ and each $j \in \{1, \dots, m\}$,

$$d(x_{k+1}, p_\epsilon) = d((1-\alpha_k)x_k \oplus \alpha_k T_{r(k)}(x_k), p_\delta)$$

$$\le (1-\alpha_k)d(x_k, p_\delta) + \alpha_k d(T_{r(k)}(x_k), p_\delta)$$

$$\le (1-\alpha_k)d(x_k, p_\delta) + \alpha_k d(x_k, p_\delta) + \delta$$

$$\le d(x_k, p_\delta) + \delta,$$

$$d(T_j(x_k), p_\delta) \le d(x_k, p_\delta) + \delta,$$

$$d(x_k, p_\delta) \le d(x_0, p_\delta) + k\delta,$$

$$d(T_j(x_k), p_\delta) \le d(x_0, p_\delta) + (k+1)\delta,$$

$$d(x_k, \theta) \le d(x_k, p_\delta) + d(p_\delta, \theta) \le d(x_0, p_\delta) + k\delta + d(p_\delta, \theta)$$

$$\le d(x_0, \theta) + d(\theta, p_\delta) + d(p_\delta, \theta) + k\delta \le 3M_0 + k\delta,$$

$$d(T_j(x_k), \theta) \le d(T_j(x_k), p_\delta) + d(p_\delta, \theta)$$

$$\le d(x_0, p_\delta) + d(p_\delta, \theta) + (k+1)\delta \le 3M_0 + (k+1)\delta.$$

Since δ is any element of $(0, 1)$ we conclude that for each integer $k \ge 0$ and each $j \in \{1, \dots, m\}$,

$$d(x_k, \theta) \le 3M_0, d(T_j(x_k), \theta) \le 3M_0.$$

Proposition 7.17 is proved.

Lemma 7.18 *Assume that $\gamma \in (0, 1)$, $n \ge 0$ is an integer,*

$$d(x_n, T_{r(n)}(x_n)) \ge \gamma, \tag{7.83}$$

$$\delta \in (0, 4^{-1}\gamma\alpha_n(1-\alpha_n)\eta(4M_0+1, \gamma(4M_0+1)^{-1})) \tag{7.84}$$

and that p_δ is as guaranteed by (A2). Then

$$d(x_{n+1}, p_\delta) \le d(x_n, p_\delta) - 4^{-1}\gamma\eta(4M_0 + 1, \gamma(4M_0 + 1)^{-1})\alpha_n(1 - \alpha_n).$$

Proof By (A2) and (7.83) and (7.84),

$$\gamma \le d(x_n, T_{r(n)}(x_n))$$

$$\le d(x_n, p_\delta) + d(p_\delta, T_{r(n)}(p_\delta)) \le d(x_n, p_\delta) + d(x_n, p_\delta) + \delta$$

and

$$d(x_n, p_\delta) \ge 2^{-1}(\gamma - \delta) \ge 4^{-1}\gamma. \tag{7.85}$$

Proposition 7.17, (A2), and (7.83) imply that

$$d(x_n, T_{r(n)}(x_n)) \ge \gamma \ge \gamma(4M_0 + 1)^{-1}(\delta + d(x_n, p_\delta)). \tag{7.86}$$

In view of (A2),

$$d(p_\delta, T_{r(n)}(x_n)) \le \delta + d(x_n, p_\delta). \tag{7.87}$$

In view of (7.86) and (7.87), we apply Lemma 7.1 with

$$x = x_n, \ y = T_{r(n)}(x_n), \ a = p_\delta,$$

$$r = \delta + d(x_n, p_\delta), \ \epsilon = \gamma(4M_0 + 1)^{-1}, \ s = 4M_0 + 1$$

and combining with (7.82), (7.84), and (7.85) obtain that

$$d(x_{n+1}, p_\delta) = d((1 - \alpha_n)x_n \oplus \alpha_n T_{r(n)}(x_n), p_\delta)$$

$$\le (1 - 2\alpha_n(1 - \alpha_n)\eta(4M_0 + 1, \gamma(4M_0 + 1)^{-1}))(\delta + d(x_n, p_\delta))$$

$$\le \delta + d(x_n, p_\delta) - 2d(x_n, p_\delta)\alpha_n(1 - \alpha_n)\eta(4M_0 + 1, \gamma(4M_0 + 1)^{-1})$$

$$\le \delta + d(x_n, p_\delta) - 2^{-1}\alpha_n(1 - \alpha_n)\gamma\eta(4M_0 + 1, \gamma(4M_0 + 1)^{-1})$$

$$\le d(x_n, p_\delta) - 4^{-1}\alpha_n(1 - \alpha_n)\gamma\eta(4M_0 + 1, \gamma(4M_0 + 1)^{-1}).$$

Lemma 7.18 is proved.

7.9 Proof of Theorem 7.16

Choose a positive number δ satisfying

$$\delta < (8\bar{N})^{-1}\alpha_i(1-\alpha_i)\gamma\eta(4M_0+1, \gamma(4M_0+1)^{-1}), \ i = 0, \dots, \bar{N}Q. \tag{7.88}$$

Let p_δ be as guaranteed by (A2). We show that there exists an integer $n \in \{0, \dots, Q\}$ such that

$$d(x_i, T_{r(i)}(x_i)) \leq \gamma, \ i \in \{n\bar{N}, \dots, (n+1)\bar{N} - 1.$$

Assume the contrary. Then for each integer $n \in \{0, \dots, Q\}$,

$$\max\{d(x_i, T_{r(i)}(x_i)) : i \in \{n\bar{N}, \dots, (n+1)\bar{N} - 1\}\} > \gamma. \tag{7.89}$$

Let $n \in \{0, \dots, Q\}$. Then (7.89) holds. Proposition 2.1, (A2) and (7.82) imply that for each $i \in \{n\bar{N}, \dots, (n+1)\bar{N} - 1\}$,

$$\begin{aligned} d(x_{i+1}, p_\delta) &\leq d((1-\alpha_i)x_i \oplus \alpha_i T_{r(i)}(x_i), p_\delta) \\ &\leq (1-\alpha_i)d(x_i, p_\delta) + \alpha_i d(T_{r(i)}(x_i), p_\delta) \\ &\leq d(x_i, p_\delta) + \delta. \end{aligned} \tag{7.90}$$

By (7.89), there exists

$$j \in \{n\bar{N}, \dots, (n+1)\bar{N} - 1\} \tag{7.91}$$

such that

$$d(x_j, T_{r(j)}(x_j)) > \gamma. \tag{7.92}$$

Lemma 7.18 and (7.88), (7.92) imply that

$$d(x_{j+1}, p_\delta) \leq d(x_j, p_\delta) - 4^{-1}\gamma\eta(4M_0+1, \gamma(4M_0+1)^{-1})\alpha_j(1-\alpha_j). \tag{7.93}$$

It follows from (7.88), (7.90), (7.91), and (7.93) that

$$d(x_{n\bar{N}}, p_\delta) - d(x_{(n+1)\bar{N}}, p_\delta)$$

$$\sum_{i=n\bar{N}}^{(n+1)\bar{N}-1} (d(x_i, p_\delta) - d(x_{i+1}, p_\delta))$$

$$\geq -\delta\bar{N} + d(x_j, p_\delta) - d(x_{j+1}, p_\delta)$$

$$\geq 4^{-1}\gamma\eta(4M_0 + 1, \gamma(4M_0 + 1)^{-1})\alpha_j(1 - \alpha_j) - \delta\bar{N}$$

$$\geq 8^{-1}\gamma\eta(4M_0 + 1, \gamma(4M_0 + 1)^{-1})\alpha_j(1 - \alpha_j). \tag{7.94}$$

By (A2) and (7.81),

$$2M_0 \geq d(x_0, p_\delta) \geq d(x_0, p_\delta) - d(x_{(Q+1)\bar{N}}, p_\delta)$$

$$= \sum_{n=0}^{Q}(d(x_{n\bar{N}}, p_\delta) - d(x_{(n+1)\bar{N}}, p_\delta)$$

$$\geq 8^{-1}\gamma\eta(4M_0 + 1, \gamma(4M_0 + 1)^{-1})\sum_{n=0}^{Q}\min\{\alpha_i(1 - \alpha_i) :$$

$$i \in \{n\bar{N}, \ldots, (n + 1)\bar{N} - 1\}\}$$

and

$$\sum_{n=0}^{Q}\min\{\alpha_i(1 - \alpha_i) : \ i \in \{n\bar{N}, \ldots, (n + 1)\bar{N} - 1\}\}$$

$$\leq 16M_0\gamma^{-1}\eta(4M_0 + 1, \gamma(4M_0 + 1)^{-1}))^{-1}.$$

This contradicts our choice of Q. The contradiction we have reached proves that there exists $n \in \{0, \ldots, Q\}$ such that

$$d(x_i, T_{r(i)}(x_i)) \leq \gamma, \ i \in \{n\bar{N}, \ldots, (n + 1)\bar{N} - 1\}. \tag{7.95}$$

Proposition 2.1 and (7.82) and (7.95) imply that for each $i \in \{n\bar{N}, \ldots, (n+1)\bar{N}-1\}$,

$$d(x_i, x_{i+1}) = d(x_i, (1 - \alpha_i)x_i \oplus \alpha_i T_{r(i)}(x_i)) \leq \alpha_i d(x_i, T_{r(i)}(x_i)) \leq \gamma.$$

This implies that for each $i, j \in \{n\bar{N}, \ldots, (n + 1)\bar{N} - 1\}$,

$$d(x_i, x_j) \leq \bar{N}\gamma. \tag{7.96}$$

Let $k \in \{n\bar{N}, \ldots, (n + 1)\bar{N}\}$ and $s \in \{1, \ldots, m\}$. Equation (7.79) implies that there exists

$$i \in \{n\bar{N}, \ldots, (n + 1)\bar{N} - 1\}$$

such that

$$r(i) = s.$$

By (7.95) and (7.96),

$$d(x_i, T_s(x_i)) \le \gamma,$$

$$d(x_k, x_i) \le \bar{N}\gamma = \epsilon.$$

Theorem 7.16 is proved.

7.10 Approximate Common Fixed Points of Semigroups

In this section based on [21], we study the common fixed point set of nonexpansive semigroups of nonexpansive mappings. Assume that $(\mathcal{M}, d)$ is a metric space. Assume that there exists a family $\mathcal{F}$ of metric segments such that each two points $x, y \in \mathcal{M}$ are endpoints of a unique metric segment $[x, y] \in \mathcal{F}$ ($[x, y]$ is an isometric image of the real line interval $[0, d(x, y)]$). We will denote by $(1-\beta)x \oplus \beta y$ the unique point z of $[x, y]$ which satisfies

$$d(x, z) = \beta d(x, y) \text{ and } d(x, y) = (1-\beta)d(x, y),$$

where $\beta \in [0, 1]$. Such metric spaces are usually called convex metric spaces. If for all $p, x, y \in \mathcal{M}$,

$$d(2^{-1}p \oplus 2^{-1}x, 2^{-1}p \oplus 2^{-1}y) \le 2^{-1}d(x, y),$$

then $\mathcal{M}$ is a hyperbolic metric space [162]. We say that a subset C of a hyperbolic space $\mathcal{M}$ is convex if $[x, y] \subset C$ for each $x, y \in C$.

Assume that $(\mathcal{M}, d)$ is a hyperbolic metric space. We say that $\mathcal{M}$ is uniformly convex (in short, UC) if for each $a \in \mathcal{M}$, each $r > 0$ and each $\epsilon > 0$,

$$\delta(r, \epsilon) = \inf\{1 - r^{-1}d(2^{-1}x \oplus 2^{-1}y, a) \colon d(x, a) \le r,\ d(y, a) \le r,\ d(x, y) \ge r\epsilon\} > 0.$$

In this section we assume that $(\mathcal{M}, d)$ is a hyperbolic metric space, and if $(\mathcal{M}, d)$ is uniformly convex, then for each $s \ge 0$, each $\epsilon > 0$, there exists $\eta(s, \epsilon) > 0$ depending on s and ϵ such that

$$\delta(r, \epsilon) > \eta(s, \epsilon) > 0 \text{ for every } r > s.$$

The hyperbolic metric space $(\mathcal{M}, d)$ is strictly convex if for all $a, x, y \in \mathcal{M}$ satisfying

$$d(a, x) = d(a, y) = d(a\lambda x \oplus (1 - \lambda)y$$

we have $x = y$.

Note that for every fixed r, $\delta(r, 0) = 0$, $\delta(r, \epsilon)$ is an increasing function of ϵ, if $r_1 \le r_2$, then

$$1 - r_2 r_1^{-1}(1 - \delta(r_2, \epsilon r_1 r_2^{-1})) \le \delta(r_1, \epsilon)$$

and that if $(\mathcal{M}, d)$ is uniformly convex, then $(\mathcal{M}, d)$ is strictly convex [99, 101].

The next useful lemma was proved in [99, 101].

Lemma 7.19 *Assume that $(\mathcal{M}, d)$ is uniformly convex and that $\{C_n\}_{n=1}^{\infty}$ be a sequence of nonempty, convex, bounded, closed, and nonincreasing sets in $\mathcal{M}$. Let $x \in \mathcal{M}$ satisfy*

$$0 < d = \lim_{n\to\infty} d(x, C_n) < \infty.$$

Assume that $x_n \in C_n$ for every natural number n and $\lim_{n\to\infty} d(x, x_n) = d$. Then $\{x_n\}_{n=1}^{\infty}$ is a Cauchy sequence.

We say that a hyperbolic metric space $(\mathcal{M}, d)$ has property (R) if every nonincreasing sequence of nonempty, convex, bounded, and closed sets has a nonempty intersection. For the proof of the following two results, see also [99, 101].

Theorem 7.20 *Assume that $(\mathcal{M}, d)$ is complete and uniformly convex and that $C \subset \mathcal{M}$ is nonempty, closed and convex. Then for each $x \in \mathcal{M}$, there exists a unique $x_0 \in C$ such that $d(x, x_0) = d(x, C)$.*

Theorem 7.21 *Every complete and uniformly convex space $(\mathcal{M}, d)$ has the property (R).*

Assume that $(\mathcal{M}, d)$ is a hyperbolic metric space and C is its nonempty subset. We say that a family $\mathcal{F} = \{T_t : t \in [0, \infty)\}$ of self-mappings of C is a nonexpansive semigroup on C if the following conditions hold:

(i) $T_0(x) = x$ for all $x \in C$;
(ii) $T_{t+s}(x) = T_t(T_s(x))$ for each $x \in C$ and all $t, s \in [0, \infty)$;
(iii) for each $t \ge 0$ and each $x, y \in C$, we have

$$d(T_s(x), T_s(y)) \le d(x, y).$$

Define

$$\mathrm{Fix}(\mathcal{F}) = \cap\{\mathrm{Fix}(T_t) : t \in [0, \infty)\}.$$

The following result was obtained in [21].

Lemma 7.22 *Assume that $(\mathcal{M}, d)$ is a hyperbolic metric space which is strictly convex and that C is its nonempty subset.Assume that S, T and nonexpansive self-mappings of C such that for each $x, y \in C$,*

$$d(T(x), T(y)) \leq d(x, y), \ d(T(x), T(y)) \leq d(x, y),$$

$$Fix(T) \cap Fix(S) \neq \emptyset.$$

Then for each $\lambda \in (0, 1)$,

$$U(x) = \lambda S(x) \oplus (1 - \lambda)T(x), \ x \in C$$

is a nonexpansive mapping and

$$Fix(U) = Fix(T) \cap Fix(S).$$

Proof Clearly,

$$\mathrm{Fix}(T) \cap \mathrm{Fix}(S) \subset \mathrm{Fix}(U).$$

We show that

$$\mathrm{Fix}(U) \subset \mathrm{Fix}(T) \cap \mathrm{Fix}(S).$$

Assume that $x \in \mathrm{Fix}(U)$. Fix

$$a \in \mathrm{Fix}(T) \cap \mathrm{Fix}(S).$$

It is not difficult to see that

$$d(a, x) = d(a, \lambda S(x) \oplus (1 - \lambda)T(x)) \leq \lambda d(a, S(x)) + (1 - \lambda)d(a, T(x))$$

and

$$d(a, x) = d(a, \lambda S(x) \oplus (1 - \lambda)T(x)) \leq \lambda d(a, x) + (1 - \lambda)d(a, x) = d(a, x).$$

This implies that

$$d(a, S(x)) = d(a, T(x)) = d(a, x).$$

The strict convexity of M is implies that $S(x) = T(x)$. Together with the relation

$$x = \lambda S(x) \oplus (1 - \lambda)T(x)$$

this implies that $x \in \mathrm{Fix}(T) \cap \mathrm{Fix}(S)$. Lemma 7.22 is proved.

Assume that $(\mathcal{M}, d)$ is a metric space and C is its nonempty and closed subset. We say that a family $\mathcal{F} = \{T_t : t \in [0, \infty)\}$ of mappings from C to M is continuous if for any $x \in C$, the mapping $t \to T_t(x)$, $t \in [0, \infty)$ is continuous.

It is not difficult to see that the following result is true.

Proposition 7.23 *Assume that $(\mathcal{M}, d)$ is a metric space, C is its nonempty and closed subset, and $\mathcal{F} = \{T_t : t \in [0, \infty)\}$ is a semigroup of mappings from C into M which is continuous on C. Let A be a dense subset of $[0, \infty)$. Then $Fix(\mathcal{F}) = \cap\{Fix(T_t) : t \in A\}$.*

Denote by $\mathbf{Z}$ the set of all integers. In this section we use the following well-known result [21].

Lemma 7.24 *Let G be a nonempty additive subgroup of R^1. Then G is either dense in R^1 or there exists $a > 0$ such that $G = a \cdot \mathbf{Z} = \{an : n \in \mathbf{Z}\}$. Therefore, if $\alpha, \beta \in R^1$ and $\alpha\beta^{-1}$ is irrational, then the set*

$$G(\alpha, \beta) = \{\alpha n + \beta m : n, m\mathbf{Z}\}$$

is dense in R^1.

The next result was proved in [21].

Proposition 7.25 *Assume that $(\mathcal{M}, d)$ is a metric space, C is its nonempty and closed subset, and $\mathcal{F} = \{T_t : t \in [0, \infty)\}$ is a semigroup of mappings from C into M. Let $\alpha, \beta \in (0, \infty)$. Then*

$$\cap \{Fix(T_a) : a \in G_+(\alpha, \beta)\} = Fix(T_\alpha) \cap Fix(T_\beta),$$

where $G_+(\alpha, \beta) = \{\alpha n + \beta m : n, m \in \mathbf{Z}\} \cap [0, \infty)$.

Proof Clearly,

$$\cap \{\mathrm{Fix}(T_a) : a \in G_+(\alpha, \beta)\} \subset \mathrm{Fix}(T_\alpha) \cap \mathrm{Fix}(T_\beta).$$

Assume that

$$x \in \mathrm{Fix}(T_\alpha) \cap \mathrm{Fix}(T_\beta)$$

and that $a \in G_+(\alpha, \beta)$. There exist $n, m \in \mathbf{Z}$ such that $a = n\alpha + m\beta$. If $n, m > 0$, then

$$T_a(x) = T_{n\alpha+m\beta}(x) = T_{n\alpha}(T_{m\beta}(x)) = T_\alpha^n(T_\beta^m(x)) = x.$$

Otherwise, assume that $a = n\alpha - \tilde{m}\beta$, where $n, \tilde{m}$ are positive. It is clear that $a + \tilde{m}\beta = n\alpha$ and

$$T_a(x) = T_a(T_\beta^\beta(x)) = T_{a+\tilde{m}\beta}(x) = T_{n\alpha}(x) = T_a^n(x) = x.$$

Proposition 7.25 is proved.

Propositions 7.23 and 7.25 imply the following result.

Theorem 7.26 *Assume that* $(\mathcal{M}, d)$ *is a metric space,* C *is its nonempty and closed subset, and* $\mathcal{F} = \{T_t : t \in [0, \infty)\}$ *is a semigroup of mappings from* C *into* M *which is continuous on* C*. Let* $\alpha, \beta \in (0, \infty)$ *such that* $\alpha\beta^{-1}$ *is irrational. Then*

$$Fix(\mathcal{F}) = Fix(T_\alpha) \cap Fix(T_\beta).$$

Lemma 7.22 implies the following result.

Theorem 7.27 *Assume that* $(\mathcal{M}, d)$ *is a hyperbolic metric space which is strictly convex,* C *is its nonempty and closed subset, and* $\mathcal{F} = \{T_t : t \in [0, \infty)\}$ *is a semigroup of nonexpansive mappings from* C *into* M *which is continuous on* C*. Let* $\alpha, \beta \in (0, \infty)$ *be such that* $\alpha\beta^{-1}$ *is irrational. Then for every* $\lambda \in (0, 1)$*,*

$$Fix(\mathcal{F}) = Fix(\lambda T_\alpha \oplus (1 - \lambda)T_\beta).$$

Now we discuss the approximation of common fixed points of semigroups. Assume that $T : C \to C$ us a nonexpansive mapping and $\sigma \in (0, 1)$. The Mann iteration process generated by T and σ is defined by the iteration formula

$$x_{n+1} = \sigma T(x_n) \oplus (1 - \sigma)x_n, \tag{7.97}$$

where $x_1 \in X$.

For the proof of the following result, see [99, 101].

Lemma 7.28 *Assume that* $(\mathcal{M}, d)$ *is a uniformly convex hyperbolic metric space,* $r \geq 0$, $\sigma \in (0, 1)$ *and that*

$$\limsup_{n\to\infty} d(x_n, a) \leq r, \ \limsup_{n\to\infty} d(y_n, a) \leq r$$

and

$$\limsup_{n\to\infty} d(a, \sigma x_n \oplus (1 - \sigma)x_n, (1 - \sigma)y_n) = r.$$

Then $\lim_{n\to\infty} d(x_n, y_n) = 0$.

The next result was proved in [21].

Lemma 7.29 *Assume that* $(\mathcal{M}, d)$ *is a uniformly convex hyperbolic metric space and* C *is its nonempty, bounded, convex, and closed subset. Let* $T : C \to C$ *be nonexpansive,* $\sigma \in (0, 1)$ *and* x_n *be defined by* (7.97)*. Assume that* ω *is a fixed point og* T*. Then there exists* $\lim_{n\to\infty} d(x_n, \omega)$.

Proof We have

$$d(x_{n+1}, \omega) = d(\sigma T(x_n) \oplus (1-\sigma)x_n, \omega) \le \sigma d(T(x_n), \omega) + (1-\sigma)d(x_n, \omega)$$

$$= \sigma d(T(x_n), T(\omega)) + (1-\sigma)d(x_n, \omega)$$

$$\le \sigma(x_n, \omega) + (1-\sigma)d(x_n, \omega)$$

for any integer $n \ge 1$. This completes the proof of Lemma 7.29.

Assume that $(\mathcal{M}, d)$ is a metric space and C is its nonempty subset. Let $\{x_n\}_{n=1}^\infty$ be a bounded sequence in $\mathcal{M}$. Define $r(\cdot, \{x_n\}_{n=1}^\infty) : C \to [0, \infty)$ by

$$r(x, \{x_n\}_{n=1}^\infty) = \limsup_{n\to\infty} d(x, x_n).$$

The asymptotic radius ρ_C of $\{x_n\}_{n=1}^\infty$ with respect to C is

$$\rho_C = \inf\{r(x, \{x_n\}_{n=1}^\infty) : x \in C\}.$$

Set $\rho = \rho_{\mathcal{M}}$.

A point $\xi \in C$ is an asymptotic center of $\{x_n\}_{n=1}^\infty$ with respect to C if

$$\rho(\xi, \{x_n\}_{n=1}^\infty) = \rho(C, \{x_n\}_{n=1}^\infty) = \min\{r(x, \{x_n\}_{n=1}^\infty) : x \in C\}.$$

The set of all asymptotic centers of $\{x_n\}_{n=1}^\infty$ with respect to C is denoted by $A(C, \{x_n\}_{n=1}^\infty)$. Set

$$A(\{x_n\}_{n=1}^\infty) = A(\mathcal{M}, \{x_n\}_{n=1}^\infty).$$

Let $(\mathcal{M}, d)$ be a metric space. A bounded sequence $\{x_n\}_{n=1}^\infty$ in $\mathcal{M}$ Δ-converges to $x \in \mathcal{M}$ if x is the unique asymptotic center of every subsequence of $\{x_n\}_{n=1}^\infty$. We write

$$x = \Delta - \lim_{n\to\infty} x_n$$

whenever $\{x_n\}_{n=1}^\infty$ Δ-converges to x.

For the proofs of the following results see [99].

Lemma 7.30 *Assume that $(\mathcal{M}, d)$ is a uniformly convex hyperbolic metric space and C is its nonempty, convex, and closed subset. The every bounded sequence $\{x_n\}_{n=1}^\infty$ in $\mathcal{M}$ has a unique asymptotic center with respect to C.*

Lemma 7.31 *Assume that $(\mathcal{M}, d)$ is a uniformly convex hyperbolic metric space and C is its nonempty, convex, and closed subset. Let $\{x_n\}_{n=1}^\infty$ be a bounded sequence in C such that $A(\{x_n\}_{n=1}^\infty) = y$ in and $r(\{x_n\}_{n=1}^\infty) = \rho$. If $\{y_n\}_{n=1}^\infty \subset C$ and $\lim_{n\to\infty} r(y_n, \{x_k\}_{k=1}^\infty) = \rho$, then $\lim_{n\to\infty} y_n = y$.*

For the proof of the following result see [79].

Lemma 7.32 *Assume that $(\mathcal{M}, d)$ is a complete uniformly convex hyperbolic metric space and C is its nonempty, convex, and closed subset. Let $T : C \to C$ be a nonexpansive mapping and $\{x_n\}_{n=1}^{\infty} \subset X$ satisfy $\lim_{n\to\infty} d(x_n, T(x_n)) = 0$. If $x \in C$ is the asymptotic center of $\{x_n\}_{n=1}^{\infty}$ with respect to C, then $T(x) = x$. In particular, if $\{x_n\}_{n=1}^{\infty} \subset X$ satisfy $\lim_{n\to\infty} d(x_n, T(x_n)) = 0$ and $x = \Delta - \lim_{n\to\infty} x_n$, then $T(x) = x$.*

The next result was obtained in [21].

Theorem 7.33 *Assume that $(\mathcal{M}, d)$ is a uniformly convex hyperbolic metric space and C is its nonempty, convex, and closed subset. Let $T : C \to C$ be a nonexpansive mapping which has a fixed point, $\sigma \in (0, 1)$ $x_1 \in C$ and the sequence $\{x_n\}_{n=1}^{\infty} \subset X$ satisfy* (4.97). *Then $\lim_{n\to\infty} d(x_n, T(x_n)) = 0$.*

Proof Assume that ω is a fixed point of T. In view of Lemma 7.29, there exists

$$r = \lim_{n\to\infty} d(x_n, \omega). \tag{7.98}$$

Without loss of generality, we may assume that $r > 0$. Clearly,

$$\limsup_{n\to\infty} d(T(x_n), \omega) = \limsup_{n\to\infty} d(T(x_n), T(\omega)) \le \limsup_{n\to\infty} d(x_n, \omega) = r \tag{7.99}$$

and

$$\begin{aligned} r = \limsup_{n\to\infty} d(x_n, \omega) &= \limsup_{n\to\infty} d(\sigma T(x_n) \oplus (1-\sigma)x_n, \omega) \\ &\le \limsup_{n\to\infty} (\sigma d(T(x_n), \omega) + (1-\sigma)d(x_n, \omega)) \\ \le \limsup_{n\to\infty} (\sigma d(x_n, \omega) &+ (1-\sigma)d(x_n, \omega)) = \limsup_{n\to\infty} \sigma d(x_n, \omega) = r. \end{aligned}$$

Hence

$$\limsup_{n\to\infty} d(\sigma T(x_n) \oplus (1-\sigma)x_n, \omega) = r.$$

In view of Lemma 7.28 we conclude that $\lim_{n\to\infty} d(x_n, T(x_n)) = 0$.

The following two results were obtained in [21].

Theorem 7.34 *Assume that $(\mathcal{M}, d)$ is a uniformly convex hyperbolic complete metric space and C is its nonempty, convex, and closed subset. Let $T : C \to C$ be a nonexpansive mapping which has a fixed point, $\sigma \in (0, 1)$ $x_1 \in C$ and the sequence $\{x_n\}_{n=1}^{\infty} \subset X$ satisfy* (4.97). *Then $\{x_n\}_{n=1}^{\infty}$ Δ-converges to x which is a fixed point of T.*

Proof Note that $\mathrm{Fix}(T)$ is convex and closed [99]. Theorem 4.33 implies that $d(x_n, T(x_n)) = 0$. Let x be the unique asymptotic center of $\{x_n\}_{n=1}^{\infty}$. Lemma 7.32 implies that $T(x) = x$. Let us show that the sequence $\{x_n\}_{n=1}^{\infty}$ Δ-converges to x.

Let $\{x_{n_i}\}_{i=1}^{\infty}$ be a subsequence of $\{x_n\}_{n=1}^{\infty}$ and z be its unique asymptotic center. Since

$$\lim_{i\to\infty} d(x_{n_i}, T(x_{n_i})) = 0$$

we obtain that $z = T(z)$. Hence

$$\limsup_{i\to\infty} d(x_{n_i}, z) \le \limsup_{i\to\infty} d(x_{n_i}, x).$$

Since x, z are fixed points of T, we get

$$\limsup_{i\to\infty} d(x_{n_i}, z) = \lim_{n\to\infty} d(x_n, z) \text{ and } \limsup_{i\to\infty} d(x_{n_i}, x) = \lim_{n\to\infty} d(x_n, x).$$

Since X is the unique asymptotic center of $\{x_n\}_{n=1}^{\infty}$ we get $x = z$. This proves that $\{x_n\}_{n=1}^{\infty}$ Δ-converges to x.

Theorem 7.35 *Assume that $(\mathcal{M}, d)$ is a uniformly convex hyperbolic complete metric space and C is its nonempty, convex, and closed subset. Let $\mathcal{F} = \{T_t : t \in [0, \infty)\}$ is a continuous semigroup of nonexpansive mappings from C into C which has a common fixed point. Let $\alpha, \beta \in (0, \infty)$ be such that $\alpha\beta^{-1}$ is irrational and $\lambda, \mu \in (0, 1)$ satisfy $\lambda + \mu < 1$. Assume that $x_1 \in C$ and that the sequence $\{x_n\}_{n=1}^{\infty}$ is defined by*

$$x_{n+1} = (\mu + \lambda)(\mu T_\alpha(x_n) \oplus \lambda T_\beta(x_n)) \oplus (1 - \mu - \lambda)x_n \tag{7.100}$$

for every integer $n \ge 1$. Then $\{x_n\}_{n=1}^{\infty}$ Δ-converges to a common fixed point of the semigroup $\mathcal{F}$.

Proof Set

$$S = \mu(\mu + \lambda)^{-1} T_\alpha \oplus \lambda(\mu + \lambda)^{-1} T_\beta.$$

It is not difficult to see that the map S is nonexpansive and $\mathrm{Fix}(\mathcal{F}) \subset \mathrm{Fix}(S)$. Let $x_1 \in C$ and $\{x_n\}_{n=1}^{\infty}$ be the sequence generated by (7.100). Then

$$x_{n+1} = \sigma S(x_n) + (1 - \sigma)x_n,$$

where $\sigma = \mu+\lambda \in (0, 1)$. Clearly, $\{x_n\}_{n=1}^{\infty}$ is the same sequence generated by (7.97) for the map S. Theorem 7.34 implies that the sequence $\{x_n\}_{n=1}^{\infty}$ Δ-converges to some $x \in \mathrm{Fix}(S)$. By Theorem 7.27, $\mathrm{Fix}(\mathcal{F}) =\subset \mathrm{Fix}(S)$ and $x \in \mathrm{Fix}(\mathcal{F})$. Theorem 7.35 is proved.

Chapter 8
Perov Contraction Mappings

In this chapter, we study contraction mappings of Perov type in a generalized complete metric space. We show that these mappings possess a unique fixed point which attracts all (inexact) iterates of the mapping uniformly on bounded sets. We study the convergence of exact iterates as well as the asymptotic behavior of inexact iterates under the presence of computational errors.

8.1 Contractive Mappings of Perov Type

In this chapter, we study contraction mappings of Perov type in a generalized complete metric space. Our goal is to show that analogs of the results which hold for Banach contractions and Rakotch contractions hold for Perov contraction mappings too which is quite natural. In this section, we consider a contraction mapping of Perov type which maps a closed subset of a generalized complete metric space into the space. The study of this class of mappings is an important topic in the fixed point theory [146, 147, 149, 198]. We show the existence of a unique fixed point which attracts all (inexact) iterates of the mapping uniformly on bounded sets.

Let R^n be an n-dimensional Euclidean space. In other words,

$$R^n = \{x = (x_1, \dots, x_n) : x_i \in R^1, \ i = 1, \dots, n\}.$$

Let

$$R^n_+ = \{x = (x_1, \dots, x_n) \in R^n : x_i \geq 0, \ i = 1, \dots, n\}$$

and $e = (1, 1, \dots, 1) \in R^n$. We say that $x = (x_1, \dots, x_n), \ y = (y_1, \dots, y_n) \in R^n$ satisfy $x \leq y$ if $x_i \leq y_i$ for all $i = 1, \dots, n$. For each $x = (x_1, \dots, x_n) \in R^n$ set

A. J. Zaslavski, *Approximate Fixed Points of Nonexpansive Mappings*,
Developments in Mathematics 80, https://doi.org/10.1007/978-3-031-70710-0_8

$$\|x\|_1 = \sum_{i=1}^{n} |x_i|.$$

Let Y be a nonempty set and let $S : Y \to Y$. We denote by S^0 the identity mapping in Y, set $S^1 = S$ and for every integer $i \geq 0$ define

$$S^{i+1} = S \circ S^i.$$

We suppose that the sum over an empty set is zero.

Assume that X is a nonempty set and a function $d : X \times X \to R^n_+$ satisfies for each $x, y, z \in X$,

$$d(x, y) = 0 \text{ if and only if } x = y,$$

$$d(x, y) = d(y, x),$$

$$d(x, z) \leq d(x, y) + d(y, z).$$

The pair (X, d) is called a generalized metric space and d is called a generalized metric [146, 147, 149, 198].

For all $x, y \in X$ set

$$d(x, y) = (d_1(x, y), \ldots, d_n(x, y)).$$

Fix $\theta \in X$. For each $x \in X$ and each $r \in R^n_+$ set,

$$B(x, r) = \{y \in X : \ d(x, y) \leq r\}.$$

We say that a sequence $\{x_i\}_{i=0}^{\infty} \subset X$ converges to $x_* \in X$ if

$$\lim_{i \to \infty} \|d(x_i, x_*)\|_1 = 0.$$

We say that $\{x_i\}_{i=0}^{\infty} \subset X$ is a Cauchy sequence if for each $\epsilon > 0$ there exists a natural number $n(\epsilon)$ such that for each pair of integers $p, m \geq n(\epsilon)$

$$\|d(x_p, x_m)\|_1 \leq \epsilon.$$

The generalized metric space (X, d) is complete if every Cauchy sequence converges.

For all $x, y \in X$ set

$$\tilde{d}(x, y) = \|d(x, y)\|_1 = d_1(x, y) + \cdots + d_n(x, y).$$

Clearly, $(X, \tilde{d})$ is a metric space and convergence in (X, d) is equivalent to the convergence in $(X, \tilde{d})$.

We assume that the metric space $(X, \tilde{d})$ is complete.

Assume that $A : R^n_+ \to R^n_+$ and that the following properties hold:

(i) $A(0) = 0$ and A is continuous at zero;
(ii) for each $z_1, z_2 \in R^n_+$ satisfying $0 \le z_1 \le z_2$,

$$A(z_1) \le A(z_2);$$

(iii) for each $z_1, z_2 \in R^n_+$,

$$A(z_1 + z_2) \le A(z_1 + z_2);$$

(iv) $A^k(e) \to 0$ as $k \to \infty$;
(v) $A(\lambda z) = \lambda A(z)$ for each $\lambda \ge 0$ and each $z \in R^n_+$.

Assume that K is a nonempty closed subset of X and $T : K \to X$ satisfies for each $x, y \in K$,

$$d(T(x), T(y)) \le A(d(x, y)). \tag{8.1}$$

It is natural to call the mapping T as a generalized contraction [156, 168]. This class of mappings was introduced in [146]. A particular case when A is a linear mapping was introduced in [147].

It is easy to see that the following auxiliary result holds.

Lemma 8.1 *Let $x, y \in K$, $m \ge 1$ be an integer and let $T^m(x), T^m(y)$ exist. Then*

$$d(T^m(x), T^m(y)) \le A^m(d(x, y)).$$

Set

$$\Delta_1 = \sup\{\|A^i(e)\|_1 : i = 1, 2, \dots\} \tag{8.2}$$

(see property (iv)).

We will prove the following results obtained in [226]. The first and the second results provide sufficient conditions for the existences of a unique fixed point of T, while the third and fourth results shows the uniform convergence of iterates of T to its unique fixed point.

Theorem 8.2 *Assume that for each $\epsilon > 0$ there exists $x_\epsilon \in K$ such that*

$$d(x_\epsilon, T(x_\epsilon)) \le \epsilon e.$$

Then the following assertions hold:

1. *There exists a unique point* $x_T \in K$ *such that* $T(x_T) = x_T$.
2. *For each* $\epsilon > 0$ *there exists* $\delta > 0$ *such that if* $x \in K$ *satisfies* $d(x, T(x)) \le \delta e$, *then* $d(x, x_T) \le \epsilon e$.

Theorem 8.3 *Assume that* $c > 0$ *and that for each integer* $m \ge 1$ *there exists* $\{x_i^{(m)}\}_{i=0}^m \subset K$ *such that*

$$d(x_0^{(m)}, x_1^{(m)}) \le ce \tag{8.3}$$

and that for each integer $i \in \{0, \dots, m-1\}$,

$$d(x_{i+1}^{(m)}, T(x_i^{(m)})) \le m^{-1}e. \tag{8.4}$$

Then there exists a unique point $x_T \in K$ *such that* $T(x_T) = x_T$.

Theorem 8.4 *Assume that* $x_T \in K$ *satisfies* $T(x_T) = x_T$ *and* $c, \epsilon > 0$. *Then there exist* $\delta \in (0, \epsilon)$ *and a natural number* n_0 *such that for each integer* $m \ge n_0$ *and each sequence* $\{x_i\}_{i=0}^m \subset K$ *which satisfies*

$$d(x_0, x_T) \le ce,$$

$$d(x_{i+1}, T(x_i)) \le \delta e, \ i = 0, \dots, m-1$$

the inequality

$$d(x_i, x_T) \le \epsilon e$$

holds for all integers $i = n_0, \dots, m$.

Theorem 8.5 *Assume that* $x_T \in K$ *satisfies* $T(x_T) = x_T$ *and* $\epsilon > 0$. *Then there exists* $\delta \in (0, \epsilon)$ *such that for each integer* $k \ge 1$ *and each sequence* $\{x_i\}_{i=0}^k \subset K$ *which satisfies*

$$d(x_0, x_T) \le \delta e,$$

$$d(x_{i+1}, T(x_i)) \le \delta e, \ i = 0, \dots, k-1$$

the inequality

$$d(x_i, x_T) \le \epsilon e$$

holds for all integers $i = 0, \dots, k$.

8.2 Proof of Theorem 8.2

By property (iv), there exists a natural number $k_* > 4$ such that for each integer $k \geq k_*$,

$$A^k(e) \leq (8n)^{-1}e. \tag{8.5}$$

Proposition 8.6 *Let $\epsilon \in (0, 1)$, a positive number δ satisfy*

$$\delta(2k_* + 1)\sum_{i=0}^{k_*-1} A^i(e) \leq (4n)^{-1}\epsilon e \tag{8.6}$$

and let $\{x_i\}_{i=0}^{k_*}, \{y_i\}_{i=0}^{k_*} \subset K$ satisfy

$$d(x_0, T(x_0)) \leq \delta e, \ d(y_0, T(y_0)) \leq \delta e \tag{8.7}$$

and for each $i = 0, \dots, k_* - 1$

$$d(x_{i+1}, T(x_i)) \leq \delta e, \ d(y_{i+1}, T(y_i)) \leq \delta e. \tag{8.8}$$

Then

$$d(x_0, y_0) \leq \epsilon e.$$

Proof Assume that the proposition is not true. Then

$$\tilde{d}(x_0, y_0) > \epsilon. \tag{8.9}$$

In view of (8.7) and (8.8),

$$d(x_0, x_1) \leq d(x_0, T(x_0)) + d(T(x_0), x_1) \leq 2\delta e,$$

$$d(y_0, y_1) \leq d(y_0, T(y_0)) + d(T(y_0), y_1) \leq 2\delta e. \tag{8.10}$$

By (8.1), (8.8), and (8.10),

$$d(x_1, x_2) \leq d(x_1, T(x_0)) + d(T(x_0), T(x_1)) + d(T(x_1), x_2)$$

$$\leq 2\delta e + A(d(x_0, x_1)) \leq 2\delta e + A(2\delta e),$$

$$d(y_1, y_2) \leq d(y_1, T(y_0)) + d(T(y_0), T(y_1)) + d(T(y_1), y_2)$$

$$\leq 2\delta e + A(d(y_0, y_1)) \leq 2\delta e + A(2\delta e). \tag{8.11}$$

We show that for all $p = 0, \dots, k_* - 1$,

$$d(x_p, x_{p+1}) \leq \sum_{i=0}^{p} A^i(2\delta e), \tag{8.12}$$

$$d(y_p, y_{p+1}) \leq \sum_{i=0}^{p} A^i(2\delta e). \tag{8.13}$$

In view of (8.10) and (8.11), Eqs. (8.12) and (8.13) hold for $p = 0, 1$.

Assume that $p \in \{1, \dots, k_* - 2\}$ and that (8.12) and (8.13) hold. It follows from (8.1), (8.8), and (8.12) that

$$\begin{aligned} d(x_{p+1}, x_{p+2}) &\leq d(x_{p+1}, T(x_p)) + d(T(x_p), T(x_{p+1})) + d(T(x_{p+1}), x_{p+2}) \\ &\leq 2\delta e + A(d(x_p, x_{p+1})) \\ &\leq 2\delta e + A(\sum_{i=0}^{p} A^i(2\delta e)) \leq \sum_{i=0}^{p+1} A^i(2\delta e). \end{aligned}$$

Analogously, we show that

$$d(y_{p+1}, y_{p+2}) \leq \sum_{i=0}^{p+1} A^i(2\delta e).$$

Thus the assumption made for p also holds for $p + 1$. Thus we showed that (8.12) and (8.13) hold for all $p = 0, \dots, k_* - 1$. Equations (8.12) and (8.13) imply that

$$d(x_0, x_{k_*}) \leq \sum_{p=0}^{k_*-1} d(x_p, x_{p+1}) \leq \sum_{p=0}^{k_*-1} (\sum_{i=0}^{p} A^i(2\delta e)) \leq k_* \sum_{i=0}^{k_*-1} A^i(2\delta e) \tag{8.14}$$

and analogously

$$d(y_0, y_{k_*}) \leq k_* \sum_{i=0}^{k_*-1} A^i(2\delta e) \tag{8.15}$$

By (8.1) and (8.8),

$$\begin{aligned} d(x_1, y_1) &\leq d(x_1, T(x_0)) + d(T(x_0), T(y_0)) + d(T(y_0), y_1) \\ &\leq 2\delta e + A(d(x_0, y_0)). \end{aligned} \tag{8.16}$$

We show that for all $p = 1, \dots, k_*$,

$$d(x_p, y_p) \leq A^p(d(x_0, y_0)) + \sum_{i=0}^{p-1} A^i(2\delta e). \tag{8.17}$$

In view of (8.16), Eq. (8.17) holds for $p = 1$.

Assume that $p \in \{1, \dots, k_* - 1\}$ and (8.17) holds. It follows from (8.1), (8.8), and (8.17) that

$$d(x_{p+1}, y_{p+1}) \leq d(x_{p+1}, T(x_p)) + d(T(x_p), T(y_p)) + d(T(y_p), y_{p+1})$$

$$\leq 2\delta e + A(d(x_p, y_p)) \leq A^{p+1}(d(x_0, y_0)) + \sum_{i=0}^{p} A^i(2\delta e)$$

and (8.17) is true for $p+1$ too. Thus (8.17) holds for all integers $p = 0, \dots, k_*$ and

$$d(x_{k_*}, y_{k_*}) \leq A^{k_*}(d(x_0, y_0)) + \sum_{i=0}^{k_*-1} A^i(2\delta e). \tag{8.18}$$

By (8.5), (8.6), (8.14), (8.15), and (8.18),

$$d(x_0, y_0) \leq d(x_0, x_{k_*}) + d(x_{k_*}, y_{k_*}) + d(y_{k_*}, y_0)$$

$$\leq 2k_* \sum_{i=0}^{k_*-1} A^i(2\delta e) + \sum_{i=0}^{k_*-1} A^i(2\delta e) + A^{k_*}(d(x_0, y_0))$$

$$\leq 2\delta(2k_* + 1) \sum_{i=0}^{k_*-1} A^i(e) + \tilde{d}(x_0, y_0) A^{k_*}(e)$$

$$\leq (2n)^{-1}\epsilon e + (4n)^{-1}\tilde{d}(x_0, y_0)e,$$

$$\tilde{d}(x_0, y_0) \leq 2^{-1}\epsilon + 4^{-1}\tilde{d}(x_0, y_0)$$

and $\tilde{d}(x_0, y_0) \leq \epsilon$. Proposition 8.6 is proved.

Proof of Theorem 8.2 For each integer $k \geq 1$ there exists $x_k \in K$ such that

$$d(x_k, T(x_k)) \leq k^{-1}e. \tag{8.19}$$

By Proposition 8.6, $\{x_k\}_{k=1}^{\infty}$ is a Cauchy sequence in $(X, \tilde{d})$. There exists

$$x_T = \lim_{k \to \infty} x_k. \tag{8.20}$$

By (8.19) and (8.20), for each integer $k \geq 1$,

$$d(x_T, T(x_T)) \leq d(x_T, x_k) + d(x_k, T(x_k)) + d(T(x_k), T(x_T))$$

$$\leq d(x_T, x_k) + d(x_k, T(x_k)) + A(d(x_k, T(x_k)) \to 0$$

as $k \to \infty$. Thus $x_T = T(x_T)$. Now Assertion 2 follows from Proposition 8.6. The uniqueness of the fixed point of T follows from Assertion 2. Theorem 8.2 is proved.

8.3 Proof of Theorem 8.3

Let $\epsilon \in (0, 1)$. In view of Theorem 8.2, it is sufficient to show that there exists $x_\epsilon \in K$ such that

$$d(x_k, T(x_k)) \leq \epsilon e.$$

Property (iv) implies that there exists a natural number $n_0 > 4$ such that

$$cA^{n_0-1}(e) < 8^{-1}\epsilon e. \tag{8.21}$$

Choose an integer

$$m > n_0 + 2$$

such that

$$m^{-1} \sum_{i=0}^{n_0-1} A^i(e) < 8^{-1}\epsilon e. \tag{8.22}$$

We show that for all integers $p = 1, \ldots, m-1$,

$$d(x_p^{(m)}, T(x_p^{(m)})) \leq cA^p(e) + 2\sum_{i=0}^{p-1} A^i(m^{-1}e) - m^{-1}e. \tag{8.23}$$

By (8.1)–(8.4),

$$d(x_1^{(m)}, T(x_1^{(m)})) \leq d(x_1^{(m)}, T(x_0^{(m)})) + d(T(x_0^{(m)}), T(x_1^{(m)}))$$

$$\leq m^{-1}e + A(d(x_0^{(m)}, x_1^{(m)})) \leq m^{-1}e + A(ce)$$

and (8.23) holds for $p = 1$.

Assume that $p \in \{1, \dots, m-2\}$ and that (8.23) holds. By (8.1), (8.4), and (8.23),

$$d(x_{p+1}^{(m)}, T(x_{p+1}^{(m)})) \le d(x_{p+1}^{(m)}, T(x_p^{(m)})) + d(T(x_p^{(m)}), T(x_{p+1}^{(m)}))$$

$$\le m^{-1}e + A(d(x_p^{(m)}, x_{p+1}^{(m)}))$$

$$\le m^{-1}e + A(d(x_p^{(m)}, T(x_p^{(m)})) + d(T(x_p^{(m)}), x_{p+1}^{(m)}))$$

$$\le m^{-1}e + A(d(T(x_p^{(m)}), x_p^{(m)}) + m^{-1}e)$$

$$\le m^{-1}e + A(2\sum_{i=0}^{p-1} A^i(m^{-1}e)) + A^{p+1}(ce)$$

$$= cA^{p+1}(e) + 2\sum_{i=0}^{p} A^i(m^{-1}e) - m^{-1}e$$

and (8.23) holds for $p+1$ too. Thus (8.23) holds for all $p = 1, \dots, m-1$. In particular for $p = n_0 - 1$, in view of (8.21) and (8.22),

$$d(x_{n_0-1}^{(m)}, T(x_{n_0-1}^{(m)})) \le cA^{n_0-1}(e) + 2\sum_{i=0}^{n_0-1} A^i(m^{-1}e) < 8^{-1}\epsilon e + 8^{-1}\epsilon e < \epsilon e.$$

This completes the proof of Theorem 8.3.

8.4 Auxiliary Results for Theorems 8.4 and 8.5

Lemma 8.7 *Assume that $x_T \in K$ satisfies*

$$T(x_T) = x_T, \tag{8.24}$$

$\delta \in (0, 1)$, $m \ge 1$ *is an integer and* $\{x_i\}_{i=0}^m \subset K$ *satisfies*

$$d(x_{i+1}, T(x_i)) \le \delta,\ i = 0, \dots, m-1. \tag{8.25}$$

Then for all $p = 0, \dots, m$,

$$d(x_p, x_T) \le A^p(d(x_0, x_T)) + \sum_{i=0}^{p-1} A^i(\delta e). \tag{8.26}$$

Proof Clearly, (8.26) holds for $p = 0$. By (8.1), (8.24), and (8.25),

$$d(x_1, x_T) \le d(x_T, T(x_0)) + d(T(x_0), x_1) \le A(d(x_T, x_0)) + \delta e$$

and (8.26) holds for $p = 1$.

Assume that $p \in \{0, \dots, m-1\}$ and (8.26) holds. It follows from (8.1), (8.25), and (8.26) that

$$d(x_{p+1}, x_T) \le d(x_{p+1}, T(x_p)) + d(T(x_p), x_T)$$

$$\le \delta e + A(A^p(d(x_0, x_T)) + \sum_{i=0}^{p-1} A^i(\delta e))$$

$$\le A^{p+1}(d(x_0, x_T)) + \sum_{i=0}^{p} A^i(\delta e)$$

and (8.26) holds for $p + 1$. This completes the proof of Lemma 8.7.

Lemma 8.8 *Assume that $x_T \in K$ satisfies*

$$T(x_T) = x_T,$$

$c > 0$ and $\epsilon \in (0, 1)$. Then there exist a natural number k and $\delta \in (0, \epsilon)$ such that for each finite sequence $\{x_i\}_{i=0}^k \subset K$ which satisfies

$$d(x_0, x_T) \le ce, \tag{8.27}$$

$$d(x_{i+1}, T(x_i)) \le \delta e, \ i = 0, \dots, k-1 \tag{8.28}$$

the inequality

$$d(x_k, x_T) \le \epsilon e.$$

holds.

Proof Property (iv) implies that there exists a natural number k such that

$$cA^k(e) < 4^{-1}\epsilon e. \tag{8.29}$$

Choose a positive number $\delta \in (0, \epsilon)$ such that

$$\delta \sum_{i=0}^{k-1} A^i(e) \le 4^{-1}\epsilon e. \tag{8.30}$$

Assume that $\{x_i\}_{i=0}^k \subset K$ satisfies (8.27) and (8.28). By Lemma 8.7 and (8.27)–(8.30),

$$d(x_k, x_T) \leq A^k(d(x_0, x_T)) + \sum_{i=0}^{k-1} A^i(\delta e)$$

$$\leq A^k(ce) + \delta \sum_{i=0}^{k-1} A^i(e) \leq \epsilon e.$$

Lemma 8.8 is proved.

8.5 Proofs of Theorem 8.4 and 8.5

Proof of Theorem 8.5 We may assume that $\epsilon \in (0, 1)$. Properties (iv) and (v) imply that there exists

$$\epsilon_1 \in (0, \epsilon)$$

such that

$$A^i(\epsilon_1 e) \leq 8^{-1}\epsilon e, \; i = 0, 1, \ldots. \tag{8.31}$$

By property (iv), there exists an integer $k_0 > 1$ such that

$$A^{k_0}(e) \leq 8^{-1}\epsilon_1 e. \tag{8.32}$$

Choose $\delta \in (0, \epsilon_1)$ such that

$$\delta \sum_{i=0}^{k_0} A^i(e) \leq 8^{-1}\epsilon_1 e. \tag{8.33}$$

Assume that $m \geq 1$ is an integer, $\{x_i\}_{i=0}^m \subset K$,

$$d(x_0, x_T) \leq \delta e, \tag{8.34}$$

and that for all $i = 0, \ldots, m-1$,

$$d(x_{i+1}, T(x_i)) \leq \delta. \tag{8.35}$$

Assume that $j \in \{0, \ldots, m\}$,

$$j + k_0 \leq m \tag{8.36}$$

and that

$$d(x_{j_0}, x_T) \leq e. \tag{8.37}$$

Set

$$y_i = x_{i+j_0},\ i = 0, \ldots, k_0. \tag{8.38}$$

Lemma 8.7 and Eqs. (8.32), (8.33), (8.37), and (8.38) imply that

$$d(x_{j+k_0}, x_T) = d(y_{k_0}, x_T) \leq A^{k_0}(d(y_{k_0}, x_T))$$

$$+ \sum_{i=0}^{k_0-1} A^i(\delta e) \leq 8^{-1}\epsilon_1 e + 8^{-1}\epsilon_1 e.$$

Thus we have shown that the following property holds:

(a) if $j \in \{0, \ldots, m\}$ satisfies (8.36) and (8.37), then $d(x_{j+k_0}, x_T) \leq 4^{-1}\epsilon_1 e$.
Property (a) and (8.34) imply that the following property holds:
(b) if an integer $s \geq 0$ satisfies $sk_0 \leq m$, then $d(x_{sk_0}, x_T) \leq 4^{-1}\epsilon_1 e$.

Assume that $p \in \{1, \ldots, m\}$. There exists an integer $s \geq 0$ such that

$$sk_0 \leq p < (s+1)k_0. \tag{8.39}$$

Property (b) implies that

$$d(x_{sk_0}, x_T) \leq 4^{-1}\epsilon_1 e. \tag{8.40}$$

Lemma 8.7 and (8.31), (8.33), (8.35), (8.39), and (8.46) imply that

$$d(x_p, x_T) \leq A^{p-sk_0}(d(x_{sk_0}, x_T)) + \sum_{i=0}^{k_0} A^i(\delta e)$$

$$\leq 4^{-1}\epsilon_1 A^{p-s_0}(e) + (\epsilon_1/8)e \leq \epsilon.$$

Theorem 8.5 is proved.

Theorem 8.4 follows from Theorem 8.5 and Lemma 8.8.

8.6 Convergence of Inexact Iterates of Perov Nonexpansive Mappings

In [30] it was shown that if any exact orbit of a nonexpansive mapping converges to its fixed point, then this convergence property also holds for its inexact orbits with summable errors. We discuss an analog of this result for inexact iterates of a contraction mapping of Perov-type acting on a generalized complete metric space and also prove a convergence property for such operators which maps a closed ball in a generalized complete metric space into the space. These results were obtained in [220].

Assume that (X, d) is a generalized metric space and d is its generalized metric. For all $x, y \in X$ set

$$d(x, y) = (d_1(x, y), \dots, d_n(x, y)).$$

Fix $\theta \in X$. For all $x, y \in X$ set

$$\tilde{d}(x, y) = \|d(x, y)\|_1 = d_1(x, y) + \dots + d_n(x, y).$$

Clearly, $(X, \tilde{d})$ is a metric space and convergence in (X, d) is equivalent to the convergence in $(X, \tilde{d})$.

We assume that the metric space $(X, \tilde{d})$ is complete and continue to assume that $A : R^n_+ \to R^n_+$ and that the following properties hold:

(i) $A(0) = 0$ and A is continuous at zero;
(ii) for each $z_1, z_2 \in R^n_+$ satisfying $0 \le z_1 \le z_2$,

$$A(z_1) \le A(z_2);$$

(iii) for each $z_1, z_2 \in R^n_+$,

$$A(z_1 + z_2) \le A(z_1 + z_2);$$

8.7 Contraction Mappings on a Closed Ball

Let $a \in X$, $r \in R^n_+$ and let

$$T : B(a, r) \to X$$

satisfy for each $x, y \in X$,

$$d(T(x), T(y)) \le A(d(x, y)). \tag{8.41}$$

Assume that

$$(A^0 + A + \cdots + A^i + \cdots)(d(F(a), a)) \leq r. \tag{8.42}$$

The following result was stated in [146].

Theorem 8.9 *A sequence $T^i(a)$, $i = 0, 1, \ldots$ is well-defined, $d(T^i(a), a) \leq r$ for all integers $i \geq 0$. Moreover,for each integer $k \geq 1$,*

$$d(a, T^k(a)) \leq (\sum_{i=0}^{k-1} A^i)(d(a, T(a)))$$

and there exists

$$x_* = \lim_{i \to \infty} T^i(a)$$

such that $T(x_*) = x_*$.

Proof In view of (8.42),

$$d(a, T(a)) \leq r,$$

$$T(a) \in B(a, r). \tag{8.43}$$

Thus $T^2(a)$ is defined. By (8.41) and (8.42),

$$d(a, T^2(a)) \leq d(a, T(a)) + d(T(a), T^2(a))$$

$$\leq d(a, T(a)) + A(d(a, T(a))) \leq r. \tag{8.44}$$

We show that for each integer $k \geq 1$,

$$d(a, T^k(a)) \leq (\sum_{i=0}^{k-1} A^i)(d(a, T(a))). \tag{8.45}$$

In view of (8.43) and (8.44), Eq. (8.45) holds for $k = 1, 2$.

Assume that $k \geq 1$ is an integer, $T^k(a)$ is defined, and (8.45) holds. Then

$$T^k(a) \in B(a, r),$$

$T^{k+1}(a)$ is defined and by (8.41) and (8.45)

$$d(a, T^{k+1}(a)) \leq d(a, T(a)) + d(T(a), T^{k+1}(a))$$

$$\leq d(a, T(a)) + A(d(a, T^k(a)))$$

$$\leq d(a, T(a)) + A((\sum_{i=0}^{k-1} A^i)(d(a, T(a))))$$

$$\leq d(a, T(a)) + \sum_{i=0}^{k-1} A^{i+1}(d(a, T(a))) = \sum_{i=0}^{k} A^i(d(a, T(a))).$$

Therefore we showed by induction that for all integers $k \geq 1$,

$$T^k(a) \in B(a, r)$$

and (8.45) is true. By (8.45) and (8.42), for each pair of integers $p > k \geq 1$,

$$d(T^k(a), T^p(a)) \leq \sum_{i=k}^{p-1} d(T^i(a), T^{i+1}(a)) \leq \sum_{i=k}^{p-1} A^i(d(a, T(a))),$$

$$\|d(T^k(a), T^p(a))\|_1 \leq \sum_{i=k}^{p-1} \|A^i(d(a, T(a)))\|_1 \to 0$$

as $k \to \infty$. Thus $\{T^i(a)\}_{i=1}^{\infty}$ is a Cauchy sequence and there exists

$$x_* = \lim_{i \to \infty} T^i(a) \in B(a, r).$$

Clearly, $T(x_*) = x_*$.

8.8 Inexact Orbits of the Contraction Mappings on the Ball

We continue to use all the assumptions, definitions, and notation introduced in Sect. 8.8. Assume that

$$r_i > 0, \ i = 1, \dots, m,$$

$$\delta \in R^m_+$$

and

$$\sum_{i=0}^{\infty} A^i(d(a, T(a))) + \sum_{i=0}^{\infty} A^i(\delta e) \leq r. \tag{8.46}$$

By Theorem 8.9, there exists

$$x_* = \lim_{i\to\infty} T^i(a).$$

In this section we study the behavior of inexact iterates of the mapping T under the presence of a computational error δ.

Theorem 8.10 *Assume that*

$$x_0 \in B(a, \delta e), \tag{8.47}$$

and that for each integer $i \ge 0$ if

$$x_i \in B(a, r)$$

is defined then

$$x_{i+1} \in B(T(x_i), \delta e) \tag{8.48}$$

Then for each integer $k \ge 1$,

$$x_k \in B(a, r)$$

is defined

$$d(T^k(a), x_k) \le \sum_{i=0}^{k} A^i(\delta e),$$

$$d(a, x_k) \le \sum_{i=0}^{k} A^i(\delta e) + \sum_{i=0}^{k-1} A^i(d(a, T(a)))$$

and

$$d(x_*, x_k) \le \sum_{i=0}^{k} A^i(\delta e) + \sum_{i=0}^{\infty} A^{k+i}(d(a, T(a))).$$

Proof Clearly,

$$x_0 \in B(a, r).$$

In view of (8.41) and (8.48),

$$d(T(a), x_1) \le d(T(a), T(x_0)) + d(T(x_0), x_1)$$

$$\leq A(d(a, x_0)) + \delta e \leq \delta e + A(\delta e). \tag{8.49}$$

By (8.46) and (8.49),

$$d(a, x_1) \leq d(a, T(x_0)) + d(T(x_0), x_1)$$

$$\leq d(a, T(a)) + d(T(a), x_1) \leq d(a, T(a)) + (A^0 + A)(\delta e) \leq r. \tag{8.50}$$

Thus $x_1 \in B(a, r)$ and x_2 is well-defined. Theorem 8.9 and Eqs. (8.41), (8.48), and (8.49) imply that

$$d(T^2(a), x_2) \leq d(T^2(a), T(x_1)) + d(T(x_1), x_2)$$

$$\leq A(T(a), x_1) + \delta e$$

$$\leq A(\delta e) + A^2(\delta e) + \delta e. \tag{8.51}$$

Theorem 8.9 and (8.51) imply that

$$d(a, x_2) \leq d(a, T^2(a)) + d(T^2(a), x_2)$$

$$\leq d(a, T(a)) + A(d(a, T(a))) + \delta e + A(\delta e) + A^2\delta e \leq r. \tag{8.52}$$

In view of (8.46) and (8.52), $x_2 \in B(a, r)$ and x_3 is well-defined. We show by induction that for all integers $k \geq 1$,

$$x_k \in B(a, r)$$

is well-defined

$$d(T^k(a), x_k) \leq \sum_{i=0}^{k} A^i(\delta e), \tag{8.53}$$

$$d(a, x_k) \leq \sum_{i=0}^{k} A^i(\delta e) + \sum_{i=0}^{k-1} A^i(d(a, T(a))) \leq r. \tag{8.54}$$

By (8.49)–(8.52), Eqs. (8.53) and (8.54) are true for $k = 1, 2$.

Assume that $k \geq 1$ is an integer,

$$x_k \in B(a, r)$$

is defined, and (8.53) and (8.54) hold. It follows from (8.41), (8.48), and (8.53) that

$$d(T^{k+1}(a), x_{k+1}) \le d(T^{k+1}(a), T(x_k)) + d(T(x_k), x_{k+1})$$
$$\le d(T(T^k(a)), T(x_k)) + d(T(x_k), x_{k+1})$$
$$\le A(d(T^k(a), x_k)) + \delta e$$
$$\le \delta e + A(\sum_{i=0}^{k} A^i(\delta e)) \le \sum_{i=0}^{k+1} A^i(\delta e) \tag{8.55}$$

and (8.53) holds for $k + 1$ too.

By Theorem 8.9 and (8.55),

$$d(a, x_{k+1}) \le d(a, T^{k+1}(a)) + d(T^{k+1}(a), x_{k+1})$$
$$\le (\sum_{i=0}^{k} A^i)(d(a, T(a))) + (\sum_{i=0}^{k+1} A^i)(\delta e) \tag{8.56}$$

and (8.54) holds for $k + 1$ too. In view of (8.46) and (8.56),

$$x_{k+1} \in B(a, r).$$

Clearly, the assumption made for k also holds for $k + 1$. Therefore we showed that for all integers $k \ge 1$

$$x_k \in B(a, r)$$

and (8.53) and (8.54) hold.

Let $k \ge 1$ be an integer. Then

$$d(T^k(a), x_*) = \lim_{p\to\infty} d(T^k(a), T^k(a)).$$

By (8.41), for each integer $p > k$,

$$d(T^k(a), T^p(a)) \le \sum_{i=0}^{p-k-1} d(T^{k+i}(a), T^{k+i+1}(a))$$
$$\le \sum_{i=0}^{p-k-1} A^{k+i}(d(a, T(a))).$$

Then

$$d(T^k(a), x_*) \le \sum_{i=0}^{\infty} A^{k+i}(d(a, T(a))).$$

Together with (8.54) this implies that

$$d(x_k, x_*) \le \sum_{i=0}^{\infty} A^{k+i}(d(a, T(a))) + \sum_{i=0}^{k} A^i(\delta e).$$

Theorem 8.10 is proved.

8.9 Inexact Orbits with Summable Computational Errors

We continue to use all the assumptions, definitions, and notation introduced in Sect. 8.8. Suppose also that the following assumptions hold:

(iv) there exists $D_0 > 1$ such that

$$A^i(e) \le D_0 e, \ i = 0, 1, \ldots \tag{8.57}$$

(v) $A(\lambda z) = \lambda A(z)$ for all $z \in R^n_+$ and all $\lambda \ge 0$.

Assume that a mapping $T : X \to X$ satisfies for each $x, y \in X$,

$$d(T(x), T(y)) \le A(d(x, y)). \tag{8.58}$$

For each $x \in X$ and each $C \subset X$ set

$$\tilde{d}(x, C) = \inf\{\tilde{d}(x, y) : \ y \in C\}.$$

We will prove the following three results obtained in [220] which describe the asymptotic behavior of inexact iterates under the presence of summable computational errors.

Theorem 8.11 *Assume that for each $x \in X$ the sequence $\{T^i(x)\}_{i=1}^{\infty}$ converges and that a sequence $\{x_i\}_{i=0}^{\infty} \subset X$ satisfies*

$$\sum_{i=0}^{\infty} \tilde{d}(x_{i+1}, T(x_i)) < \infty.$$

Then the sequence $\{x_i\}_{i=0}^{\infty}$ converges to a fixed point of T.

Theorem 8.12 *Assume that F is a nonempty closed set in X and that for each $x \in X$,*

$$\lim_{i\to\infty} \tilde{d}(T^i(x), F) = 0$$

and that a sequence $\{x_i\}_{i=0}^{\infty} \subset X$ *satisfies*

$$\sum_{i=0}^{\infty} \tilde{d}(x_{i+1}, T(x_i)) < \infty.$$

Then

$$\lim_{i\to\infty} \tilde{d}(x_i(x), F) = 0.$$

Theorem 8.13 *Assume that for each* $x \in X$ *there exists a nonempty compact set* $E(x) \subset X$ *such that*

$$\lim_{i\to\infty} \tilde{d}(T^i(x), E(x)) = 0$$

and that a sequence $\{x_i\}_{i=0}^{\infty} \subset X$ *satisfies*

$$\sum_{i=0}^{\infty} \tilde{d}(x_{i+1}, T(x_i)) < \infty.$$

Then there exists a nonempty compact set $E \subset X$ *such that*

$$\lim_{i\to\infty} \tilde{d}(x_i, E) = 0.$$

8.10 An Auxiliary Result for Theorem 8.11–8.13

Lemma 8.14 *Assume that* $\{x_i\}_{i=0}^{\infty} \subset X$ *satisfies*

$$\sum_{i=0}^{\infty} \|d(x_{i+1}, T(x_i))\|_1 < \infty,$$

k *is a natural number and that*

$$y_k = x_k, \tag{8.59}$$

$$y_{i+1} = T(y_i) \text{ for all integers } i \geq k. \tag{8.60}$$

Then for all integers $p > k$,

$$d(x_p, y_p) \leq \sum_{i=k}^{p-1} A^{p-i-1}(d(x_{i+1}, T(x_i))), \tag{8.61}$$

$$d(x_p, y_p) \leq \sum_{i=k}^{p-1} A^{p-i-1}(\|d(x_{i+1}, T(x_i))\|_1 e) \leq \sum_{i=k}^{p-1} \|d(x_{i+1}, T(x_i))\|_1 D_0 e.$$

Proof By (8.59) and (8.60),

$$d(y_{k+1}, x_{k+1}) = d(x_{k+1}, T(x_k))$$

and (8.61) holds for $p = k + 1$.

Assume that $p > k$ is an integer and (8.63) holds. It follows from (8.58), (8.60), and (8.61) that

$$d(x_{p+1}, y_{p+1}) \leq d(T(y_p), T(x_p)) + d(T(x_p), x_{p+1})$$

$$\leq A(d(y_p, x_p)) + d(T(x_p), x_{p+1})$$

$$\leq d(T(x_p), x_{p+1}) + A(\sum_{i=k}^{p-1} A^{p-i-1}(d(x_{i+1}, T(x_i))))$$

$$\leq \sum_{i=k}^{p} A^{p-i}(d(x_{i+1}, T(x_i)))$$

and the assumption made for p also holds for $p + 1$. Thus by induction, we showed that (8.61) holds for all integers $p > k$. The final equation in the statement of the theorem easily follows from (8.61). Lemma 8.14 is proved.

8.11 Proof of Theorem 8.11

Let

$$\epsilon \in (0, 1).$$

There exists is a natural number $k > 1$ for which

$$\sum_{i=k}^{\infty} \|d(x_{i+1}, T(x_i))\|_1 < (\epsilon/2)(D_0 + 1)^{-1} n^{-1}. \tag{8.62}$$

Set

$$y_k = x_k, \ \ y_{i+1} = T(y_i) \text{ for all integers } i \geq k. \tag{8.63}$$

Clearly, there exists

$$\lim_{i \to \infty} y_i = y_* \tag{8.64}$$

and

$$T(y_*) = y_*. \tag{8.65}$$

In view of (8.64), there exists an integer $p_1 > k$ such that

$$\|d(y_i, y_*)\|_1 \leq (4n)^{-1}\epsilon \text{ for all integers } i \geq p_1. \tag{8.66}$$

Lemma 8.14 and Eq. (8.62) imply that for all integers $p > k$,

$$d(x_p, y_p) \leq \sum_{i=k}^{p-1} \|d(x_{i+1}, T(x_i))\|_1 D_0 e,$$

$$\|d(x_p, y_p)\|_1 \leq \sum_{i=k}^{p-1} \|d(x_{i+1}, T(x_i))\|_1 D_0 n < \epsilon/2. \tag{8.67}$$

By (8.66) and (8.67), for all integers $i > p_1$,

$$\|d(x_p, y_*)\|_1 \leq \|d(x_p, y_p)\|_1 + \|d(y_p, y_*)\|_1 \leq \epsilon.$$

Since ϵ is any number from the interval $(0, 1)$, we conclude that $\{x_i\}_{i=0}^{\infty}$ is a Cauchy sequence and there exists

$$x_* = \lim_{i \to \infty} x_i.$$

It is easy to see that

$$\|d(x_*, y_*)\|_1 \leq \epsilon.$$

Since ϵ is any number from the interval $(0, 1)$ we conclude that

$$T(x_*) = x_*.$$

Theorem 8.11 is proved.

8.12 Proof of Theorem 8.12

Let

$$\epsilon > 0.$$

There exists is a natural number k for which

$$\sum_{i=k}^{\infty} \|d(x_{i+1}, T(x_i))\|_1 < (\epsilon/4)(D_0 + 1)^{-1}n^{-1}. \tag{8.68}$$

Set

$$y_k = x_k, \;\; y_{i+1} = T(y_i) \text{ for all integers } i \geq k. \tag{8.69}$$

Clearly, by (8.69),

$$\lim_{i \to \infty} \tilde{d}(y_i, F) = 0.$$

Lemma 8.14 and (8.68) imply that for all integers $i > k$,

$$d(y_i, x_i) < \epsilon/4. \tag{8.70}$$

There exists an integer $p_0 > k$ such that

$$\tilde{d}(y_i, F) < \epsilon/4 \text{ for all integers } i \geq p_0. \tag{8.71}$$

By (8.70) and (8.71), for all integers $i \geq p_0$,

$$\tilde{d}(x_i, F) \leq \tilde{d}(x_i, y_i) + \tilde{d}(y_i, F) \leq \epsilon.$$

Theorem 8.12 is proved.

8.13 Proof of Theorem 8.13

Let

$$\epsilon > 0.$$

There exists is a natural number k for which

$$\sum_{i=k}^{\infty} \|d(x_{i+1}, T(x_i))\|_1 < (\epsilon/4)(D_0 + 1)^{-1}n^{-1}. \tag{8.72}$$

Set

$$y_k = x_k, \ y_{i+1} = T(y_i) \text{ for all integers } i \geq k. \tag{8.73}$$

By (8.73), there exists a nonempty compact set $E_0 \subset X$ such that

$$\tilde{d}(y_i, E_0) \to 0 \text{ as } i \to \infty. \tag{8.74}$$

By (8.74), there exists an integer $p_0 > k$ such that

$$\tilde{d}(y_i, E_0) \leq \epsilon/4 \text{ for all integers } i \geq p_0. \tag{8.75}$$

Lemma 8.14 and (8.72) and (8.73) imply that for all integers $i \geq k$,

$$\tilde{d}(x_i, y_i) \leq \epsilon/4. \tag{8.76}$$

It follows from (8.75) and (8.76) that

$$\tilde{d}(x_i, E) \leq \epsilon/2 \text{ for all integers } i \geq p_0.$$

It is not difficult to see that there exists a finite set

$$\{z_1, \ldots, z_q\} \subset \{x_i : i = 0, 1, \ldots\}$$

such that

$$\{x_i\}_{i=0}^{\infty} \subset \cup_{i=1}^{q} B(z_i, \epsilon/2).$$

Since ϵ is any positive number, we conclude that there exists a nonempty compact set $E \subset X$ such that

$$\lim_{i \to \infty} \tilde{d}(x_i, E) = 0.$$

Theorem 8.13 is proved.

8.14 Existence and Convergence Results

Recall that for each $z \in R^1$,

$$\lfloor z \rfloor = \max\{i : i \text{ is an integer and } i \le z\}$$

and that the sum over empty set is zero. We continue to use the notation and definitions of Sect. 8.2. In particular we assume that (X, d) is a generalized metric space equipped with the generalized metric $d : X \times X \to R^n_+$. For all $x, y \in X$ set

$$d(x, y) = (d_1(x, y), \dots, d_n(x, y)).$$

Recall that for all $x, y \in X$,

$$\tilde{d}(x, y) = \|d(x, y)\|_1 = d_1(x, y) + \dots + d_n(x, y)$$

and $(X, \tilde{d})$ is a metric space.

We assume that the metric space $(X, \tilde{d})$ is complete.

Denote by $\mathcal{A}$ the set of all mappings $A : R^n_+ \to R^n_+$ such that the following properties hold:

(i) $A(0) = 0$;

(ii) for each $z_1, z_2 \in R^n_+$ satisfying $0 \le z_1 \le z_2$,

$$A(z_1) \le A(z_2);$$

(iii) for each $z_1, z_2 \in R^n_+$,

$$A(z_1 + z_2) \le A(z_1 + z_2);$$

(iv) for each $z \in R^n_+$, $A^t(z) \to 0$ as $t \to \infty$.

Let $A_1, A_2 \in \mathcal{A}$. We say that $A_1 \le A_2$ if $A_1(z) \le A_2(z)$ for all $z \in R^n_+$.

Assume that for each $t \in (0, \infty)$, $A(t) \in \mathcal{A}$,

$$\text{if } 0 < t_1 < t_2, \text{ then } A(t_2) \le A(t_1) \tag{8.77}$$

and for each $z \in R^n_+$ there exists

$$\tilde{A}(z) = \lim_{t \to 0^+} A(t)(z). \tag{8.78}$$

Clearly,

$$\tilde{A}(0) = 0,$$

if $z_1, z_2 \in R^n_+$ and $z_1 \le z_2$, then

$$\tilde{A}(z_1) \le \tilde{A}(z_2)$$

and for all $z_1, z_2 \in R^n_+$,

$$\tilde{A}(z_1 + z_2) \le \tilde{A}(z_1) + \tilde{A}(z_2).$$

We suppose that the following assumptions holds:

(v) the set $\{\tilde{A}^i(e) : i = 0, 1, \dots, \}$ is bounded;
(vi) for each $\epsilon > 0$ there exists $\delta > 0$ such that $A^i(\delta e) \le \epsilon e$ for all integers $i \ge 0$. (Note that if all the mappings $A(\cdot)$ are positively homogeneous, then property (vi) follows from property (v).)

Set

$$\Delta_* = \sup\{\|\tilde{A}^i(e)\|_1 : i = 0, 1, \dots\} \tag{8.79}$$

and

$$A(0) = \tilde{A}. \tag{8.80}$$

Assume that $T : X \to X$ and that for each $x, y \in X$,

$$d(T(x), T(y)) \le A(\tilde{d}(x, y))(d(x, y)). \tag{8.81}$$

It is natural to call the mapping T as a generalized contractive mapping [156, 168].

We prove the following results obtained in [224].

Theorem 8.15 *Let $\epsilon > 0$. Then there exists $\delta > 0$ such that for each $x, y \in X$ satisfying*

$$\|d(x, T(x))\|_1 \le \delta, \ \|d(y, T(y))\|_1 \le \delta$$

the inequality $\|d(x, y)\|_1 \le \epsilon$ holds.

Theorem 8.16 *1. There exists a unique point $x_* \in X$ such that $T(x_*) = x_*$.*
2. For each $\epsilon > 0$ there exists $\delta > 0$ such that if $x \in X$ satisfies $\tilde{d}(x, T(x)) \le \delta$, then $\tilde{d}(x, x_*) \le \epsilon$.

Theorem 8.17 *1. Let $x_* \in X$ satisfy $T(x_*) = x_*$. Then for each $x \in X$, the sequence $\{T^i(x)\}_{i=1}^\infty$ converges to x_* uniformly on all bounded subsets of X.*
2. Let $\epsilon, M > 0$. Then there exist $\delta > 0$ and a natural number n_0 such that for each integer $m \ge n_0$ and each sequence $\{x_i\}_{i=0}^m \subset X$ satisfying

$$\tilde{d}(x_0, x_*) \le M,$$

$$\tilde{d}(x_{i+1}, T(x_i)) \le \delta, \ i = 0, \dots, m-1$$

the inequality $\tilde{d}(x_i, x_*) \le \epsilon$ *holds for all integers* $i = n_0, \ldots, m-1$.

Particular cases of the mapping T satisfying (8.81) are studied in the literature. See [146, 147, 149, 198] and the references mentioned therein. A class of mappings satisfying (8.81) with a constant function $A(\cdot)$ which does not depend on t and such that $A(t)$ is a linear map was introduced in [147], while condition (8.81) with a nonlinear map $A(t)$ (which also does not depend on t) was introduced in [146]. A mapping T satisfying (8.81) with the constant function $A(t)$ which does not depend on t can be considered as a generalized strict contraction. In the previous sections of this chapter, we also consider the case with the constant function $A(t)$ which does not depend on t. From now on, we consider the mapping T satisfying (8.81) with the function $A(t)$ which do depends on t.

In the sequel we use the following auxiliary result

Lemma 8.18 *Let* $z_1, z_2 \in X$. *Then for every integer* $m \ge 1$,

$$d(T^m(z_1), T^m(z_2)) \le \tilde{A}^m(d(z_1, z_2)).$$

Proof By (8.77), (8.78), and (8.81),

$$d(T(z_1), T(z_2)) \le A(\tilde{d}(z_1, z_2))(d(z_1, z_2)) \le \tilde{A}(d(z_1, z_2)).$$

By induction, using the relation above, we can easily show that for every integer $m \ge 1$,

$$d(T^m(z_1), T^m(z_2)) \le \tilde{A}^m(d(z_1, z_2)).$$

Lemma 8.18 is proved.

8.15 Proof of Theorem 8.15

We may assume without loss of generality that $\epsilon < 1$. By property (iv), there exists a natural number k_0 such that

$$\|A^i(\epsilon/2)(e)\|_1 \le 4^{-1}\epsilon \text{ for all integers } i \ge k_0. \tag{8.82}$$

Choose a positive number

$$\delta_1 < 8^{-1}\epsilon(k_0+1)^{-1}n^{-1}. \tag{8.83}$$

Property (v) implies that there exists $\delta \in (0, \delta_1)$ such that

$$\tilde{A}^i(\delta e) \le \delta_1 e \text{ for each integer } i \ge 0. \tag{8.84}$$

Assume that $x, y \in X$ and

$$\|d(x, T(x))\|_1 \leq \delta, \ \|d(y, T(y))\|_1 \leq \delta. \tag{8.85}$$

We show that $\|d(x, y)\|_1 \leq \epsilon$. Assume the contrary. Then

$$\|d(x, y)\|_1 > \epsilon. \tag{8.86}$$

Lemma 8.18 and (8.84) and (8.85) imply that for all integers $i \geq 0$

$$d(T^i(x), T^{i+1}(x)) \leq \tilde{A}^i(d(x, T(x)) \leq \tilde{A}^i(\delta e) \leq \delta_1 e, \tag{8.87}$$

$$d(T^i(y), T^{i+1}(y)) \leq \tilde{A}^i(d(y, T(y)) \leq \tilde{A}^i(\delta e) \leq \delta_1 e. \tag{8.88}$$

We show that there exists $i \in \{0, \ldots, k_0 + 1\}$ such that

$$\|d(T^i(x), T^i(y))\|_1 \leq \epsilon/2.$$

Assume the contrary. Then for all integers $i \in \{0, \ldots, k_0 + 1\}$,

$$\|d(T^i(x), T^i(y))\|_1 > \epsilon/2. \tag{8.89}$$

By (8.1), (8.77), and (8.89), for all integers $i = 0, \ldots, k_0$,

$$\begin{aligned} d(T^{i+1}(x), T^{i+1}(y)) &\leq A(\tilde{d}(T^i(x), T^i(y)))(d(T^i(x), T^i(y))) \\ &\leq A(\epsilon/2)(d(T^i(x), T^i(y))). \end{aligned} \tag{8.90}$$

Since (8.90) holds for all $i = 0, \ldots, k_0$, we conclude that

$$d(T^{k_0+1}(x), T^{k_0+1}(y)) \leq A(\epsilon/2)^{k_0+1}(d(x, y)). \tag{8.91}$$

Clearly,

$$d(x, y) \leq \|d(x, y)\|_1 e. \tag{8.92}$$

Equations (8.91) and (8.92) imply that

$$d(T^{k_0+1}(x), T^{k_0+1}(y)) \leq A(\epsilon/2)^{k_0+1}(\|d(x, y)\|_1 e). \tag{8.93}$$

In view of (8.86), there exists an integer $p \geq 1$, such that

$$p\epsilon \leq \|d(x, y)\|_1 e < (p + 1)\epsilon. \tag{8.94}$$

Set

$$r_0 = \|d(x, y)\|_1 - p\epsilon. \tag{8.95}$$

It follows from (8.94) and (8.95) that

$$\|d(x, y)\|_1 e = p\epsilon e + r_0 e. \tag{8.96}$$

Properties (iii) and (v) and Eqs. (8.94)–(8.96) imply that

$$A(\epsilon/2)^{k_0+1}(\|d(x, y)\|_1 e) = A(\epsilon/2)^{k_0+1}(p\epsilon e + r_0 e)$$

$$\leq (p+1)A(\epsilon/2)^{k_0+1}(\epsilon e) \leq 2pA(\epsilon/2)^{k_0+1}(\epsilon e)$$

$$\leq 2\|d(x, y)\|_1 \epsilon^{-1} A(\epsilon/2)^{k_0+1}(\epsilon e). \tag{8.97}$$

By (8.93) and (8.97),

$$d(T^{k_0+1}(x), T^{k_0+1}(y)) \leq 2\|d(x, y)\|_1 \epsilon^{-1} A(\epsilon/2)^{k_0+1}(e). \tag{8.98}$$

By (8.82) and (8.98),

$$\|d(T^{k_0+1}(x), T^{k_0+1}(y))\|_1 \leq 2\|d(x, y)\|_1 \epsilon^{-1} \|A(\epsilon/2)^{k_0+1}(e)\|_1$$

$$\leq 2^{-1}\|d(x, y)\|_1. \tag{8.99}$$

It follows from (8.87) and (8.88) that

$$d(x, T^{k_0+1}(x)) \leq \sum_{i=0}^{k} d(T^i(x), T^{i+1}(x)) \leq (k_0+1)\delta_1 e, \tag{8.100}$$

$$d(y, T^{k_0+1}(y)) \leq \sum_{i=0}^{k} d(T^i(y), T^{i+1}(y)) \leq (k_0+1)\delta_1 e. \tag{8.101}$$

Clearly,

$$d(x, y) \leq d(x, T^{k_0+1}(x)) + d(T^{k_0+1}(x), T^{k_0+1}(y)) + d(T^{k_0+1}(y), y). \tag{8.102}$$

By (8.86) and (8.99)–(8.102),

$$\epsilon < \|d(x, y)\|_1$$

$$\leq 2\|(k_0+1)\delta_1 e\|_1 + \|d(T^{k_0+1}(x), T^{k_0+1}(y))\|_1 \leq 2(k_0+1)\delta_1 n + 2^{-1}\|d(x, y)\|_1$$

and

$$2^{-1}\epsilon < 2^{-1}\|d(x,y)\|_1 \le 2(k_0+1)\delta_1 n.$$

This contradicts (8.83). The contradiction we have reached proves that there exists $i \in \{0, \dots, k+1\}$ such that

$$\|d(T^i(x), T^i(y))\|_1 \le \epsilon/2. \tag{8.103}$$

Clearly,

$$d(x,y) \le d(x, T^i(x)) + d(T^i(x), T^i(y)) + d(T^i(y), y). \tag{8.104}$$

Equations (8.86) and (8.104) imply that

$$\begin{aligned} \epsilon &< \|d(x,y)\|_1 \\ &\le \|d(x, T^i(x))\|_1 + \|d(T^i(x), T^i(y))\|_1 + \|d(T^i(y), y)\|_1. \end{aligned} \tag{8.105}$$

By (8.87) and (8.88),

$$\begin{aligned} d(x, T^i(x)) &\le \sum \{d(T^j(x), T^{j+1}(x)) :\ j \in \{0, \dots, i\},\ j < i\} \\ &\le i\delta_1 e \le \delta_1(k_0+1)e, \end{aligned} \tag{8.106}$$

$$\begin{aligned} d(y, T^i(y)) &\le \sum \{d(T^j(y), T^{j+1}(y)) :\ j \in \{0, \dots, i\},\ j < i\} \\ &\le i\delta_1 e \le \delta_1(k_0+1)e. \end{aligned} \tag{8.107}$$

By (8.83), (8.103), and (8.105)–(8.107),

$$\epsilon < \|d(x,y)\|_1 < \epsilon/2 + 2(k_0+1)\delta_1 n < \epsilon.$$

The contradiction we have reached proves that

$$\|d(x,y)\|_1 \le \epsilon.$$

Theorem 8.15 is proved.

8.16 Proof of Theorem 8.16

Let $\epsilon > 0$. We show that there exists $x_\epsilon \in X$ such that $\|d(T(x_\epsilon), x_\epsilon)\|_1 \le \epsilon$.

Fix $x_0 \in X$ and consider a sequence$\{T^i(x_0)\}_{i=1}^{\infty}$. Lemma 8.18, property (ii), and Eqs. (8.77) and (8.79) imply that for each integer $i \ge 0$,

$$d(T^i(x_0), T^{i+1}(x_0)) \le \tilde{A}^i(d(x_0, T(x_0)))$$

$$\le \tilde{A}^i(\|d(x_0, T(x_0))\|_1 e) \le \tilde{A}^i((\lfloor \|d(x_0, T(x_0))\|_1 \rfloor + 1)e) \tag{8.108}$$

$$\le (\lfloor \|d(x_0, T(x_0))\|_1 \rfloor + 1)\tilde{A}^i(e) \le (\lfloor d(x_0, T(x_0)) \rfloor + 1)\Delta_* e.$$

We show that there exists an integer $k \ge 0$ such that

$$\|d(T^k(x_0), T^{k+1}(x_0))\|_1 \le \epsilon.$$

Assume the contrary. Then for each integer $k \ge 0$,

$$\|d(T^k(x_0), T^{k+1}(x_0))\|_1 > \epsilon. \tag{8.109}$$

Property (ii) and Eqs. (8.77), (8.81), and (8.109) imply that for each integer $i \ge 0$,

$$d(T^{i+1}(x_0), T^{i+2}(x_0)) \le A(\|d(T^i(x_0), T^{i+1}(x_0))\|_1)(d(T^i(x_0), T^{i+1}(x_0)))$$

$$\le A(\epsilon)(d(T^i(x_0), T^{i+1}(x_0))).$$

Together with property (iv), this implies that for each integer $k \ge 0$,

$$d(T^k(x_0), T^{k+1}(x_0)) \le A(\epsilon)^k(d(x_0, T(x_0))) \to 0 \text{ as } k \to \infty.$$

This contradicts (8.109). The contradiction we have reached proves that there exists an integer $k \ge 0$ such that

$$\|d(T^k(x_0), T^{k+1}(x_0))\|_1 \le \epsilon.$$

Therefore for each integer $k \ge 1$, there exists $x_k \in X$ such that

$$\|d(x_k, T(x_k))\|_1 \le 1/k. \tag{8.110}$$

Theorem 8.15 and (8.110) imply that $\{x_k\}_{k=1}^{\infty}$ is a Cauchy sequence and there exists

$$x_* = \lim_{k \to \infty} x_k. \tag{8.111}$$

Property (iii) and Eqs. (8.77), (8.78), (8.81), (8.110), and (8.111) imply that for each integer $k \ge 1$,

$$d(T(x_*), x_*) \le d(T(x_*), T(x_k)) + d(T(x_k), x_k) + d(x_k, x_*)$$

$$\le d(T(x_k), x_k) + d(x_k, x_*) + \tilde{A}(d(x_k, x_*)) \to 0 \text{ as } k \to \infty$$

and $T(x_*) = x_*$. Now Assertion 2 follows from Theorem 8.15 and implies the uniqueness of x_*. Theorem 8.16 is proved.

8.17 Proof of Theorem 8.17

Let $\epsilon > 0$ and M be a natural number. Property (vi) implies that there exists a number $\delta \in (0, \epsilon)$ such that

$$\tilde{A}^i(\delta e) \leq \epsilon e \text{ for all integers } i \geq 0. \tag{8.112}$$

Property (iv) implies that there exists a natural number k_0 such that

$$\|A(\delta)^{k_0}(e)\|_1 \leq \delta n^{-2}M^{-1}. \tag{8.113}$$

Let $x \in X$ satisfy

$$\|d(x, x_*)\|_1 \leq M. \tag{8.114}$$

We show that there exists an integer $i \in \{0, \dots, k_0 + 1\}$ such that

$$\|d(T^i(x), x_*)\| \leq \delta.$$

Assume the contrary. Then for all integers $i = 0, \dots, k_0$,

$$\|d(T^i(x), x_*)\|_1 > \delta. \tag{8.115}$$

It follows from equality $T(x_*) = x_*$ and Eqs. (8.77) and (8.81) that for all integers $i = 0, \dots, k_*$,

$$d(T^{i+1}(x), x_*) \leq A(\|d(T^i(x), x_*)\|_1)(d(T^i(x), x_*)) \leq A(\delta)(d(T^i(x), x_*)).$$

Combined with (8.113) and (8.114) and property (iii), this implies that

$$d(T^{k_0}(x), x_*) \leq A(\delta)^{k_0}(d(x, x_*)) \leq A(\delta)^{k_0}(nMe)$$

$$\leq nMA(\delta)^{k_0}(e) < \delta n^{-1}e$$

and

$$\|d(T^{k_0}(x), x_*)\|_1 < \delta,$$

a contradiction (see (8.115)). Therefore there exists an integer $i \in \{0, \dots, k_0 + 1\}$ such that

$$\|d(T^i(x), x_*)\|_1 \leq \delta$$

and

$$d(T^i(x), x_*) \leq \delta e.$$

Together with (8.77), (8.78), (8.81), and (8.112) this implies that for all integers $j > i$,

$$d(T^j(x), x_*) = d(T^{j-i}(T^i(x)), x_*) \leq \tilde{A}^{j-i}(d(T^i(x), x_*)) \leq \epsilon e.$$

Thus Assertion 1 is proved.

The mapping T has a unique fixed point, and all its iterates converge to this fixed point uniformly on bounded sets. In view of (8.77), (8.78), and (8.81) and property (v), T is bounded on bounded sets. Property (v) implies that T is uniformly continuous on X. Now Assertion 2 follows from Theorem 2.65 of [168] which was obtained in [31]. Theorem 8.17 is proved.

Chapter 9
Cyclical Mappings

In this chapter, we study cyclical mappings acting in a complete metric space. We study the asymptotic behavior of (inexact) iterates of cyclical mapping of contractive type. Using the Baire generic approach, it is shown that a typical cyclical mapping is of contractive type. More precisely, we consider a complete metric space of cyclical nonexpansive mappings acting on a union of two sets in a complete hyperbolic space. If these sets have a nonempty intersection, then using Baire category approach, we show that for most mappings, their iterates converge to a unique fixed point belonging to the intersection of the sets. If these two sets have an empty intersection, then using Baire category approach, we show for most mappings the distance between iterates converge to the distance between these two sets.

9.1 Cyclical Contractive Mappings

In this chapter, we study cyclical contraction mappings in a complete metric space. Our goal is to show that analogs of the results that hold for nonexpansive mappings in their usual sense hold for cyclical contraction mappings too which is quite natural.

Let (X, ρ) be a complete metric space and A and B be nonempty and closed subsets of X. For each $x \in K$ and each $r > 0$ set,

$$B(x, r) = \{y \in X : \ \rho(x, y) \le r\}.$$

Assume that $\phi : [0, \infty) \to [0, 1]$ is a decreasing function such that

$$\phi(t) < 1, \ t \in (0, \infty) \tag{9.1}$$

and $T : A \cup B \to X$ is a mapping such that for any point $a \in A$ and any point $b \in B$, we have

A. J. Zaslavski, *Approximate Fixed Points of Nonexpansive Mappings*,
Developments in Mathematics 80, https://doi.org/10.1007/978-3-031-70710-0_9

$$\rho(T(a), T(b)) \leq \phi(\rho(a, b))\rho(a, b). \tag{9.2}$$

The mapping T is called contractive [168].

In the paper [109], it was shown that if the function ϕ is a constant and $T(A) \subset B$ and $T(B) \subset A$, then the map T has a unique fixed point belonging to the union $A \cap B$. In this chapter, we present generalizations of this result and prove the following theorems obtained in [222].

Theorem 9.1 *Assume that for every positive number ϵ there are two points $x_{\epsilon,1} \in A$ and $x_{\epsilon,2} \in B$ for which*

$$\rho(x_{\epsilon,i}, T(x_{\epsilon,i})) \leq \epsilon, \ i = 1, 2.$$

Then the following assertions hold:

1. *There is a unique point $x_T \in A \cap B$ that satisfies $T(x_T) = x_T$.*
2. *For any positive ϵ, there is a positive number δ such that for every point $x \in A \cup B$ satisfying $\rho(x, T(x)) \leq \delta$, the inequality $\rho(x, x_T) \leq \epsilon$ is true.*

Theorem 9.2 *Suppose that M is a real positive number and that for every $n \in \{1, 2, \ldots\}$ there is*

$$\{x_i^{(n)}\}_{i=0}^{2n-1} \subset A \cup B \tag{9.3}$$

for which the following equations are true:

$$\rho(x_0^{(n)}, x_1^{(n)}) \leq M, \tag{9.4}$$

$$x_{2i}^{(n)} \in A, \ i = 0, \ldots, n-1, \tag{9.5}$$

$$x_{2i+1}^{(n)} \in B, \ i = 0, \ldots, n-1, \tag{9.6}$$

$$\max\{\rho(T(x_i^{(n)}), x_{i+1}^{(n)}) : \ i = 0, \ldots, 2n-1\} \leq 1/n. \tag{9.7}$$

Then there exists $x_* \in A \cap B$ such that $T(x_*) = x_*$ and x_* is a unique fixed point of T.

The next result easily follows from Theorem 3.12 of [168].

Theorem 9.3 *Suppose that a point $x_T \in A \cap B$ satisfies $T(x_T) = x_T$. Then for every pair of real numbers $\epsilon, M > 0$ there is a real number $\delta > 0$ and $k \in \{1, 2, \ldots\}$ such that for any natural number $n > k$ and any sequence $\{x_i\}_{i=0}^{n-1} \subset A \cup B$ for which*

$$\rho(x_0, x_T) \leq M,$$

$$\rho(x_{i+1}, T(x_i)) \le \delta, \ i = 0, \dots, n-1$$

the inequality $\rho(x_i, x_T) \le \epsilon$ *is valid for all* $i \in \{k, \dots, n\}$.

9.2 An Auxiliary Result

Lemma 9.4 *Let* $\epsilon \in (0, 1)$,

$$\delta \in (0, 2^{-1}\epsilon(1 - \phi(\epsilon))) \tag{9.8}$$

and

$$x \in A, \ y \in B \tag{9.9}$$

satisfy

$$\rho(x, T(x)), \ \rho(y, T(y)) \le \delta. \tag{9.10}$$

Then $\rho(x, y) \le \epsilon$.

Proof Suppose that the inequality

$$\rho(x, y) > \epsilon \tag{9.11}$$

is true. Then (9.2) and (9.4) imply that

$$\rho(T(x), T(y)) \le \phi(\rho(x, y))\rho(x, y) \le \phi(\epsilon)\rho(x, y). \tag{9.12}$$

By (9.10) and (9.12), we have

$$\begin{aligned}\rho(x, y) &\le \rho(x, T(x)) + \rho(T(x), T(y)) + \rho(T(y), y) \\ &\le 2\delta + \phi(\epsilon)\rho(x, y).\end{aligned} \tag{9.13}$$

In view of (9.11) and (9.13), we have

$$\epsilon(1 - \phi(\epsilon)) \le \rho(x, y)(1 - \phi(\epsilon)) \le 2\delta$$

and

$$\delta \ge 2^{-1}\epsilon(1 - \phi(\epsilon)).$$

This contradicts (9.8) and proves Lemma 9.4.

Corollary 9.5 *Suppose that a point $x_* \in A \cap B$ satisfies $T(x_*) = x_*$, $\epsilon \in (0, 1)$ and that*

$$\delta \in (0, 2^{-1}\epsilon(1 - \phi(\epsilon))).$$

Then for any point $x \in A \cup B$ for which $\rho(x, T(x)) \leq \delta$ the relation $\rho(x, x_*) \leq \epsilon$ is true.

9.3 Proof of Theorem 9.1

For any natural number n, there is a point

$$x_{n,1} \in A, \ x_{n,2} \in B \tag{9.14}$$

for which

$$\rho(x_{n,i}, T(x_{n,i})) \leq 1/n, \ i = 1, 2. \tag{9.15}$$

Let $\epsilon \in (0, 1)$ be given and choose

$$\delta \in (0, 2^{-1}\epsilon(1 - \phi(\epsilon))). \tag{9.16}$$

Lemma 9.4 and Eqs. (9.14)–(9.16) imply that for every pair of positive integers $n, m > \delta^{-1}$, we have

$$\rho(x_{n,1}, x_{m,2}), \ \rho(x_{n,2}, x_{m,1}) \leq \epsilon. \tag{9.17}$$

In view of Eq. (9.17), for every pair of positive integers $n, m > \delta^{-1}$, we have

$$\rho(x_{n,1}, x_{m,1}), \ \rho(x_{n,2}, x_{m,2}) \leq 2\epsilon. \tag{9.18}$$

Since ϵ is an arbitrary number of $(0, 1)$, we conclude that $\{x_{n,i}\}_{n=1}^{\infty}$, $i = 1, 2$ are Cauchy sequences that have a limit:

$$x_* = \lim_{n \to \infty} x_{n,1} = \lim_{n \to \infty} x_{n,2} \in A \cap B. \tag{9.19}$$

By (9.1), (9.2), (9.14), (9.15), and (9.19), for any positive integer n, we have

$$\rho(T(x_*), T(x_{n,1})) \leq \rho(x_*, x_{n,1}) \to 0 \text{ as } n \to \infty$$

and

$$\rho(T(x_*), x_{n,1}) \leq \rho(T(x_*), T(x_{n,1})) + \rho(T(x_{n,1}), x_{n,1}) \to 0 \text{ as } n \to \infty.$$

Thus, $T(x_*) = x_*$. The uniqueness of the fixed point is obvious.
The second assertion follows now from Corollary 9.5.

9.4 Proof of Theorem 9.2

By Theorem 9.1 we need to show that for any positive real number ϵ, there is a point

$$x_\epsilon \in A, \ y_\epsilon \in B$$

for which

$$\rho(x_\epsilon, T(x_\epsilon)) \leq \epsilon, \ \rho(y_\epsilon, T(y_\epsilon)) \leq \epsilon.$$

Let $\epsilon \in (0, 1)$ be given. Choose a positive integer n for which

$$n^{-1} < 16^{-1}\epsilon(1 - \phi(4^{-1}\epsilon))(M + 1)^{-1}. \tag{9.20}$$

There is a finite sequence

$$\{x_i\}_{i=0}^{2n-1} \subset A \cup B$$

which satisfies

$$x_{2i} \in A, \ x_{2i+1} \in B, \ i = 0, \dots, n-1, \tag{9.21}$$

$$\rho(x_0, x_1) \leq M, \tag{9.22}$$

$$\max\{\rho(T(x_i), x_{i+1}) : \ i = 0, \dots, 2n-1\} \leq 1/n. \tag{9.23}$$

We claim that there is $i \in \{0, \dots, 2n-2\}$ for which

$$\rho(T(x_i), x_i) \leq \epsilon/2.$$

Assume the contrary. Then for any number $i \in \{0, \dots, 2n-2\}$, we have

$$\rho(T(x_i), x_i) > \epsilon/2. \tag{9.24}$$

In view of (9.20), (9.23), and (9.24),

$$\rho(x_i, x_{i+1}) \geq \rho(T(x_i), x_i) - \rho(T(x_i), x_{i+1}) > \epsilon/2 - n^{-1} > \epsilon/4. \tag{9.25}$$

It follows from (9.2) and (9.25) that

$$\rho(T(x_i), T(x_{i+1})) \le \phi(\rho(x_i, x_{i+1}))\rho(x_i, x_{i+1}) \le \phi(\epsilon/4)\rho(x_i, x_{i+1}). \tag{9.26}$$

By (9.23) and (9.26),

$$\rho(x_{i+1}, x_{i+2}) \le \rho(x_{i+1}, T(x_i)) + \rho(T(x_i), T(x_{i+1})) + \rho(T(x_{i+1}), x_{i+2})$$
$$\le 2n^{-1} + \phi(\epsilon/4)\rho(x_i, x_{i+1}). \tag{9.27}$$

It follows from (9.8) and (9.27) that

$$\rho(x_i, x_{i+1}) - \rho(x_{i+1}, x_{i+2})$$
$$\ge (1 - \phi(\epsilon/4))\rho(x_i, x_{i+1}) - 2n^{-1}$$
$$\ge (1 - \phi(\epsilon/4))\epsilon/4 - 2/n \ge 8^{-1}\epsilon(1 - \phi(4^{-1}\epsilon)). \tag{9.28}$$

Equations (9.22) and (9.28) imply that

$$M \ge \rho(x_0, x_1) \ge \rho(x_0, x_1) - \rho(x_{2n-2}, x_{2n-1})$$
$$= \sum_{i=0}^{2n-2} (\rho(x_i, x_{i+1}) - \rho(x_{i+1}, x_{i+2}))$$
$$\ge (2n - 1)8^{-1}\epsilon(1 - \phi(4^{-1}\epsilon))$$

and

$$n \le 2n - 1 \le 8M\epsilon^{-1}(1 - \phi(4^{-1}\epsilon))^{-1}.$$

This contradicts (9.20). The contradiction we have reached implies there is a number $i \in \{0, \dots, 2n - 2\}$ for which

$$\rho(T(x_i), x_i) \le \epsilon/2. \tag{9.29}$$

By (9.1), (9.20), (9.23), and (9.29), we have

$$\rho(T(x_{i+1}), x_{i+1}) \le \rho(T(x_{i+1}), T(x_i)) + \rho(T(x_i), x_{i+1})$$
$$\le \rho(x_{i+1}, x_i) + n^{-1} \le \rho(x_i, T(x_i)) + \rho(T(x_i), x_{i+1}) + n^{-1} \le \epsilon/2 + 2n^{-1} < \epsilon.$$

Now Theorem 9.2 follows from Theorem 9.1.

9.5 Inexact Iterates of Cyclic Mappings with Summable Errors

In [30] it was shown that if any exact orbit of a nonexpansive mapping converges to its fixed point, then this convergence property also holds for its inexact orbits with summable errors. Here we present an analog of this result for inexact iterates of cyclic nonexpansive mappings obtained in [225].

Assume that (X, ρ) is a complete metric space and $A, B \subset X$ are nonempty, closed sets such that

$$A \cap B = \emptyset \tag{9.30}$$

and a mapping $T : A \cup B \to A \cup B$ satisfies

$$T(A) \subset B \text{ and } T(B) \subset A \tag{9.31}$$

and

$$\rho(T(x), T(y)) \leq \rho(x, y) \text{ for all } x, y \in A \cup B. \tag{9.32}$$

Set

$$\rho(A, B) = \inf\{\rho(x, y) : \ x \in A, \ \ y \in B\}.$$

We prove the following results obtained in [225].

Theorem 9.6 *Assume that for each $x \in A \cup B$ the sequence $\{T^n(x)\}_{n=1}^{\infty}$ satisfies*

$$\lim_{i \to \infty} \rho(T^i(x), T_{i+1}(x)) = \rho(A, B)$$

and that a sequence $\{x_i\}_{i=0}^{\infty} \subset A \cup B$ satisfies

$$\sum_{i=0}^{\infty} \rho(x_{i+1}, T(x_i)) < \infty \tag{9.33}$$

and for each integer $i \geq 0$, if $x_i \in A$, then $x_{i+1} \in B$ and if $x_i \in B$, then $x_{i+1} \in A$. Then

$$\lim_{i \to \infty} \rho(x_i, x_{i+1}) = \rho(A, B).$$

Theorem 9.7 *Assume that for each $x \in A \cup B$, there exists $\bar{x} \in A \cup B$ such that*

$$\rho(A, B) = \rho(\bar{x}, T(\bar{x})),$$

$$\lim_{i\to\infty} \rho(T^{2i}(x), \bar{x}) = 0, \ \lim_{i\to\infty} \rho(T^{2i+1}(x), T(\bar{x})) = 0$$

and that a sequence $\{x_i\}_{i=0}^{\infty} \subset A \cup B$ *satisfies*

$$\sum_{i=0}^{\infty} \rho(x_{i+1}, T(x_i)) < \infty \tag{9.34}$$

and for each integer $i \geq 0$, *if* $x_i \in A$, *then* $x_{i+1} \in B$ *and if* $x_i \in B$, *then* $x_{i+1} \in A$. *Then there exists* $\bar{x} \in A \cup B$ *such that*

$$\rho(A, B) = \rho(\bar{x}, T(\bar{x}))$$

and

$$\lim_{i\to\infty} \rho(x_{2i}, \bar{x}) = 0, \ \lim_{i\to\infty} \rho(x_{2i+1}, T(\bar{x})) = 0.$$

9.6 An Auxiliary Result for Theorems 9.6 and 9.7

Lemma 9.8 *Assume that* $\{x_i\}_{i=0}^{\infty} \subset A \cup B$ *satisfies*

$$\sum_{i=0}^{\infty} \rho(x_{i+1}, T(x_i)) < \infty,$$

for each integer $i \geq 0$, *if* $x_i \in A$, *then* $x_{i+1} \in B$ *and if* $x_i \in B$, *then* $x_{i+1} \in A$, n_0 *is a natural number and that*

$$y_{n_0} = x_{n_0}, \tag{9.35}$$

$$y_{i+1} = T(y_i) \text{ for all integers } i \geq n_0. \tag{9.36}$$

Then for all integers $n \geq n_0 + 1$,

$$\rho(x_n, y_n) \leq \sum_{i=n_0}^{n-1} \rho(x_{i+1}, T(x_i)). \tag{9.37}$$

Proof By (9.36),

$$\rho(x_{n_0+1}, y_{n_0+1}) = \rho(x_{n_0+1}, T(x_{n_0}))$$

and (9.37) holds for $n = n_0 + 1$.

Assume that $n \geq n_0 + 1$ is an integer and (9.37) holds. It follows from (9.32) and (9.37) that

$$\rho(x_{n+1}, y_{n+1}) \leq \rho(x_{n+1}, T(x_n)) + \rho(T(x_n), T(y_n))$$

$$\leq \rho(x_{n+1}, T(x_n)) + \rho(x_n, y_n)$$

$$\leq \rho(x_{n+1}, T(x_n)) + \sum_{i=n_0}^{n-1} \rho(x_{i+1}, T(x_i))$$

$$= \sum_{i=n_0}^{n} \rho(x_{i+1}, T(x_i))$$

and the assumption made for n also holds for $n + 1$. Thus by induction, we showed that (9.37) holds for all integers $n \geq n_0 + 1$. Lemma 9.8 is proved.

9.7 Proof of Theorem 9.6

Let

$$\epsilon \in (0, 1).$$

By (9.33), there exists a natural number n_0 for which

$$\sum_{i=n_0}^{\infty} \rho(x_{i+1}, T(x_i)) < \epsilon/8. \tag{9.38}$$

Set

$$y_{n_0} = x_{n_0}, \tag{9.39}$$

$$y_{i+1} = T(y_i) \text{ for all integers } i \geq n_0. \tag{9.40}$$

Lemma 9.8 and Eqs. (9.37)–(9.40) imply that for all integers $n > n_0$,

$$\rho(x_n, y_n) \leq \sum_{i=n_0}^{n-1} \rho(x_{i+1}, T(x_i)) < \epsilon/8. \tag{9.41}$$

By (9.37) and (9.40), there exists an integer $n_1 > n_0$ such that for all integers $n \geq n_1$,

$$\rho(y_n, y_{n+1}) \leq \rho(A, B) + \epsilon/4.$$

Together with (9.41), this implies that for all integers $n \geq n_1$,

$$\rho(x_n, n_{n+1}) \leq \rho(A, B) + \epsilon/4 + \epsilon/4.$$

Theorem 9.6 is proved.

9.8 Proof of Theorem 9.7

Let

$$\epsilon \in (0, 1).$$

By (9.34), there exists a natural number n_0 for which (9.38) is true. Define y_i, $i = n_0, n_0 + 1, \ldots$ by (9.39) and (9.40). Lemma 9.8 and Eqs. (9.37)–(9.40) imply that for all integers $n \geq n_0 + 1$,

$$\rho(x_n, y_n) \leq \sum_{i=n_0}^{n-1} \rho(x_{i+1}, T(x_i)) < \epsilon/8. \tag{9.42}$$

By (9.38)–(9.40), there exist $\bar{x} \in A$ and an integer $n_1 > n_0$ such that

$$\rho(A, B) = \rho(\bar{x}, T(\bar{x}))$$

and for all integers $i \geq n_1$,

$$\rho(y_{2i}, \bar{x}), \ \rho(y_{2i+1}, T(\bar{x})) \leq \epsilon/8.$$

Together with (9.42), this implies that for all integers $i \geq n_1$,

$$\rho(x_{2i}, \bar{x}), \ \rho(x_{2i+1}, T(\bar{x})) \leq \epsilon/8 + \epsilon/4.$$

Since ϵ is an arbitrary element of the interval (0, 1), we obtained that $\{x_{2i}\}_{i=0}^{\infty}$ and $\{x_{2i+1}\}_{i=0}^{\infty}$ are Cauchy sequences. There exists

$$\lim_{i\to\infty} x_{2i}, \ \lim_{i\to\infty} x_{2i+1},$$

$$T(\lim_{i\to\infty} x_{2i}) = \lim_{i\to\infty} x_{2i+1},$$

$$\rho(\lim_{i\to\infty} x_{2i}, \bar{x}), \ \rho(\lim_{i\to\infty} x_{2i+1}, \bar{x}) \le \epsilon$$

and

$$\rho(\lim_{i\to\infty} x_{2i}, \lim_{i\to\infty} x_{2i+1}) \ge \rho(\bar{x}, T(\bar{x})) - \epsilon \ge \rho(A, B) - \epsilon.$$

Since ϵ is an arbitrary element of the interval $(0, 1)$, this completes the proof of Theorem 9.7.

9.9 Generic Existence of Fixed Points

We consider a complete metric space of cyclical nonlinear mappings defined on a bounded closed subset of a complete W-hyperbolic space. Using the Baire category approach, we show that most mappings possess a unique fixed point that attracts all the iterates. This result was obtained in [229].

We use the notation, definitions, and assumptions of Sect. 2.1 and assume that (X, ρ, W) is a W-hyperbolic space that is denoted by X for simplicity. We think of $W(x, y, \lambda)$ as a convex combinations of the points $x, y \in X$ with the coefficients $1-\lambda,\ \lambda$ and use the notation

$$W(x, y, \lambda) := (1-\lambda)x \oplus \lambda y, \ x, y \in X, \ \lambda \in [0, 1].$$

By Proposition 2.1, for each $x, y, w, z \in X$ and each $\lambda, \tilde{\lambda} \in [0, 1]$,

$$\rho((1-\lambda)x \oplus \lambda z, (1-\tilde{\lambda})y \oplus \tilde{\lambda} w)$$
$$\le (1-\lambda)\rho(x, y) + \lambda\rho(z, w) + |\lambda - \tilde{\lambda}|\rho(y, w). \tag{9.43}$$

It should be mentioned that the result stated below (Theorem 9.10) was obtained in [229] for hyperbolic spaces, a particular case of a W-hyperbolic space, but the proof remains the same.

Let A and B be nonempty, bounded, and closed subsets of X such that

$$A \cap B \ne \emptyset.$$

For each $x \in K$ and each $r > 0$, set

$$B(x, r) = \{y \in X : \ \rho(x, y) \le r\}.$$

Denote by $\mathcal{M}$ the set of all mappings $T : A \cup B \to A \cup B$ such that

$$T(A) \subset B, \ T(B) \subset A, \tag{9.44}$$

$$\rho(T(x), T(y)) \le \rho(x, y) \text{ for each } x \in A \text{ and each } y \in B. \tag{9.45}$$

The following result was obtained in [109].

Theorem 9.9 *Assume that $T \in \mathcal{M}$, $\alpha \in (0, 1)$ and that*

$$\rho(T(x), T(y)) \le \alpha\rho(x, y) \text{ for each } x \in A \text{ and each } y \in B.$$

Then T has a unique fixed point in $A \cap B$.

For each $T_1, T_2 \in \mathcal{M}$, define

$$d(T_1, T_2) = \sup\{\rho(T_1(x), T_2(x)) : \ x \in A \cup B\}.$$

Clearly, the metric space $(\mathcal{M}, d)$ is complete. Set

$$D = \sup\{\rho(z_1, z_2) : \ z_1, z_2 \in A \cup B\}.$$

We prove the following result obtained in [229].

Theorem 9.10 *Assume that there exists*

$$x_* \in A \cap B \tag{9.46}$$

such that for each $a \in A$, each $b \in B$, and each $\gamma \in (0, 1)$,

$$\gamma x_* \oplus (1 - \gamma)a \in A \text{ and } \gamma x_* \oplus (1 - \gamma)b \in B. \tag{9.47}$$

Then there exists a set $\mathcal{F} \subset \mathcal{M}$ which is a countable intersection of open everywhere dense sets in $(\mathcal{M}, d)$ such that for each $T \in \mathcal{F}$, the following assertions hold:

1. *There exists a unique $x_T \in A \cap B$ such that*

$$T(x_T) = x_T.$$

2. *For each $\epsilon > 0$, there exist an integer $n_\epsilon \ge 1$ and a neighborhood $\mathcal{U}$ of T in $(\mathcal{M}, d)$ such that for each $S \in \mathcal{U}$, each $x \in A \cup B$, and each integer $n \ge n_\epsilon$,*

$$\rho(S^n(x), x_T) \le \epsilon.$$

9.10 Proof of Theorem 9.10

Let $\gamma \in (0, 1)$ and $T \in \mathcal{M}$. For each $x \in A \cup B$, set

$$T_\gamma(x) = \gamma x_* \oplus (1 - \gamma)T(x). \tag{9.48}$$

By (9.44), (9.47), and (9.48),

$$T_\gamma(A) \subset B, \ T_\gamma(B) \subset A. \tag{9.49}$$

Assume that $a \in A$ and $b \in B$. By (9.43), (9.45), and (9.48),

$$\rho(T_\gamma(a), T_\gamma(b)) = \rho(\gamma x_* \oplus (1 - \gamma)T(a), \gamma x_* \oplus (1 - \gamma)T(b))$$

$$\leq (1 - \gamma)\rho(T(a), T(b)) \leq (1 - \gamma)\rho(a, b). \tag{9.50}$$

Theorem 9.9 and (9.50) imply that there exists

$$x_{T,\gamma} \in A \cap B \tag{9.51}$$

such that

$$T_\gamma(x_{T,\gamma}) = x_{T,\gamma}. \tag{9.52}$$

In view of (9.43) and (9.48),

$$d(T_\gamma, T) = \sup\{\rho(T_\gamma(x), T(x)) : \ x \in A \cup B\}$$

$$\sup\{\rho(\gamma x_* \oplus (1 - \gamma)T(x), T(x)) : \ x \in A \cup B\}$$

$$\leq \sup\{\gamma\rho(x_*, T(x)) : \ x \in A \cup B\} \leq \gamma D. \tag{9.53}$$

Define

$$E = \{T_\gamma : \ T \in \mathcal{M}, \ \gamma \in (0, 1)\}. \tag{9.54}$$

Equations (9.53) and (9.54) imply that E is an everywhere dense set in $(\mathcal{M}, d)$.

Let $T \in \mathcal{M}$, $\gamma \in (0, 1)$ and n be a natural number. There exists a natural number $m(\gamma)$ such that

$$(1 - \gamma)^{m(\gamma)}(D + 1) < (4n)^{-1}. \tag{9.55}$$

Choose a positive number

$$\delta < \gamma(4n)^{-1}. \tag{9.56}$$

Assume that $S \in \mathcal{M}$ satisfies

$$d(T_\gamma, S) \le \delta. \tag{9.57}$$

It follows from (9.50)–(9.52) and (9.57) that for each $x \in A \cup B$,

$$\begin{aligned}\rho(S(x), x_{T,\gamma}) &\le \rho(S(x), T_\gamma(x)) + \rho(T_\gamma(x), x_{T,\gamma})\\ &\le \delta + (1-\gamma)\rho(x, x_{T,\gamma}).\end{aligned} \tag{9.58}$$

Let $x \in A \cup B$. We show that for all integers $i \ge 1$,

$$\rho(S^i(x), x_{T,\gamma}) \le (1-\gamma)^i \rho(x, x_{T,\gamma}) + \delta(\sum_{j=0}^{i-1}(1-\gamma)^j). \tag{9.59}$$

In view of (9.58), Eq. (9.59) holds for $i = 1$.

Assume that $i \ge 1$ is an integer and (9.59) holds. By (9.50)–(9.52), (9.57), and (9.59),

$$\begin{aligned}\rho(S^{i+1}(x), x_{T,\gamma}) &\le \rho(S(S^i(x)), T_\gamma(S^i(x))) + \rho(T_\gamma(S^i(x)), x_{T,\gamma})\\ &\le \delta + (1-\gamma)[(1-\gamma)^i \rho(x, x_{T,\gamma}) + \delta(\sum_{j=0}^{i-1}(1-\gamma)^j)]\\ &= \delta + (1-\gamma)^{i+1}\rho(x, x_{T,\gamma}) + \delta(\sum_{j=0}^{i}(1-\gamma)^j).\end{aligned}$$

Thus, (9.59) holds for $i+1$ too. Thus, (9.59) holds for all integers $i \ge 1$ and in view of (9.55) and (9.56),

$$\rho(S^{m(\gamma)}(x), x_{T,\gamma}) \le (1-\gamma)^{m(\gamma)} D + \delta\gamma^{-1} < (2n)^{-1} \tag{9.60}$$

for all $x \in A \cup B$.

Let $W(T, \gamma, n)$ be an open neighborhood of T_γ in $(\mathcal{M}, d)$ such that

$$W(T, \gamma, n) \subset \{S \in \mathcal{M} : d(S, T_\gamma) < \delta\}. \tag{9.61}$$

In view of (9.60) and (9.61), the following property holds:

(a) For each $S \in W(T, \gamma, n)$ and each $x \in A \cup B$,

$$\rho(S^{m(\gamma)}(x), x_{T,\gamma}) \le (2n)^{-1}. \tag{9.62}$$

Define

$$\mathcal{F} = \cap_{n=1}^{\infty} \cup \{W(T, \gamma, n) : T \in \mathcal{M}, \gamma \in (0, 1), n = 1, 2, \ldots\}. \tag{9.63}$$

Clearly, $\mathcal{F}$ is a countable intersection of open everywhere dense sets in $(\mathcal{M}, d)$.

Assume that

$$Q \in \mathcal{F}. \tag{9.64}$$

By (9.63) and (9.64), for every integer $n \ge 1$, there exist

$$T_n \in \mathcal{M}, \ \gamma_n \in (0, 1)$$

such that

$$Q \in W(T_n, \gamma_n, n). \tag{9.65}$$

Property (a) and (9.65) imply that

$$Q^{m(\gamma)}(A \cup B) \subset B(x_{T_n,\gamma_n}, (2n)^{-1}). \tag{9.66}$$

Clearly,

$$Q^i(A \cup B) \subset B(x_{T_n,\gamma_n}, (2n)^{-1})$$

for all integers $i \ge m(\gamma)$. Since n is any natural number, we obtain that for every $x \in A \cup B$, $\{Q^i(x)\}_{i=1}^{\infty}$ is a Cauchy sequence, and there exists

$$\lim_{i\to\infty} Q^i(x) \in A \cap B.$$

In view of (9.66),

$$\lim_{i\to\infty} Q^i(x) \in B(x_{T_n,\gamma_n}, (2n)^{-1}).$$

for all $x \in A \cup B$. Since n is any natural number, we obtain that

$$\lim_{i\to\infty} Q^i(x_1) = \lim_{i\to\infty} Q^i(x_2)$$

for all $x_1, x_2 \in A \cup B$.

Define

$$x_Q = \lim_{i \to \infty} Q^i(x), \ x \in A \cup B. \tag{9.67}$$

By (9.45) and (9.57), for every $x \in A \cup B$,

$$Q(\lim_{i \to \infty} Q^i(x)) = \lim_{i \to \infty} Q^{i+1}(x) = \lim_{i \to \infty} Q^i(x)$$

and

$$Q(x_Q) = x_Q.$$

Clearly, x_Q is a unique fixed point of Q.

Let $\epsilon \in (0, 1)$. Choose a natural number

$$n > 4/\epsilon. \tag{9.68}$$

Assume that

$$S \in W(T_n, \gamma_n, n). \tag{9.69}$$

By (9.66) and (9.67),

$$x_Q \in B(x_{T_n, \gamma_n)}, (2n)^{-1}). \tag{9.70}$$

It follows from (9.62) and (9.68)–(9.70) that

$$S^{m(\gamma)}(A \cup B) \subset B(x_{T_n, \gamma_n}, (2n)^{-1}) \subset B(x_Q, n^{-1}) \subset B(x_Q, \epsilon).$$

This completes the proof of Theorem 9.10.

9.11 Generic Results

Now we consider a complete metric space of cyclical nonexpansive mappings acting on a union of two sets in a complete hyperbolic space with an empty intersection. Using the Baire category approach, we show for most mappings, the distance between iterates converges to the distance between these two sets. This result was obtained in [223].

Let (X, ρ, W) be a complete W-hyperbolic space and A and B be nonempty, bounded, and closed subsets of X such that

$$A \cap B = \emptyset \tag{9.71}$$

and let

$$x_A \in A,\ x_B \in B \tag{9.72}$$

satisfy

$$\rho(x_A, x_B) = \rho(A, B) := \inf\{\rho(a, b) :\ a \in A,\ b \in B\}. \tag{9.73}$$

Assume that for every pair of points $a \in A$ and $b \in B$ and every number $\gamma \in (0, 1)$,

$$\gamma x_A \oplus (1 - \gamma)a \in A \tag{9.74}$$

$$\gamma x_B \oplus (1 - \gamma)b \in B. \tag{9.75}$$

Denote by $\mathcal{M}$ the collection of all operators

$$T : A \cup B \to A \cup B \tag{9.76}$$

such that

$$T(A) \subset B,\ T(B) \subset A \tag{9.77}$$

and

$$\rho(T(x), T(y)) \le \rho(x, y) \text{ for all } x \in A \text{ and all } y \in B. \tag{9.78}$$

For every pair of operators $T_1, T_2 \in \mathcal{M}$, define

$$d(T_1, T_2) = \sup\{\rho(T_1(x), T_2(x)) :\ x \in A \cup B\}. \tag{9.79}$$

Evidently, $(\mathcal{M}, d)$ is a complete metric space. Set

$$D = \sup\{\rho(a, b) :\ a, b \in A \cup B\}. \tag{9.80}$$

We will prove the following result obtained in [223].

Theorem 9.11 *There exists a set $\mathcal{F} \subset \mathcal{M}$ which is a countable intersection of open everywhere dense sets in $(\mathcal{M}, d)$ such that the following assertion holds.*

For every mapping $S \in \mathcal{F}$ and every positive number ϵ, there exist an integer $n_0 \ge 1$ and $\delta > 0$ such that for every sequence $\{x_i\}_{i=0}^{\infty} \subset A \cup B$ which satisfies for every nonnegative integer i,

$$\text{if } x_i \in A, \text{ then } x_{i+1} \in B \text{ and if } x_i \in B \text{ then } x_{i+1} \in A$$

and

$$\rho(S(x_i), x_{i+1}) \le \delta$$

the inequality

$$\rho(x_n, x_{n+1}) \le \rho(A, B) + \epsilon$$

holds for all integers $n \ge n_0$.

Theorem 9.11 easily implies the following result.

Theorem 9.12 *Assume that for every positive number* ϵ*, there exists a positive number* δ *such that for every pair of points* $x \in A$ *and* $y \in B$ *which satisfy*

$$\rho(x, y) \le \rho(A, B) + \delta,$$

there exist a pair of points $\widehat{x} \in A$ *and* $\widehat{y} \in B$ *such that*

$$\rho(\widehat{x}, \widehat{y}) = \rho(A, B)$$

and

$$\max\{\rho(x, \widehat{x}),\ \rho(y, \widehat{y})\} \le \epsilon.$$

Let a set $\mathcal{F}$ *be as guaranteed by Theorem 9.11. Then the following assertion holds.*

For every mapping $S \in \mathcal{F}$ *and every positive number* ϵ*, there exist an integer* $n_0 \ge 1$ *and* $\delta > 0$ *such that for every sequence* $\{x_i\}_{i=0}^{\infty} \subset A \cup B$ *that satisfies for every integer* $i \ge 0$,

$$\text{if } x_i \in A, \text{ then } x_{i+1} \in B \text{ and if } x_i \in B \text{ then } x_{i+1} \in A$$

and

$$\rho(S(x_i), x_{i+1}) \le \delta$$

and for every integer $n \ge n_0$*, there exist a pair of points* $\widehat{x},\ \widehat{y} \in A \cup B$ *for which*

$$\rho(\widehat{x}, \widehat{y}) = \rho(A, B),$$

$$\max\{\rho(x_n, \widehat{x}),\ \rho(x_{n+1}, \widehat{y})\} \le \epsilon$$

if $x_n \in A$*, then* $\widehat{x} \in A$*,* $\widehat{y} \in B$ *and and if* $x_n \in B$*, then* $\widehat{x} \in B$*,* $\widehat{y} \in A$*.*

9.12 Auxiliary Results for Theorem 9.11

Let a map $T \in \mathcal{M}$ and a number $\gamma \in (0, 1)$ be given. Set

$$T_\gamma(a) = (1-\gamma)T(a) \oplus \gamma x_B, \ a \in A, \tag{9.81}$$

$$T_\gamma(b) = (1-\gamma)T(b) \oplus \gamma x_A, \ b \in B. \tag{9.82}$$

In view of (9.74), (9.75), (9.77), (9.81), and (9.82), we have

$$T_\gamma(A) \subset B, \ T_\gamma(B) \subset A. \tag{9.83}$$

It follows from Eqs. (9.79), (9.81), and (9.82) that for every pair of points $a \in A$ and $b \in B$, we have

$$\rho(T_\gamma(a), T(a)) = \rho((1-\gamma)T(a) \oplus \gamma x_B, T(a)) \le \gamma\rho(x_B, T(a)) \le \gamma D, \tag{9.84}$$

$$\rho(T_\gamma(b), T(b)) = \rho((1-\gamma)T(b) \oplus \gamma x_A, T(b)) \le \gamma\rho(x_A, T(b)) \le \gamma D. \tag{9.85}$$

By (9.78), (9.81), and (9.82), for every pair of points $a \in A$ and $b \in B$, we have

$$\begin{aligned} \rho(T_\gamma(a), T_\gamma(b)) &= \rho((1-\gamma)T(a) \oplus \gamma x_B, \ (1-\gamma)T(b) \oplus \gamma x_A) \\ &\le (1-\gamma)\rho(T(a), T(b)) + \gamma\rho(x_A, x_B) \\ &\le (1-\gamma)\rho(a, b) + \gamma\rho(x_A, x_B). \end{aligned} \tag{9.86}$$

In view of Eqs. (9.85) and (9.86), the set

$$\{T_\gamma : T \in \mathcal{M}, \ \gamma \in (0, 1)\}$$

is an everywhere dense set in the space $(\mathcal{M}, d)$.

Lemma 9.13 *Let $T \in \mathcal{M}$, $\gamma \in (0, 1)$, $\delta \ge 0$ and let sequences*

$$\{x_i\}_{i=0}^\infty, \ \{y_i\}_{i=0}^\infty \subset A \cup B$$

satisfy

$$x_0 \in A, \ y_0 \in B, \tag{9.87}$$

for every nonnegative integer i,

$$x_{2i} \in A, \ x_{2i+1} \in B, \tag{9.88}$$

$$y_{2i} \in B, \ y_{2i+1} \in A, \tag{9.89}$$

$$\rho(T_\gamma(x_i), x_{i+1}) \le \delta, \tag{9.90}$$

$$\rho(T_\gamma(y_i), y_{i+1}) \le \delta. \tag{9.91}$$

Then for every natural number $n \geq 1$,

$$\rho(x_n, y_n) \leq (1-\gamma)^n \rho(x_0, y_0) + (\gamma\rho(x_A, x_B) + 2\delta)\sum_{i=0}^{n-1}(1-\gamma)^i. \tag{9.92}$$

Proof By Eqs. (9.86), (9.90), and (9.91), we have

$$\rho(x_1, y_1) \leq \rho(x_1, T_\gamma(x_0)) + \rho(T_\gamma(x_0), T_\gamma(y_0)) + \rho(T_\gamma(y_0), y_1)$$

$$\leq 2\delta + (1-\gamma)\rho(x_0, y_0) + \gamma\rho(x_A, x_B)$$

and Eq. (9.92) is true for $n = 1$.

Assume that n is a natural number and that Eq. (9.92) holds. In view of (9.86) and (9.88)–(9.91), we have

$$\rho(x_{n+1}, y_{n+1}) \leq \rho(x_{n+1}, T_\gamma(x_n)) + \rho(T_\gamma(x_n), T_\gamma(y_n)) + \rho(T_\gamma(y_n), y_{n+1})$$

$$\leq 2\delta + \rho(T_\gamma(x_n), T_\gamma(y_n))$$

$$\leq 2\delta + \gamma\rho(x_A, x_B)$$

$$+ (1-\gamma)[(1-\gamma)^n \rho(x_0, y_0) + (\gamma\rho(x_A, x_B) + 2\delta)\sum_{i=0}^{n-1}(1-\gamma)^i]$$

$$\leq (1-\gamma)^{n+1}\rho(x_0, y_0) + (\gamma\rho(x_A, x_B) + 2\delta)\sum_{i=0}^{n}(1-\gamma)^i$$

and Eq. (9.92) is true for $n + 1$ too. Therefore, Eq. (9.92) is valid for all natural numbers n. This completes the proof of Lemma 9.13.

Lemma 9.14 *Let $T \in \mathcal{M}$, $\gamma \in (0, 1)$, $\epsilon \in (0, 1)$, and*

$$\delta \in (0, \epsilon\gamma/8). \tag{9.93}$$

Then there exists an integer $n_0 \geq 1$ such that for every pair of sequences,

$$\{x_i\}_{i=0}^{\infty}, \ \{y_i\}_{i=0}^{\infty} \subset A \cup B$$

which satisfy

$$x_0 \in A, \ \ y_0 \in B, \tag{9.94}$$

for every nonnegative integer i,

$$x_{2i} \in A, \; x_{2i+1} \in B, \tag{9.95}$$

$$y_{2i} \in B, \; y_{2i+1} \in A, \tag{9.96}$$

$$\max\{\rho(T_\gamma(x_i), x_{i+1}), \; \rho(T_\gamma(y_i), y_{i+1})\} \le \delta \tag{9.97}$$

the inequality

$$\rho(x_n, y_n) \le \rho(x_A, x_B) + \epsilon$$

holds for every integer $n \ge n_0$*,*

Proof Choose an integer $n_0 \ge 1$ for which

$$(1-\gamma)^{n_0} D \le \epsilon/8. \tag{9.98}$$

Assume that $n \ge n_0$ is an integer and sequences

$$\{x_i\}_{i=0}^{\infty}, \; \{y_i\}_{i=0}^{\infty} \subset A \cup B$$

satisfy Eqs. (9.94)–(9.97). It follows from Lemma 9.13 and Eqs. (9.79) and (9.93)–(9.97) that

$$\rho(x_n, y_n) \le (1-\gamma)^n \rho(x_0, y_0) + (\gamma\rho(x_A, x_B) + 2\delta)\sum_{i=0}^{n-1}(1-\gamma)^i$$

$$\le (1-\gamma)^{n_0} D + 2\delta/\gamma + \rho(x_A, x_B)$$

$$\le \rho(x_A, x_B) + \epsilon/2.$$

Lemma 9.14 is proved.

Corollary 9.15 *Assume that the assumptions of Lemma 9.14 hold and that an integer* $n_0 \ge 1$ *is as guaranteed by Lemma 9.14. Then for every pair of sequences,*

$$\{x_i\}_{i=0}^{\infty}, \; \{y_i\}_{i=0}^{\infty} \subset A \cup B$$

which satisfy Eqs. (9.94)–(9.97) *and every integer* $n \ge n_0$*,*

$$\rho(x_n, x_{n+1}) \le \rho(x_A, x_B) + \epsilon$$

and

$$\rho(y_n, y_{n+1}) \le \rho(x_A, x_B) + \epsilon.$$

9.13 Proof of Theorem 9.11

Let $T \in \mathcal{M}$, $\gamma \in (0, 1)$ and k be a natural number. Put

$$\delta(T, \gamma, k) = 32^{-1}k^{-1}\gamma.$$

In view of Lemma 9.14, there exists an integer $n(T, \gamma, k) \geq 1$ such that Lemma 9.14 and Corollary 9.15 hold with

$$\epsilon = (2k)^{-1}, \; n_0 = n(T, \gamma, k), \; \delta = \delta(T, \gamma, k).$$

More precisely, the following property holds:

(a) For every integer $n \geq n(T, \gamma, k)$, every pair of sequences

$$\{x_i\}_{i=0}^{\infty}, \; \{y_i\}_{i=0}^{\infty} \subset A \cup B$$

which satisfy

$$x_0 \in A, \; y_0 \in B,$$

for every nonnegative integer i,

$$x_{2i} \in A, \; x_{2i+1} \in B,$$

$$y_{2i} \in B, \; y_{2i+1} \in A,$$

$$\max\{\rho(T_\gamma(x_i), x_{i+1}), \; \rho(T_\gamma(y_i), y_{i+1})\} \leq \delta(T, \gamma, k)$$

we have

$$\rho(x_n, y_n) \leq \rho(x_A, x_B) + (2k)^{-1},$$

$$\rho(x_n, x_{n+1}) \leq \rho(x_A, x_B) + (2k)^{-1}$$

and

$$\rho(y_n, y_{n+1}) \leq \rho(x_A, x_B) + (2k)^{-1}.$$

There exists an open neighborhood $\mathcal{U}(T, \gamma, k)$ of the mapping T_γ in the space $(\mathcal{M}, d)$ such that

$$\mathcal{U}(T, \gamma, k) \subset \{S \in \mathcal{M} : \; d(S, T_\gamma) < 2^{-1}\delta(T, \gamma, k)\}. \tag{9.99}$$

By Eq. (9.99) and property (a), the following property holds:

(b) For every operator $S \in \mathcal{U}(T, \gamma, k)$ and every sequence $\{x_i\}_{i=0}^{\infty} \subset A \cup B$ which satisfies for every nonnegative integer i,

$$\text{if } x_i \in A, \text{ then } x_{i+1} \in B \text{ and if } x_i \in B, \text{ then } x_{i+1} \in A,$$

$$\rho(S(x_i), x_{i+1}) \le \delta(T, \gamma, k)/2$$

we have

$$\rho(x_n, x_{n+1}) \le \rho(x_A, x_B) + (2k)^{-1}$$

for every integer $n \ge n(T, \gamma, k)$.

Define

$$\mathcal{F} = \cap_{k=1}^{\infty} \cup \{\mathcal{U}(T, \gamma, k) : T \in \mathcal{M}, \gamma \in (0, 1)\}. \tag{9.100}$$

It is clear that $\mathcal{F}$ is a countable intersection of open everywhere dense sets in $(\mathcal{M}, d)$.

Assume that

$$S \in \mathcal{F}, \ \epsilon \in (0, 1). \tag{9.101}$$

Choose an integer

$$k > 8\epsilon^{-1}.$$

In view of Eqs. (9.100) and (9.101), there exist $T \in \mathcal{M}$ and $\gamma \in (0, 1)$ such that

$$S \in \mathcal{U}(T, \gamma, k).$$

Now it is clear that property (b) holds and implies the validity of Theorem 9.11.

9.14 The First Porosity Result

Let X be a complete metric space. According to Baire's theorem, the intersection of every countable collection of open dense subsets of X is dense in X. This rather simple, yet powerful, result has found many applications. In particular, given a property that elements of X may have, it is of interest to determine whether this property is generic, that is, whether the set of elements which do enjoy this property contains a countable intersection of open dense sets. Such an approach, when a certain property is investigated for the whole space X and not just for a single point in X, has already been successfully applied in many areas of analysis (see, e.g.,

[61, 62, 144, 145, 153, 154, 168, 197, 200] and the references mentioned there). According to the generic approach, we say that a property holds for a generic (typical) element of a complete metric space (or the property holds generically) if the set of all elements of the metric space possessing this property contains a G_δ everywhere dense subset of the metric space which is a countable intersection of open everywhere dense sets.

Our results involves the notion of porosity which we now recall [144, 145, 165, 168, 210].

Let (Y, d) be a complete metric space. We denote by $B(y, r)$ the closed ball of center $y \in Y$ and radius $r > 0$. A subset $E \subset Y$ is called porous (with respect to the metric d) if there exist $\alpha \in (0, 1)$ and $r_0 > 0$ such that for each $r \in (0, r_0]$ and each $y \in Y$, there exists $z \in Y$ for which

$$B(z, \alpha r) \subset B(y, r) \setminus E.$$

A subset of the space Y is called σ-porous (with respect to d) if it is a countable union of porous subsets of Y.

Since porous sets are nowhere dense, all σ-porous sets are of the first Baire category. If Y is a finite-dimensional Euclidean space, then σ-porous sets are of Lebesgue measure 0. In fact, the class of σ-porous sets in such a space is much smaller than the class of sets that has Lebesgue measure 0 and is of the Baire first category. Also, every Banach space contains a set of the first Baire category which is not σ-porous.

To point out the difference between porous and nowhere dense sets, note that if $E \subset Y$ is nowhere dense, $y \in Y$, and $r > 0$, then there is a point $z \in Y$ and a number $s > 0$ such that $B(z, s) \subset B(y, r) \setminus E$. If, however, E is also porous, then for small enough r, we can choose $s = \alpha r$, where $\alpha \in (0, 1)$ is a constant that depends only on E.

Now we present three results obtained in [227]. In the first porosity result, we show that most cyclical nonexpansive mappings are contractive when the intersection of our two sets is not empty. In the second porosity theorem, we demonstrate that an analogous result is true in the case when the intersection is empty. Our third result shows that for every complete metric space, the distance between iterates of a contractive mapping converges to the distance between these two sets when the intersection of these two sets is empty.

Let (X, ρ, W) be a complete W-hyperbolic space and A and B be nonempty, bounded, and closed subsets of X such that

$$A \cap B \neq \emptyset.$$

For each $x \in K$ and each $r > 0$, set

$$B(x, r) = \{y \in X : \rho(x, y) \leq r\}.$$

Denote by $\mathcal{M}$ the set of all mappings $T : A \cup B \to A \cup B$ such that

$$T(A) \subset B,\ T(B) \subset A, \tag{9.102}$$

$$\rho(T(x), T(y)) \le \rho(x, y) \text{ for each } x \in A \text{ and each } y \in B. \tag{9.103}$$

For each $T_1, T_2 \in \mathcal{M}$, define

$$d(T_1, T_2) = \sup\{\rho(T_1(x), T_2(x)) :\ x \in A \cup B\}.$$

Clearly, the metric space $(\mathcal{M}, d)$ is complete. Set

$$D = \sup\{\rho(z_1, z_2) :\ z_1, z_2 \in A \cup B\}.$$

A mapping $T : A \cup B \to A \cup B$ is called contractive (in the sense of Rakotch) [156, 168] if there exists a decreasing function $\phi : [0, \infty) \to [0, 1]$ such that

$$\phi(t) < 1,\ t \in (0, \infty)$$

and for any point $a \in A$ and any point $b \in B$, we have

$$\rho(T(a), T(b)) \le \phi(\rho(a, b))\rho(a, b).$$

Theorems 9.1–9.3 imply the following theorem.

Theorem 9.16 *Assume that a mapping $T : A \cup B \to A \cup B$ is contractive. Then there exists $x_* \in A \cap B$ such that $T(x_*) = x_*$ and x_* is a unique fixed point of T. Moreover, for each $\epsilon > 0$, there exists $\delta > 0$ and a natural number k such that for every integer $n > k$ and any sequence $\{x_i\}_{i=0}^{n-1} \subset A \cup B$ which satisfies*

$$\rho(x_{i+1}, T(x_i)) \le \delta,\ i = 0, \ldots, n-1$$

the inequality $\rho(x_i, x_T) \le \epsilon$ is valid for all $i \in \{k, \ldots, n\}$.

Denote by $\mathcal{F}$ the set of all contractive mappings $T \in \mathcal{M}$. Now we state our first porosity result that is proved in the next section.

Theorem 9.17 *Assume that there exists*

$$x_* \in A \cap B$$

such that for each $a \in A$, each $b \in B$, and each $\gamma \in (0, 1)$,

$$(1-\gamma)a \oplus \gamma x_* \in A \text{ and } (1-\gamma)b \oplus \gamma x_* \in B.$$

Then the set $\mathcal{M} \setminus \mathcal{F}$ is σ-porous in $(\mathcal{M}, d)$.

9.15 Proof of Theorem 9.16

Let $\gamma \in (0, 1)$ and $T \in \mathcal{M}$. For each $a \in A$ and each $b \in B$, set

$$T_\gamma(a) = (1-\gamma)T(a) \oplus \gamma x_* \in A, \;\; T_\gamma(b) = (1-\gamma)T(b) \oplus \gamma x_* \in B. \quad (9.104)$$

By (9.43), (9.103), and (9.104), for each $a \in A$ and each $b \in B$,

$$\rho(T_\gamma(a), T_\gamma(b)) = \rho((1-\gamma)T(a) \oplus \gamma x_*, (1-\gamma)T(b) \oplus \gamma x_*)$$
$$\le (1-\gamma)\rho(T(a), T(b)) \le (1-\gamma)\rho(a, b) \quad (9.105)$$

and

$$\rho(T_\gamma(a), T(a)) = \rho((1-\gamma)T(a) \oplus \gamma x_*, T(a)) \le \gamma\rho(T(a), x_*) \le \gamma D, \quad (9.106)$$
$$\rho(T_\gamma(b), T(b)) = \rho((1-\gamma)T(b) \oplus \gamma x_*, T(b)) \le \gamma\rho(T(b), x_*) \le \gamma D. \quad (9.107)$$

Fix

$$\theta_1 \in A, \;\; \theta_2 \in B.$$

For each integer $n \ge 1$, denote by $\mathcal{F}_n$ the set of all $T \in \mathcal{M}$ such that

$$\sup\{\rho(T(a), T(b))\rho(a, b)^{-1} :$$
$$a \in A, \;\; b \in B \text{ and } \rho(a, b) \ge \min\{D/n, \;\; \rho(\theta_1, \theta_2)\}\} < 1. \quad (9.108)$$

Define

$$\mathcal{F} = \cap_{n=1}^{\infty} \mathcal{F}_n. \quad (9.109)$$

It is not difficult to see that every mapping $T \in \mathcal{F}_n$ is contractive. Indeed, if $T \in \mathcal{F}$, then we put $\phi(0) = 1$, and for every $t > 0$, define

$$\phi(t) = \sup\{\rho(T(a), T(b))\rho(a, b)^{-1} :$$
$$a \in A, \;\; b \in B \text{ and } \rho(a, b) \ge \min\{t, \;\; \rho(\theta_1, \theta_2)\}\} < 1$$

in view of (9.108), and now it is not difficult to see that for every $a \in A$ and every $b \in B$, we have

$$\rho(T(a), T(b)) \le \phi(\rho(a, b))\rho(a, b).$$

In order to complete the proof, it is sufficient to show that for every integer $n \geq 1$, the set $\mathcal{M} \setminus \mathcal{F}_n$ is porous in $(\mathcal{M}, d)$.

Let $n \geq 1$ be an integer. Define

$$\alpha = 16^{-1}(D+1)^{-1} \min\{D/n, \rho(\theta_1, \theta_2)\}. \tag{9.110}$$

Let $T \in \mathcal{M}$ and $r \in (0, 1]$. Set

$$\gamma = 4^{-1}r(D+1)^{-1} \tag{9.111}$$

and consider the mapping T_γ defined by (9.104). By (9.106), (9.107), and (9.111),

$$d(T_\gamma, T) \leq \gamma D \leq r/4. \tag{9.112}$$

Assume that $S \in \mathcal{M}$ satisfies

$$d(S, T_\gamma) \leq \alpha r. \tag{9.113}$$

In view of (9.110), (9.112), and (9.113),

$$d(S, T) \leq d(S, T_\gamma) + d(T_\gamma, T) \leq \alpha r + r/4 \leq r. \tag{9.114}$$

It follows from (9.105), (9.110), (9.111), and (9.113) that for each $a \in A$ and each $b \in B$ satisfying

$$\rho(a, b) \geq \min\{D/n, \ \rho(\theta_1, \theta_2)\},$$

we have

$$\rho(S(a), S(b))\rho(a, b)^{-1}$$

$$\leq (\rho(S(a), T_\gamma(a)) + \rho(T_\gamma(a), T_\gamma(b)) + \rho(T_\gamma(b), S(b)))\rho(a, b)^{-1}$$

$$\leq (2\alpha r + (1-\gamma)\rho(a, b))\rho(a, b)^{-1}$$

$$\leq 2\alpha r\rho(a, b)^{-1} + (1-\gamma)$$

$$\leq (1-\gamma) + 2\alpha r \min\{D/n, \ \rho(\theta_1, \theta_2)\}^{-1}$$

$$\leq 1 - 4^{-1}r(D+1)^{-1} + 8^{-1}r(D+1)^{-1} \leq 1 - 8^{-1}r(D+1)^{-1}.$$

Hence,

$$S \in \mathcal{F}_n.$$

Together with (9.114), this implies that

$$\{S \in \mathcal{M} : d(S, T_\gamma) \le \alpha r\} \subset \{S \in \mathcal{M} : d(S, T) \le r\} \cap \mathcal{F}_n.$$

Therefore, the set $\mathcal{M} \setminus \mathcal{F}_n$ is porous. This completes the proof of Theorem 9.17.

9.16 The Second Porosity Result

Let (X, ρ, W) be a complete W-hyperbolic space and A and B be nonempty, bounded, and closed subsets of X such that

$$A \cap B = \emptyset \tag{9.115}$$

and let

$$x_A \in A, \ x_B \in B \tag{9.116}$$

satisfy

$$\rho(x_A, x_B) = \rho(A, B) := \inf\{\rho(a, b) : \ a \in A, \ b \in B\}. \tag{9.117}$$

Assume that for each $a \in A$, each $b \in B$, and each $\gamma \in (0, 1)$,

$$\gamma x_A \oplus (1 - \gamma)a \in A \tag{9.118}$$

$$\gamma x_B \oplus (1 - \gamma)b \in B. \tag{9.119}$$

We continue to consider the space $\mathcal{M}$ of all mappings

$$T : A \cup B \to A \cup B \tag{9.120}$$

such that

$$T(A) \subset B, \ T(B) \subset A \tag{9.121}$$

and

$$\rho(T(x), T(y)) \le \rho(x, y) \text{ for all } x \in A \text{ and all } y \in B \tag{9.122}$$

equipped with the metric

$$d(T_1, T_2) = \sup\{\rho(T_1(x), T_2(x)) : \ x \in A \cup B\}, \ T_1, T_2 \in \mathcal{M}. \tag{9.123}$$

Note that $(\mathcal{M}, d)$ is a complete metric space and recall that

$$D = \sup\{\rho(a, b) : a, b \in A \cup B\}. \tag{9.124}$$

A mapping $T : A \cup B \to A \cup B$ is called contractive (in the sense of Rakotch) [156, 168] if there exists a decreasing function $\phi : [0, \infty) \to [0, 1]$ such that

$$\phi(t) < 1, \ t \in (0, \infty) \tag{9.125}$$

and for any point $a \in A$ and any point $b \in B$, we have

$$\rho(T(a), T(b)) - \rho(A, B) \le \phi(\rho(a, b) - \rho(A, B))(\rho(a, b) - \rho(A, B)). \tag{9.126}$$

Denote by $\tilde{\mathcal{F}}$ the set of all contractive mappings $T \in \mathcal{M}$. Now we prove our second main result.

Theorem 9.18 *The set $\mathcal{M} \setminus \tilde{\mathcal{F}}$ is σ-porous in $(\mathcal{M}, d)$.*

Proof We may assume without loss of generality that

$$\tilde{D} := \sup\{\rho(a, b) : a \in A, \ b \in B\} - \rho(A, B) > 0.$$

Let $T \in \mathcal{M}$ and $\gamma \in (0, 1)$. Define

$$T_\gamma(a) = (1 - \gamma)T(a) \oplus \gamma x_B, \ a \in A, \tag{9.127}$$

$$T_\gamma(b) = (1 - \gamma)T(b) \oplus \gamma x_A, \ b \in B. \tag{9.128}$$

By (9.118), (9.119), (9.121), (9.127), and (9.128),

$$T_\gamma(A) \subset B, \ T_\gamma(B) \subset A. \tag{9.129}$$

It follows from (9.43), (9.124), (9.127), and (9.128) that each $a \in A$ and each $b \in B$,

$$\rho(T_\gamma(a), T(a)) = \rho((1 - \gamma)T(a) \oplus \gamma x_B, T(a)) \le \gamma\rho(x_B, T(a)) \le \gamma D, \tag{9.130}$$

$$\rho(T_\gamma(b), T(b)) = \rho((1 - \gamma)T(b) \oplus \gamma x_A, T(b)) \le \gamma\rho(x_A, T(b)) \le \gamma D. \tag{9.131}$$

In view of (9.43), (9.122), (9.127), and (9.128), for each $a \in A$ and each $b \in B$,

$$\rho(T_\gamma(a), T_\gamma(b)) = \rho((1 - \gamma)T(a) \oplus \gamma x_B, \ (1 - \gamma)T(b) \oplus \gamma x_A)$$

$$\le (1 - \gamma)\rho(T(a), T(b)) + \gamma\rho(x_A, x_B)$$

$$\leq (1-\gamma)\rho(a,b) + \gamma\rho(x_A, x_B)$$

and

$$\rho(a,b) - \rho(T_\gamma(a), T_\gamma(b)) \geq \gamma(\rho(a,b) - \rho(x_A, x_B)). \tag{9.132}$$

Let $n \geq 2$ be an integer. Define

$$\Omega_n = \{(a,b) \in A \times B : \ \rho(a,b) - \rho(x_A, x_B) \geq \tilde{D}/n\}. \tag{9.133}$$

Clearly,

$$\Omega_n \neq \emptyset.$$

It follows from (9.132) and (9.133) that for each $(a,b) \in \Omega_n$,

$$\rho(a,b) - \rho(T_\gamma(a), T_\gamma(b)) \geq \gamma(\rho(a,b) - \rho(x_A, x_B)) \geq \gamma\tilde{D}/n. \tag{9.134}$$

Denote by $\mathcal{M}_n$ the set of all $S \in \mathcal{M}$ such that

$$\inf\{\rho(a,b) - \rho(S(a), S(b))(\rho(a,b) - \rho(A,B))^{-1} : \ (a,b) \in \Omega_n\} > 0. \tag{9.135}$$

We show that the set $\mathcal{M} \setminus \mathcal{M}_n$ is porous in $(\mathcal{M}, d)$. Set

$$\alpha = 8^{-1}(D+1)^{-1}\tilde{D}/n. \tag{9.136}$$

Let $T \in \mathcal{M}$ and $r \in (0, 1]$. Set

$$\gamma = 2^{-1}r(D+1)^{-1} \tag{9.137}$$

and consider the mapping T_γ defined by (9.127) and (9.128). By (9.130), (9.131), and (9.137),

$$d(T_\gamma, T) \leq \gamma D \leq 2^{-1}r. \tag{9.138}$$

It follows from (9.132) and (9.137) that for each $(a,b) \in \Omega_n$,

$$(\rho(a,b) - \rho(T_\gamma(a), T_\gamma(b))) \geq \gamma(\rho(a,b) - \rho(A,B))$$

$$= 2^{-1}r(D+1)^{-1}(\rho(a,b) - \rho(A,B)). \tag{9.139}$$

Assume that $S \in \mathcal{M}$ satisfies

$$d(S, T_\gamma) \leq \alpha r. \tag{9.140}$$

In view of (9.136), (9.138), and (9.140),

$$d(S, T) \le \alpha r + r/2 \le 3r/4. \tag{9.141}$$

Let

$$(a, b) \in \Omega_n. \tag{9.142}$$

It follows from (9.133), (9.136), (9.139), (9.140), and (9.142) that

$$\begin{aligned}
&\rho(a, b) - \rho(S(a), S(b)) \\
&\ge \rho(a, b) - \rho(T_\gamma(a), T_\gamma(b)) - \rho(T_\gamma(a), S(a)) - \rho(T_\gamma(b), S(b)) \\
&\ge 2^{-1} r(D + 1)^{-1}(\rho(a, b) - \rho(A, B)) - 2\alpha r \\
&\ge 2^{-1} r(D + 1)^{-1}(\rho(a, b) - \rho(A, B) - 4(D + 1)\alpha) \\
&\ge 2^{-1} r(D + 1)^{-1}(\rho(a, b) - \rho(A, B) - 2^{-1}\tilde{D}/n) \\
&\ge 4^{-1} r(D + 1)^{-1}(\rho(a, b) - \rho(A, B)).
\end{aligned} \tag{9.143}$$

In view of (9.143),

$$S \in \mathcal{M}_n.$$

Together with (9.141), this implies that

$$\{S \in \mathcal{M} : d(S, T_\gamma) \le \alpha r\} \subset \{S \in \mathcal{M} : d(S, T) \le r\} \cap \mathcal{M}_n.$$

Therefore, the set $\mathcal{M} \setminus \mathcal{M}_n$ is porous. Set

$$\mathcal{F} = \cap_{n=2}^{\infty} \mathcal{M}_n. \tag{9.144}$$

Clearly, $\mathcal{M} \setminus \mathcal{F}$ is σ-porous in $(\mathcal{M}, d)$.

Assume that

$$S \in \mathcal{F}. \tag{9.145}$$

Equations (9.135), (9.144), and (9.145) imply that for every integer $n \ge 2$, $S \in \mathcal{M}_n$ and there exists $\gamma_n \in (0, 1)$ such that

$$(\rho(a, b) - \rho(S(a), S(b)))(\rho(a, b) - \rho(A, B))^{-1} > \gamma_n, \ (a, b) \in \Omega_n. \tag{9.146}$$

In view of (9.146), for every integer $n \ge 2$ and every $(a, b) \in \Omega_n$, we have

$$\rho(S(a), S(b)) < \rho(a, b) - \gamma_n(\rho(a, b) - \rho(A, B)),$$

$$\rho(S(a), S(b)) - \rho(A, B) < \rho(a, b) - \rho(A, B) - \gamma_n(\rho(a, b) - \rho(A, B))$$
$$< (1 - \gamma_n)(\rho(a, b) - \rho(A, B))$$

and

$$\sup\{(\rho(S(a), S(b)) - \rho(A, B))(\rho(a, b) - \rho(A, B))^{-1} : (a, b) \in \Omega_n\} < 1. \tag{9.147}$$

Since n is an arbitrary integer satisfying $n \geq 2$, it follows from (9.133) and (9.147) that for every $\epsilon > 0$, we have

$$\phi(\epsilon) := \sup\{(\rho(S(a), S(b)) - \rho(A, B))(\rho(a, b) - \rho(A, B))^{-1} :$$
$$(a, b) \in A \times B, \ \rho(a, b) - \rho(A, B) \geq \epsilon\} < 1.$$

(Here the supremum over an empty set is zero.) Set $\phi(0) = 1$. It is easy to see that the function $\phi : [0, \infty) \to [0, 1]$ is decreasing and that for each $(a, b) \in A \times B$, we have,

$$\rho(S(a), S(b)) - \rho(A, B) \leq \phi(\rho(a, b) - \rho(A, B))(\rho(a, b) - \rho(A, B)).$$

Thus, the mapping S is contractive. Theorem 9.18 is proved.

9.17 The Third Result

Let (X, ρ) be a complete metric space and A and B be nonempty and closed subsets of X such that

$$A \cap B = \emptyset, \tag{9.148}$$

$$\rho(A, B) := \inf\{\rho(a, b) : \ a \in A, \ b \in B\} \tag{9.149}$$

and let a mapping $T : A \cup B \to A \cup B$ satisfies

$$T(A) \subset B, \ T(B) \subset A. \tag{9.150}$$

Assume that there exists a decreasing function $\phi : [0, \infty) \to [0, 1]$ such that

$$\phi(t) < 1, \ t \in (0, \infty)$$

and that for any point $a \in A$ and any point $b \in B$, we have

$$\rho(T(a), T(b)) - \rho(A, B) \le \phi(\rho(a, b) - \rho(A, B))(\rho(a, b) - \rho(A, B)). \quad (9.151)$$

In other words, the mapping T is contractive (in the sense of Rakotch) [156, 168].

Fix $\theta \in A \cup B$. We prove the following theorem.

Theorem 9.19 *Let $M > 0$ and $\epsilon \in (0, 1)$. Then there exist a natural number n_0 and $\delta > 0$ such that for each sequence $\{x_i\}_{i=0}^{\infty}$ which satisfies*

$$\rho(x_0, \theta) \le M$$

and for each integer $i \ge 0$,

$$\text{if } x_i \in A, \text{ then } x_{i+1} \in B \text{ and if } x_i \in B, \text{ then } x_{i+1} \in A$$

and

$$\rho(x_{i+1}, T(x_i)) \le \delta$$

the inequality $\rho(x_i, x_{i+1}) \le \rho(A, B) + \epsilon$ holds for all integers $n \ge n_0$.

Proof Fix

$$x_A \in A, \ x_B \in B \quad (9.152)$$

satisfying

$$\rho(x_A, x_B) \le \rho(A, B) + 1. \quad (9.153)$$

Choose

$$\delta \in (0, 8^{-1}(1 - \phi(\epsilon/2))\epsilon). \quad (9.154)$$

Assume that

$$z_0 \in A \cup B, \ \rho(z_0, \theta) \le M, \ z_1 \in X, \ \rho(z_1, T(z_0)) \le 1. \quad (9.155)$$

In view of (9.155),

$$\rho(z_0, x_A) \le \rho(z_0, \theta) + \rho(\theta, x_A) \le M + \rho(\theta, x_A) \quad (9.156)$$

and

$$\rho(z_0, x_B) \le \rho(z_0, \theta) + \rho(\theta, x_B) \le M + \rho(\theta, x_B). \quad (9.157)$$

If $z_0 \in A$, then by (9.151), (9.155), and (9.157),

$$\rho(z_0, T(z_0)) \le \rho(z_0, x_B) + \rho(x_B, T(x_B)) + \rho(T(x_B), T(z_0))$$
$$\le 2\rho(z_0, x_B) + \rho(x_B, T(x_B))$$
$$\le 2M + 2\rho(\theta, x_B) + \rho(x_B, T(x_B))$$

and

$$\rho(z_0, z_1) \le 1 + 2M + 2\rho(\theta, x_B) + \rho(x_B, T(x_B)). \tag{9.158}$$

If $z_0 \in B$, then by (9.151), (9.155), and (9.156),

$$\rho(z_0, T(z_0)) \le \rho(z_0, x_A) + \rho(x_A, T(x_A)) + \rho(T(x_A), T(z_0))$$
$$\le 2M + 2\rho(\theta, x_A) + \rho(x_A, T(x_A))$$

and

$$\rho(z_0, z_1) \le 1 + 2M + 2\rho(\theta, x_A) + \rho(x_A, T(x_A)).$$

Thus (see (9.158)) in both cases,

$$\rho(z_0, z_1) \le c_1 := 1 + 2M + 2\rho(\theta, x_A) + 2\rho(\theta, x_B) + \rho(x_A, T(x_A)) + \rho(x_B, T(x_B)). \tag{9.159}$$

Hence, we showed that the following property holds:

(i) Equation (9.155) implies (9.159).

Choose an integer

$$n_0 > 4c_1\epsilon^{-1}(1 - \phi(\epsilon/2))^{-1} + 4. \tag{9.160}$$

Assume that $\{x_i\}_{i=0}^{\infty} \subset A \cup B$ satisfies for each integer $i \ge 0$,

$$\text{if } x_i \in A, \text{ then } x_{i+1} \in B \text{ and if } x_i \in B, \text{ then } x_{i+1} \in A, \tag{9.161}$$

$$\rho(x_0, \theta) \le M \tag{9.162}$$

and for each integer $i \ge 0$,

$$\rho(x_{i+1}, T(x_i)) \le \delta. \tag{9.163}$$

We show that there exists an integer $j \in [0, n_0]$ such that

$$\rho(x_j, x_{j+1}) \le \rho(A, B) + \epsilon/2.$$

Assume the contrary. Then

$$\rho(x_i, x_{i+1}) > \rho(A, B) + \epsilon/2, \ i = 0, \dots, n_0. \tag{9.164}$$

By (9.151), (9.161), and (9.164) and monotonicity of ϕ,

$$\rho(T(x_i), T(x_{i+1})) - \rho(A, B)$$

$$\le (\rho(x_i, x_{i+1}) - \rho(A, B))\phi(\rho(x_i, x_{i+1}) - \rho(A, B))$$

$$\le \phi(\epsilon/2)(\rho(x_i, x_{i+1}) - \rho(A, B)),$$

$$\rho(x_i, x_{i+1}) - \rho(T(x_i), T(x_{i+1}))$$

$$= \rho(x_i, x_{i+1}) - \rho(A, B) - (\rho(T(x_i), T(x_{i+1})) - \rho(A, B))$$

$$\ge (1 - \phi(\epsilon/2))(\rho(x_i, x_{i+1}) - \rho(A, B)) \ge (1 - \phi(\epsilon/2))\epsilon/2$$

and together with (9.154) and (9.163), this implies that

$$\rho(x_i, x_{i+1}) - \rho(x_{i+1}, x_{i+2})$$

$$\ge \rho(x_i, x_{i+1}) - \rho(T(x_i), T(x_{i+1})) - \rho(T(x_i), x_{i+1}) - \rho(T(x_{i+1}), x_{i+2})$$

$$\ge (1 - \phi(\epsilon/2))\epsilon/2 - 2\delta \ge (1 - \phi(\epsilon/2))\epsilon/4. \tag{9.165}$$

Property (i) and Eqs. (9.162), (9.163), and (9.165) imply that

$$c_1 \ge \rho(x_0, x_1) \ge \rho(x_0, x_1) - \rho(x_{n_0}, x_{n_0+1})$$

$$= \sum_{i=0}^{n_0-1} (\rho(x_i, x_{i+1}) - \rho(x_{i+1}, x_{i+2})) \ge 4^{-1}(1 - \phi(\epsilon/2))\epsilon n_0$$

and

$$n_0 \le 4c_1(1 - \phi(\epsilon/2))^{-1}\epsilon^{-1}.$$

This contradicts (9.160). The contradiction we have reached proves that there exists an integer

$$j \in [0, n_0] \tag{9.166}$$

such that

$$\rho(x_j, x_{j+1}) \le \rho(A, B) + \epsilon/2. \tag{9.167}$$

In order to complete the proof of the theorem, it is sufficient to show that

$$\rho(x_i, x_{i+1}) \le \rho(A, B) + \epsilon$$

for all integers $i \ge j$.

Assume the contrary. Then there is an integer $k > j$ such that

$$\rho(x_k, x_{k+1}) > \rho(A, B) + \epsilon. \tag{9.168}$$

We may assume without loss of generality that for all integers $i \in \{j, \dots k - 1\}$,

$$\rho(x_i, x_{i+1}) \le \rho(A, B) + \epsilon$$

and in particular

$$\rho(x_{k-1}, x_k) \le \rho(A, B) + \epsilon. \tag{9.169}$$

There are two cases:

$$\rho(x_{k-1}, x_k) \le \rho(A, B) + \epsilon/2; \tag{9.170}$$

$$\rho(x_{k-1}, x_k) > \rho(A, B) + \epsilon/2. \tag{9.171}$$

Assume that (9.170) holds. Then by (9.151), (9.154), (9.163), and (9.170),

$$\rho(x_k, x_{k+1}) \le \rho(x_k, T(x_{k-1})) + \rho(T(x_{k-1}), T(x_k)) + \rho(T(x_k), x_{k+1})$$

$$\le 2\delta + \rho(T(x_{k-1}), T(x_k)) \le 2\delta + \rho(x_{k-1}, x_k)$$

$$\le \rho(A, B) + \epsilon/2 + 2\delta \le \rho(A, B) + \epsilon.$$

This contradicts (9.168). The contradiction we have reached proves that (9.170) holds. By (9.151), (9.171), and monotonicity of ϕ,

$$\rho(T(x_{k-1}), T(x_k)) - \rho(A, B)$$

$$\le (\rho(x_{k-1}, x_k) - \rho(A, B))\phi(\rho(x_{k-1}, x_k) - \rho(A, B)),$$

$$\rho(x_k, x_{k-1}) - \rho(T(x_{k-1}), T(x_k))$$

$$= \rho(x_{k-1}, x_k) - \rho(A, B) - (\rho(T(x_{k-1}), T(x_k)) - \rho(A, B))$$

$$\geq (\rho(x_{k-1}, x_k) - \rho(A, B))(1 - \phi(\rho(x_{k-1}, x_k) - \rho(A, B)))$$

$$\geq (\rho(x_{k-1}, x_k) - \rho(A, B))(1 - \phi(\epsilon/2))$$

$$\geq (1 - \phi(\epsilon/2))\epsilon/2. \tag{9.172}$$

It follows from (9.154), (9.163), and (9.172) that

$$\rho(x_k, x_{k+1}) \leq \rho(x_k), T(x_{k-1})) + \rho(T(x_{k-1}), T(x_k)) + \rho(T(x_k), x_{k+1})$$

$$\leq 2\delta + (1 - \phi(\epsilon/2))\epsilon/2 + \rho(A, B) < \rho(A, B) + \epsilon.$$

This contradicts (9.168). The contradiction we have reached completes the proof of Theorem 9.19.

Chapter 10
Monotone Nonexpansive Mappings

In this chapter, we study monotone nonexpansive mappings. First, we consider three classes of monotone contractive mappings defined on a complete metric space, and for each mapping in one of these classes, we establish the existence of a unique fixed point that attracts all iterates. Then it is shown that the convergence of iterates of a monotone nonexpansive mapping is stable in the presence of summable computational errors. We consider monotone nonexpansive self-mappings of a bounded closed subset of a Banach space, and using the Baire category approach, we show that for most of monotone nonexpansive mappings, the fixed point problem is well-posed.

10.1 Monotone Contractive Mappings

The class of monotone nonexpansive mappings has been the subject of a rapidly growing area of research [7, 77]. We consider three classes of monotone contractive mappings defined on a complete metric space. For each mapping in one of these classes, we establish the existence of a unique fixed point that attracts all iterates. These results were obtained in [170].

Let (X, ρ) be a complete metric space equipped with a partial order $\leq$. That is, for all points $x, y, z \in X$, we have

$$x \leq x,$$

$$\text{if } x \leq y, \;\; y \leq x, \;\; \text{then } x = y,$$

and

$$\text{if } x \leq y, \;\; y \leq z, \;\; \text{then } x \leq z.$$

A. J. Zaslavski, *Approximate Fixed Points of Nonexpansive Mappings*,
Developments in Mathematics 80, https://doi.org/10.1007/978-3-031-70710-0_10

We also assume that

$$\{(x, y) \in X \times X : x \le y\}$$

is a closed subset of $X \times X$.

Let K be a nonempty closed subset of X which is not a singleton. Let $x_K \in K$ and assume that at least one of the following relations holds:

$$x_K \le x \text{ for all } x \in K \tag{10.1}$$

or

$$x \le x_K \text{ for all } x \in K. \tag{10.2}$$

For each $x \in K$ and each $r > 0$, set

$$B(x, r) := \{y \in X : \rho(x, y) \le r\}.$$

Let $T : K \to X$. Denote by T^0 the identity operator $I : K \to K$, that is, $I(x) = x,\ x \in K$. Suppose that the graph of T

$$\mathrm{graph}(T) = \{(x, T(x)) : x \in K\}$$

is a closed subset of $X \times X$,

$$T^i(x_K) \in K \text{ for all integers } i \ge 1 \tag{10.3}$$

and

$$T(x) \le T(y) \text{ for all } x, y \in K \text{ such that } x \le y. \tag{10.4}$$

Our goal is to show that analogs of many results obtained in our previous research for nonexpansive mappings hold for monotone nonexpansive operators too. For these operators, relation (10.4) is true only if $x \le y$. Therefore, in order to obtain such analogs, we should guarantee that for our sequences of iterates, the inequality $x_n \le x_{n+1}$ holds for any integer $n \ge 0$. In order to meet this goal, we need assumptions like (10.1) or (10.2).

We prove three theorems obtained in [170] regarding the existence of a unique fixed point of such a mapping T under three different contractivity assumptions. In the first result, we use contractivity in the sense of Rakotch [156], the second is in the spirit of Boyd and Wong [23], while in the third result, T is a contractive mapping in the sense of Matkowski [128].

Theorem 10.1 *Assume that* $\phi : [0, \infty) \to [0, 1]$ *is a decreasing function:*

$$\phi(t) < 1 \text{ for all } t > 0 \tag{10.5}$$

and that for all $x, y \in K$ satisfying $x \le y$, we have

$$\rho(T(x), T(y)) \le \phi(\rho(x, y))\rho(x, y). \tag{10.6}$$

Then the sequence $\{T^i(x_K)\}_{i=1}^{\infty}$ converges, $\lim_{i\to\infty} T^i(x_K)$ is the unique fixed point of the mapping T, and the following assertion holds.

Let M and ϵ be positive. Then there exists a natural number n_0 such that if

$$x \in K \cap B(x_K, M),$$

$n > n_0$ is an integer and $T^n(x)$ is defined, then

$$\rho(T^i(x), \lim_{j\to\infty} T^j(x_K)) \le \epsilon$$

for all $i = n_0 + 1, \dots, n$.

Theorem 10.2 *Assume that the function $\phi : [0, \infty) \to [0, \infty)$ is upper semicontinuous:*

$$\phi(t) < t \text{ for all } t > 0 \tag{10.7}$$

and that for all $x, y \in K$ satisfying $x \le y$, we have

$$\rho(T(x), T(y)) \le \phi(\rho(x, y)). \tag{10.8}$$

Then the sequence $\{T^i(x_K)\}_{i=1}^{\infty}$ converges, $\lim_{i\to\infty} T^i(x_K)$ is the unique fixed point of the mapping T, and the following assertion holds.

Let M and ϵ be positive. Then there exists a natural number n_0 such that if

$$x \in K \cap B(x_K, M),$$

$n > n_0$ is an integer and $T^n(x)$ is defined, then

$$\rho(T^i(x), \lim_{j\to\infty} T^j(x_K)) \le \epsilon$$

for all $i = n_0 + 1, \dots, n$.

Theorem 10.3 *Assume that $\phi : [0, \infty) \to [0, \infty)$ is an increasing function:*

$$\lim_{n\to\infty} \phi^n(t) = 0 \text{ for all } t > 0 \tag{10.9}$$

and that for all $x, y \in K$ satisfying $x \le y$, we have

$$\rho(T(x), T(y)) \le \phi(\rho(x, y)). \tag{10.10}$$

Then the sequence $\{T^i(x_K)\}_{i=1}^{\infty}$ *converges,* $\lim_{i\to\infty} T^i(x_K)$ *is the unique fixed point of the mapping* T*, and the following assertion holds.*

Let M *and* ϵ *be positive. Then there exists a natural number* n_0 *such that if*

$$x \in K \cap B(x_K, M),$$

$n > n_0$ *is an integer and* $T^n(x)$ *is defined, then*

$$\rho(T^i(x), \lim_{j\to\infty} T^j(x_K)) \le \epsilon$$

for all $i = n_0 + 1, \ldots, n$.

10.2 Proof of Theorem 10.1

If (10.1) holds, then

$$T^0(x_K) \le T^1(x_K)$$

and in view of (10.4), for all integers $i \ge 0$, we have

$$T^i(x_K) \le T^{i+1}(x_K). \tag{10.11}$$

If (10.2) holds, then

$$T^1(x_K) \le T^0(x_K)$$

and in view of (10.4), for all integers $i \ge 0$, we have

$$T^{i+1}(x_K) \le T^i(x_K). \tag{10.12}$$

By (10.6), (10.11), and (10.12), for all integers $i \ge 0$,

$$\rho(T^i(x_K), T^{i+1}(x_K)) \ge \rho(T^{i+1}(x_K), T^{i+2}(x_K)). \tag{10.13}$$

We claim that

$$\lim_{i\to\infty} \rho(T^i(x_K), T^{i+1}(x_K)) = 0. \tag{10.14}$$

Suppose to the contrary that (10.14) does not hold. Then in view of (10.13), there exists a number $r > 0$ such that

$$\rho(T^i(x_K), T^{i+1}(x_K)) > r \text{ for all integers } i \geq 0. \tag{10.15}$$

Since the function ϕ is decreasing, it follows from (10.6), (10.11), (10.12), and (10.15) that for all integers $i \geq 0$,

$$\rho(T^i(x_K), T^{i+1}(x_K)) - \rho(T^{i+1}(x_K), T^{i+2}(x_K))$$

$$\geq \rho(T^i(x_K), T^{i+1}(x_K)) - \phi(\rho(T^i(x_K), T^{i+1}(x_K)))\rho(T^i(x_K), T^{i+1}(x_K))$$

$$\geq \rho(T^i(x_K), T^{i+1}(x_K))(1 - \phi(\rho(T^i(x_K), T^{i+1}(x_K))))$$

$$\geq \rho(T^i(x_K), T^{i+1}(x_K))(1 - \phi(r)) \geq r(1 - \phi(r)).$$

This implies that for every natural number n,

$$\rho(x_K, T^1(x_K))$$

$$\geq \rho(x_K, T^1(x_K)) - \rho(T^n(x_K), T^{n+1}(x_K))$$

$$= \sum_{i=0}^{n-1}[\rho(T^i(x_K), T^{i+1}(x_K)) - \rho(T^{i+1}(x_K), T^{i+2}(x_K))]$$

$$\geq nr(1 - \phi(r)) \to \infty \text{ as } n \to \infty.$$

The contradiction we have reached proves that (10.14) does hold.

Next, we prove that $\{T^i(x_K)\}_{i=1}^{\infty}$ is a Cauchy sequence. To this end, let $\epsilon > 0$. By (10.14), there exists a natural number n_0 such that for all integers $i \geq n_0$,

$$\rho(T^i(x_K), T^{i+1}(x_K)) \leq (\epsilon/4)(1 - \phi(\epsilon)). \tag{10.16}$$

Assume that

$$n_2 > n_1 \geq n_0 \tag{10.17}$$

are integers. We now show that

$$\rho(T^{n_1}(x_K), T^{n_2}(x_K)) \leq \epsilon.$$

Suppose to the contrary that

$$\rho(T^{n_1}(x_K), T^{n_2}(x_K)) > \epsilon. \tag{10.18}$$

By (10.6), (10.11), (10.12), and (10.18),

$$\rho(T^{n_1+1}(x_K), T^{n_2+1}(x_K))$$

$$\leq \phi(\rho(T^{n_1}(x_K), T^{n_2}(x_K)))\rho(T^{n_1}(x_K), T^{n_2}(x_K))$$

$$\leq \phi(\epsilon)\rho(T^{n_1}(x_K), T^{n_2}(x_K)). \tag{10.19}$$

In view of (10.19),

$$\phi(\epsilon)\rho(T^{n_1}(x_K), T^{n_2}(x_K)) \geq \rho(T^{n_1+1}(x_K), T^{n_2+1}(x_K))$$

$$\geq \rho(T^{n_1}(x_K), T^{n_2}(x_K)) - \rho(T^{n_1}(x_K), T^{n_1+1}(x_K)) - \rho(T^{n_2}(x_K), T^{n_2+1}(x_K)). \tag{10.20}$$

It follows from (10.6)–(10.8) and (10.20) that

$$(1-\phi(\epsilon))\epsilon/2 \geq \rho(T^{n_1}(x_K), T^{n_1+1}(x_K)) + \rho(T^{n_2}(x_K), T^{n_2+1}(x_K))$$

$$\geq (1-\phi(\epsilon))\rho(T^{n_1}(x_K), T^{n_2}(x_K)) \geq \epsilon(1-\phi(\epsilon)),$$

which is a contradiction. The contradiction we have reached proves that

$$\rho(T^{n_1}(x_K), T^{n_2}(x_K)) \leq \epsilon$$

for all integers $n_1, n_2 \geq n_0$. Thus, $\{T^i(x_K)\}_{i=1}^{\infty}$ is indeed a Cauchy sequence, as claimed, and there exists

$$\widehat{x} = \lim_{i\to\infty} T^i(x_K). \tag{10.21}$$

Since graph(T) is a closed set, (10.21) implies that

$$T(\widehat{x}) = \widehat{x}.$$

Next we show that $\widehat{x}$ is the unique fixed point of T.

To this end, assume that $z \in K$ and

$$T(z) = z. \tag{10.22}$$

If (10.1) holds, then

$$T^i(x_K) \leq z \text{ for all integers } i \geq 1$$

and

$$\widehat{x} \leq z,$$

and if (10.2) holds, then

$$z \le T^i(x_K) \text{ for all integers } i \ge 1$$

and

$$z \le \widehat{x}.$$

In both cases, (10.6) implies that

$$\rho(z, \widehat{x}) \le \phi(\rho(z, \widehat{x}))\rho(z, \widehat{x}).$$

If $z \ne \widehat{x}$, then we have reached a contradiction. Therefore,

$$z = \widehat{x},$$

as claimed.

Now let M and ϵ be positive. There exists a natural number n_0 such that

$$n_0 > M((1 - \phi(\epsilon/2))(\epsilon/2))^{-1} \tag{10.23}$$

and

$$\rho(T^i(x_K), \widehat{x}) \le \epsilon/2 \text{ for all integers } i \ge n_0. \tag{10.24}$$

Assume that

$$x \in B(x_K, M) \cap K, \tag{10.25}$$

$n > n_0$ is an integer and that $T^n(x)$ is defined.

If (10.1) holds, then by (10.6),

$$T^i(x_K) \le T^i(x) \text{ for all integers } i = 1, \dots, n \tag{10.26}$$

and if (10.2) holds, then by (10.6),

$$T^i(x) \le T^i(x_K) \text{ for all integers } i = 1, \dots, n. \tag{10.27}$$

In both cases, (10.6) implies that for all $i = 0, \dots, n-1$,

$$\rho(T^{i+1}(x), T^{i+1}(x_K)) \le \phi(\rho(T^i(x), T^i(x_K)))\rho(T^i(x), T^i(x_K)). \tag{10.28}$$

We now show that there exists an integer $i \in [0, n_0]$ such that

$$\rho(T^i(x), T^i(x_K)) \le \epsilon/2.$$

Suppose to the contrary that this does not hold. Then for all $i = 0, \dots, n_0$,

$$\rho(T^i(x), T^i(x_K)) > \epsilon/2. \tag{10.29}$$

In view of (10.28) and (10.29), for all $i = 0, \dots, n_0$,

$$\rho(T^{i+1}(x), T^{i+1}(x_K)) \le \phi(\rho(T^i(x), T^i(x_K)))\rho(T^i(x), T^i(x_K))$$

$$\le \phi(\epsilon/2)\rho(T^i(x), T^i(x_K)).$$

By (10.19), (10.25), and the equation above, we have

$$M \ge \rho(x, x_K)$$

$$\ge \rho(x, x_K) - \rho(T^{n_0}(x), T^0(x_K))$$

$$= \sum_{i=0}^{n_0-1} [\rho(T^i(x), T^i(x_K)) - \rho(T^{i+1}(x), T^{i+1}(x_K))]$$

$$\ge \sum_{i=0}^{n_0-1} (1 - \phi(\epsilon/2))\rho(T^i(x), T^i(x_K))$$

$$\ge n_0(1 - \phi(\epsilon/2))\epsilon/2$$

and

$$n_0 \le M(1 - \phi(\epsilon/2))^{-1}(\epsilon/2)^{-1}.$$

This inequality contradicts (10.23). The contradiction we have reached proves that there indeed exists an integer $i_0 \in \{0, \dots, n_0\}$ such that

$$\rho(T^{i_0}(x), T^{i_0}(x_K)) \le \epsilon/2. \tag{10.30}$$

Assume now that an integer i satisfies

$$i_0 \le i \le n.$$

In view of (10.28) and (10.30),

$$\rho(T^i(x), T^i(x_K)) \le \epsilon/2.$$

If $i \ge n_0$, we have by (10.24) and the relation above,

$$\rho(\widehat{x}, T^i(x)) \le \rho(\widehat{x}, T^i(x_K)) + \rho(T^i(x), T^i(x_K)) \le \epsilon.$$

This completes the proof of Theorem 10.1.

10.3 Proof of Theorem 10.2

If (10.1) holds, then

$$T^0(x_K) \le T^1(x_K)$$

and in view of (10.4), for all integers $i \ge 0$,

$$T^i(x_K) \le T^{i+1}(x_K). \tag{10.31}$$

If (10.2) holds, then

$$T^1(x_K) \le T^0(x_K)$$

and in view of (10.4), for all integers $i \ge 0$,

$$T^{i+1}(x_K) \le T^i(x_K). \tag{10.32}$$

By (10.7), (10.8), (10.31), and (10.32), for all integers $i \ge 0$,

$$\rho(T^{i+1}(x_K), T^{i+2}(x_K)) \le \phi(\rho(T^i(x_K), T^{i+1}(x_K))) \le \rho(T^i(x_K), T^{i+1}(x_K)). \tag{10.33}$$

We claim that

$$\lim_{i\to\infty} \rho(T^i(x_K), T^{i+1}(x_K)) = 0.$$

Suppose to the contrary that this does not hold. Then in view of (10.33), there exists a number $r > 0$ such that

$$\rho(T^i(x_K), T^{i+1}(x_K)) > r \text{ for all integers } i \ge 0. \tag{10.34}$$

Since the function $t - \phi(t)$ is positive for all $t > 0$ and lower semicontinuous, there exists a number $\gamma > 0$ such that

$$t - \phi(t) > \gamma \text{ for all } t \in [r/4, \rho(T^0(x_K), T^1(x_K)) + 1]. \tag{10.35}$$

It now follows from (10.33)–(10.35) that for all integers $i \ge 0$,

$$\rho(T^{i+1}(x_K), T^{i+2}(x_K)) \le \phi(\rho(T^i(x_K), T^{i+1}(x_K)))$$
$$\le \rho(T^i(x_K), T^{i+1}(x_K)) - \gamma. \tag{10.36}$$

By (10.36), for all integers $n \ge 1$, we have

$$\rho(T^0(x_K), T^1(x_K)) \ge \rho(T^0(x_K), T^1(x_K)) - \rho(T^n(x_K), T^{n+1}(x_K))$$
$$= \sum_{i=0}^{n-1} [\rho(T^i(x_K), T^{i+1}(x_K)) - \rho(T^{i+1}(x_K), T^{i+2}(x_K))]$$
$$\ge \gamma n \to \infty \text{ as } n \to \infty.$$

The contradiction we have reached proves that

$$\lim_{i\to\infty} \rho(T^i(x_K), T^{i+1}(x_K)) = 0, \tag{10.37}$$

as claimed. Next, we prove that $\{T^i(x_K)\}_{i=1}^{\infty}$ is a Cauchy sequence. To this end, let $\delta > 0$. We show that there exists a natural number n_0 such that for each pair of integers $i, j \ge n_0$, we have

$$\rho(T^i(x_K), T^j(x_K)) \le \delta.$$

Suppose to the contrary that this does not hold. Then for each natural number k, there exist integers i_k, j_k such that

$$k \le i_k < j_k$$

and

$$\rho(T^{i_k}(x_K), T^{j_k}(x_K)) > \delta. \tag{10.38}$$

We may assume without any loss of generality that for each natural number k, the following property holds:

(P1) If an integer j satisfies $i_k \le j < j_k$, then

$$\rho(T^{i_k}(x_K), T^j(x_K)) \le \delta. \tag{10.39}$$

Let k be a natural number. By (10.38) and (10.39),

$$\delta < \rho(T^{i_k}(x_K), T^{j_k}(x_K))$$
$$\le \rho(T^{j_k}(x_K), T^{j_k-1}(x_K)) + \rho(T^{j_k-1}(x_K), T^{i_k}(x_K))$$

$$\leq \rho(T^{j_k}(x_K), T^{j_k-1}(x_K)) + \delta. \tag{10.40}$$

In view of (10.37),

$$\lim_{k\to\infty} \rho(T^{j_k}(x_K), T^{j_k-1}(x_K)) = 0. \tag{10.41}$$

It follows from (10.40) and (10.41) that

$$\lim_{k\to\infty} \rho(T^{i_k}(x_K), T^{j_k}(x_K)) = \delta. \tag{10.42}$$

By (10.33) and (10.38),

$$\delta < \rho(T^{i_k}(x_K), T^{j_k}(x_K))$$

$$\leq \rho(T^{i_k}(x_K), T^{i_k+1}(x_K)) + \rho(T^{i_k+1}(x_K), T^{j_k+1}(x_K)) + \rho(T^{j_k+1}(x_K), T^{j_k}(x_K))$$

$$\leq \phi(\rho(T^{i_k}(x_K), T^{j_k}(x_K))) + \rho(T^{i_k}(x_K), T^{i_k+1}(x_K)) + \rho(T^{j_k+1}(x_K), T^{j_k}(x_K)). \tag{10.43}$$

By (10.37), (10.42), and (10.43),

$$\delta = \lim_{k\to\infty} \rho(T^{i_k}(x_K), T^{j_k}(x_K))$$

$$\leq \liminf_{k\to\infty} \phi(\rho(T^{i_k}(x_K), T^{j_k}(x_K))) = \phi(\delta)$$

so that

$$\delta \leq \phi(\delta).$$

The contradiction we have reached proves that $\{T^i(x_K)\}_{i=1}^{\infty}$ is indeed a Cauchy sequence, as claimed, and there exists

$$\widehat{x} = \lim_{i\to\infty} T^i(x_K).$$

Since graph(T) is a closed set, we have

$$T(\widehat{x}) = \widehat{x}. \tag{10.44}$$

Now we show that $\widehat{x}$ is the unique fixed point of T.

To this end, assume that $z \in K$ and

$$T(z) = z. \tag{10.45}$$

If (10.1) holds, then

$$T^i(x_K) \le z \text{ for all integers } i \ge 1$$

and

$$\widehat{x} \le z,$$

and if (10.2) holds, then

$$z \le T^i(x_K) \text{ for all integers } i \ge 1$$

and

$$z \le \widehat{x}.$$

In both cases,

$$\rho(z, \widehat{x}) \le \phi(\rho(z, \widehat{x})).$$

This implies that

$$z = \widehat{x},$$

as claimed. Let M and ϵ be positive. Since the function $t - \phi(t)$ is positive for all $t > 0$ and lower semicontinuous, there exists a number $\gamma > 0$ such that

$$t - \phi(t) > \gamma \text{ for all } t \in [\epsilon/4, M + \epsilon + 1]. \tag{10.46}$$

There also exists a natural number n_0 such that

$$n_0 > M\gamma^{-1} \tag{10.47}$$

and

$$\rho(T^i(x_K), \widehat{x}) \le \epsilon/2 \text{ for all integers } i \ge n_0. \tag{10.48}$$

Assume that

$$x \in B(x_K, M), \tag{10.49}$$

$n > n_0$ is an integer and that $T^n(x)$ is defined.

If (10.1) holds, then by (10.4),

$$T^i(x_K) \le T^i(x) \text{ for all integers } i = 1, \dots, n \tag{10.50}$$

and if (10.2) holds, then by (10.4),

$$T^i(x) \le T^i(x_K) \text{ for all integers } i = 1, \ldots, n. \tag{10.51}$$

In both cases, inequality (10.10) implies that for all $i = 0, \ldots, n-1$,

$$\rho(T^{i+1}(x), T^{i+1}(x_K)) \le \phi(\rho(T^i(x), T^i(x_K))). \tag{10.52}$$

We now show that there exists an integer $i \in [0, n_0]$ such that

$$\rho(T^i(x), T^i(x_K)) \le \epsilon/2.$$

Suppose to the contrary that this does not hold. Then in view of (10.49) and (10.52), we have for all $i = 0, \ldots, n_0$,

$$M \ge \rho(x, x_K) \ge \rho(T^i(x), T^i(x_K)) > \epsilon/2. \tag{10.53}$$

By (10.46) and (10.53), for all $i = 0, \ldots, n_0$,

$$\rho(T^i(x), T^i(x_K)) - \phi(\rho(T^i(x), T^i(x_K))) > \gamma.$$

When combined with (10.52), this inequality implies that

$$\rho(T^i(x), T^i(x_K)) - \rho(T^{i+1}(x), T^{i+1}(x_K)) > \gamma. \tag{10.54}$$

By (10.49) and (10.54),

$$M \ge \rho(x, x_K)$$

$$\ge \rho(x, x_K) - \rho(T^{n_0}(x), T^0(x_K))$$

$$= \sum_{i=0}^{n_0-1} [\rho(T^i(x), T^i(x_K)) - \rho(T^{i+1}(x), T^{i+1}(x_K))] \ge \gamma n_0$$

and

$$n_0 \le M\gamma^{-1}.$$

This contradicts (10.47). The contradiction we have reached proves that there indeed exists $i_0 \in \{0, \ldots, n_0\}$ such that

$$\rho(T^{i_0}(x), T^{i_0}(x_K)) \le \epsilon/2. \tag{10.55}$$

It follows from (10.52) and (10.55) that for all integers $i \in \{i_0, \ldots, n\}$,

$$\rho(T^i(x), T^i(x_K)) \le \epsilon/2$$

and since $i \ge n_0$, we have by (10.48),

$$\rho(\widehat{x}, T^i(x)) \le \rho(\widehat{x}, T^i(x_K)) + \rho(T^i(x), T^i(x_K)) < \epsilon.$$

This completes the proof of Theorem 10.2.

10.4 Proof of Theorem 10.3

If (10.1) holds, then

$$T^0(x_K) \le T^1(x_K)$$

and for all integers $i \ge 0$,

$$T^i(x_K) \le T^{i+1}(x_K). \tag{10.56}$$

If (10.2) holds, then

$$T^1(x_K) \le T^0(x_K)$$

and for all integers $i \ge 0$,

$$T^{i+1}(x_K) \le T^i(x_K). \tag{10.57}$$

Since $\phi^n(t) \to 0$ as $n \to \infty$ for all $t > 0$ and ϕ is increasing, we have

$$\phi(t) < t \text{ for all } t > 0. \tag{10.58}$$

By (10.9), (10.10), and (10.58), for all integers $i \ge 0$,

$$\rho(T^{i+1}(x_K), T^{i+2}(x_K)) \le \phi(\rho(T^i(x_K), T^{i+1}(x_K))) \le \rho(T^i(x_K), T^{i+1}(x_K)) \tag{10.59}$$

and for all integers $i \ge 1$,

$$\rho(T^i(x_K), T^{i+1}(x_K)) \le \phi^i(\rho(x_K, T^1(x_K))) \to 0 \text{ as } i \to \infty.$$

Thus,

$$\lim_{i\to\infty} \rho(T^i(x_K), T^{i+1}(x_K)) = 0. \tag{10.60}$$

We claim that $\{T^i(x_K)\}_{i=1}^{\infty}$ is a Cauchy sequence. Let $\delta > 0$. In view of (10.58),

$$\phi(\delta) < \delta. \tag{10.61}$$

By (10.60) and (10.61), there exists a natural number i_0 such that

$$\rho(T^i(x_K), T^{i+1}(x_K)) \le \delta - \phi(\delta) \tag{10.62}$$

for all integers $i \ge i_0$.

Now we show that

$$\rho(T^i(x_K), T^j(x_K)) \le \delta$$

for all integers $j > i \ge i_0$.

To this end, assume that $i \ge i_0$ is an integer:

$$x \in K \cap B(T^i(x_K), \delta) \tag{10.63}$$

and

$$\text{either } x \le T^i(x_K) \text{ or } T^i(x_K) \le x. \tag{10.64}$$

It follows from (10.58), (10.62), and (10.64) that

$$\rho(T(x), T^i(x_K)) \le \rho(T(x), T^{i+1}(x_K)) + \rho(T^{i+1}(x_K), T^i(x_K))$$

$$\le \phi(\rho(x, T^i(x_K))) + \delta - \phi(\delta) \le \delta.$$

Therefore,

$$T(K \cap B(T^i(x_K), \delta)) \cap (\{x \in K : T^i(x_K) \le x\} \cup \{x \in K : x \le T^i(x_K)\}$$

$$\subset B(T^i(x_K), \delta). \tag{10.65}$$

By (10.65), if (10.1) holds, then

$$T^{i_1}(x_K) \le T^{i_2}(x_K) \text{ for all integers } i_2 \ge i_1 \ge 0$$

and

$$T^j(x_K) \in K \cap B(T^i(x_K), \delta)$$

for all integers $j > i$.

By (10.65), if (10.2) holds, then

$$T^{i_2}(x_K) \le T^{i_1}(x_K) \text{ for all integers } i_2 \ge i_1 \ge 0$$

and

$$T^j(x_K) \in K \cap B(T^i(x_K), \delta)$$

for all integers $j > i$.

Thus, in both cases,

$$\rho(T^i(x_K), T^j(x_K)) \le \delta$$

for all integers $j > i \ge i_0$, as claimed. Therefore, $\{T^i(x_K)\}_{i=1}^{\infty}$ is indeed a Cauchy sequence and there exists

$$\widehat{x} = \lim_{i \to \infty} T^i(x_K). \tag{10.66}$$

Since graph(T) is a closed set, we have

$$T(\widehat{x}) = \widehat{x}.$$

Next we show that $\widehat{x}$ is the unique fixed point of T.

To this end, assume that $z \in K$ and

$$T(z) = z.$$

If (10.1) holds, then

$$T^i(x_K) \le z \text{ for all integers } i \ge 1$$

and

$$\widehat{x} \le z,$$

and if (10.2) holds, then

$$z \le T^i(x_K) \text{ for all integers } i \ge 1$$

and

$$z \le \widehat{x}.$$

In both cases,

$$\rho(z, \widehat{x}) \le \phi(\rho(z, \widehat{x})).$$

This implies that

$$z = \widehat{x}.$$

Now let M and ϵ be positive. By (10.9) and (10.66), there exists a natural number n_0 such that

$$\phi^{n_0}(M) < \epsilon/2 \tag{10.67}$$

and

$$\rho(T^i(x_K), \widehat{x}) \le \epsilon/2 \text{ for all integers } i \ge n_0. \tag{10.68}$$

Assume that

$$x \in B(x_K, M) \cap K, \tag{10.69}$$

$n > n_0$ is an integer and $T^n(x)$ is defined.

If (10.1) holds, then again by (10.4),

$$T^i(x_K) \le T^i(x) \text{ for all integers } i = 1, \dots, n$$

and if (10.2) holds, then by (10.4),

$$T^i(x) \le T^i(x_K) \text{ for all integers } i = 1, \dots, n.$$

In both cases, inequality (10.10) implies that for all $i = 0, \dots, n-1$,

$$\rho(T^{i+1}(x), T^{i+1}(x_K)) \le \phi(\rho(T^i(x), T^i(x_K))). \tag{10.70}$$

It follows from (10.67)–(10.70) that

$$\rho(T^{n_0}(x), T^{n_0}(x_K)) \le \phi^{n_0}(\rho(x, x_K)) \le \phi^{n_0}(M) < \epsilon/2.$$

By (10.68), (10.70), and the relation above, for $i = n_0, \dots, n$,

$$\rho(T^i(x), T^i(x_K)) \le \epsilon/2$$

and

$$\rho(\widehat{x}, T^i(x)) \le \rho(\widehat{x}, T^i(x_K)) + \rho(T^i(x), T^i(x_K)) < \epsilon.$$

This completes the proof of Theorem 10.3.

10.5 Inexact Iterates of Monotone Nonexpansive Mappings with Summable Errors

In [30] it was shown that the convergence of iterates of a nonexpansive self-mapping of a complete metric space is stable in the presence of summable computational errors. We prove such results for monotone nonexpansive mappings which were obtained in [177].

Assume that (X, ρ) is a complete metric space equipped with a partial order $\leq$ such that $x \leq x$ for each $x \in X$; if $x, y \in X$ satisfy $x \leq y$ and $y \leq x$, then $x = y$, and if $x, y, z \in X$ satisfy $x \leq y$ and $y \leq z$, then $x \leq z$.

Assume that a mapping $T : X \to X$ satisfies

$$T(x) \leq T(y) \text{ for each } x, y \in X \text{ such that } x \leq y \tag{10.71}$$

and

$$\rho(T(x), T(y)) \leq \rho(x, y) \text{ for each } x, y \in X \text{ such that } x \leq y. \tag{10.72}$$

We will prove the following three results obtained in [177] which describe the asymptotic behavior of inexact iterates under the presence of computational errors.

Theorem 10.4 *Assume that for each $x \in X$, the sequence $\{T^i(x)\}_{i=1}^{\infty}$ converges. Let a sequence $\{x_i\}_{i=0}^{\infty}$ satisfy the conditions*

$$\sum_{i=0}^{\infty} \rho(x_{i+1}, T(x_i)) < \infty \tag{10.73}$$

and

$$x_{i+1} \geq T(x_i) \text{ for each integer } i \geq 0. \tag{10.74}$$

Then the sequence $\{x_i\}_{i=0}^{\infty}$ converges. If, in addition, T is continuous, then its limit is a fixed point of T.

Theorem 10.4 is proved in Sect. 10.8.

Theorem 10.5 *Assume that F is a nonempty subset of X, that for each point $x \in X$,*

$$\lim_{i \to \infty} \rho(T^i(x), F) = 0$$

and that a sequence $\{x_i\}_{i=0}^{\infty}$ satisfies (10.73) and (10.74). Then

$$\lim_{i \to \infty} \rho(x_i, F) = 0.$$

Theorem 10.5 is proved in Sect. 10.9.

Theorem 10.6 *Assume that for each point $x \in X$, there exists a compact set $E(x) \subset X$ such that*

$$\lim_{i\to\infty} \rho(T^i(x), E(x)) = 0$$

and that a sequence $\{x_i\}_{i=0}^{\infty}$ satisfies (10.73) *and* (10.74). *Then there exists a compact set $E \subset X$ such that* $\lim_{i\to\infty} \rho(x_i, E) = 0$.

Theorem 10.6 is proved in Sect. 10.9.

10.6 An Auxiliary Result for Theorems 10.4–10.6

Lemma 10.7 *Assume that a sequence $\{x_i\}_{i=0}^{\infty}$ satisfies*

$$\sum_{i=0}^{\infty} \rho(x_{i+1}, T(x_i)) < \infty \tag{10.75}$$

and

$$x_{i+1} \geq T(x_i) \text{ for each integer } i \geq 0, \tag{10.76}$$

that $n_0 \geq 0$ is an integer and that

$$y_{n_0} = x_{n_0}$$

and

$$y_{i+1} = T(y_i) \text{ for every integer } i > n_0. \tag{10.77}$$

Then for each integer $n > n_0$, we have

$$x_n \geq y_n \tag{10.78}$$

and

$$\rho(x_n, y_n) \leq \sum_{i=n_0+1}^{n} \rho(x_i, T(x_{i-1})). \tag{10.79}$$

Proof In view of (10.76) and (10.77),

$$x_{n_0+1} \geq T(x_{n_0}) = T(y_{n_0}) = y_{n_0+1},$$

$$\rho(y_{n_0+1}, x_{n_0+1}) = \rho(T(y_{n_0}), x_{n_0+1}) = \rho(x_{n_0+1}, T(x_{n_0})),$$

and relations (10.78) and (10.79) hold with $n = n_0 + 1$.

Assume now that $n > n_0$ is an integer and that relations (10.78) and (10.79) hold. By (10.77), we have

$$\rho(x_{n+1}, y_{n+1}) \leq \rho(x_{n+1}, T(x_n)) + \rho(T(x_n), T(y_n)). \tag{10.80}$$

In view of (10.78),

$$\rho(T(x_n), T(y_n)) \leq \rho(x_n, y_n). \tag{10.81}$$

Relations (10.79) and (10.81) imply that

$$\rho(x_{n+1}, y_{n+1}) \leq \rho(x_{n+1}, T(x_n)) + \rho(x_n, y_n) \leq \sum_{i=n_0+1}^{n+1} \rho(x_i, T(x_{i-1})).$$

It follows from (10.72) and (10.76)–(10.79) that

$$x_{n+1} \geq T(x_n) \geq T(y_n) = y_{n+1}.$$

Thus, (10.78) and (10.79) hold for $n + 1$ too. This means that the assumption made for n also holds for $n + 1$. Lemma 10.7 is proved.

10.7 Proof of Theorem 10.4

Given $\epsilon > 0$, there exists an integer $n_0 \geq 1$ such that

$$\sum_{i=n_0}^{\infty} \rho(x_i, T(x_{i-1})) < \epsilon/2. \tag{10.82}$$

Set

$$y_{n_0} = x_{n_0}$$

and

$$y_{i+1} = T(y_i) \text{ for each integer } i \geq n_0. \tag{10.83}$$

Lemma 10.7 and relations (10.82) and (10.83) imply that for each integer $n > n_0$, we have

$$\rho(x_n, y_n) \le \epsilon/2. \tag{10.84}$$

In view of (10.83), there exists the limit

$$y_* = \lim_{n\to\infty} y_n. \tag{10.85}$$

By (10.84) and (10.85), for all sufficiently large natural numbers n,

$$\rho(x_n, y_*) \le \rho(x_n, y_n) + \rho(y_n, y_*) \le \epsilon.$$

Thus, $\{x_n\}_{n=0}^{\infty}$ is a Cauchy sequence, and there exists

$$x_* = \lim_{n\to\infty} x_n.$$

Clearly, if T is continuous, then x_* is a fixed point of T. This completes the proof of Theorem 10.4.

10.8 Proof of Theorem 10.5

Given $\epsilon > 0$, there exists an integer $n_0 \ge 1$ such that (10.82) holds. Define a sequence $\{y_i\}_{i=n_0}^{\infty}$ by (10.83). Lemma 10.7 and relations (10.82) and (10.83) imply that for each integer $n > n_0$, we have

$$\rho(x_n, y_n) \le \epsilon/2.$$

In view of (10.83) and the above inequality, for every sufficiently large natural number n, we have

$$\rho(x_n, F) \le \rho(x_n, y_n) + \rho(y_n, F) < \epsilon.$$

This completes the proof of Theorem 10.5.

10.9 Proof of Theorem 10.6

Given $\epsilon > 0$, there exists an integer $n_0 \ge 1$ such that (10.6) holds. Define a sequence $\{y_i\}_{i=n_0}^{\infty}$ by (10.83). Lemma 10.7 and relations (10.82) and (10.83) imply that for each integer $n > n_0$, we have

$$\rho(x_n, y_n) \leq \epsilon/2.$$

In view of (10.83), there exists a compact set $E_0 \subset X$ such that

$$\lim_{n\to\infty} \rho(y_n, E_0) = 0.$$

Clearly, for every sufficiently large natural number $n > n_0$, we have

$$\rho(x_n, E_0) \leq \rho(x_n, y_n) + \rho(y_n, E_0) < \epsilon.$$

Thus, we have shown that there exists a compact set E_0 such that

$$\rho(x_n, E_0) < \epsilon$$

for each sufficiently large natural number $n > n_0$. We may assume that E_0 is finite. This implies that each subsequence of $\{x_i\}_{i=0}^{\infty}$ has a convergent subsequence. Denote by E the set of all limit points of the sequence $\{x_i\}_{i=0}^{\infty}$. It is not difficult to see that E is compact and that

$$\lim_{i\to\infty} \rho(x_i, E) = 0.$$

This completes the proof of Theorem 10.6.

10.10 Well-Posedness of the Fixed Point Problem for Monotone Nonexpansive Maps

Now we study the well-posedness of the fixed point problem for monotone nonexpansive mappings and present the results obtained in [172].

Let (X, ρ) be a complete metric space equipped with an order $'' \leq''$ such that for all $x, y, z \in X$,

$$x \leq x,$$

$$\text{if } x \leq y,\ \ y \leq x,\ \text{ then } x = y,$$

$$\text{if } x \leq y,\ \ y \leq z,\ \text{ then } x \leq z.$$

We suppose that

$$\{(x, y) \in X \times X : \ x \leq y\}$$

is a closed subset of $X \times X$ and that for all $x, y, z \in X$ satisfying $x \le y \le z$, we have

$$\rho(x, y) \le \rho(x, z). \tag{10.86}$$

Let K be a nonempty closed subset of X which is not a singleton and $x_K \in K$ satisfy

$$x_K \le x \text{ for all } x \in K. \tag{10.87}$$

Denote by $\mathcal{A}$ the set of all continuous mappings $T : K \to K$ such that

$$T(x) \le T(y) \text{ for all } x, y \in K \text{ such that } x \le y. \tag{10.88}$$

The set $\mathcal{A}$ is equipped with the uniformity determined by the base

$$\mathcal{U}(n, \epsilon) = \{(T_1, T_2) \in \mathcal{A} \times \mathcal{A} :$$

$$\rho(T_1(x), T_2(x)) \le \epsilon \text{ for all } x \in B(x_K, n)\},$$

where $n, \epsilon > 0$. It is not difficult to see that the uniform space $\mathcal{A}$ is metrizable and complete.

Theorem 10.8 *Let $T \in \mathcal{A}$, for all $x, y \in K$ such that $x \le y$,*

$$\rho(T(x), T(y)) \le \rho(x, y), \tag{10.89}$$

$\bar{x}, x_* \in K$ and

$$T(x_*) = x_*. \tag{10.90}$$

Assume that for each $\epsilon > 0$, there exists a natural number n_0 such that for each integer $n \ge n_0$ and each $x \in K$ satisfying $x_K \le x \le \bar{x}$,

$$\rho(T^n(x), x_*) \le \epsilon.$$

Then for each $\epsilon > 0$, there exist a neighborhood $\mathcal{U}$ of T in $\mathcal{A}$ and a natural number n_0 such that for each mapping $S \in \mathcal{U}$ and each $x \in K$ satisfying $x_K \le x \le \bar{x}$,

$$\rho(S^{n_0}(x), x_*) \le \epsilon.$$

10.11 Proof of Theorem 10.8

Let $\epsilon \in (0, 1)$. There exists a natural number n_0 such that the following property holds:

(i) For each integer $n \geq n_0$ and each $x \in K$ satisfying $x \leq \bar{x}$, we have

$$\rho(T^n(x), x_*) \leq \epsilon/8.$$

Set $T^0(x) = x$ for all $x \in K$ and

$$\epsilon_{n_0} = \epsilon/8. \tag{10.91}$$

By induction define a finite sequence of positive numbers $\{\epsilon_i\}_{i=0}^{n_0}$ such that for each integer $i \in \{0, \dots, n_0 - 1\}$,

$$\epsilon_i < \epsilon_{i+1}/8, \tag{10.92}$$

$$\rho(T(T^i(x_K)), T(z)) \leq \epsilon_{i+1}/8 \text{ for each } z \in K \cap B(T^i(x_K), \epsilon_i) \tag{10.93}$$

and

$$\rho(T(T^i(\bar{x})), T(z)) \leq \epsilon_{i+1}/8 \text{ for each } z \in K \cap B(T^i(\bar{x}), \epsilon_i). \tag{10.94}$$

Set

$$\mathcal{U} = \{S \in \mathcal{A} : \rho(T(z), S(z)) \leq \epsilon_0/4$$
$$\text{for all } z \in K \cap (\cup_{i=0}^{n_0} B(T^i(x_K), 1) \cup \cup_{i=0}^{n_0} B(T^i(\bar{x}), 1)). \tag{10.95}$$

Let

$$S \in \mathcal{U} \tag{10.96}$$

and

$$z \in \{\bar{x}, \ x_K\}.$$

We show that for all $i = 0, \dots, n_0$,

$$\rho(S^i(z), T^i(z)) \leq \epsilon_i. \tag{10.97}$$

Clearly, (10.97) holds for $i = 0$. By (10.92), (10.95), and (10.96), Eq. (10.97) holds for $i = 1$.

Assume that an integer i satisfies $0 \le i < n_0$ and (10.97) holds. It follows from (10.91), (10.92), and (10.97) that

$$S^i(z) \in K \cap (\cup_{j=0}^{n_0} B(T^j(x_K), 1) \cup \cup_{j=0}^{n_0} B(T^j(\bar{x}), 1)). \tag{10.98}$$

By (10.96) and (10.98),

$$\rho(S^{i+1}(z), T(S^i(z))) \le \epsilon_0/4. \tag{10.99}$$

In view of (10.94) and (10.97),

$$\rho(T(S^i(z)), T(S^i(z))) \le \epsilon_{i+1}/8. \tag{10.100}$$

It follows from (10.92), (10.99), and (10.100) that

$$\rho(S^{i+1}(z), T^{i+1}(z)) \le \epsilon_{i+1}/8 + \epsilon_0/4 \le \epsilon_{i+1}.$$

Thus, (10.97) holds for all $n = 0, \dots, n_0$ and

$$\rho(S^{n_0}(z), T^{n_0}(z)) \le \epsilon_{n_0} = \epsilon/8. \tag{10.101}$$

Property (i) and (10.107) imply that

$$\rho(S^{n_0}(z), x_*) \le \epsilon/4$$

and

$$\rho(S^{n_0}(x_K), x_*) \le \epsilon/4, \ \ \rho(S^{n_0}(\bar{x}), x_*) \le \epsilon/4. \tag{10.102}$$

Let $x \in K$ satisfy

$$x_K \le x \le \bar{x}. \tag{10.103}$$

By (10.88) and (10.103),

$$S^{n_0}(x_K) \le S^{n_0}(x) \le S^{n_0}(\bar{x}). \tag{10.104}$$

It follows from (10.86), (10.102), and (10.104) that

$$\rho(S^{n_0}(x), S^{n_0}(x_K)) \le \rho(S^{n_0}(x_K), S^{n_0}(\bar{x})) \le \epsilon/2.$$

Together with (10.102), this implies that

$$\rho(S^{n_0}(x), x_*) \le \epsilon.$$

Theorem 10.8 is proved.

Theorem 10.8 easily implies the following result.

Theorem 10.9 *Let there exist $\bar{x} \in X$ such that*

$$x \leq \bar{x} \text{ for all } x \in K,$$

$T \in \mathcal{A}$, for all $x, y \in K$ such that $x \leq y$,

$$\rho(T(x), T(y)) \leq \rho(x, y),$$

$x_* \in K$ and

$$T(x_*) = x_*.$$

Assume that $T^n(x) \to x_*$ as $n \to \infty$ uniformly over $x \in K$. Then for each $\epsilon > 0$, there exist a neighborhood $\mathcal{U}$ of T in $\mathcal{A}$ and a natural number n_0 such that for each mapping $S \in \mathcal{U}$ and each $x \in K$ and each integer $n \geq n_0$,

$$\rho(S^n(x), x_*) \leq \epsilon.$$

10.12 A Generic Well-Posedness Result

Let $(X, \|\cdot\|)$ be a Banach space ordered by a closed convex cone $X_+ \subset X$ satisfying

$$X_+ \cap (-X_+) = \{0\}.$$

Note that for $x, y \in X$,

$$x \leq y \text{ if and only if } y - x \in X_+.$$

Suppose that

$$\|x\| \leq \|y\| \text{ for all } x, y \in X_+ \text{ satisfying } x \leq y. \tag{10.105}$$

Let K be a nonempty bounded closed subset of X,

$$\bar{x}_K, \widehat{x}_K \in K,$$

$$\bar{x}_K \leq x \leq \widehat{x}_K \text{ for all } x \in K, \tag{10.106}$$

$\theta \in K$ and for each $x \in K$,

$$\{\gamma x + (1-\gamma)\theta : \gamma \in [0, 1]\} \subset K. \tag{10.107}$$

Set

$$d(K) = \sup\{\|x - y\| : x, y \in K\}. \tag{10.108}$$

Denote by $\mathcal{M}$ the set of all continuous mappings $T : K \to K$ such that

$$T(x) \le T(y) \text{ for all } x, y \in K \text{ satisfying } x \le y \tag{10.109}$$

and

$$\|T(x) - T(y)\| \le \|x - y\| \tag{10.110}$$

for all $x, y \in K$ satisfying $x \le y$.

For each $T_1, T_2 \in \mathcal{M}$, set

$$d(T_1, T_2) = \sup\{\|T_1(x) - T_2(x)\| : x \in K\}. \tag{10.111}$$

Clearly, $(\mathcal{M}, d)$ is a complete metric space.

Let $T \in \mathcal{M}$. We say that the fixed point problem (FPP) is well-posed for T is there exists a unique point $x_T \in K$ such that

$$T(x_T) = x_T$$

and the following property holds:

(WP) For each $\epsilon > 0$, there exists a natural number n_ϵ such that for each $x \in K$,

$$\|x_T - T^{n_\epsilon}(x)\| \le \epsilon.$$

We prove the following result obtained in [172].

Theorem 10.10 *There exists a set $\mathcal{F} \subset \mathcal{M}$ which is a countable intersection of open everywhere dense sets in $\mathcal{M}$ such that for each $T \in \mathcal{F}$, FPP is well-posed.*

Proof Let $A \in \mathcal{M}$ and $\gamma \in (0, 1)$. Define

$$A_\gamma(x) = (1-\gamma)A(x) + \gamma\theta. \tag{10.112}$$

Clearly, $A_\gamma : K \to K$ is a continuous mapping; if $x, y \in K$ satisfy $x \le y$, then

$$A_\gamma(x) \le A_\gamma(y), \tag{10.113}$$

$$\|A_\gamma(x) - A_\gamma(y)\|$$

$$= \|(1-\gamma)A(x) + \gamma\theta - (1-\gamma)A(y) - \gamma\theta\|$$

$$= (1-\gamma)\|A(x) - A(y)\| \le (1-\gamma)\|x-y\| \tag{10.114}$$

and

$$d(A, A_\gamma) = \sup\{\|A(x) - A_\gamma(x)\| : x \in K\}$$

$$= \sup\{A(x) - (1-\gamma)A(x) - \gamma\theta : x \in K\} \le \gamma d(K). \tag{10.115}$$

In view of (10.115),

$$\{A_\gamma : A \in \mathcal{M} \text{ and } \gamma \in (0,1)\}$$

in an everywhere dense set in $\mathcal{M}$.

For $A \in \mathcal{M}$, set $A^0(x) = x$ for all $x \in K$.

Let $A \in \mathcal{M}$ and $\gamma \in (0, 1)$. We show that FPP is well-posed for A_γ. By (10.106), (10.110), and (10.114), for all $x, y \in K$ satisfying $x \le y$ and every integer $n \ge 0$,

$$(A_\gamma)^n(\bar{x}_K) \le (A_\gamma)^n(x) \le (A_\gamma)^n(\widehat{x}_K),$$

$$\|(A_\gamma)^{n+1}(x) - (A_\gamma)^{n+1}(y)\| \le (1-\gamma)\|(A_\gamma)^n(x) - (A_\gamma)^n(y)\|.$$

This implies that for all integers $n \ge 0$ and all $x \in K$,

$$\|(A_\gamma)^n(x) - (A_\gamma)^n(\bar{x}_K)\| \le (1-\gamma)^n\|x - \bar{x}_K\| \le (1-\gamma)^n d(K). \tag{10.116}$$

Let $x \in K$. Equation (10.116) implies that $\{(A_\gamma)^n(\bar{x}_K)\}_{n=0}^\infty$ is a Cauchy sequence and there exists

$$x_* = \lim_{n\to\infty} (A_\gamma)^n(\bar{x}_K). \tag{10.117}$$

In view of (10.116) and (10.117), FPP is well-posed for A_γ.

Let $A \in \mathcal{M}$, $\gamma \in (0, 1)$ and k be a natural number. There exists a unique point $x_{A,\gamma} \in K$ such that

$$A_\gamma(x_{A,\gamma}) = x_{A,\gamma}. \tag{10.118}$$

Theorem 10.9 and (10.118) imply that that there exist a natural number $n(A, \gamma, k)$ and an open neighborhood $\mathcal{U}(A, \gamma, k)$ of A_γ in $\mathcal{M}$ such that the following property holds:

(P) For each mapping $S \in \mathcal{U}(A, \gamma, k)$, each $x \in K$ and each integer $i \ge n(A, \gamma, k)$,

$$\|S^i(x) - x_{A,\gamma}\| \le 1/k.$$

Set

$$\mathcal{F} = \cap_{p=1}^{\infty} \cup \{\mathcal{U}(A, \gamma, k) : \; A \in \mathcal{M}, \; \gamma \in (0, 1), \; \text{ an integer } k \ge p\}. \tag{10.119}$$

Clearly, $\mathcal{F}$ is a countable intersection of open everywhere dense subsets of the metric space $(\mathcal{M}, d)$.

Let

$$T \in \mathcal{F}, \; \epsilon \in (0, 1). \tag{10.120}$$

Choose a natural number p such that

$$4p^{-1} < \epsilon.$$

By (10.119) and (10.120), there exist

$$A \in \mathcal{M}, \; \gamma \in (0, 1) \text{ and an integer } k \ge p$$

such that

$$T \in \mathcal{U}(A, \gamma, k). \tag{10.121}$$

It follows from (10.121), the choice of p and property (P) that for each $x \in K$ and each integer $i \ge n(A, \gamma, k)$,

$$\|T^i(x) - x_{A,\gamma}\| \le (k)^{-1} \le p^{-1} < \epsilon/4.$$

Since $\epsilon \in (0, 1)$ is arbitrary, for every $x \in K$, $\{T^i(x)\}_{i=1}^{\infty}$ is a Cauchy sequence and there exists

$$\lim_{i \to \infty} T^i(x)$$

which satisfies

$$\|\lim_{i \to \infty} T^i(x) - x_{A,\gamma}\| \le \epsilon/4.$$

for all $x \in K$. This implies that for each $x_1, x_2 \in K$,

$$\lim_{i \to \infty} T^i(x_1) = \lim_{i \to \infty} T^i(x_2).$$

Therefore, there exists a unique point $x_T \in K$ such that

$$T(x_T) = x_T,$$

$$\|x_T - x_{A,\gamma}\| \leq \epsilon/4$$

and for each $x \in K$ and each integer $i \geq n(A, \gamma, k)$,

$$\|x_T - T^i(x)\| \leq \epsilon.$$

Theorem 10.10 is proved.

Chapter 11
Uniformly Locally Contractive Mappings

In this chapter, we study uniformly locally contractive mappings. We show that a uniformly locally contractive mapping has a fixed point, that its fixed point problem is well posed, and that inexact iterates of such a mapping converge to its unique fixed point, uniformly on bounded sets. Using the porosity notion, we also show that most uniformly locally nonexpansive mappings are, in fact, uniformly locally contractive. We extend these results to uniformly locally contractive non-self-mappings defined on a closed subset of the metric space.

11.1 Properties of Uniformly Locally Contractive Mappings

The results of this section were obtained in [175].

Assume that (X, ρ) is a complete metric space, $\Delta > 0$, and that the following assumption holds.

(A) For each $x, y \in X$, there exist an integer $q \geq 1$ and points $x_i \in X$, $i = 0, \dots, q$, such that

$$x_0 = x,\ x_q = y,\ \rho(x_i, x_{i+1}) \leq \Delta,\ i \in \{0, \dots, q\} \setminus \{q\}.$$

For each $x, y \in X$, define

$$\rho_1(x, y) := \inf\{\sum_{i=0}^{q-1} \rho(x_i, x_{i+1}) :\ q \geq 1 \text{ is an integer},$$

$$x_i \in X,\ i = 0, \dots, q,\ x_0 = x,\ x_q = y,\ \rho(x_i, x_{i+1}) \leq \Delta,\ i \in \{0, \dots, q\} \setminus \{q\}\}. \tag{11.1}$$

A. J. Zaslavski, *Approximate Fixed Points of Nonexpansive Mappings*,
Developments in Mathematics 80, https://doi.org/10.1007/978-3-031-70710-0_11

It follows from assumption (A) and Eq. (11.1) that for each $x, y, z \in X$, $\rho_1(x, y)$ is finite,

$$\rho(x, y) \le \rho_1(x, y) < \infty, \tag{11.2}$$

$$\text{if } \rho(x, y) \le \Delta, \text{ then } \rho_1(x, y) = \rho(x, y), \tag{11.3}$$

$$\rho_1(x, y) = \rho_1(y, x),$$

$$\rho_1(x, x) = 0,$$

$$\text{if } \rho_1(x, y) = 0, \text{ then } \rho(x, y) = 0 \text{ and } x = y,$$

and

$$\rho_1(x, z) \le \rho_1(x, y) + \rho_1(y, z).$$

Clearly, ρ_1 is a metric on X. This metric plays an important role in our study.

Let $T : X \to X$ be a self-mapping of X.

We assume that for each $x, y \in X$ satisfying $\rho(x, y) \le \Delta$, the inequality

$$\rho(T(x), T(y)) \le \rho(x, y) \tag{11.4}$$

holds. In view of (11.1) and (11.4), for each $x, y \in X$,

$$\rho_1(T(x), T(y)) \le \rho_1(x, y). \tag{11.5}$$

Such a mapping T is said to be uniformly locally nonexpansive.

We remark in passing that the smaller class of uniformly local (strict) contractions was introduced in [75], while the larger class of locally nonexpansive mappings was studied in [25]. The work [75] also contains an example of a uniformly local (strict) contraction which is not nonexpansive.

Our goal is to show that analogs of many results obtained in our previous research for nonexpansive mappings hold for uniformly locally nonexpansive operators. For these operators, relation (11.4) is true only if $\rho(x, y) \le \Delta$. This makes our study more difficult and less understood than in the case of nonexpansive operators. To overcome this difficulty we introduced the quasi-metric ρ_1 (see (11.1)). It turns out that the estimations that are main ingredients in our proofs for nonexpansive mappings have corresponding analogs for uniformly locally nonexpansive operators when the metric ρ is replaced by quasi-metric ρ_1.

Assume that $\phi : [0, \Delta] \to [0, 1]$ is a decreasing function,

$$\phi(t) < 1 \text{ for all } t \in (0, \Delta], \tag{11.6}$$

and for each $x, y \in X$ satisfying $\rho(x, y) \le \Delta$, we have

$$\rho(T(x), T(y)) \le \phi(\rho(x, y))\rho(x, y). \tag{11.7}$$

We prove the following theorems obtained in [175]. The first of them shows the well-posedness of the fixed point problem for the mapping T.

Theorem 11.1 *Given $\epsilon > 0$, there exists $\delta > 0$ such that for each $x, y \in X$ satisfying*

$$\max\{\rho(T(x), x),\ \rho(T(y), y)\} \le \delta,$$

the inequality

$$\rho(x, y) \le \epsilon$$

holds.

Theorem 11.2 *Let $\epsilon, M > 0$. Then there exist a number $\delta \in (0, \epsilon)$ and an integer $n_0 \ge 1$ such that for each integer $n > n_0$ and each finite sequence $\{x_i\}_{i=0}^n \subset X$ which satisfies*

$$\rho_1(x_0, x_1) \le M$$

and

$$\rho(x_{i+1}, T(x_i)) \le \delta,\ i = 0, \dots, n-1,$$

the inequality

$$\rho(x_{i+1}, x_i) \le \epsilon$$

holds for all $i = n_0, \dots, n-1$.

Theorems 11.1 and 11.2 imply that there exists a unique $x_* \in K$ such that

$$T(x_*) = x_*.$$

Note that the existence of the unique fixed point was obtained in [155] (see also [158]).

Theorems 11.1 and 11.2 easily imply the following convergence result for inexact iterates of the mapping T.

Theorem 11.3 *Let the point $x_* \in X$ satisfy*

$$T(x_*) = x_*$$

and let $\epsilon, M > 0$. Then there exist a number $\delta > 0$ and an integer $n_0 \ge 1$ such that for each integer $n > n_0$ and each finite sequence $\{x_i\}_{i=0}^n \subset X$ which satisfies

$$\rho_1(x_0, x_1) \le M$$

and

$$\rho(x_{i+1}, T(x_i)) \le \delta, \ i = 0, \dots, n-1,$$

the inequality

$$\rho(x_i, x_*) \le \epsilon$$

holds for all $i = n_0, \dots, n-1$.

11.2 Auxiliary Results for Theorems 11.1 and 11.2

Lemma 11.4 *Let* $x, y \in X$ *satisfy*

$$\rho(x, y) > \Delta. \tag{11.8}$$

Then

$$\rho_1(x, y) = \inf\{\sum_{i=0}^{q-1} \rho(x_i, x_{i+1}) : \ q \ge 1 \text{ is an integer},$$

$$x_i \in X, \ i = 0, \dots, q, \ x_0 = x, \ x_q = y, \ \rho(x_i, x_{i+1}) \le \Delta, \ i \in \{0, \dots, q-1\},$$

$$\max\{\rho(x_i, x_{i+1}) : \ i = 0, \dots, q-1\} \ge \Delta/4\}. \tag{11.9}$$

Proof Assume that $q \ge 1$ is an integer, $x_i \in X, \ i = 0, \dots, q$,

$$x_0 = x, \ x_q = y, \tag{11.10}$$

$$\rho(x_i, x_{i+1}) \le \Delta, \ i \in \{0, \dots, q-1\}, \tag{11.11}$$

and

$$\max\{\rho(x_i, x_{i+1}) : \ i = 0, \dots, q-1\} < \Delta/4. \tag{11.12}$$

By (11.8), (11.10), and (11.12), we have

$$q \ge 4$$

and

$$\sum_{i=0}^{q-1} \rho(x_i, x_{i+1}) \geq \rho(x, y) > \Delta. \tag{11.13}$$

In view of (11.12) and (11.13), there exists a natural number

$$j \in \{2, \ldots, q-1\}$$

such that

$$\rho(x_0, x_j) < \Delta/2 \tag{11.14}$$

and

$$\rho(x_0, x_{j+1}) \in [\Delta/2, 3\Delta/4]. \tag{11.15}$$

Set

$$\widehat{x_0} = x_0,\ \widehat{x_1} = x_j,\ \widehat{x_i} = x_{j+i-1},\ i = 2, \ldots, q-j+1. \tag{11.16}$$

By (11.10), (11.11), (11.16), and the inequality $j \geq 2$, we have

$$q - j + 1 \leq q - 1$$

and

$$\widehat{x_0} = x,\ \widehat{x}_{q-j+1} = x_q = y.$$

It follows from (11.9), (11.14), and (11.16) that

$$\rho(\widehat{x_i}, \widehat{x}_{i+1}) \leq \Delta,\ i \in \{0, \ldots, q-j\}$$

and

$$\sum_{i=0}^{q-j} \rho(\widehat{x_i}, \widehat{x}_{i+1}) \leq \sum_{i=0}^{q-1} \rho(x_i, x_{i+1}).$$

Equations (11.12), (11.15), and (11.16) imply that

$$\rho(\widehat{x_0}, \widehat{x_1}) = \rho(x_0, x_j) \geq \rho(x_0, x_{j+1}) - \rho(x_j, x_{j+1}) \geq \Delta/2 - \Delta/4.$$

Thus, for every integer $q \geq 1$ and every finite sequence $x_i \in X,\ i = 0, \ldots, q$, satisfying (11.10)–(11.12), there exist an integer $\widehat{q} \geq 1$ and a finite sequence $\widehat{x_i} \in X,\ i = 0, \ldots, \widehat{q}$, such that

$$\widehat{x_0} = x_0, \ \widehat{x_{\widehat{q}}} = y,$$

$$\rho(\widehat{x_i}, \widehat{x_{i+1}}) \le \Delta, \ i \in \{0, \dots, \widehat{q} - 1\},$$

$$\max\{\rho(\widehat{x_i}, \widehat{x_{i+1}}) : \ i = 0, \dots, \widehat{q} - 1\} \ge \Delta/4$$

and

$$\sum_{i=0}^{\widehat{q}} \rho(\widehat{x_i}, \widehat{x_{i+1}}) \le \sum_{i=0}^{q-1} \rho(x_i, x_{i+1}).$$

This completes the proof of Lemma 11.4.

Lemma 11.5 *Let $x, y \in X$ satisfy*

$$\rho(x, y) > \Delta. \tag{11.17}$$

Then

$$\rho_1(T(x), T(y)) \le \rho_1(x, y) - 4^{-1}\Delta(1 - \phi(\Delta/4)).$$

Proof Let $q \ge 1$ be an integer,

$$x_i \in X, \ i = 0, \dots, q, \ x_0 = x, \ x_q = y,$$

$$\rho(x_i, x_{i+1}) \le \Delta, \ i \in \{0, \dots, q - 1\}, \tag{11.18}$$

and

$$\max\{\rho(x_i, x_{i+1}) : \ i = 0, \dots, q - 1\} \ge \Delta/4. \tag{11.19}$$

In view of (11.19), there exists a natural number $j \in \{0, \dots, q - 1\}$ such that

$$\rho(x_j, x_{j+1}) \ge \Delta/4. \tag{11.20}$$

By (11.6), (11.7), and (11.18), we have, for all integers $i = 0, \dots, q - 1$,

$$\rho(T(x_i), T(x_{i+1})) \le \phi(\rho(x_i, x_{i+1}))\rho(x_i, x_{i+1}) \le \rho(x_i, x_{i+1}). \tag{11.21}$$

Equations (11.7), (11.18), and (11.20) imply that

$$\rho(T(x_j), T(x_{j+1})) \le \phi(\rho(x_j, x_{j+1}))\rho(x_j, x_{j+1}) \le \phi(\Delta/4)\rho(x_j, x_{j+1})$$

and

$$\rho(x_j, x_{j+1}) - \rho(T(x_j), T(x_{j+1})) \ge \rho(x_j, x_{j+1})(1 - \phi(\Delta/4))$$

$$\ge 4^{-1}\Delta(1 - \phi(\Delta/4)). \tag{11.22}$$

It follows from (11.7), (11.21), and (11.22) that

$$\sum_{i=0}^{q-1} \rho(x_i, x_{i+1}) \ge \sum_{i=0}^{q-1} \rho(T(x_i), T(x_{i+1})) + 4^{-1}\Delta(1 - \phi(\Delta/4)). \tag{11.23}$$

Lemma 11.4 and Eqs. (11.21) and (11.23) imply that

$$\rho_1(T(x), T(y)) \le \inf\{\sum_{i=0}^{q-1} \rho(x_i, x_{i+1}) - 4^{-1}\Delta(1 - \phi(\Delta/4)) :$$

$$q \ge 1 \text{ is an integer},$$

$$x_i \in X,\ i = 0, \dots, q,\ x_0 = x,\ x_q = y,\ \rho(x_i, x_{i+1}) \le \Delta,\ i \in \{0, \dots, q-1\},$$

$$\max\{\rho(x_i, x_{i+1}) :\ i = 0, \dots, q-1\} \ge \Delta/4\}$$

$$= \rho_1(x, y) - 4^{-1}\Delta(1 - \phi(\Delta/4)).$$

Lemma 11.5 is proved.

11.3 Proof of Theorem 11.1

We may assume that $\epsilon < \min\{1, \Delta\}$. Choose a positive number

$$\delta < 8^{-1}\epsilon(1 - \phi(\epsilon/4)). \tag{11.24}$$

Assume that $x, y \in X$ satisfy

$$\rho(x, T(x)),\ \rho(y, T(y)) \le \delta. \tag{11.25}$$

By (11.3), (11.24), and (11.25),

$$\rho_1(x, y) \le \rho_1(x, T(x)) + \rho_1(T(x), T(y)) + \rho_1(T(y), y)$$

$$= \rho_1(T(x), T(y)) + \rho(x, T(x)) + \rho(y, T(y)) \le 2\delta + \rho_1(T(x), T(y)). \tag{11.26}$$

In order to complete the proof of the theorem, it is sufficient to show that

$$\rho(x, y) \le \epsilon.$$

Suppose to the contrary that

$$\rho(x, y) > \epsilon. \tag{11.27}$$

In view of (11.2) and (11.27),

$$\rho_1(x, y) \ge \rho(x, y) > \epsilon. \tag{11.28}$$

There are two cases:

$$\rho(x, y) \le \Delta; \tag{11.29}$$

$$\rho(x, y) > \Delta. \tag{11.30}$$

Assume that (11.29) holds. Then Eqs. (11.3) and (11.29) imply that

$$\rho_1(x, y) = \rho(x, y) \le \Delta. \tag{11.31}$$

It follows from (11.7), (11.27), and (11.31) that

$$\rho(T(x), T(y)) \le \phi(\rho(x, y))\rho(x, y) \le \phi(\epsilon)\rho(x, y). \tag{11.32}$$

By (11.24), (11.27), and (11.32), we have

$$\rho(x, y) \le 2\delta + \rho(T(x), T(y)) \le 2\delta + \phi(\epsilon)\rho(x, y)$$

and

$$2\delta \ge \rho(x, y)(1 - \phi(\epsilon)) \ge \epsilon(1 - \phi(\epsilon)).$$

This, however, contradicts (11.25). The contradiction we have reached proves that (11.30) holds. Lemma 11.5 and (11.37) imply that

$$\rho_1(T(x), T(y)) \le \rho_1(x, y) - 4^{-1}\Delta(1 - \phi(\Delta/4)). \tag{11.33}$$

It follows from Eqs. (11.3), (11.24), (11.25), and (12.34) that

$$\begin{aligned}\rho_1(x, y) &\le 2\delta + \rho_1(T(x), T(y))\\ &\le 2\delta + \rho_1(x, y) - 4^{-1}\Delta(1 - \phi(\Delta/4))\end{aligned}$$

and

$$\delta \geq 8^{-1}\Delta(1-\phi(\Delta/4)).$$

This, however, contradicts (11.25). The contradiction we have reached completes the proof of Theorem 11.1.

11.4 Proof of Theorem 11.2

Without any loss of generality, we may assume that $M > 1$ and

$$\epsilon < \min\{1, \Delta/4\}.$$

Choose a positive number

$$\delta < 8^{-1}\epsilon(1-\phi(\epsilon/2)) \tag{11.34}$$

and a natural number

$$n_0 > 1 + 8M\epsilon^{-1}(1-\phi(\epsilon/2))^{-1}. \tag{11.35}$$

Assume that $n > n_0$ is an integer and that a finite sequence $\{x_i\}_{i=0}^{n} \subset X$ satisfies

$$\rho_1(x_0, x_1) \leq M \tag{11.36}$$

and

$$\rho(x_{i+1}, T(x_i)) \leq \delta,\ i = 0, \dots, n-1. \tag{11.37}$$

First, we show that there exists an integer $i \in \{0, \dots, n_0 - 1\}$ such that

$$\rho(x_i, x_{i+1}) \leq \epsilon/2.$$

Suppose to the contrary that no such integer exists. Then,

$$\rho(x_i, x_{i+1}) > \epsilon/2,\ i = 0, \dots, n_0 - 1. \tag{11.38}$$

Let $i \in \{0, \dots, n_0 - 1\}$. Assume first that

$$\rho(x_i, x_{i+1}) \leq \Delta.$$

Then, in view of (11.3), (11.7), (11.34), (11.37), and (11.38), we have

$$\rho_1(x_{i+2}, x_{i+1}) \le \rho_1(x_{i+2}, T(x_{i+1})) + \rho_1(T(x_{i+1}), T(x_i)) + \rho_1(T(x_i), x_{i+1})$$
$$\le 2\delta + \rho_1(T(x_{i+1}), T(x_i)) = 2\delta + \rho(T(x_{i+1}), T(x_i))$$
$$\le 2\delta + \phi(\rho(x_{i+1}, x_i))\rho(x_{i+1}, x_i)$$
$$\le 2\delta + \phi(\epsilon/2)\rho(x_{i+1}, x_i). \tag{11.39}$$

By (11.34), (11.38), and (11.39), we also have

$$\rho_1(x_i, x_{i+1}) - \rho_1(x_{i+1}, x_{i+2}) \ge (1 - \phi(\epsilon/2))\rho(x_{i+1}, x_i) - 2\delta$$
$$\ge (1 - \phi(\epsilon/2))\epsilon/2 - 2\delta \ge (1 - \phi(\epsilon/2))\epsilon/4. \tag{11.40}$$

Now assume that

$$\rho(x_i, x_{i+1}) > \Delta. \tag{11.41}$$

Then Lemma 11.5 and (11.41) imply that

$$\rho_1(T(x_i), T(x_{i+1})) \le \rho_1(x_i, x_{i+1}) - (1 - \phi(\Delta/4))\Delta/4. \tag{11.42}$$

It follows from (11.3), (11.37), (11.39), and (11.42) that

$$\rho_1(x_{i+2}, x_{i+1}) \le \rho_1(x_{i+1}, T(x_i)) + \rho_1(T(x_i), T(x_{i+1})) + \rho_1(T(x_{i+1}), x_{i+2})$$
$$\le 2\delta + \rho_1(T(x_{i+1}), T(x_i)) \le 2\delta + \rho_1(x_i, x_{i+1}) - 4^{-1}\Delta(1 - \phi(\Delta/4))$$
$$\le 2\delta + \rho_1(x_i, x_{i+1}) - 2^{-1}\epsilon(1 - \phi(\epsilon/2))$$
$$\le \rho_1(x_i, x_{i+1}) - 4^{-1}\epsilon(1 - \phi(\epsilon/2)).$$

Thus, in both cases (see (11.40)), we have

$$\rho_1(x_i, x_{i+1}) - \rho_1(x_{i+1}, x_{i+2}) \ge (1 - \phi(\epsilon/2))\epsilon/4. \tag{11.43}$$

Equations (11.36) and (11.43) imply that

$$M \ge \rho_1(x_0, x_1) \ge \rho_1(x_0, x_1) - \rho_1(x_{n_0}, x_{n_0+1})$$
$$= \sum_{i=0}^{n_0-1} (\rho_1(x_i, x_{i+1}) - \rho_1(x_{i+1}, x_{i+2}))$$
$$\ge n_0(1 - \phi(\epsilon/2))\epsilon/4$$

and

$$n_0 \leq 4M(1 - \phi(\epsilon/2))^{-1}\epsilon^{-1}.$$

This, however, contradicts (11.35). The contradiction we have reached proves that after all, there exists an integer $j \in \{0, \ldots, n_0 - 1\}$ such that

$$\rho(x_j, x_{j+1}) \leq \epsilon/2. \tag{11.44}$$

We claim that for all integers $i \in \{j, \ldots, n - 1\}$,

$$\rho(x_i, x_{i+1}) \leq \epsilon.$$

Suppose to the contrary that this does not hold. Then there exists an integer $k \in \{j, \ldots, n - 1\}$ such that

$$\rho(x_k, x_{k+1}) > \epsilon. \tag{11.45}$$

By (11.44) and (11.45),

$$k > j. \tag{11.46}$$

We may assume without any loss of generality that

$$\rho(x_i, x_{i+1}) \leq \epsilon, \;\; i = j, \ldots, k - 1,$$

and, in particular,

$$\rho(x_{k-1}, x_k) \leq \epsilon. \tag{11.47}$$

There are two cases:

$$\rho(x_{k-1}, x_k) \leq \epsilon/2; \tag{11.48}$$

$$\rho(x_{k-1}, x_k) > \epsilon/2. \tag{11.49}$$

Assume that (11.48) holds. By (11.7), (11.34), (11.37), and (11.38),

$$\rho(x_k, x_{k+1}) \leq \rho(x_k, T(x_{k-1})) + \rho(T(x_{k-1}), T(x_k)) + \rho(T(x_k), x_{k+1})$$

$$\leq 2\delta + \rho(x_k, x_{k-1}) \leq 2\delta + \epsilon/2 < \epsilon.$$

This, however, contradicts (11.45). Therefore, (11.49) holds.

It now follows from (11.7), (11.34), (11.37), (11.47), and (11.49) that

$$\rho(x_{k+1}, x_k) \le \rho(x_{k+1}, T(x_k)) + \rho(T(x_k), T(x_{k-1})) + \rho(T(x_{k-1}), x_k)$$

$$\le 2\delta + \rho(x_k, x_{k-1})\phi(\rho(x_k, x_{k-1})) \le 2\delta + \phi(\epsilon/2)\epsilon \le \epsilon.$$

This contradicts (11.45) once again. The contradiction we have reached proves that for all integers $i \in \{j, \dots, n-1\}$, we have

$$\rho(x_i, x_{i+1}) \le \epsilon.$$

This completes the proof of Theorem 11.2.

11.5 The First Porosity Result

Let (Y, d) be a complete metric space. We denote by $B(y, r)$ the closed ball of center $y \in Y$ and radius $r > 0$. Recall that a subset $E \subset Y$ is called porous (with respect to the metric d) if there exist $\alpha \in (0, 1)$ and $r_0 > 0$ such that for each $r \in (0, r_0]$ and each $y \in Y$, there exists $z \in Y$ for which

$$B(z, \alpha r) \subset B(y, r) \setminus E.$$

A subset of the space Y is called σ-porous (with respect to d) if it is a countable union of porous subsets of Y.

Using the porosity notion, we now show that most uniformly locally nonexpansive mappings are, in fact, uniformly locally contractive. For mappings that are nonexpansive and contractive, such a result was obtained in [166].

We continue to use the definitions, notations, and assumptions introduced in Sect. 11.1. Assume that

$$\mathrm{diam}(X) := \sup\{\rho_1(x, y) : x, y \in X\} < \infty.$$

We call a mapping $T : X \to X$ Δ-contractive if there exists a decreasing function

$$\phi : [0, \Delta] \to [0, 1]$$

such that

$$\phi(t) < 1, \ t \in (0, \Delta], \tag{11.50}$$

and for each $x, y \in X$ satisfying $\rho(x, y) \le \Delta$,

$$\rho(T(x), T(y)) \le \phi(\rho(x, y))\rho(x, y). \tag{11.51}$$

Denote by $\mathcal{F}$ the set of all Δ-contractive self-mappings of X.

Also, denote by $\mathfrak{M}$ the set of all mappings $T : X \to X$ such that for each $x, y \in X$ satisfying $\rho(x, y) \le \Delta$,

$$\rho(T(x), T(y)) \le \rho(x, y).$$

Assume that $\theta \in X$ and that for each $\gamma \in (0, 1)$ and each $x \in X$, there is a point

$$(1-\gamma)x \oplus \gamma\theta \in X$$

such that for each $x, y \in X$ and each $\gamma \in (0, 1)$,

$$\rho((1-\gamma)x \oplus \gamma\theta,\ (1-\gamma)y \oplus \gamma\theta) \le (1-\gamma)\rho(x, y) \tag{11.52}$$

and

$$\rho((1-\gamma)x \oplus \gamma\theta, x) \le \gamma\rho(x, \theta). \tag{11.53}$$

Note that this assumption indeed holds if X is a convex subset of a hyperbolic space in the sense of [85, 168].

For each $S, T \in \mathfrak{M}$, set

$$d(S, T) := \sup\{\rho(S(x), T(x)) :\ x \in X\}.$$

It is not difficult to see that $(\mathfrak{M}, d)$ is a complete metric space.

We prove the following result obtained in [175].

Theorem 11.6 *The set $\mathfrak{M} \setminus \mathcal{F}$ is σ-porous in $(\mathfrak{M}, d)$.*

Proof For every natural number n, define

$$\mathcal{F}_n = \{T \in \mathfrak{M} :$$

$$\sup\{\rho(T(x), T(y))\rho(x, y)^{-1} :\ x, y \in X,\ \rho(x, y) \in [n^{-1}\Delta, \Delta]\} < 1\}. \tag{11.54}$$

It is not difficult to see that

$$\mathcal{F} = \cap_{n=1}^{\infty} \mathcal{F}_n. \tag{11.55}$$

In view of (11.55),

$$\mathfrak{M} \setminus \mathcal{F} = \cup_{n=1}^{\infty} (\mathfrak{M} \setminus \mathcal{F}). \tag{11.56}$$

Let $n \ge 1$ be an integer. By (11.55) and (11.56), in order to complete the proof of the theorem, it is sufficient to show that the set $\mathfrak{M} \setminus \mathcal{F}_n$ is σ-porous in $(\mathfrak{M}, d)$. To this end, define

$$\alpha := (8n)^{-1} \min\{1, \Delta\}(\mathrm{diam}(X) + 1)^{-1}. \tag{11.57}$$

Let $r \in (0, 1]$ and $T \in \mathfrak{M}$. Define

$$\gamma := 2^{-1} r(\mathrm{diam}(X) + 1)^{-1} \tag{11.58}$$

and

$$T_\gamma(x) := (1 - \gamma)T(x) \oplus \gamma\theta, \ x \in X. \tag{11.59}$$

Let $x, y \in X$ satisfy

$$\rho(x, y) \leq \Delta. \tag{11.60}$$

In view of (11.52), (11.54), and (11.59), we have

$$\rho(T_\gamma(x), T_\gamma(y)) \leq (1 - \gamma)\rho(T(x), T(y)) \leq (1 - \gamma)\rho(x, y). \tag{11.61}$$

By (11.50) and (11.59),

$$\rho(T_\gamma(x), T(x)) = \rho((1 - \gamma)T(x) \oplus \gamma\theta, T(x)) \leq \gamma\rho(\theta, T(x)) \leq \gamma\mathrm{diam}(X). \tag{11.62}$$

In view of (11.61),

$$T_\gamma \in \mathfrak{M}.$$

Assume now that $S \in \mathfrak{M}$ satisfies

$$d(S, T_\gamma) \leq \alpha r.$$

Then Eqs. (11.57), (11.58), (11.61), and the relation above imply that for each $x, y \in X$ satisfying

$$\rho(x, y) \in [n^{-1}\Delta, \Delta],$$

we have

$$\rho(S(x), S(y))\rho(x, y)^{-1} \leq (\rho(T_\gamma(x), T_\gamma(y)) + 2d(S, T_\gamma))\rho(x, y)^{-1}$$
$$\leq ((1 - \gamma)\rho(x, y) + 2\alpha r)\rho(x, y)^{-1} \leq (1 - \gamma) + 2\alpha r n\Delta^{-1}$$
$$\leq (1 - \gamma) + 4^{-1} r(\mathrm{diam}(X) + 1)^{-1} = 1 - 4^{-1} r(\mathrm{diam}(X) + 1)^{-1}.$$

Therefore,

$$S \in \mathcal{F}_n$$

and

$$\{S \in \mathfrak{M} : d(S, T_\gamma) \le \alpha r\} \subset \mathcal{F}_n.$$

By (11.57), (11.58), and (11.62),

$$\{S \in \mathfrak{M} : d(S, T_\gamma) \le \alpha r\} \subset \{S \in \mathfrak{M} : d(S, T) \le \alpha r + \gamma \mathrm{diam}(X)\}$$

$$\subset \{S \in \mathfrak{M} : d(S, T) \le r\}.$$

In view of the above relations, the set $\mathfrak{M} \setminus \mathcal{F}_n$ is indeed σ-porous in $(\mathfrak{M}, d)$. This completes the proof of Theorem 11.6.

11.6 Uniformly Locally Contractive Non-self-mappings

Now we extend the results of the previous sections to uniformly locally contractive non-self-mappings defined on a closed subset of the metric space. These extensions were obtained in [176].

We use the notation, definitions, and assumptions of Sect. 11.1. In particular, we assume that assumption (A) holds which we recall.

Assume that (X, ρ) is a complete metric space, $\Delta > 0$, and that the following assumption holds.

(A) For each $x, y \in X$, there exist an integer $q \ge 1$ and points $x_i \in X$, $i = 0, \dots, q$, such that

$$x_0 = x,\ x_q = y,\ \rho(x_i, x_{i+1}) \le \Delta,\ i \in \{0, \dots, q\} \setminus \{q\}.$$

Recall that for each $x, y \in X$,

$$\rho_1(x, y) := \inf\{\sum_{i=0}^{q-1} \rho(x_i, x_{i+1}) :\ q \ge 1 \text{ is an integer},$$

$$x_i \in X,\ i = 0, \dots, q,\ x_0 = x,\ x_q = y,\ \rho(x_i, x_{i+1}) \le \Delta,\ i \in \{0, \dots, q\} \setminus \{q\}\}.$$

Clearly, ρ_1 is a metric on X.

Assume that $K \subset X$ is a nonempty closed set and that the following assumption holds.

(B) For each $x, y \in K$,

$$\rho_1(x, y) := \inf\{\sum_{i=0}^{q-1} \rho(x_i, x_{i+1}) : \ q \geq 1 \text{ is an integer,}$$

$$x_i \in K, \ i = 0, \dots, q, \ x_0 = x, \ x_q = y, \ \rho(x_i, x_{i+1}) \leq \Delta, \ i \in \{0, \dots, q\} \setminus \{q\}\}.$$

Let $T : K \to X$ be a mapping. We assume that for each $x, y \in X$ satisfying $\rho(x, y) \leq \Delta$, the inequality

$$\rho(T(x), T(y)) \leq \rho(x, y) \tag{11.63}$$

holds. In view of (B) and (11.63), for each $x, y \in X$, we have

$$\rho_1(T(x), T(y)) \leq \rho_1(x, y).$$

Such a mapping T is said to be uniformly locally nonexpansive.
Assume that $\phi : [0, \Delta] \to [0, 1]$ is a decreasing function,

$$\phi(t) < 1 \text{ for all } t \in (0, \Delta], \tag{11.64}$$

and that for each $x, y \in K$ satisfying $\rho(x, y) \leq \Delta$, we have

$$\rho(T(x), T(y)) \leq \phi(\rho(x, y))\rho(x, y). \tag{11.65}$$

We prove the following theorems obtained in [176]. The first of them shows the well-posedness of the fixed point problem for the mapping T.

Theorem 11.7 *Given $\epsilon > 0$, there exists $\delta > 0$ such that for each $x, y \in K$ satisfying*

$$\max\{\rho(T(x), x), \ \rho(T(y), y)\} \leq \delta,$$

the inequality

$$\rho(x, y) \leq \epsilon$$

holds.

Theorem 11.8 *Let $\epsilon, M > 0$ be given. Then there exist a number $\delta \in (0, \epsilon)$ and an integer $n_0 \geq 1$ such that for each integer $n > n_0$ and each finite sequence $\{x_i\}_{i=0}^{n} \subset K$ which satisfies*

$$\rho_1(x_0, x_1) \leq M$$

and

$$\rho(x_{i+1}, T(x_i)) \le \delta,\ i = 0, \dots, n-1,$$

the inequality

$$\rho(x_i, x_{i+1}) \le \epsilon$$

holds for all $i = n_0, \dots, n-1$.

Theorems 11.7 and 11.8 easily imply the following two results regarding the existence of a unique fixed point and the convergence of inexact iterates.

Theorem 11.9 *Assume that for each integer* $n \ge 1$*, there exists a finite sequence* $\{x^{(n)}\}_{i=0}^{n} \subset K$ *such that*

$$\sup\{\rho_1(x_0^{(n)}, x_1^{(n)}) : n = 1, 2, \dots\} < \infty$$

and

$$\lim_{n\to\infty} \sup\{\rho(x_{i+1}^{(n)}, Tx_i^{(n)}) : i = 0, \dots, n-1\} = 0.$$

Then there exists a unique point $x_* \in K$ *such that* $T(x_*) = x_*$.

Theorem 11.10 *Let the point* $x_* \in K$ *satisfy*

$$T(x_*) = x_*$$

and let $\epsilon, M > 0$ *be given. Then there exist a number* $\delta > 0$ *and an integer* $n_0 \ge 1$ *such that for each integer* $n > n_0$ *and each finite sequence* $\{x_i\}_{i=0}^{n} \subset K$ *which satisfies*

$$\rho_1(x_0, x_1) \le M$$

and

$$\rho(x_{i+1}, T(x_i)) \le \delta,\ i = 0, \dots, n-1,$$

the inequality

$$\rho(x_i, x_*) \le \epsilon$$

holds for all $i = n_0, \dots, n$.

11.7 Auxiliary Results for Theorems 11.7 and 11.8

For the proof of the following result, see Lemma 11.4.

Lemma 11.11 *Let $x, y \in K$ satisfy*

$$\rho(x, y) > \Delta,$$

let $q \geq 1$ be an integer, $x_i \in K,\ i = 0, \ldots, q,$

$$x_0 = x,\ x_q = y,$$

$$\rho(x_i, x_{i+1}) \leq \Delta,\ i \in \{0, \ldots, q-1\},$$

and let

$$\max\{\rho(x_i, x_{i+1}) : i = 0, \ldots, q-1\} < \Delta/4.$$

Then there exist an integer $\widehat{q} \geq 1$ and a finite sequence $\widehat{x}_i \in K,\ i = 0, \ldots, \widehat{q}$, such that

$$\widehat{x}_0 = x_0,\ \widehat{x}_{\widehat{q}} = y,$$

$$\{\widehat{x}_i : i = 0, \ldots, \widehat{q}\} \subset \{x_i : i = 0, \ldots, q\},$$

$$\rho(\widehat{x}_i, \widehat{x}_{i+1}) \leq \Delta,\ i \in \{0, \ldots, \widehat{q}-1\},$$

$$\max\{\rho(\widehat{x}_i, \widehat{x}_{i+1}) : i = 0, \ldots, \widehat{q}-1\} \geq \Delta/4$$

and

$$\sum_{i=0}^{\widehat{q}} \rho(\widehat{x}_i, \widehat{x}_{i+1}) \leq \sum_{i=0}^{q-1} \rho(x_i, x_{i+1}).$$

Lemma 11.11 and assumption (B) yield the following result.

Lemma 11.12 *Let $x, y \in K$ satisfy*

$$\rho(x, y) > \Delta.$$

Then,

$$\rho_1(x, y) = \inf\{\sum_{i=0}^{q-1} \rho(x_i, x_{i+1}) :\ q \geq 1 \textit{ is an integer},$$

$$x_i \in K,\ i = 0, \dots, q,\ x_0 = x,\ x_q = y,\ \rho(x_i, x_{i+1}) \le \Delta,\ i \in \{0, \dots, q-1\},$$

and

$$\max\{\rho(x_i, x_{i+1}) :\ i = 0, \dots, q-1\} \ge \Delta/4\}.$$

Lemma 11.13 *Let $x, y \in X$ satisfy*

$$\rho(x, y) > \Delta.$$

Then,

$$\rho_1(T(x), T(y)) \le \rho_1(x, y) - 4^{-1}\Delta(1 - \phi(\Delta/4)).$$

Proof Lemma 11.12 and (11.65) imply that

$$\rho_1(T(x), T(y)) \le \inf\{\sum_{i=0}^{q-1} \rho(T(x_i), T(x_{i+1})) :\ q \ge 1 \text{ is an integer},$$

$$x_i \in K,\ i = 0, \dots, q,\ x_0 = x,\ x_q = y,\ \rho(x_i, x_{i+1}) \le \Delta,\ i \in \{0, \dots, q-1\},$$

$$\max\{\rho(x_i, x_{i+1}) :\ i = 0, \dots, q-1\} \ge \Delta/4\}$$

$$\le \inf\{\sum_{i=0}^{q-1} \rho(x_i, x_{i+1}) - 4^{-1}\Delta(1 - \phi(\Delta/4)) :\ q \ge 1 \text{ is an integer},$$

$$x_i \in K,\ i = 0, \dots, q,\ x_0 = x,\ x_q = y,\ \rho(x_i, x_{i+1}) \le \Delta,\ i \in \{0, \dots, q-1\},$$

$$\max\{\rho(x_i, x_{i+1}) :\ i = 0, \dots, q-1\} \ge \Delta/4\}$$

$$= \rho_1(x, y) - 4^{-1}\Delta(1 - \phi(\Delta/4)).$$

This completes the proof of Lemma 11.13.

11.8 Proof of Theorem 11.7

We may assume that

$$\epsilon < \min\{1, \Delta\}. \tag{11.66}$$

Choose a positive number

$$\delta < 8^{-1}\epsilon(1 - \phi(\epsilon/4)) \tag{11.67}$$

and assume that $x, y \in K$ satisfy

$$\rho(x, T(x)),\ \rho(y, T(y)) \leq \delta. \tag{11.68}$$

Using (11.66)–(11.68), we see that

$$\rho_1(x, y) \leq \rho_1(T(x), T(y)) + 2\delta. \tag{11.69}$$

If $\rho(x, y) > \Delta$, then Lemma 11.12, (11.66) and (11.67) imply that

$$\rho_1(x, y) \leq \rho_1(T(x), T(y)) + 2\delta$$

$$\leq \rho_1(x, y) - 4^{-1}\Delta(1 - \phi(\Delta/4)) + 2\delta < \rho_1(x, y).$$

The contradiction we have reached shows that

$$\rho(x, y) \leq \Delta.$$

If $\rho(x, y) > \epsilon$, then (11.63), (11.65), (11.67), (11.69), (11.70), and the relation above imply that

$$\rho(x, y) \leq \rho(T(x), T(y)) + 2\delta$$

$$\leq 2\delta + \rho(x, y) - (\rho(x, y) - \rho(T(x), T(y)))$$

$$\leq 2\delta + \rho(x, y) - \rho(x, y)(1 - \phi(\rho(x, y)))$$

$$\leq 2\delta + \rho(x, y) - \epsilon(1 - \phi(\epsilon)) < \rho(x, y).$$

The contradiction we have reached yields that $\rho(x, y) \leq \epsilon$, as asserted. This completes the proof of Theorem 11.7.

11.9 Proof of Theorem 11.8

Without any loss of generality, we may assume that $M > 1$ and

$$\epsilon < \min\{1, \Delta/4\}.$$

Choose a positive number

$$\delta < 8^{-1}\epsilon(1 - \phi(\epsilon/2)) \tag{11.70}$$

and a natural number

$$n_0 > 1 + 8M\epsilon^{-1}(1 - \phi(\epsilon/2))^{-1}. \tag{11.71}$$

Assume that $n > n_0$ is an integer and that a finite sequence $\{x_i\}_{i=0}^{n} \subset K$ satisfies

$$\rho_1(x_0, x_1) \le M \tag{11.72}$$

and

$$\rho(x_{i+1}, T(x_i)) \le \delta,\ i = 0, \dots, n-1. \tag{11.73}$$

Let $i \in \{0, \dots, n_0 - 2\}$. In view of (11.70) and (11.73), we have

$$\begin{aligned} \rho_1(x_{i+2}, x_{i+1}) &\le \rho_1(x_{i+2}, T(x_{i+1})) + \rho_1(T(x_{i+1}), T(x_i)) + \rho_1(T(x_i), x_{i+1}) \\ &\le 2\delta + \rho_1(T(x_{i+1}), T(x_i)). \end{aligned} \tag{11.74}$$

If $\rho(x_i, x_{i+1}) \le \Delta$, then it follows from (11.63), (11.65), and (11.74) that

$$\begin{aligned} \rho_1(x_{i+2}, x_{i+1}) &\le 2\delta + \rho(T(x_{i+1}), T(x_i)) \\ &\le 2\delta + \phi(\rho(x_i, x_{i+1}))\rho(x_{i+1}, x_i) \\ &\le 2\delta + \phi(\rho_1(x_i, x_{i+1}))\rho_1(x_{i+1}, x_i). \end{aligned} \tag{11.75}$$

If $\rho(x_i, x_{i+1}) > \Delta$, then Lemma 11.12, (11.65), and (11.74) imply that

$$\begin{aligned} \rho_1(x_{i+2}, x_{i+1}) &\le \rho_1(T(x_i), T(x_{i+1})) + 2\delta \\ &\le 2\delta + \rho_1(x_i, x_{i+1}) - (\rho_1(x_i, x_{i+1}) - \rho_1(T(x_i), T(x_{i+1}))) \\ &\le 2\delta + \rho_1(x_i, x_{i+1}) - (1 - \phi(\Delta))\Delta. \end{aligned} \tag{11.76}$$

If $\rho(x_i, x_{i+1}) > \epsilon/2$, then relations (11.70), (11.75), and (11.76) imply that

$$\begin{aligned} &\rho_1(x_i, x_{i+1}) - \rho_1(x_{i+2}, x_{i+1}) \\ &\ge \min\{(1 - \phi(\epsilon/2))\epsilon/2 - 2\delta,\ (1 - \phi(\Delta))\Delta - 2\delta\} \\ &= (1 - \phi(\epsilon/2))\epsilon/2 - 2\delta \ge (1 - \phi(\epsilon/2))\epsilon/4. \end{aligned}$$

If $\rho(x_i, x_{i+1}) \leq \epsilon/2$, then relations (11.63), (11.70), and (11.75) imply that

$$\rho_1(x_{i+1}, x_{i+2}) \leq 2\delta + \rho(x_i, x_{i+1}) \leq \epsilon.$$

Therefore, we have shown that the following property holds:

(C) For each $i \in \{0, \ldots, n_0 - 2\}$, if $\rho(x_i, x_{i+1}) > \epsilon/2$, then

$$\rho_1(x_i, x_{i+1}) - \rho_1(x_{i+2}, x_{i+1}) \geq (1 - \phi(\epsilon/2))\epsilon/4 \tag{11.77}$$

and if $\rho(x_i, x_{i+1}) \leq \epsilon/2$, then

$$\rho_1(x_{i+1}, x_{i+2}) \leq \epsilon. \tag{11.78}$$

We now claim that there exists an integer $i \in \{0, \ldots, n_0 - 1\}$ such that

$$\rho(x_i, x_{i+1}) \leq \epsilon/2. \tag{11.79}$$

Suppose to the contrary that no such integer exists. Then by (C) and (11.72), for all $i \in \{0, \ldots, n_0 - 1\}$, (11.77) holds,

$$M \geq \rho_1(x_0, x_1) \geq \sum_{i=0}^{n_0-1} (\rho_1(x_i, x_{i+1}) - \rho_1(x_{i+1}, x_{i+2}))$$

$$\geq n_0(1 - \phi(\epsilon/2))\epsilon/4$$

and

$$n_0 \leq 4M(1 - \phi(\epsilon/2))^{-1}\epsilon^{-1}.$$

This, however, contradicts (11.71). The contradiction we have reached shows that after all, there does exist an integer $j \in \{0, \ldots, n_0 - 1\}$ such that

$$\rho(x_j, x_{j+1}) \leq \epsilon/2,$$

as claimed.

Property (C) implies that if an integer $i \in \{j, \ldots, n - 1\} \setminus \{n - 1\}$ satisfies

$$\rho(x_i, x_{i+1}) \leq \epsilon,$$

then $\rho(x_{i+1}, x_{i+2}) \leq \epsilon$ too. Therefore, it follows that $\rho(x_i, x_{i+1}) \leq \epsilon$ for all $i \in \{j, \ldots, n - 1\}$, as asserted. This completes the proof of Theorem 11.8.

11.10 The Second Porosity Result

Let (X, ρ, W) be a complete W-hyperbolic space and let K be a nonempty, closed, and bounded subset of (X, ρ). Assume that

$$\Delta > 0,\ \theta \in K$$

and that for each $\gamma \in (0, 1)$ and each $x \in K$,

$$\gamma\theta \oplus (1-\gamma)x \in K. \tag{11.80}$$

Set

$$\text{diam}(K) = \sup\{\rho(x, y) :\ x, y \in K\}. \tag{11.81}$$

It follows from (11.80) and the boundedness of K in (X, ρ) that (A) holds for $X = K$ and that

$$\sup\{\rho_1(x, y) :\ x, y \in K\} < \infty.$$

Denote by $\mathcal{A}$ the set of all mappings $T : K \to X$ such that

$$\rho(T(x), T(y)) \le \rho(x, y) \text{ for each } x, y \in K \text{ satisfying } \rho(x, y) \le \Delta. \tag{11.82}$$

For each $A, B \in \mathcal{A}$, define

$$d(A, B) := \sup\{\rho(A(x), B(x)) :\ x \in K\}.$$

Since K is bounded in (X, ρ_1), $d(A, B)$ is finite for each $A, B \in \mathcal{A}$.

It is not difficult to see that $(\mathcal{A}, d)$ is a complete metric space.

Recall that a mapping $T \in \mathcal{A}$ is a uniformly locally contraction if there exists a decreasing function $\phi : [0, \Delta] \to [0, 1]$ such that

$$\phi(t) < 1 \text{ for all } t \in (0, \Delta],$$

and that for each $x, y \in K$ satisfying $\rho(x, y) \le \Delta$, we have

$$\rho(T(x), T(y)) \le \phi(\rho(x, y))\rho(x, y).$$

Denote by $\mathcal{F}$ the set of all uniformly locally contractions $A \in \mathcal{A}$. We prove the following result obtained in [178].

Theorem 11.14 *The set $\mathcal{A} \setminus \mathcal{F}$ is σ-porous.*

We may assume that

$$\Delta \le \mathrm{diam}(K).$$

Let $A \in \mathcal{A}$ and $\gamma \in (0, 1)$. Define

$$A_\gamma(x) = (1-\gamma)A(x) \oplus \gamma\theta, \ x \in K. \tag{11.83}$$

In view of Proposition 2.1 and (11.83), for each $x, y \in K$ satisfying $\rho(x, y) \le \Delta$,

$$\rho(A_\gamma(x), A_\gamma(y)) \le (1-\gamma)\rho(A(x), A(y)) \le (1-\gamma)\rho(x, y), \tag{11.84}$$

and by (11.80), for each $x \in K$,

$$\rho(A(x), A_\gamma(x)) = \rho(A(x), (1-\gamma)A(x) \oplus \gamma\theta)$$

$$\le \gamma\rho(A(x), \theta) \le \gamma\rho(A(x), A(\theta)) + \gamma\rho(A(\theta), \theta) \le \rho(x, \theta) + \gamma\rho(A(\theta), \theta). \tag{11.85}$$

By (11.85),

$$d(A, A_\gamma) \le \gamma\mathrm{diam}(K) + \gamma\rho(A(\theta), \theta). \tag{11.86}$$

Theorem 11.14 will follow easily from the following two propositions.
For each integer $n \ge 1$, denote by $\mathcal{F}_n$ the set of all $A \in \mathcal{A}$ such that

$$\sup\{\rho(A(x), A(y))\rho(x, y)^{-1} :$$

$$x, y \in K \text{ and } \Delta \ge \rho(x, y) \ge \Delta n^{-1}(\mathrm{diam}(K) + 1)^{-1}\} < 1. \tag{11.87}$$

Set

$$\mathcal{F} := \cap_{n=1}^{\infty} \mathcal{F}_n. \tag{11.88}$$

Proposition 11.15 *Assume that $A \in \mathcal{F}$. Then there exists a decreasing function $\phi : [0, \Delta] \to [0, 1]$ such that*

$$\phi(t) < 1 \textit{ for all } t \in (0, \Delta],$$

and that for each $x, y \in K$ satisfying $\rho(x, y) \le \Delta$,

$$\rho(A(x), A(y)) \le \phi(\rho(x, y))\rho(x, y).$$

Proof Set $\phi(0) = 1$. For each $t \in (0, \Delta]$, define

$$\phi(t) := \sup\{\rho(A(x), A(y))\rho(x, y)^{-1} :$$
$$x, y \in K \text{ and } \Delta \geq \rho(x, y) \geq t\}. \tag{11.89}$$

By (11.79)–(11.89),

$$\phi(t) < 1, \ t \in (0, \Delta].$$

It is clear that $\phi : [0, \Delta] \to [0, 1]$ is a decreasing function. Let $x, y \in K$ satisfy

$$0 < \rho(x, y) \leq \Delta.$$

Set $t = \rho(x, y)$. By (11.89),

$$\phi(\rho(x, y)) = \phi(t) \geq \rho(A(x), A(y))\rho(x, y)^{-1}.$$

Proposition 11.15 is proved.

Proposition 11.16 *Let n be a natural number. Then $\mathcal{A}\setminus\mathcal{F}_n$ is a porous set in $(\mathcal{A}, d)$.*

Proof Set

$$\alpha = \Delta(8n)^{-1}(\text{diam}(K) + 1 + \rho(A(\theta), \theta))^{-2}. \tag{11.90}$$

Assume that

$$A \in \mathcal{A}, \ r \in (0, 1].$$

Set

$$\gamma = 2^{-1}(\text{diam}(K) + 1 + \rho(A(\theta), \theta))^{-1}r. \tag{11.91}$$

Define

$$A_\gamma(x) := (1 - \gamma)A(x) \oplus \gamma\theta, \ x \in K.$$

In view of (11.86) and (11.91),

$$d(A_\gamma, A) \leq \gamma\text{diam}(K) + \gamma\rho(A(\theta), \theta) \leq r/2. \tag{11.92}$$

Assume now that $B \in \mathcal{A}$ satisfies

$$d(B, A_\gamma) \leq \alpha r. \tag{11.93}$$

It follows from (11.90), (11.92), and (11.93) that

$$d(A,B) \le d(A,A_\gamma) + d(A_\gamma,B)$$

$$\le \alpha r + r/2 \le r. \tag{11.94}$$

By (11.83), (11.84), and (11.93), for each $x, y \in K$ satisfying

$$\Delta \ge \rho(x,y) \ge \Delta n^{-1}(\mathrm{diam}(K)+1)^{-1} \tag{11.95}$$

we have

$$\rho(B(x),B(y)) \le \rho(A_\gamma(x),A_\gamma(y)) + 2\alpha r$$

$$\le (1-\gamma)\rho(x,y) + 2\alpha r$$

and in view of (11.90), (11.91), and (11.95),

$$\rho(B(x),B(y))\rho(x,y)^{-1} \le 1-\gamma+2\alpha r\rho(x,y)^{-1}$$

$$\le 1-\gamma+2\alpha r\Delta^{-1}n(\mathrm{diam}(K)+1)$$

$$\le 1-2^{-1}(\mathrm{diam}(K)+1+\rho(A(\theta),\theta))^{-1}r + 2r\alpha n(\mathrm{diam}(K)+1)\Delta^{-1}$$

$$\le 1-2^{-1}(\mathrm{diam}(K)+1+\rho(A(\theta),\theta))^{-1}r$$

$$+4^{-1}(\mathrm{diam}(K)+1+\rho(A(\theta),\theta))^{-1}r$$

$$= 1-4^{-1}(\mathrm{diam}(K)+1+\rho(A(\theta),\theta))^{-1}r.$$

Thus,

$$B \in \mathcal{F}_n$$

and

$$\{B \in \mathcal{A} : d(B,A_\gamma) \le \alpha r\} \subset \{B \in \mathcal{A} : d(A,B) \le r\} \cap \mathcal{F}_n.$$

Therefore $\mathcal{A} \setminus \mathcal{F}_n$ is a porous set. This completes the proof of Proposition 11.16.

Chapter 12
Set-Valued Mappings

In this chapter, we study iteration processes associated with a given set-valued mapping which generate its approximate fixed point and in some cases converge to a fixed point of the mapping. We consider these processes under the presence of computational errors which always present in practice. Our results are proved for various classes of set-valued mappings including strict contractions and the Mizoguchi-Takahashi class of mappings. We consider spaces of nonexpansive mappings, acting on a closed subset of a complete metric space, which is equipped with the topology of uniform convergence on bounded sets. We show the existence of an open and everywhere dense subset in this space such that every its element possesses an approximate fixed point, which is stable under small perturbations. The results of Sects. 12.12–12.15 are new.

12.1 Approximate Fixed Points

Let (X, ρ) be a complete metric space. For each $x \in X$ and each $r > 0$, set

$$B(x, r) = \{y \in X : \ \rho(x, y) \le r\}, \ B^0(x, r) = \{y \in X : \ \rho(x, y) < r\}.$$

For each $x \in X$ and each $A \subset X$, set

$$\rho(x, A) = \inf\{\rho(x, y) : \ y \in A\}.$$

Note that we assume that the infimum over empty set is ∞. For each pair of nonempty sets $A, B \subset X$, set

$$H(A, B) = \max\{\sup\{\rho(x, B) : \ x \in A\}, \ \sup\{\rho(y, A) : \ y \in B\}\}. \tag{12.1}$$

A. J. Zaslavski, *Approximate Fixed Points of Nonexpansive Mappings*,
Developments in Mathematics 80, https://doi.org/10.1007/978-3-031-70710-0_12

Assume that $T : X \to 2^X \setminus \{\emptyset\}$, $T(x)$ is closed for each $x \in X$, $c \in [0, 1)$ and that for each $x, y \in X$,

$$H(T(x), T(y)) \le c\rho(x, y). \tag{12.2}$$

We are interested to solve the problem

$$\text{Find } z \in X \text{ such that } z \in T(z).$$

In practice we can obtain only an approximate solution of this problem $z \in X$ such that $\rho(z, T(z))$ is small.

More precisely, in order to meet this goal, we use the following algorithm.

Initialization: select an arbitrary point $x_0 \in X$ and a small positive constant δ.

Iterative step: given a current iteration point x_n, calculate the next iteration point x_{n+1} such that

$$B(x_{n+1}, \delta) \cap \{\xi \in T(x_n) :$$

$$\rho(\xi, x_n) \le \delta + \rho(x_n, T(x_n))\} \ne \emptyset.$$

Clearly, at each iterative step, x_{n+1} is an approximate solution of the problem

$$\rho(\xi, x_n) \to \min, \ \xi \in T(x_n).$$

Since the space X is not compact, a solution of the problem above does not exist in general.

The following results were proved in [62] where iterative schemes for approximating fixed points of closed-valued strict contractions in metric spaces were considered.

Theorem 12.1 *Assume that $x_0 \in X$, $\{\epsilon_i\}_{i=0}^{\infty} \subset (0, \infty)$, $\sum_{i=0}^{\infty} \epsilon_i < \infty$ and that for each integer $i \ge 0$,*

$$x_{i+1} \in T(x_i), \ \rho(x_i, x_{i+1}) \le \rho(x_i, T(x_i)) + \epsilon_i. \tag{12.3}$$

Then, $\{x_i\}_{i=0}^{\infty}$ converges to a fixed point of T.

Proof We show that $\{x_i\}_{i=0}^{\infty}$ is a Cauchy sequence. By (12.1) and (12.3), for each integer $i \ge 0$,

$$\begin{aligned} \rho(x_{i+1}, x_{i+2}) &\le \rho(x_{i+1}, T(x_{i+1})) + \epsilon_{i+1} \\ &\le H(T(x_i), T(x_{i+1})) + \epsilon_{i+1} \\ &\le c\rho(x_i, x_{i+1}) + \epsilon_{i+1}. \end{aligned} \tag{12.4}$$

By (12.4),

$$\rho(x_1, x_2) \le c\rho(x_0, x_1) + \epsilon_1$$

and

$$\rho(x_2, x_3) \le c\rho(x_1, x_2) + \epsilon_2$$

$$\le c^2\rho(x_0, x_1) + c\epsilon_1 + \epsilon_2. \tag{12.5}$$

Now we use induction to show that for each integer $n \ge 1$,

$$\rho(x_n, x_{n+1}) \le c^n \rho(x_0, x_1) + \sum_{i=0}^{n-1} c^i \epsilon_{n-i}. \tag{12.6}$$

In view of (12.5), inequality (12.6) is valid for $n = 1, 2$.

Assume that $k \ge 1$ is an integer and that (12.6) holds for $n = k$. When combined with (12.4), this implies that

$$\rho(x_{k+1}, x_{k+2}) \le c\rho(x_k, x_{k+1}) + \epsilon_{k+1}$$

$$\le c^{k+1}\rho(x_0, x_1) + \sum_{i=0}^{k-1} c^{i+1}\epsilon_{k-i} + \epsilon_{k+1}$$

$$= c^{k+1}\rho(x_0, x_1) + \sum_{i=0}^{k} c^i \epsilon_{k+1-i}.$$

Thus, (12.6) holds with $n = k + 1$ and therefore (12.6) holds for all integers $n \ge 1$. By (12.6),

$$\sum_{n=1}^{\infty} \rho(x_n, x_{n+1}) \le \sum_{n=1}^{\infty}(c^n \rho(x_0, x_1) + \sum_{i=0}^{n} c^{n-i}\epsilon_i)$$

$$\le \rho(x_0, x_1)\sum_{n=1}^{\infty} c^n + \sum_{i=1}^{\infty}(\sum_{j=0}^{\infty} c^j)\epsilon_i < \infty.$$

Thus, $\{x_n\}_{n=0}^{\infty}$ is indeed a Cauchy sequence and there exists

$$x_* = \lim_{n\to\infty} x_n. \tag{12.7}$$

We claim that $x_* \in T(x_*)$. Let $\epsilon > 0$ be given. By (12.7), there is an integer $n_0 \geq 1$ such that for each integer $n \geq n_0$,

$$\rho(x_n, x_*) \leq \epsilon/8. \tag{12.8}$$

Let $n \geq n_0$ be an integer. By (12.1) and (12.8),

$$H(T(x_n), T(x_*)) \leq c\rho(x_n, x_*) \leq c\epsilon/8. \tag{12.9}$$

By (12.3),

$$x_{n+1} \in T(x_n).$$

Together with (12.9), this implies that

$$\rho(x_{n+1}, T(x_*)) \leq c\epsilon/8.$$

Hence, there is

$$y \in T(x_*) \tag{12.10}$$

such that

$$\rho(x_{n+1}, y) \leq c\epsilon/4.$$

Together with (12.8), this implies that

$$\rho(x_*, T(x_*)) \leq \rho(x_*, y) \leq \rho(x_*, x_{n+1}) + \rho(x_{n+1}, y) \leq \epsilon/8 + \epsilon/4.$$

Since ϵ is an arbitrary positive number, we conclude that

$$x_* \in T(x_*),$$

as claimed. Theorem 12.1 is proved.

The next result shows the well-posedness of the fixed point problem.

Theorem 12.2 *Let $\epsilon > 0$ be given and let*

$$\delta \in (0, 4^{-1}(1 - c)\epsilon]. \tag{12.11}$$

Then if $x \in X$ and $\rho(x, T(x)) < \delta$, then there is $\bar{x} \in X$ such that $\bar{x} \in T(\bar{x})$ and $\rho(x, \bar{x}) \leq \epsilon$.

Proof Consider

$$x \in X \text{ such that } \rho(x, T(x)) < \delta. \tag{12.12}$$

Set

$$x_0 = x. \tag{12.13}$$

By (12.12), there is

$$x_1 \in T(x_0) \tag{12.14}$$

such that

$$\rho(x_0, x_1) < \delta. \tag{12.15}$$

For each integer $n \geq 1$, choose

$$x_{n+1} \in T(x_n) \tag{12.16}$$

such that

$$\rho(x_{n+1}, x_n) \leq \rho(x_n, T(x_n))(1+c)/(2c). \tag{12.17}$$

By (12.1), (12.3), (12.16), and (12.17), for each integer $n \geq 1$,

$$\rho(x_n, x_{n+1}) \leq (1+c)H(T(x_{n-1}), T(x_n))/(2c) \leq ((1+c)/2)\rho(x_n, x_{n-1}).$$

When combined with (12.15), this implies that for each integer $n \geq 1$,

$$\rho(x_n, x_{n+1}) \leq [(1+c)/2]^n \rho(x_0, x_1) \leq [(1+c)/2]^n \delta. \tag{12.18}$$

Therefore,

$$\sum_{n=0}^{\infty} \rho(x_n, x_{n+1}) < \infty,$$

$\{x_n\}_{n=0}^{\infty}$ is a Cauchy sequence and there exists $\bar{x} \in X$ such that

$$\bar{x} = \lim_{n \to \infty} x_n. \tag{12.19}$$

Since $x_{n+1} \in T(x_n)$ for all integers $n \geq 0$, (12.19) implies that

$$\bar{x} \in T(\bar{x}).$$

By (12.11), (12.18), and (12.19),

$$\rho(x_0, \bar{x}) = \lim_{n\to\infty} \rho(x_0, x_n) \leq \sum_{n=0}^{\infty} \rho(x_i, x_{i+1})$$

$$\leq \sum_{i=0}^{\infty} [(1+c)/2]^i \delta = 2\delta/(1-c) < \epsilon/2.$$

This completes the proof of Theorem 12.2.

The next result shows the behavior of inexact iterates under the presence of nonsummable computational errors.

Theorem 12.3 *Fix $\theta \in X$. Let $\epsilon \in (0, 1]$ and $M > 0$ be given, a positive number δ satisfy*

$$\delta < 2^{-1}\epsilon(1-c], \tag{12.20}$$

and an integer $n_0 \geq 1$ satisfy

$$c^{n_0}(2M + 1 + \rho(\theta, T(\theta)) < \epsilon/2. \tag{12.21}$$

Then for each sequence $\{x_i\}_{i=0}^{\infty} \subset X$ such that $\rho(x_0, \theta) \leq M$ and such that for each integer $n \geq 0$,

$$x_{n+1} \in T(x_n) \text{ and } \rho(x_{n+1}, x_n) \leq \delta + \rho(x_n, T(x_n)),$$

we have

$$\rho(x_{n+1}, x_n) < \epsilon \text{ for all integers } n \geq n_0.$$

Proof Let

$$\{x_n\}_{n=0}^{\infty} \subset X, \ \rho(x_0, \theta) \leq M, \tag{12.22}$$

and assume that for each integer $n \geq 0$,

$$x_{n+1} \in T(x_n) \text{ and } \rho(x_{n+1}, x_n) \leq \delta + \rho(x_n, T(x_n)). \tag{12.23}$$

We now estimate $\rho(x_0, T(x_0))$. In view of (12.1) and (12.22),

$$\rho(x_0, T(x_0)) \leq \rho(x_0, \theta) + \rho(\theta, T(\theta)) + H(T(\theta), T(x_0))$$

$$\leq \rho(x_0, \theta) + \rho(\theta, T(\theta)) + \rho(\theta, x_0) \leq 2M + \rho(\theta, T(\theta)). \tag{12.24}$$

In view of (12.23) and (12.24),

$$\rho(x_0, x_1) \le \rho(x_0, T(x_0)) + \delta \le 2M + 1 + \rho(\theta, T(\theta)). \tag{12.25}$$

By (12.1) and (12.23), for each integer $n \ge 0$,

$$\rho(x_{n+2}, x_{n+1}) \le \rho(x_{n+1}, T(x_{n+1})) + \delta$$

$$\le H(T(x_n), T(x_{n+1})) + \delta \le \delta + c\rho(x_n, x_{n+1}). \tag{12.26}$$

By induction we show that for each integer $n \ge 1$,

$$\rho(x_{n+1}, x_n) \le \delta \sum_{i=0}^{n-1} c^i + c^n \rho(x_0, x_1). \tag{12.27}$$

By (12.26), inequality (12.27) holds for $n = 1$. Assume that $k \ge 1$ is an integer and that (12.27) holds with $n = k$. Then by (12.26),

$$\rho(x_{k+2}, x_{k+1}) \le c\rho(x_k, x_{k+1}) + \delta \le \delta \sum_{i=0}^{k} c^i + c^{k+1} \rho(x_0, x_1).$$

Thus, (12.27) holds with $n = k + 1$ and therefore it holds for all integers $n \ge 1$. By (12.25) and (12.27), for all natural numbers n,

$$\rho(x_{n+1}, x_n) \le \delta(1 - c)^{-1} + c^n(2M + 1 + \rho(\theta, T(\theta))). \tag{12.28}$$

Assume that $n \ge n_0$ is an integer. It follows from (12.20), (12.21), and (12.28) that

$$\rho(x_n, x_{n+1}) < \delta(1 - c)^{-1} + c^{n_0}(2M + 1 + \rho(\theta, T(\theta))) < \epsilon.$$

Theorem 12.3 is proved.

Theorems 12.2 and 12.3 imply the following additional result.

Theorem 12.4 *Let $\epsilon \in (0, 1)$ and $M > 0$ be given, a positive number δ satisfy*

$$\delta \le 8^{-1}(1 - c)^2\epsilon,$$

and a natural number n satisfy

$$c^{n_0}(2M + 1 + \rho(\theta, T(\theta)) < 8^{-1}(1 - c)\epsilon.$$

Then for each sequence $\{x_i\}_{i=0}^{\infty} \subset X$ which satisfies

$$\rho(x_0, \theta) \le M, \ x_{n+1} \in T(x_n) \text{ and } \rho(x_n, x_{n+1}) \le \rho(x_n, T(x_n)) + \delta$$

for all integers $n \ge 0$, for each integer $n \ge n_0$, there is a point $y \in X$ such that $y \in T(y)$ and $\rho(y, x_n) < \epsilon$.

The following example shows that Theorem 12.4 cannot be improved in the sense that the fixed point y, the existence of which is guaranteed by the theorem, is not, in general, the same for all integers $n \ge n_0$.

Example 12.5 Let $X = [0, 1]$, $\rho(x, y) = |x - y|$, and $T(x) = [0, 1]$ for all $x \in [0, 1]$. Let $\delta > 0$ be given. Choose a natural number k such that $1/k < \delta$. Put

$$x_0 = 0, \ x_i = i/k, \ i = 1, \dots, k,$$

$$x_{i+k} = 1 - i/k, \ i = 0, \dots, k,$$

and for all integers $p \ge 0$ and any $i \in \{0, \dots, 2k\}$, put

$$x_{2pk+i} = x_i.$$

Then $\{x_i\}_{i=0}^{\infty} \subset X$ and for any integer $i \ge 0$, we have

$$x_{i+1} \in T(x_i) \text{ and } |x_i - x_{i+1}| \le k^{-1} < \delta.$$

On the other hand, for all $x \in X$ and any integer $p \ge 0$,

$$\max\{|x - x_i| : \ i = 2kp, \dots, 2pk + 2k\} \ge 1/2.$$

Note that in the results stated above at any iterative step t, we have $x_{t+1} \in T(x_t)$. In the next sections, we generalize them in the case when x_{t+1} is just sufficiently close to $T(x_t)$. These generalizations are interesting and important since computational errors always occur in practice.

12.2 An Extension of Theorem 12.3

Let (X, ρ) be a complete metric space, $T : X \to 2^X \setminus \{\emptyset\}$, $T(x)$ be closed for each $x \in X$, $c \in [0, 1)$ and that for each $x, y \in X$,

$$H(T(x), T(y)) \le c\rho(x, y). \tag{12.29}$$

The following result was obtained in [233].

Theorem 12.6 *Let $\theta \in X$ and let $\epsilon \in (0, 1]$ and $M > 0$ be given, a positive number δ satisfy*

$$\delta < 9^{-1}\epsilon(1-c), \tag{12.30}$$

and a natural number n_0 satisfy

$$c^{n_0}(2M + 1 + \rho(\theta, T(\theta)) < \epsilon/2. \tag{12.31}$$

Then for each sequence $\{x_i\}_{i=0}^{\infty} \subset X$ such that $\rho(x_0, \theta) \le M$ and such that for each integer $n \ge 0$,

$$B(x_{n+1}, \delta) \cap \{\xi \in T(x_n) : \ \rho(\xi, x_n) \le \delta + \rho(x_n, T(x_n))\} \neq \emptyset \tag{12.32}$$

we have that for each integer $n \ge n_0$,

$$\rho(x_{n+1}, x_n) < \epsilon \text{ and } B^0(x_n, \epsilon) \cap T(x_n) \neq \emptyset.$$

Proof Let

$$\{x_n\}_{n=0}^{\infty} \subset X, \ \rho(x_0, \theta) \le M, \tag{12.33}$$

and assume that (12.32) holds for each integer $n \ge 0$. We now estimate $\rho(x_0, T(x_0))$. In view of (12.29) and (12.33),

$$\rho(x_0, T(x_0)) \le \rho(x_0, \theta) + \rho(\theta, T(\theta)) + H(T(\theta), T(x_0))$$

$$\le \rho(x_0, \theta) + \rho(\theta, T(\theta)) + \rho(\theta, x_0) \le 2M + \rho(\theta, T(\theta)). \tag{12.34}$$

In view of (12.30), (12.32), and (12.34),

$$\rho(x_0, x_1) \le \rho(x_0, T(x_0)) + 2\delta \le 2M + 1 + \rho(\theta, T(\theta)). \tag{12.35}$$

By (12.32), for each integer $n \ge 0$, there exists

$$\xi_n \in B(x_{n+1}, \delta)$$

such that

$$\rho(\xi_n, x_n) \le \delta + \rho(x_n, T(x_n)).$$

By (12.29) and the relations above, for each integer $n \ge 0$,

$$\rho(x_{n+2}, x_{n+1}) \le \rho(x_{n+1}, \xi_{n+1}) + \rho(\xi_{n+1}, x_{n+2})$$

$$\le \rho(x_{n+1}, \xi_{n+1}) + \delta$$

$$\leq \rho(x_{n+1}, T(x_{n+1})) + 2\delta$$

$$\leq \rho(x_{n+1}, \xi_n) + \rho(\xi_n, T(x_{n+1})) + 2\delta$$

$$\leq \rho(\xi_n, T(x_{n+1})) + 3\delta$$

$$\leq 3\delta + H(T(x_n), T(x_{n+1})) \leq 3\delta + c\rho(x_n, x_{n+1}). \tag{12.36}$$

Next, we show by induction that for each integer $n \geq 1$,

$$\rho(x_{n+1}, x_n) \leq 3\delta \sum_{i=0}^{n-1} c^i + c^n \rho(x_0, x_1). \tag{12.37}$$

By (12.36), inequality (12.37) holds for $n = 1$. Assume that $k \geq 1$ is an integer and that (12.37) holds with $n = k$. Then by (12.36),

$$\rho(x_{k+2}, x_{k+1}) \leq 3\delta + c\rho(x_k, x_{k+1}) + \leq 3\delta \sum_{i=0}^{k} c^i + c^{k+1} \rho(x_0, x_1).$$

Thus, (12.40) holds with $n = k + 1$ and therefore it holds for all integers $n \geq 1$. By (12.35) and (12.37), for all natural numbers n,

$$\rho(x_{n+1}, x_n) \leq 3\delta(1 - c)^{-1} + c^n(2M + 1 + \rho(\theta, T(\theta))). \tag{12.38}$$

Assume that $n \geq n_0$ is an integer. It follows from (12.30), (12.31), and (12.38) that

$$\rho(x_n, x_{n+1}) \leq 3\delta(1 - c)^{-1} + c^{n_0}(2M + 1 + \rho(\theta, T(\theta))) < \epsilon/3 + \epsilon/2$$

and in view of (12.30) and (12.32),

$$\rho(x_n, T(x_n)) \leq \rho(x_n, x_{n+1}) + \rho(x_{n+1}, T(x_n))$$

$$< \epsilon/3 + \epsilon/2 + \delta < \epsilon/3 + \epsilon/2 + \epsilon/9 < \epsilon.$$

Theorem 12.6 is proved.

Theorems 12.2 and Theorem 12.6 imply the following result.

Theorem 12.7 *Fix $\theta \in X$. Let $\epsilon \in (0, 1)$ and $M > 0$ be given, a positive number δ satisfy*

$$\delta < 36^{-1}\epsilon(1 - c)^2,$$

and a natural number n_0 satisfy

$$c^{n_0}(2M + 1 + \rho(\theta, T(\theta))) < 8^{-1}(1 - c)\epsilon.$$

Then for each sequence $\{x_i\}_{i=0}^{\infty} \subset X$ *such that* $\rho(x_0, \theta) \le M$ *and such that for each integer* $n \ge 0$,

$$B(x_{n+1}, \delta) \cap \{\xi \in T(x_n) : \ \rho(\xi, x_n) \le \delta + \rho(x_n, T(x_n))\} \ne \emptyset$$

we have that for each integer $n \ge n_0$ *there exist*

$$y \in B(x_n, \epsilon) \cap T(y).$$

12.3 An Extension of Theorem 12.1

Let (X, ρ) be a complete metric space. Assume that $T : X \to 2^X \setminus \{\emptyset\}$, $T(x)$ is closed for each $x \in X$, $c \in [0, 1)$, and that for each $x, y \in X$,

$$H(T(x), T(y)) \le c\rho(x, y). \tag{12.39}$$

The following result that was obtained in [233] describes the behavior of inexact iterates under the presence of summable computational errors.

Theorem 12.8 *Assume that* $x_0 \in X$,

$$\{\epsilon_i\}_{i=0}^{\infty} \subset (0, \infty), \ \sum_{i=0}^{\infty} \epsilon_i < \infty, \ \{r_i\}_{i=0}^{\infty} \subset (0, \infty), \ \sum_{i=0}^{\infty} r_i < \infty$$

and that for each integer $i \ge 0$, $x_i \in X$,

$$B(x_{i+1}, r_i) \cap \{\xi \in T(x_i) : \ \rho(\xi, x_i) \le \rho(x_i, T(x_i)) + \epsilon_i\} \ne \emptyset. \tag{12.40}$$

Then $\{x_i\}_{i=0}^{\infty}$ *converges to a fixed point of* T.

Proof We show that $\{x_i\}_{i=0}^{\infty}$ is a Cauchy sequence. By (12.40), for each integer $i \ge 0$, there exists

$$\xi_i \in B(x_{i+1}, r_i) \cap T(x_i) \tag{12.41}$$

such that

$$\rho(\xi_i, x_i) \le \rho(x_i, T(x_i)) + \epsilon_i. \tag{12.42}$$

Let $i \ge 0$ be an integer. By (12.39), (12.41), and (12.42),

$$\begin{aligned}\rho(x_{i+1}, x_{i+2}) &\le \rho(x_{i+1}, \xi_{i+1}) + \rho(\xi_{i+1}, x_{i+2})\\ &\le \rho(x_{i+1}, \xi_{i+1}) + r_{i+1}\\ &\le \rho(x_{i+1}, T(x_{i+1})) + \epsilon_{i+1} + r_{i+1}\\ &\le \rho(x_{i+1}, \xi_i) + \rho(\xi_i, T(x_{i+1})) + \epsilon_{i+1} + r_{i+1}\\ &\le H(T(x_i), T(x_{i+1})) + r_i + r_{i+1} + \epsilon_{i+1}\\ &\le c\rho(x_i, x_{i+1}) + r_i + r_{i+1} + \epsilon_{i+1}. \end{aligned} \tag{12.43}$$

By (12.43),

$$\rho(x_1, x_2) \le c\rho(x_0, x_1) + r_0 + r_1 + \epsilon_1$$

and

$$\begin{aligned}\rho(x_2, x_3) &\le c\rho(x_1, x_2) + r_1 + r_2 + \epsilon_2\\ &\le c^2\rho(x_0, x_1) + cr_0 + cr_1 + c\epsilon_1 + r_1 + r_2 + \epsilon_2.\end{aligned} \tag{12.44}$$

Now we use induction to show that for each integer $n \ge 1$,

$$\rho(x_n, x_{n+1}) \le c^n\rho(x_0, x_1) + \sum_{i=0}^{n-1} c^i\epsilon_{n-i} + \sum_{i=0}^{n-1} c^i r_{n-i} + \sum_{i=0}^{n-1} c^i r_{n-i-1}. \tag{12.45}$$

In view of (12.44), inequality (12.45) is valid for $n = 1, 2$.

Assume that $k \ge 1$ is an integer and that (12.45) holds for $n = k$. When combined with (12.43), this implies that

$$\begin{aligned}\rho(x_{k+1}, x_{k+2}) &\le c\rho(x_k, x_{k+1}) + r_k + r_{k+1} + \epsilon_{k+1}\\ &\le c^{k+1}\rho(x_0, x_1) + \sum_{i=0}^{k-1} c^{i+1}\epsilon_{k-i} + \sum_{i=0}^{k-1} c^{i+1} r_{k-i} + \sum_{i=0}^{k-1} c^{i+1} r_{k-i-1} + \epsilon_{k+1} + r_k + r_{k+1}\\ &= c^{k+1}\rho(x_0, x_1) + \sum_{i=0}^{k} c^i\epsilon_{k+1-i} + \sum_{i=0}^{k} c^i r_{k+1-i} + \sum_{i=0}^{k} c^i r_{k-i}.\end{aligned}$$

Thus, (12.45) holds with $n = k + 1$ and therefore (12.45) holds for all integers $n \ge 1$. By (12.45),

$$\sum_{n=1}^{\infty} \rho(x_n, x_{n+1}) \le \sum_{n=1}^{\infty} (c^n \rho(x_0, x_1) + \sum_{i=0}^{n} c^{n-i}(r_i + \epsilon_i) + \sum_{i=0}^{n-1} c^{n-i-i} r_i)$$

$$\le \rho(x_0, x_1) \sum_{n=1}^{\infty} c^n + 2 \sum_{i=0}^{\infty} (\sum_{j=0}^{\infty} c^j)(\epsilon_i + r_i) < \infty.$$

Thus, $\{x_n\}_{n=0}^{\infty}$ is indeed a Cauchy sequence and there exists

$$x_* = \lim_{n\to\infty} x_n. \tag{12.46}$$

We claim that $x_* \in T(x_*)$. Let $\epsilon > 0$ be given. By (12.46), there is an integer $n_0 \ge 1$ such that for each integer $n \ge n_0$,

$$r_n < \epsilon/8, \tag{12.47}$$

$$\rho(x_n, x_*) \le \epsilon/8. \tag{12.48}$$

Let $n \ge n_0$ be an integer. By (12.39) and (12.48),

$$H(T(x_n), T(x_*)) \le c\rho(x_n, x_*) \le c\epsilon/8. \tag{12.49}$$

By (12.41),

$$\xi_n \in T(x_n). \tag{12.50}$$

By (12.41), (12.47), (12.49), and (12.50),

$$\rho(x_{n+1}, T(x_*)) \le \rho(x_{n+1}, \xi_n) + \rho(\xi_n, T(x_*))$$

$$\le r_n + H(T(x_n, T(x_*))$$

$$\le r_n + c\epsilon/8 < \epsilon/8 + c\epsilon/8.$$

This implies that there exists

$$y \in T(x_*)$$

such that $\rho(x_{n+1}, y) \le \epsilon/4$. Together with (12.42), this implies that

$$\rho(x_*, T(x_*)) \le \rho(x_*, y) \le \rho(x_*, x_{n+1}) + \rho(x_{n+1}, y) \le \epsilon/8 + \epsilon/4.$$

Since ϵ is an arbitrary positive number, we conclude that

$$x_* \in T(x_*),$$

as claimed. Theorem 12.8 is proved.

12.4 The Mizoguchi-Takahashi Class of Mappings

In [132] N. Mizoguchi and W. Takahashi answered a question posed by S. Reich in [159] and established the existence of a fixed point for a set-valued nonexpansive mapping satisfying certain assumptions. Now we study this mapping and analyze an iterative scheme that allows us, for any initial state, to construct a sequence of iterates which converge to a fixed point of the mapping. Note that in [185] it was given an example showing that the Mizoguchi-Takahashi fixed point theorem does not reduce to the Nadler theorem. The results that are presented here were obtained in [221].

Let (X, ρ) be a complete metric space. Denote by $S(X)$ the collection of all nonempty closed bounded subsets of X.

For each $A, B \in S(X)$, set

$$H(A, B) = \max\{\sup\{\rho(x, B) : x \in A\}, \sup\{\rho(x, A) : x \in B\}\}. \quad (12.51)$$

Clearly, $(S(X), H)$ is a metric space.

Assume that a function $\phi : (0, \infty) \to [0, 1)$ satisfies for each $t \in [0, \infty)$,

$$\limsup_{r \to t^+} \phi(r) < 1. \quad (12.52)$$

Let $A : X \to S(X)$ be such that

$$H(A(x), A(y)) \le \phi(\rho(x, y))\rho(x, y) \quad (12.53)$$

for all $x, y \in X$ with $x \neq y$. In [132] it was shown that the mapping A has a fixed point. Put

$$\phi(0) = 0.$$

Then (12.53) holds for all $x, y \in X$. We prove the following result obtained in [221].

Theorem 12.9 *Let $x_0 \in X$, a sequence $\{\Delta_i\}_{i=0}^{\infty} \subset (0, \infty)$ satisfy*

$$\sum_{i=0}^{\infty} \Delta_i < \infty \quad (12.54)$$

and let a sequence $\{x_i\}_{i=0}^{\infty} \subset X$ satisfy for each integer $i \ge 0$,

$$x_{i+1} \in A(x_i), \tag{12.55}$$

$$\text{if } x_i \in A(x_i), \text{ then } x_{i+1} = x_i, \tag{12.56}$$

$$\rho(x_i, x_{i+1}) \leq \rho(x_i, A(x_i)) + \Delta_i \tag{12.57}$$

and have the following property:
(P1) If $i \geq 1$ and

$$\rho(x_i, A(x_i)) < \rho(x_{i-1}, x_i),$$

then

$$\rho(x_i, x_{i+1}) < \rho(x_{i-1}, x_i).$$

Then the sequence $\{x_i\}_{i=0}^{\infty}$ converges to a fixed point of A.

Note that for every $x_0 \in X$ and every sequence $\{\Delta_i\}_{i=0}^{\infty} \subset (0, 1)$ satisfying (12.54), there exists a sequence $\{x_i\}_{i=0}^{\infty} \subset X$ that satisfies (12.55)–(12.57) and has property (P1) for all integers $i \geq 0$.

Proof of Theorem 12.9 We may assume without loss of generality that for all integers $i \geq 0$,

$$x_i \notin A(x_i). \tag{12.58}$$

This implies that for all integers $i \geq 0$,

$$x_i \neq x_{i+1}. \tag{12.59}$$

In view of (12.57),

$$\rho(x_0, x_1) \leq \rho(x_0, A(x_0)) + \Delta_0. \tag{12.60}$$

By (12.53), (12.55), and (12.57), for each integer $i \geq 1$,

$$\begin{aligned}\rho(x_i, x_{i+1}) &\leq \rho(x_i, A(x_i)) + \Delta_i \\ &\leq H(A(x_{i-1}), A(x_i)) + \Delta_i \\ &\leq \phi(\rho(x_{i-1}, x_i))\rho(x_{i-1}, x_i) + \Delta_i.\end{aligned} \tag{12.61}$$

By (12.53), (12.55), and (12.59), for each integer $i \geq 1$,

$$\rho(x_i, A(x_i)) \leq H(A(x_{i-1}), A(x_i))$$

$$\leq \phi(\rho(x_{i-1}, x_i))\rho(x_{i-1}, x_i) < \rho(x_{i-1}, x_i). \tag{12.62}$$

It follows from (12.62) and property (P1) that for each integer $i \geq 1$,

$$\rho(x_i, x_{i+1}) < \rho(x_{i-1}, x_i). \tag{12.63}$$

In view of (12.63), there exists

$$\tau = \lim_{i\to\infty} \rho(x_i, x_{i+1})$$

$$= \inf\{\rho(x_i, x_{i+1}) : i \geq 0 \text{ is an integer}\}. \tag{12.64}$$

Set

$$\lambda = \limsup_{r\to\tau^+} \phi(r). \tag{12.65}$$

In view of (12.52) and (12.65),

$$0 \leq \lambda < 1. \tag{12.66}$$

By (12.63), (12.64), and (12.65), there exists a natural number p_0 such that for each integer $i \geq p_0$,

$$\phi(\rho(x_i, x_{i+1})) \leq 2^{-1}(1 + \lambda). \tag{12.67}$$

It follows from (12.61) and (12.67) that for each integer $i \geq p_0 + 1$,

$$\rho(x_i, x_{i+1}) \leq \phi(\rho(x_{i-1}, x_i))\rho(x_{i-1}, x_i) + \Delta_i$$

$$\leq 2^{-1}(1 + \lambda)\rho(x_{i-1}, x_i) + \Delta_i. \tag{12.68}$$

Relation (12.68) implies that

$$\rho(x_{p_0+1}, x_{p_0+2}) \leq 2^{-1}(1 + \lambda)\rho(x_{p_0}, x_{p_0+1}) + \Delta_{p_0+1}. \tag{12.69}$$

In view of (12.68) and (12.69),

$$\rho(x_{p_0+2}, x_{p_0+3}) \leq 2^{-1}(1 + \lambda)\rho(x_{p_0+1}, x_{p_0+2}) + \Delta_{p_0+2}$$

$$\leq (2^{-1}(1 + \lambda))^2\rho(x_{p_0}, x_{p_0+1}) + 2^{-1}(1 + \lambda)\Delta_{p_0+1} + \Delta_{p_0+2}. \tag{12.70}$$

We show by induction that for each integer $n \geq 1$,

$$\rho(x_{p_0+n}, x_{p_0+n+1})$$

$$\leq (2^{-1}(1+\lambda))^n \rho(x_{p_0}, x_{p_0+1}) + \sum_{i=0}^{n-1} (2^{-1}(1+\lambda))^i \Delta_{p_0+n-i}. \tag{12.71}$$

In view of (12.69) and (12.70), inequality (12.71) holds for $n = 1, 2$.

Assume that $k \geq 1$ is an integer and that (12.71) holds for $n = k$. Together with (12.68), this implies that

$$\rho(x_{p_0+k+1}, x_{p_0+k+2}) \leq 2^{-1}(1+\lambda)\rho(x_{p_0+k}, x_{p_0+k+1}) + \Delta_{p_0+k+1}$$

$$\leq (2^{-1}(1+\lambda))^{k+1} \rho(x_{p_0}, x_{p_0+1}) + \sum_{i=0}^{k-1} (2^{-1}(1+\lambda))^{i+1} \Delta_{p_0+k-i} + \Delta_{p_0+k+1}$$

$$= (2^{-1}(1+\lambda))^{k+1} \rho(x_{p_0}, x_{p_0+1}) + \sum_{i=0}^{k} (2^{-1}(1+\lambda))^i \Delta_{p_0+k+1-i}.$$

Thus, (12.71) holds for $n = k+1$. Therefore, we have shown that (12.71) holds for all integers $n \geq 1$.

By (12.66) and (12.71),

$$\sum_{n=1}^{\infty} \rho(x_{p_0+n}, x_{p_0+n+1})$$

$$\leq \sum_{n=1}^{\infty} ((2^{-1}(1+\lambda))^n \rho(x_{p_0}, x_{p_0+1}) + \sum_{i=1}^{n} (2^{-1}(1+\lambda))^{n-i} \Delta_{p_0+i}$$

$$\leq \rho(x_{p_0}, x_{p_0+1}) \sum_{n=1}^{\infty} (2^{-1}(1+\lambda))^n + \sum_{i=1}^{\infty} (\sum_{j=0}^{\infty} (2^{-1}(1+\lambda))^j) \Delta_{p_0+i}$$

$$\leq (\sum_{n=0}^{\infty} (2^{-1}(1+\lambda))^n)[\rho(x_{p_0}, x_{p_0+1}) + \sum_{i=1}^{\infty} \Delta_{p_0+i}] < \infty.$$

Thus $\{x_n\}_{n=0}^{\infty}$ is a Cauchy sequence and there exists

$$x_* = \lim_{n\to\infty} x_n. \tag{12.72}$$

We show that $x_* \in A(x_*)$. Let $\epsilon > 0$. In view of (12.72), there exists an integer $n_0 \geq 1$ such that for each integer $n \geq n_0$,

$$\rho(x_n, x_*) \leq \epsilon/4. \tag{12.73}$$

Let $n \geq n_0$ be an integer. By (12.53) and (12.73),

$$H(A(x_n), A(x_*)) \leq \rho(x_n, x_*) \leq \epsilon/4. \tag{12.74}$$

Relation (12.55) implies that

$$x_{n+1} \in A(x_n). \tag{12.75}$$

It follows from (12.74) and (12.75) that

$$\rho(x_{n+1}, A(x_*)) \leq \epsilon/4.$$

Thus, there exists

$$y \in A(x_*) \tag{12.76}$$

such that

$$\rho(x_{n+1}, y) < \epsilon/4 + \epsilon/8. \tag{12.77}$$

By (12.73), (12.76), and (12.77),

$$\rho(x_*, A(x_*)) \leq \rho(x_*, y) \leq \rho(x_*, x_{n+1}) + \rho(x_{n+1}, y) \leq \epsilon/4 + \epsilon/4 + \epsilon/8.$$

Since ϵ is an arbitrary positive number, we conclude that

$$x_* \in A(x_*).$$

Theorem 12.9 is proved.

The next theorem shows that our fixed point problem is well-posed. It was also proved in [221].

Theorem 12.10 *Let $\epsilon > 0$. Then there exists $\delta > 0$ such that for each $x \in X$ satisfying $\rho(x, A(x)) < \delta$, there exists $\bar{x} \in X$ such that $\bar{x} \in A(\bar{x})$ and $\rho(x, \bar{x}) < \epsilon$.*

Proof By (12.52), there exist $\delta_0 > 0$ and $\lambda_0 \in (0, 1)$ such that

$$\phi(t) < \lambda_0 \text{ for all } t \in (0, \delta_0]. \tag{12.78}$$

Fix

$$\lambda_1 \in (\lambda_0, 1). \tag{12.79}$$

Choose a positive number δ such that

$$\delta < \delta_0 \text{ and } \delta(1 - \lambda_1)^{-1} < \epsilon. \tag{12.80}$$

Let $x \in X$ satisfy

$$\rho(x, A(x)) < \delta. \tag{12.81}$$

Set

$$x_0 = x. \tag{12.82}$$

By (12.81) and (12.82), there exists $x_1 \in X$ such that

$$x_1 \in A(x_0), \ \ \rho(x_0, x_1) < \delta, \tag{12.83}$$

$$\text{if } x_0 \in A(x_0), \ \text{ then } x_1 = x_0. \tag{12.84}$$

By induction we define a sequence $\{x_n\}_{n=0}^{\infty} \subset X$ such that for each integer $n \geq 1$, the following properties hold:

(P2) If $x_n \in A(x_n)$, then $x_{n+1} = x_n$.

(P3) If $x_n \notin A(x_n)$, then $x_{n+1} \in A(x_n)$ and

$$\rho(x_n, x_{n+1}) < \rho(x_n, A(x_n))\lambda_1\lambda_0^{-1}. \tag{12.85}$$

(P4) If

$$\rho(x_n, A(x_n)) < \rho(x_{n-1}, x_n),$$

then

$$\rho(x_n, x_{n+1}) < \rho(x_{n-1}, x_n).$$

If $x_1 \in A(x_1)$, then in view of (12.80), (12.82), and (12.83), the assertion of the theorem holds. Thus, we may assume without loss of generality that

$$x_1 \notin A(x_1). \tag{12.86}$$

By (P3),

$$x_2 \neq x_1. \tag{12.87}$$

It follows from (12.83), (12.87), and (P2) that

$$x_1 \neq x_0. \tag{12.88}$$

Assume that an integer $k \geq 2$ and

$$x_k \neq x_{k-1}. \tag{12.89}$$

(Note that in view of (12.87), (12.89) is true for $k = 2$.) By (12.84), (12.86)–(12.89), and properties (P2) and (P3),

$$x_i \neq x_{i-1},\ x_{i-1} \notin A(x_{i-1}) \tag{12.90}$$

for all integers i satisfying $1 \le i \le k$.

Let an integer i satisfy $1 \le i \le k$. By (12.53), (12.90), (P2), and (P3),

$$\rho(x_i, A(x_i)) \le H(A(x_{i-1}), A(x_i)) \le \phi(\rho(x_{i-1}, x_i))\rho(x_{i-1}, x_i). \tag{12.91}$$

In view of (12.90) and (12.91),

$$\rho(x_i, A(x_i)) < \rho(x_{i-1}, x_i). \tag{12.92}$$

It follows from (12.92) and (P4) that

$$\rho(x_i, x_{i+1}) < \rho(x_{i-1}, x_i).$$

Since the inequality above holds for all integers i satisfying $1 \le i \le k$, we conclude using (12.83) that

$$\rho(x_k, x_{k+1}) \le \rho(x_{k-1}, x_k) \le \cdots \le \rho(x_1, x_2) \le \rho(x_0, x_1) < \delta.$$

By the relation above, (12.78), (12.80), (12.91), and property (P3), for each integer i satisfying $1 \le i < k$,

$$\rho(x_i, x_{i+1}) < \rho(x_i, A(x_i))\lambda_1\lambda_0^{-1}$$

$$\le \lambda_1\lambda_0^{-1}\phi(\rho(x_{i-1}, x_i))\rho(x_{i-1}, x_i)$$

$$\le \lambda_1\lambda_0^{-1}\lambda_0\rho(x_{i-1}, x_i) = \lambda_1\rho(x_{i-1}, x_i).$$

Together with (12.83), this implies that for each integer i satisfying $1 \le i < k$,

$$\rho(x_i, x_{i+1}) \le \lambda_1^i\rho(x_0, x_1) \le \lambda_1^i\delta. \tag{12.93}$$

Thus, we have shown that the following property holds:

(P5) If an integer $k \ge 2$ satisfies $x_k \neq x_{k-1}$, then for each integer i satisfying $1 \le i < k$, the inequality

$$\rho(x_i, x_{i+1}) \le \lambda_1^i\delta$$

holds.

There are two cases:

(a) There is an integer $k \geq 2$ such that

$$x_k \neq x_{k-1}, \ x_{k+1} = x_k.$$

(b) $x_k \neq x_{k-1}$ for all integers $k \geq 2$.

Assume that the case (a) holds and let an integer $k \geq 2$ be as guaranteed in the case (a). Then by property (P5),

$$\rho(x_i, x_{i+1}) \leq \lambda_1^i \delta, \ i = 1, \ldots, k-1. \tag{12.94}$$

In view of (P2) and (P3),

$$x_i = x_k \text{ for all integers } i \geq k, \ x_k \in A(x_k). \tag{12.95}$$

It follows from (12.80) and (12.94) that

$$\rho(x_0, x_k) \leq \sum_{i=0}^{k-1} \rho(x_i, x_{i+1}) \leq \sum_{i=0}^{\infty} \lambda_1^i \delta = \delta(1-\lambda_1)^{-1} < \epsilon.$$

Thus,

$$x_k \in A(x_k), \ \rho(x, x_k) < \epsilon$$

and in the case (a) Theorem 12.9 is proved.

Assume that the case (b) holds. By property (P5), for all integers $i \geq 1$,

$$\rho(x_i, x_{i+1}) \leq \lambda_1^i \delta. \tag{12.96}$$

In view of (12.96),

$$\sum_{i=0}^{\infty} \rho(x_i, x_{i+1}) < \infty.$$

Thus, $\{x_i\}_{i=0}^{\infty}$ is a Cauchy sequence. Let

$$\bar{x} = \lim_{i \to \infty} x_i. \tag{12.97}$$

Arguing as in the proof of Theorem 12.9, we can show that

$$\bar{x} \in A(\bar{x}).$$

By (12.80), (12.82), (12.96), and (12.97),

$$\rho(x, \bar{x}) = \rho(x_0, \bar{x}) = \lim_{n\to\infty} \rho(x_0, x_n)$$

$$\leq \sum_{i=0}^{\infty} \rho(x_i, x_{i+1}) \leq \sum_{i=0}^{\infty} \lambda_1^i \delta = \delta(1-\lambda_1)^{-1} < \epsilon.$$

This completes the proof of Theorem 12.10.

12.5 An Extension of Theorem 12.9

Let (X, ρ) be a complete metric space. We use the notation of Sect. 12.4.

Assume that a function $\phi : (0, \infty) \to [0, 1)$ satisfies for each $t \in (0, \infty)$,

$$\limsup_{r\to t^+} \phi(r) < 1. \tag{12.98}$$

Let $A : X \to S(X)$ be such that

$$H(A(x), A(y)) \leq \phi(\rho(x, y))\rho(x, y) \tag{12.99}$$

for all $x, y \in X$ with $x \neq y$. Put

$$\phi(0) = 0.$$

Then (12.99) holds for all $x, y \in X$. We prove the following extension of Theorem 12.9 which was also obtained in [221]. In Theorem 12.9, we assumed that the sequence of errors is summable. In the next result, we only assume that it converges to zero.

Theorem 12.11 *Let $x_0 \in X$, a sequence $\{\Delta_i\}_{i=0}^{\infty} \subset (0, \infty)$ satisfy*

$$\lim_{i\to\infty} \Delta_i = 0 \tag{12.100}$$

and let a sequence $\{x_i\}_{i=0}^{\infty} \subset X$ satisfy for each integer $i \geq 0$,

$$x_{i+1} \in A(x_i), \tag{12.101}$$

$$\text{if } x_i \in A(x_i), \text{ then } x_{i+1} = x_i, \tag{12.102}$$

$$\rho(x_i, x_{i+1}) \leq \rho(x_i, A(x_i)) + \Delta_i \tag{12.103}$$

and have the following property:
(P6) If $i \geq 1$ *and*

$$\rho(x_i, A(x_i)) < \rho(x_{i-1}, x_i),$$

then

$$\rho(x_i, x_{i+1}) < \rho(x_{i-1}, x_i).$$

Then $\lim_{i\to\infty} \rho(x_i, x_{i+1}) = 0$.

Proof We may assume without loss of generality that for all integers $i \geq 0$,

$$x_i \notin A(x_i). \tag{12.104}$$

This implies that for all integers $i \geq 0$,

$$x_i \neq x_{i-1}. \tag{12.105}$$

In view of (12.103),

$$\rho(x_0, x_1) \leq \rho(x_0, A(x_0)) + \Delta_0. \tag{12.106}$$

By (12.99) and (12.101), for each integer $i \geq 1$,

$$\rho(x_i, A(x_i)) \leq H(A(x_{i-1}), A(x_i)) \leq \phi(\rho(x_{i-1}, x_i))\rho(x_{i-1}, x_i). \tag{12.107}$$

It follows from (12.105) and (12.107) that for each integer $i \geq 1$,

$$\rho(x_i, A(x_i)) < \rho(x_{i-1}, x_i). \tag{12.108}$$

Property (P6) and (12.108) imply that for each integer $i \geq 1$,

$$\rho(x_i, x_{i+1}) < \rho(x_{i-1}, x_i). \tag{12.109}$$

In view of (12.109), there exists

$$\lambda_0 = \lim_{i\to\infty} \rho(x_i, x_{i+1})$$
$$= \inf\{\rho(x_i, x_{i+1}) : i \geq 0 \text{ is an integer}\}. \tag{12.110}$$

In order to complete the proof of the theorem, it is sufficient to show that $\lambda_0 = 0$. Assume the contrary. Then

$$\lambda_0 > 0. \tag{12.111}$$

By (12.103) and (12.107), for each integer $i \geq 1$,

$$\rho(x_i, x_{i+1}) \leq \rho(x_i, A(x_i)) + \Delta_i$$

$$\leq \phi(\rho(x_{i-1}, x_i))\rho(x_{i-1}, x_i) + \Delta_i. \tag{12.112}$$

In view of (12.98) and (12.111), there exist

$$\lambda_1 > \lambda_0, \ \gamma_0 \in (0, 1) \tag{12.113}$$

such that

$$\phi(t) < \gamma_0 \text{ for all } t \in (\lambda_0, \lambda_1]. \tag{12.114}$$

It follows from (12.109) and (12.110) that there exists a natural number p_0 such that for every integer $i \geq p_0$,

$$\rho(x_i, x_{i+1}) \in (\lambda_0, \lambda_1].$$

Together with (12.114), this implies that for every integer $i \geq p_0$,

$$\phi(\rho(x_i, x_{i+1})) < \gamma_0. \tag{12.115}$$

By (12.110), (12.112), (12.113), and (12.115) that for every integer $i \geq p_0 + 1$,

$$\rho(x_i, x_{i+1}) \leq \phi(\rho(x_{i-1}, x_i))\rho(x_{i-1}, x_i) + \Delta_i$$

$$\leq \gamma_0 \rho(x_{i-1}, x_i) + \Delta_i$$

$$= \rho(x_{i-1}, x_i) - (1 - \gamma_0)\rho(x_{i-1}, x_i) + \Delta_i$$

$$\leq \rho(x_{i-1}, x_i) - (1 - \gamma_0)\lambda_0 + \Delta_i. \tag{12.116}$$

In view of (12.100), there exists an integer $p_1 > p_0 + 2$ such that for each integer $i \geq p_1$,

$$\Delta_i < \lambda_0(1 - \gamma_0)/4. \tag{12.117}$$

It follows from (12.116) and (12.117) that for each integer $i \geq p_1$,

$$\rho(x_i, x_{i+1}) \leq \rho(x_{i-1}, x_i) - \lambda_0(1 - \gamma_0)/2.$$

This implies that

$$\lim_{i\to\infty} \rho(x_i, x_{i+1}) = -\infty.$$

The contradiction we have reached proves Theorem 12.4.

12.6 A Space of Single-Valued Mappings

We consider a space of nonexpansive mappings, acting on a closed subset of a complete metric space, which is equipped with the topology of uniform convergence on bounded sets. We show the existence of an open and everywhere dense subset in this space such that every element possesses an approximate fixed point, which is stable under small perturbations. These results were obtained in [219].

Let (X, ρ) be a complete metric space. Fix

$$\theta \in X.$$

We assume that there is an operation

$$(1-\gamma)x \oplus \gamma\theta, \;\; x \in X \; \gamma \in [0, 1]$$

such that for each $x, y \in X$ and each $\gamma \in (0, 1]$,

$$\rho((1-\gamma)x \oplus \gamma\theta, (1-\gamma)y \oplus \gamma\theta) \le (1-\gamma)\rho(x, y), \tag{12.118}$$

$$\rho((1-\gamma)x \oplus \gamma\theta, x) \le \gamma\rho(x, \theta). \tag{12.119}$$

Denote by $\mathcal{A}$ the set of all operators $A : X \to X$ such that

$$\rho(A(x), A(y)) \le \rho(x, y) \text{ for all } x, y \in X. \tag{12.120}$$

We equip the set $\mathcal{A}$ with the uniformity determined by the base:

$$\mathcal{U}(n) = \{(A, B) \in \mathcal{A} \times \mathcal{A} : \; \rho(A(x), B(x)) \le n^{-1} \text{ for all } x \in B(\theta, n)\}, \tag{12.121}$$

where n is a natural number. Clearly the uniform space $\mathcal{A}$ is metrizable and complete.

We prove the following result obtained in [219].

Theorem 12.12 *Let $\bar{r} > 0$, $\bar{M} > 0$, and $\epsilon \in (0, 1)$. Then there exists an open everywhere dense subset $\mathcal{F} \subset \mathcal{A}$ such that for each $B \in \mathcal{F}$ there exist $x_B \in X$, a natural number n_B, $\delta_B \in (0, \bar{r})$, and an open neighborhood $\mathcal{U}$ of B in $\mathcal{A}$ such that the following assertion holds.*

Let $C \in \mathcal{U}$, n_1 be a natural number and a sequence $\{x_i\}_{i=0}^{\infty} \subset X$ be such that

$$\rho(x_0, \theta) \le \bar{M},$$

$$\rho(x_{i+1}, C(x_i)) \le \bar{r} \text{ for all integers } i \ge 0$$

and

$$\rho(x_{i+1}, C(x_i)) \le \delta_B \text{ for all integers } i \ge n_1.$$

Then,

$$\rho(x_i, x_B) \le \epsilon \text{ for all integers } i \ge n_1 + n_B.$$

Corollary 12.13 *Let $\bar{r} > 0$, $\bar{M} > 0$, and $\epsilon \in (0, 1)$ and let an open everywhere dense subset $\mathcal{F} \subset \mathcal{A}$ be as guaranteed by Theorem 12.12. Assume that $B \in \mathcal{F}$, a point $x_B \in X$, and an open neighborhood $\mathcal{U}$ of B in $\mathcal{A}$ are as guaranteed by Theorem 12.12.*

Let $C \in \mathcal{U}$ and a sequence $\{x_i\}_{i=0}^{\infty}$ be such that

$$\rho(x_0, \theta) \le \bar{M},$$

$$\rho(x_{i+1}, C(x_i)) \le \bar{r} \text{ for all integers } i \ge 0$$

and

$$\lim_{i \to \infty} \rho(x_{i+1}, C(x_i)) = 0.$$

Then,

$$\rho(x_i, x_B) \le \epsilon \text{ for all sufficiently large natural numbers } i.$$

12.7 Proof of Theorem 12.12

Let $A \in \mathcal{A}$ and $\gamma \in (0, 1)$. Define

$$A_\gamma(x) = (1 - \gamma)A(x) \oplus \gamma\theta, \ \ x \in X. \tag{12.122}$$

Clearly, A_γ is a self-mapping of X. By (12.118), (12.120), and (12.122), for each $x, y \in X$,

$$\rho(A_\gamma(x), A_\gamma(y)) = \rho((1 - \gamma)A(x) \oplus \gamma\theta, (1 - \gamma)A(y) \oplus \gamma\theta)$$

$$\le (1 - \gamma)\rho(A(x), A(y)) \le (1 - \gamma)\rho(x, y). \tag{12.123}$$

In view of (12.123) and the Banach fixed point theorem, there exists a unique point $x(A, \gamma) \in X$ such that

$$A_\gamma(x(A, \gamma)) = x(A, \gamma). \tag{12.124}$$

Choose a positive number $M(A, \gamma)$ such that

$$M(A, \gamma) > 1 + \bar{M} + \rho(\theta, x(A, \gamma)) + (2\bar{r} + 1)\gamma^{-1}. \tag{12.125}$$

Clearly, if $x \in B(\theta, \bar{M})$, then it follows from (12.125) that

$$\rho(x, x(A, \gamma)) \le \rho(x, \theta) + \rho(\theta, x(A, \gamma))$$

$$\le \bar{M} + \rho(\theta, x(A, \gamma)) < M(A, \gamma).$$

This implies that

$$B(\theta, \bar{M}) \subset B(x(A, \gamma), M(A, \gamma)). \tag{12.126}$$

Assume that

$$x \in B(x(A, \gamma), M(A, \gamma)), \tag{12.127}$$

$$y \in X, \ \ \rho(y, A_\gamma(x)) \le 2\bar{r}. \tag{12.128}$$

By (12.123), (12.124), (12.125), (12.127), and (12.128),

$$\rho(y, x(A, \gamma)) \le \rho(y, A_\gamma(x)) + \rho(A_\gamma(x), x(A, \gamma))$$

$$\le 2\bar{r} + \rho(A_\gamma(x), A_\gamma(x(A, \gamma)))$$

$$\le 2\bar{r} + (1 - \gamma)\rho(x, x(A, \gamma))$$

$$\le 2\bar{r} + (1 - \gamma)M(A, \gamma)$$

$$= M(A, \gamma) + 2\bar{r} - \gamma M(A, \gamma) < M(A, \gamma).$$

Thus, we have shown that

$$\{y \in X : \ \text{there is } x \in B(x(A, \gamma), M(A, \gamma)) \text{ such that } \rho(y, A_\gamma(x)) \le 2\bar{r}\}$$

$$\subset B(x(A, \gamma), M(A, \gamma)). \tag{12.129}$$

Choose a number $\delta(A, \gamma) \in (0, \bar{r})$ such that

$$\delta(A, \gamma) < 4^{-1}\gamma\epsilon. \tag{12.130}$$

Denote by $\mathcal{U}(A, \gamma)$ an open neighborhood of A_γ in $\mathcal{A}$ such that

$$\mathcal{U}(A, \gamma) \subset \{C \in \mathcal{A} : \rho(C(z), A_\gamma(z)) \le \delta(A, \gamma)$$

$$\text{for all } z \in B(x(A, \gamma), M(A, \gamma))\}. \tag{12.131}$$

By (12.119), (12.120), and (12.122), for each $z \in X$,

$$\rho(A(z), A_\gamma(z)) = \rho(A(z), (1 - \gamma)A(z) \oplus \gamma\theta)$$

$$\le \gamma\rho(\theta, A(z)) \le \gamma(\rho(\theta, A(\theta)) + \rho(A(\theta), A(z)))$$

$$\le \gamma(\rho(\theta, A(\theta)) + \rho(\theta, z)).$$

This implies that for each neighborhood V of A in $\mathcal{A}$, there exists $\gamma_V \in (0, 1)$ such that for each $\gamma \in (0, \gamma_V)$, $A_\gamma \in V$. Therefore,

$$\{A_\gamma : A \in \mathcal{A}, \gamma \in (0, 1)\}$$

is an everywhere dense subset of $\mathcal{A}$.

Set

$$\mathcal{F} = \cup\{\mathcal{U}(A, \gamma) : A \in \mathcal{A}, \gamma \in (0, 1)\}. \tag{12.132}$$

Clearly, $\mathcal{F}$ is an open everywhere dense subset of $\mathcal{A}$.

Let

$$B \in \mathcal{F}. \tag{12.133}$$

By (12.132) and (12.133), there exist $A \in \mathcal{A}$ and $\gamma \in (0, 1)$ such that

$$B \in \mathcal{U}(A, \gamma). \tag{12.134}$$

Set

$$x_B = x(A, \gamma), \tag{12.135}$$

$$\delta_B = \delta(A, \gamma) \tag{12.136}$$

and choose a natural number n_B such that

$$n_B > 2M(A, \gamma)(\gamma\epsilon)^{-1}. \tag{12.137}$$

Assume that

$$C \in \mathcal{U}(A, \gamma), \tag{12.138}$$

n_1 is a natural number and that a sequence $\{x_i\}_{i=0}^{\infty} \subset X$ satisfies

$$\rho(x_0, \theta) \le \bar{M}, \tag{12.139}$$

$$\rho(x_{i+1}, C(x_i)) \le \bar{r} \text{ for all integers } i \ge 0 \tag{12.140}$$

and

$$\rho(x_{i+1}, C(x_i)) \le \delta_B \text{ for all integers } i \ge n_1. \tag{12.141}$$

In order to complete the proof, it is sufficient to show that

$$\rho(x_i, x_B) \le \epsilon \text{ for all integers } i \ge n_1 + n_B.$$

First, we show that

$$\rho(x_i, x(A, \gamma)) \le M(A, \gamma) \tag{12.142}$$

for all integers $i \ge 0$. In view of (12.126) and (12.139), inequality (12.142) holds for $i = 0$.

Assume that an integer $j \ge 0$ and (12.142) holds for $i = j$. Thus,

$$\rho(x_j, x(A, \gamma)) \le M(A, \gamma). \tag{12.143}$$

By (12.131), (12.138), and (12.143),

$$\rho(C(x_j), A_\gamma(x_j)) \le \delta(A, \gamma) < \bar{r}. \tag{12.144}$$

It follows from (12.123), (12.124), (12.125), (12.140), (12.143), and (12.144) that

$$\begin{aligned}
&\rho(x_{j+1}, x(A, \gamma)) \\
&\le \rho(x_{j+1}, C(x_j)) + \rho(C(x_j), A_\gamma(x_j)) + \rho(A_\gamma(x_j), A_\gamma x(A, \gamma)) \\
&\le \bar{r} + \bar{r} + (1 - \gamma)\rho(x_j, x(A, \gamma)) \\
&\le 2\bar{r} + (1 - \gamma)M(A, \gamma) \le M(A, \gamma).
\end{aligned}$$

Thus,

$$\rho(x_{j+1}, x(A, \gamma)) \le M(A, \gamma)$$

and (12.142) holds for $i = j + 1$. Therefore, we conclude that (12.142) holds for all integers $i \geq 0$.

We show that there exists an integer $i \in [n_1, n_1 + n_B]$ such that

$$\rho(x_i, x(A, \gamma)) \leq \epsilon.$$

Assume the contrary. Then,

$$\rho(x_i, x(A, \gamma)) > \epsilon \tag{12.145}$$

for all $i = n_1, \dots, n_1 + n_B$. Assume that an integer i satisfies

$$n_1 \leq i \leq n_1 + n_B - 1.$$

Then (12.145) holds. By (12.123), (12.124), and (12.141),

$$\begin{aligned}
&\rho(x_{i+1}, x(A, \gamma)) \\
&\leq \rho(x_{i+1}, C(x_i)) + \rho(C(x_i), A_\gamma(x_i)) + \rho(A_\gamma(x_i), x(A, \gamma)) \\
&\leq \delta_B + \rho(C(x_i), A_\gamma(x_i)) + \rho(A_\gamma(x_i), A_\gamma(x(A, \gamma))) \\
&\leq \delta_B + \rho(C(x_i), A_\gamma(x_i)) + (1 - \gamma)\rho(x_i, x(A, \gamma)).
\end{aligned} \tag{12.146}$$

It follows from (12.131), (12.136), (12.138), and (12.142) that

$$\rho(C(x_i), A_\gamma(x_i)) \leq \delta(A, \gamma) = \delta_B. \tag{12.147}$$

In view of (12.146) and (12.147),

$$\rho(x_{i+1}, x(A, \gamma)) \leq 2\delta_B + (1 - \gamma)\rho(x_i, x(A, \gamma)). \tag{12.148}$$

By (12.130), (12.136), (12.145), and (12.148),

$$\begin{aligned}
&\rho(x_i, x(A, \gamma)) - \rho(x_{i+1}, x(A, \gamma)) \\
&\geq \gamma\rho(x_i, x(A, \gamma)) - 2\delta_B > \gamma\epsilon - 2\delta(A, \gamma) > \gamma\epsilon/2.
\end{aligned}$$

Thus,

$$\rho(x_i, x(A, \gamma)) - \rho(x_{i+1}, x(A, \gamma)) > \gamma\epsilon/2$$

for every integer i satisfying $n_1 \leq i \leq n_1 + n_B - 1$. Together with (12.142), this implies that

$$M(A, \gamma) \geq \rho(x_{n_1}, x(A, \gamma)) - \rho(x_{n_1+n_B}, x(A, \gamma))$$

$$= \sum_{i=n_1}^{n_1+n_b-1} (\rho(x_i, x(A, \gamma)) - \rho(x_{i+1}, x(A, \gamma))) > n_B \gamma \epsilon / 2$$

and

$$n_B < 2(\gamma\epsilon)^{-1} M(A, \gamma).$$

This contradicts (12.137). The contradiction we have reached proves that there exists an integer

$$j \in [n_1, n_1 + n_B] \tag{12.149}$$

such that

$$\rho(x_j, x(A, \gamma)) \leq \epsilon. \tag{12.150}$$

Assume that an integer i satisfies

$$i \geq n_1, \ \rho(x_i, x(A, \gamma)) \leq \epsilon. \tag{12.151}$$

We show that

$$\rho(x_{i+1}, x(A, \gamma)) \leq \epsilon.$$

By (12.123), (12.124), (12.141), and (12.151),

$$\rho(x_{i+1}, x(A, \gamma))$$

$$\leq \rho(x_{i+1}, C(x_i)) + \rho(C(x_i), A_\gamma(x_i)) + \rho(A_\gamma(x_i), A_\gamma(x(A, \gamma)))$$

$$\leq \delta_B + \rho(C(x_i), A_\gamma(x_i)) + (1 - \gamma)\rho(x_i, x(A, \gamma)). \tag{12.152}$$

In view of (12.131), (12.138), and (12.142),

$$\rho(C(x_i), A_\gamma(x_i)) \leq \delta(A, \gamma). \tag{12.153}$$

There are two cases:

$$\rho(x_i, x(A, \gamma)) \leq \epsilon/2; \tag{12.154}$$

$$\rho(x_i, x(A, \gamma)) > \epsilon/2. \tag{12.155}$$

Assume that (12.154) holds. By (12.130), (12.136), (12.152), (12.153), and (12.154),

$$\rho(x_{i+1}, x(A, \gamma)) \le 2\delta(A, \gamma) + \rho(x_i, x(A, \gamma)) \le 2\delta(A, \gamma) + \epsilon/2 < \epsilon.$$

Assume that (12.155) holds. It follows from (12.130), (12.136), (12.151), (12.152), and (12.153) that

$$\rho(x_{i+1}, x(A, \gamma)) \le 2\delta(A, \gamma) + (1 - \gamma)\epsilon < \epsilon.$$

Thus, in the both cases,

$$\rho(x_{i+1}, x(A, \gamma)) \le \epsilon. \tag{12.156}$$

In view of (12.149) and (12.150), we have shown by induction that for all integers $i \ge j \in [n_1, n_1 + n_B]$, inequality (12.156) holds. This completes the proof of Theorem 12.12.

12.8 A Space of Set-Valued Mappings

We consider a space of nonexpansive set-valued mappings, acting on a closed subset of a complete metric space, which is equipped with the topology of uniform convergence on bounded sets. We show the existence of an open and everywhere dense subset in this space such that every element possesses an approximate fixed point, which is stable under small perturbations. These results were obtained in [228].

Let (X, ρ) be a complete metric space. Fix

$$\theta \in X.$$

We assume that there is an operation

$$(1 - \gamma)x \oplus \gamma\theta, \ x \in X \ \gamma \in [0, 1)$$

such that for each $x, y \in X$ and each $\gamma \in (0, 1]$,

$$\rho((1 - \gamma)x \oplus \gamma\theta, (1 - \gamma)y \oplus \gamma\theta) \le (1 - \gamma)\rho(x, y), \tag{12.157}$$

$$\rho((1 - \gamma)x \oplus \gamma\theta, x) \le \gamma\rho(x, \theta). \tag{12.158}$$

Denote by $S(X)$ the collection of all nonempty closed bounded subsets of X. For each $A, B \in S(X)$, set

$$H(A, B) = \max\{\sup\{\rho(x, B) : x \in A\}, \sup\{\rho(x, A) : x \in B\}\}.$$

Clearly, $(S(X), H)$ is a metric space. It is known that the metric space $(S(X), H)$ is complete.

Denote by $\mathcal{M}$ the set of all mappings $A : X \to S(X)$ such that

$$H(A(x), A(y)) \le \rho(x, y) \text{ for all } x, y \in X. \tag{12.159}$$

Fix $\theta \in K$. We equip the set $\mathcal{M}$ with the uniformity determined by the following base:

$$\mathcal{U}(n) = \{(A, B) \in \mathcal{M} \times \mathcal{M} :$$

$$H(A(x), B(x)) \le n^{-1} \text{ for all } x \in B(\theta, n)\}, \tag{12.160}$$

where n is a natural number. Clearly, the uniform space $\mathcal{M}$ is metrizable and complete.

For each $D \subset X$, denote by $\mathrm{cl}(D)$ the closure of D.

For each nonempty set $D \subset X$ and each $A \in \mathcal{M}$, set

$$A(D) = \cup\{A(x) : x \in D\}.$$

We prove the following result obtained in [228].

Theorem 12.14 *Let $\bar{r} > 0$, $\bar{M} > 0$ and $\epsilon \in (0, 1)$. Then there exists an open everywhere dense subset $\mathcal{F} \subset \mathcal{M}$ such that for each $B \in \mathcal{F}$, there exist a natural number n_B, $\delta_B \in (0, \bar{r})$, and an open neighborhood $\mathcal{U}$ of B in $\mathcal{M}$ such that the following assertion holds.*

Let $C_i \in \mathcal{U}$ for all integers $i \ge 0$, n_1 be a natural number, and a sequence $\{x_i\}_{i=0}^{\infty} \subset X$ be such that

$$\rho(x_0, \theta) \le \bar{M},$$

$$x_{i+1} \in C_i(x_i),\ \rho(x_i, x_{i+1}) \le \rho(x_i, C_i(x_i)) + \bar{r}$$

for all integers $i \ge 0$ and

$$\rho(x_i, x_{i+1}) \le \rho(x_i, C_i(x_i)) + \delta_B \text{ for all integers } i \ge n_1.$$

Then,

$$\rho(x_i, x_{i+1}) \le \epsilon \text{ for all integers } i \ge n_1 + n_B.$$

Moreover, if an integer $j \geq n_1$ satisfies $\rho(x_j, x_{j+1}) \leq \epsilon$, then

$$\rho(x_i, x_{i+1}) \leq \epsilon \text{ for all integers } i \geq j.$$

Corollary 12.15 *Let $\bar{r} > 0$, $\bar{M} > 0$ and $\epsilon \in (0, 1)$ and let an open everywhere dense subset $\mathcal{F} \subset \mathcal{M}$ be as guaranteed by Theorem 12.14. Assume that $B \in \mathcal{F}$ and an open neighborhood $\mathcal{U}$ of B in $\mathcal{M}$ is as guaranteed by Theorem 12.14.*

Let $C_i \in \mathcal{U}$ for all integers $i \geq 0$ and a sequence $\{x_i\}_{i=0}^{\infty} \subset X$ be such that

$$\rho(x_0, \theta) \leq \bar{M},$$

$$x_{i+1} \in C_i(x_i),\ \rho(x_i, x_{i+1}) \leq \rho(x_i, C_i(x_i)) + \bar{r}$$

for all integers $i \geq 0$ and

$$\lim_{i \to \infty} (\rho(x_i, x_{i+1}) - \rho(x_i, C_i(x_i))) = 0.$$

Then,

$$\rho(x_i, x_{i+1}) \leq \epsilon \text{ for all sufficiently large natural numbers } i.$$

We prove the following result obtained in [228].

Theorem 12.16 *Let $\bar{M} > 0$ and $\epsilon \in (0, 1)$. Then there exists an open everywhere dense subset $\mathcal{F} \subset \mathcal{M}$ such that for each $B \in \mathcal{F}$, there exist $x_B \in K$, a natural number n_B, and an open neighborhood $\mathcal{U}$ of B in $\mathcal{M}$ such that the following assertion holds.*

Let $C_i \in \mathcal{U}$, $i = 0, 1, \ldots$, and $x_0 \in K$ satisfy

$$\rho(x_0, \theta) \leq \bar{M}.$$

Then there exists a sequence $\{x_i\}_{i=0}^{\infty} \subset X$ such that

$$x_{i+1} \in C_i(x_i) \text{ for all integers } i \geq 0$$

and

$$\rho(x_i, x_B) \leq \epsilon \text{ for all integers } i \geq n_B.$$

Theorem 12.15 implies the following result.

Theorem 12.17 *There exists a set $\mathcal{F} \subset \mathcal{M}$ which is a countable intersection of open everywhere dense subsets of $\mathcal{M}$ such that for each $B \in \mathcal{F}$, each $M > 0$, and each $\epsilon > 0$, there exist $x_B \in K$, a natural number n_B, and an open neighborhood $\mathcal{U}$ of B in $\mathcal{M}$ such that the following assertion holds.*

Let $C_i \in \mathcal{U}$, $i = 0, 1, \ldots$, and $x_0 \in X$ satisfy

$$\rho(x_0, \theta) \le \bar{M}.$$

Then there exists a sequence $\{x_i\}_{i=0}^{\infty} \subset K$ such that

$$x_{i+1} \in C_i(x_i) \text{ for all integers } i \ge 0$$

and

$$\rho(x_i, x_B) \le \epsilon \text{ for all integers } i \ge n_B.$$

12.9 Proofs of Theorems 12.14 and 12.16

We prove Theorems 12.14 and 12.16 simultaneously. In the case of Theorem 12.16, set

$$\bar{r} = 2^{-1}. \tag{12.161}$$

We begin with the construction of the set $\mathcal{F}$.

Let $A \in \mathcal{M}$ and $\gamma \in (0, 1)$. Define

$$A_\gamma(x) = \text{cl}(\{(1-\gamma)z \oplus \gamma\theta : z \in A(x)\}), \; x \in K. \tag{12.162}$$

We show that for each $x, y \in X$,

$$H(A_\gamma(x), A_\gamma(y)) \le (1-\gamma)\rho(x, y). \tag{12.163}$$

Let $x, y \in X$. It is sufficient to show that for each

$$\xi \in \{(1-\gamma)z \oplus \gamma\theta : z \in A(x)\},$$

we have

$$\rho(\xi, A_\gamma(y)) \le (1-\gamma)\rho(x, y).$$

Assume that $z \in A(x)$ and $\epsilon > 0$. By (12.159), there exists $u \in A(y)$ such that

$$\rho(z, u) \le \rho(z, A(y)) + \epsilon$$

$$\le H(A(x), A(y)) + \epsilon \le \rho(x, y) + \epsilon.$$

Then,

$$(1-\gamma)u \oplus \theta \in A_\gamma(y),$$

$$\rho((1-\gamma)z \oplus \gamma\theta, A_\gamma(y))$$

$$\leq \rho((1-\gamma)z \oplus \gamma\theta, ((1-\gamma)u \oplus \gamma\theta))$$

$$\leq (1-\gamma)\rho(z, u) \leq (1-\gamma)\rho(x, y) + \epsilon.$$

Since ϵ is any positive number, we conclude that

$$\rho((1-\gamma)z \oplus \gamma\theta, A_\gamma(y)) \leq (1-\gamma)\rho(x, y).$$

This implies (12.163). Fix

$$M_0(A, \gamma) > 1 + \gamma^{-1}(H(A(\theta), \{\theta\}) + 1.$$

We show that

$$A_\gamma(B(\theta, M_0(A, \gamma))) \subset B(\theta, M_0(A, \gamma) - 1). \tag{12.164}$$

Let

$$x \in B(\theta, M_0(A, \gamma)),$$

$$z \in A_\gamma(x) = \text{cl}(\{(1-\gamma)\xi \oplus \gamma\theta : \ \xi \in A(x)\}.$$

In order to prove (12.164), it is sufficient to show that

$$B(z, \theta) \leq M_0(A, \gamma) - 1.$$

We may assume without loss of generality that there exists

$$\xi \in A(x)$$

such that

$$z = (1-\gamma)\xi \oplus \gamma\theta.$$

By the relation above, the choice of $M_0(A, \gamma)$, (12.158) and (12.159),

$$\rho(z, \theta) = \rho((1-\gamma)\xi \oplus \gamma\theta, \theta)$$

$$\leq (1-\gamma)\rho(\xi, \theta) \leq (1-\gamma)\rho(\xi, A(\theta)) + H(A(\theta), \{\theta\})$$

$$\leq (1-\gamma)(H(A(x), A(\theta)) + H(A(\theta), \{\theta\})$$

$$\leq (1-\gamma)(\rho(x,\theta) + H(A(\theta), \{\theta\})$$

$$\leq (1-\gamma)(M_0(A,\gamma) + H(A(\theta), \{\theta\})$$

$$\leq M_0(A,\gamma) - 1 + 1 - \gamma(M_0(A,\gamma) - H(A(\theta), \{\theta\}) \leq M_0(A,\gamma) - 1.$$

Thus, (12.164) holds.

In view of (12.163), (12.164), and the Nadler fixed point theorem [137], there exists

$$x(A,\gamma) \in B(\theta, M_0(A,\gamma)) \tag{12.165}$$

such that

$$x(A,\gamma) \in A_\gamma(x(A,\gamma)). \tag{12.166}$$

It follows from (12.164)–(12.166) that

$$x(A,\gamma) \in B(\theta, M_0(A,\gamma) - 1). \tag{12.167}$$

Choose a positive number $M(A,\gamma)$ such that

$$M(A,\gamma) > 2M_0(A,\gamma) + 2\bar{M} + 2$$

$$+ (\sup\{\rho(\xi_1,\xi_2) : \xi_1, \xi_2 \in A_\gamma(x(A,\gamma))\} + 2)\gamma^{-1}. \tag{12.168}$$

By (12.167) and (12.168),

$$B(\theta, \bar{M}) \subset B(x(A,\gamma), \bar{M} + M_0(A,\gamma) - 1)$$

$$\subset B(x(A,\gamma), M(A,\gamma) - 1). \tag{12.169}$$

We show that

$$A_\gamma(B(x(A,\gamma), M(A,\gamma))) \subset B(x(A,\gamma), M(A,\gamma) - 1). \tag{12.170}$$

Let

$$z \in A_\gamma(B(x(A,\gamma), M(A,\gamma))). \tag{12.171}$$

In view of (12.171), there exists

$$y \in B(x(A,\gamma), M(A,\gamma)) \tag{12.172}$$

such that

$$z \in A_\gamma(y). \tag{12.173}$$

By (12.163) and (12.173),

$$\rho(z, A_\gamma(x(A,\gamma))) \le H(A_\gamma(y), A_\gamma(x(A,\gamma)))$$
$$\le (1-\gamma)\rho(y, x(A,\gamma)). \tag{12.174}$$

It follows from (12.166), (12.168), (12.172), and (12.174) that

$$\rho(z, x(A,\gamma))$$
$$\le \rho(z, A_\gamma(x(A,\gamma))) + \sup\{\rho(\xi_1,\xi_2) :\ \xi_1,\xi_2 \in A_\gamma(x(A,\gamma))\}$$
$$\le (1-\gamma)\rho(y, x(A,\gamma)) + \sup\{\rho(\xi_1,\xi_2) :\ \xi_1,\xi_2 \in A_\gamma(x(A,\gamma))\}$$
$$\le (1-\gamma)M(A,\gamma) + \sup\{\rho(\xi_1,\xi_2) :\ \xi_1,\xi_2 \in A_\gamma(x(A,\gamma))\}$$
$$= M(A,\gamma) - \gamma M(A,\gamma) + \sup\{\rho(\xi_1,\xi_2) :\ \xi_1,\xi_2 \in A_\gamma(x(A,\gamma))\}$$
$$< M(A,\gamma) - 1.$$

Thus, (12.170) holds.

Choose $\delta(A,\gamma) \in (0,\bar{r})$ such that

$$\delta(A,\gamma) < 12^{-1}\gamma\epsilon. \tag{12.175}$$

There exists an open neighborhood $\mathcal{U}(A,\gamma)$ of A_γ in $\mathcal{M}$ such that

$$\mathcal{U}(A,\gamma) \subset \{C \in \mathcal{M} :\ H(C(z), A_\gamma(z)) \le \delta(A,\gamma)$$
$$\text{for all } z \in B(x(A,\gamma), M(A,\gamma))\}. \tag{12.176}$$

By (12.162), for each $z \in X$,

$$H(A(z), A_\gamma(z)) = H(A(z), \mathrm{cl}(\{(1-\gamma)\xi \oplus \gamma\theta :\ \xi \in A(z)\}))$$
$$\le \gamma \sup\{\rho(\theta,\xi) :\ \xi \in A(z)\}. \tag{12.177}$$

For each $z \in K$ and each $\xi \in A(z)$,

$$\rho(\theta,\xi)$$

$$\le \rho(\xi, A(\theta)) + \sup\{\rho(\eta_1, \eta_2) : \eta_1, \eta_2 \in A(\theta)\}$$

$$+ \sup\{\rho(\theta, \eta) : \eta \in A(\theta)\}$$

$$\le \sup\{\rho(\eta_1, \eta_2) : \eta_1, \eta_2 \in A(\theta)\}$$

$$+ \sup\{\rho(\theta, \eta) : \eta \in A(\theta)\} + H(A(z), A(\theta))$$

$$\le \sup\{\rho(\theta, \eta) : \eta \in A(\theta)\} + \rho(z, \theta)$$

$$+ \sup\{\rho(\eta_1, \eta_2) : \eta_1, \eta_2 \in A(\theta)\}$$

$$\le \rho(z, \theta) + 3 \sup\{\rho(\theta, \eta) : \eta \in A(\theta)\}. \tag{12.178}$$

By (12.177) and (12.178), for each $z \in X$,

$$H(A(z), A_\gamma(z)) \le \gamma(\rho(z, \theta) + 3 \sup\{\rho(\theta, \eta) : \eta \in A(\theta)\}).$$

By the relation above, for each neighborhood V of A in $\mathcal{M}$, there exists $\gamma_V \in (0, 1)$ such that for each $\gamma \in (0, \gamma_V]$, $A_\gamma \in V$. Therefore,

$$\{A_\gamma : A \in \mathcal{M}, \ \gamma \in (0, 1)\}$$

is an everywhere dense subset of $\mathcal{M}$. Set

$$\mathcal{F} = \cup\{\mathcal{U}(A, \gamma) : A \in \mathcal{M}, \ \gamma \in (0, 1)\}. \tag{12.179}$$

Clearly, $\mathcal{F}$ is an open everywhere dense subset of $\mathcal{M}$.

Completion of the Proof of Theorem 12.14 Assume that

$$B \in \mathcal{F}. \tag{12.180}$$

By (12.179) and (12.180), there exist $A \in \mathcal{M}$ and $\gamma \in (0, 1)$ such that

$$B \in \mathcal{U}(A, \gamma). \tag{12.181}$$

Set

$$\delta_B = \delta(A, \gamma) \tag{12.182}$$

and choose a natural number n_B such that

$$n_B > 4M(A, \gamma)(\gamma\epsilon)^{-1} + 4. \tag{12.183}$$

Assume that

$$C_i \in \mathcal{U}(A, \gamma), \ i = 0, 1, \ldots, \tag{12.184}$$

n_1 is a natural number and that a sequence $\{x_i\}_{i=0}^{\infty} \subset K$ satisfies

$$\rho(x_0, \theta) \le \bar{M}, \tag{12.185}$$

$$x_{i+1} \in C_i(x_i), \ \rho(x_i, x_{i+1}) \le \rho(x_i, C_i(x_i)) + \bar{r} \tag{12.186}$$

for all integers $i \ge 0$ and

$$\rho(x_i, x_{i+1}) \le \rho(x_i, C_i(x_i)) + \delta_B \text{ for all integers } i \ge n_1. \tag{12.187}$$

In order to complete the proof, it is sufficient to show the following:

$$\rho(x_i, x_{i+1}) \le \epsilon \text{ for all integers } i \ge n_1 + n_B;$$

if an integer $j \ge n_1$ satisfies $\rho(x_j, x_{j+1}) \le \epsilon$, then

$$\rho(x_{j+1}, x_{j+2}) \le \epsilon.$$

First, we show that

$$x_i \in B(x(A, \gamma), M(A, \gamma) - 1/2) \tag{12.188}$$

for all integers $i \ge 0$. In view of (12.169) and (12.185), inequality (12.188) holds for $i = 0$.

Assume that an integer $j \ge 0$ and (12.188) holds with $i = j$. By (12.170) and (12.188) with $i = j$,

$$A_\gamma(x_j) \subset B(x(A, \gamma), M(A, \gamma) - 1). \tag{12.189}$$

In view of (12.176), (12.184), and (12.188) (with $i = j$),

$$H(C_j(x_j), A_\gamma(x_j)) \le \delta(A, \gamma).$$

Together with (12.186), this implies that

$$\rho(x_{j+1}, A_\gamma(x_j)) \le H(C_j(x_j), A_\gamma(x_j)) \le \delta(A, \gamma).$$

Combined with (12.175) and (12.189), this implies that

$$\rho(x_{j+1}, B(x(A, \gamma), M(A, \gamma) - 1)) \le \delta(A, \gamma) < 1/4$$

and

$$\rho(x_{j+1}, x(A, \gamma)) < M(A, \gamma) - 1/2.$$

Thus, (12.188) holds for $i = j + 1$. Therefore, we have shown by induction that (12.188) holds for all integers $i \geq 0$.

We show that there exists an integer $i \in [n_1, n_1 + n_B]$ such that

$$\rho(x_i, x_{i+1}) \leq \epsilon.$$

Assume the contrary. Then,

$$\rho(x_i, x_{i+1}) > \epsilon \tag{12.190}$$

for all $i = n_1, \ldots, n_1 + n_B$. Assume that an integer i satisfies

$$i \in \{n_1, \ldots, n_1 + n_B - 1\}. \tag{12.191}$$

In view of (12.187) and (12.191),

$$\rho(x_{i+1}, x_{i+2}) \leq \rho(x_{i+1}, C_{i+1}(x_{i+1})) + \delta_B. \tag{12.192}$$

By (12.176), (12.182), (12.184), and (12.188),

$$H(C_j(x_j), A_\gamma(x_j)) \leq \delta(A, \gamma) = \delta_B \tag{12.193}$$

for all integers $j \geq 0$. It follows from (12.163), (12.186), and (12.193) that

$$\begin{aligned}
\rho(x_{i+1}, C_{i+1}(x_{i+1})) &\leq H(C_i(x_i), C_{i+1}(x_{i+1})) \\
&\leq H(C_i(x_i), A_\gamma(x_i)) + H(A_\gamma(x_i), A_\gamma(x_{i+1})) + H(A_\gamma(x_{i+1}), C_{i+1}(x_{i+1})) \\
&\leq 2\delta(A, \gamma) + H(A_\gamma(x_i), A_\gamma(x_{i+1})) \\
&\leq 2\delta(A, \gamma) + (1 - \gamma)\rho(x_i, x_{i+1}).
\end{aligned} \tag{12.194}$$

By (12.182), (12.192), and (12.194),

$$\rho(x_{i+1}, x_{i+2}) \leq 3\delta(A, \gamma) + (1 - \gamma)\rho(x_i, x_{i+1}).$$

Together with (12.175), (12.190), and (12.191), this implies that

$$\begin{aligned}
&\rho(x_i, x_{i+1}) - \rho(x_{i+1}, x_{i+2}) \\
&\geq \gamma\rho(x_i, x_{i+1}) - 3\delta(A, \gamma)
\end{aligned}$$

$$\geq \gamma\epsilon - 3\delta(A, \gamma) \geq \gamma\epsilon/2$$

and

$$\rho(x_i, x_{i+1}) - \rho(x_{i+1}, x_{i+2}) \geq 2^{-1}\gamma\epsilon.$$

Since the inequality above holds for all integers i satisfying (12.191), it follows from (12.188) that

$$2M(A, \gamma) \geq \rho(x_{n_1}, x_{n_1+1}) - \rho(x_{n_1+n_B-1}, x_{n_1+n_B})$$

$$= \sum_{i=n_1}^{n_1+n_B-2} (\rho(x_i, x_{i+1}) - \rho(x_{i+1}, x_{i+2})) \geq (n_B - 1)\gamma\epsilon/2$$

and

$$n_B \leq 4(\gamma\epsilon)^{-1}M(A, \gamma) + 1.$$

This contradicts (12.183). The contradiction we have reached proves that there exists an integer

$$j \in [n_1, n_1 + n_B] \tag{12.195}$$

such that

$$\rho(x_j, x_{j+1}) \leq \epsilon. \tag{12.196}$$

Assume that an integer j satisfies

$$j \geq n_1, \ \rho(x_j, x_{j+1}) \leq \epsilon. \tag{12.197}$$

By (12.163), (12.175), (12.176), (12.184), (12.186)–(12.188), (12.193), and (12.197),

$$\rho(x_{j+1}, x_{j+2}) \leq \rho(x_{j+1}, C_{j+1}(x_{j+1})) + \delta_B$$

$$\leq \delta_B + H(C_j(x_j), C_{j+1}(x_{j+1}))$$

$$\leq \delta_B + H(C_j(x_j), A_\gamma(x_j)) + H(A_\gamma(x_j), A_\gamma(x_{j+1})) + H(A_\gamma(x_{j+1}), C_{j+1}(x_{j+1}))$$

$$\leq 3\delta(A, \gamma) + (1 - \gamma)\rho(x_j, x_{j+1})$$

$$\leq 3\delta(A, \gamma) + (1 - \gamma)\epsilon \leq \epsilon.$$

Thus, if an integer j satisfies (12.197), then $\rho(x_{j+1}, x_{j+2}) \leq \epsilon$. Combined with (12.195) and (12.196), this completes the proof of Theorem 12.14.

Completion of the Proof of Theorem 12.16 Let

$$B \in \mathcal{F}. \tag{12.198}$$

By (12.179) and (12.198), there exist $A \in \mathcal{M}$ and $\gamma \in (0, 1)$ such that

$$B \in \mathcal{U}(A, \gamma). \tag{12.199}$$

Set

$$x_B = x(A, \gamma) \tag{12.200}$$

and choose a natural number n_B such that

$$n_B > 4 + 2M(A, \gamma)(\gamma\epsilon)^{-1}. \tag{12.201}$$

Assume that

$$C_i \in \mathcal{U}(A, \gamma),\ i = 0, 1, \ldots, \tag{12.202}$$

$$x_0 \in B(\theta, \bar{M}). \tag{12.203}$$

By induction define a sequence $\{x_i\}_{i=0}^{\infty} \subset K$ such that for each integer $i \geq 0$,

$$x_{i+1} \in C_i(x_i), \tag{12.204}$$

$$\rho(x(A, \gamma), x_{i+1}) \leq \rho(x(A, \gamma), C_i(x_i)) + 4^{-1}\delta(A, \gamma). \tag{12.205}$$

In order to complete the proof, it is sufficient to show that

$$\rho(x_i, x(A, \gamma)) \leq \epsilon \text{ for all integers } i \geq n_B.$$

First, we show that

$$\rho(x_i, x(A, \gamma)) \leq M(A, \gamma) \tag{12.206}$$

for all integers $i \geq 0$. Note that in view of (12.169) and (12.203), inequality (12.206) holds for $i = 0$.

Assume that an integer $j \geq 0$ and that

$$\rho(x_j, x(A, \gamma)) \leq M(A, \gamma). \tag{12.207}$$

By (12.264), (12.166), and (12.205),

$$\rho(x_{j+1}, x(A, \gamma)) \le \rho(x(A, \gamma), C_j(x_j)) + 4^{-1}\delta(A, \gamma)$$

$$\le \rho(x(A, \gamma), A_\gamma(x_j)) + H(A_\gamma(x_j), C_j(x_j)) + 4^{-1}\delta(A, \gamma)$$

$$\le H(A_\gamma(x(A, \gamma)), A_\gamma(x_j)) + H(A_\gamma(x_j), C_j(x_j)) + 4^{-1}\delta(A, \gamma)$$

$$\le (1 - \gamma)\rho(x(A, \gamma), x_j) + H(A_\gamma(x_j), C_j(x_j)) + 4^{-1}\delta(A, \gamma). \tag{12.208}$$

In view of (12.176), (12.202), and (12.207),

$$H(A_\gamma(x_j), C_j(x_j)) \le \delta(A, \gamma).$$

Together with (12.208), this implies that

$$\rho(x_{j+1}, x(A, \gamma)) \le (1 - \gamma)\rho(x(A, \gamma), x_j) + 2\delta(A, \gamma).$$

Thus, we have shown that the following property holds:
(P) If an integer $j \ge 0$ satisfies

$$\rho(x_j, x(A, \gamma)) \le M(A, \gamma),$$

then

$$\rho(x_{j+1}, x(A, \gamma)) \le (1 - \gamma)\rho(x(A, \gamma), x_j) + 2\delta(A, \gamma). \tag{12.209}$$

Assume that an integer $j \ge 0$ and that (12.207) holds. Then by property (P), (12.168), (12.175), and (12.207),

$$\rho(x_{j+1}, x(A, \gamma)) \le (1 - \gamma)M(A, \gamma) + 2\delta(A, \gamma)$$

$$< (1 - \gamma)M(A, \gamma) + 1 < M(A, \gamma)$$

and (12.206) holds with $i = j + 1$. Therefore, we have shown by induction that (12.206) holds for all integers $i \ge 0$. Together with property (P), this implies that (12.209) holds for all integers $j \ge 0$.

We show that there exists an integer $j \in [0, n_B]$ such that

$$\rho(x_j, x(A, \gamma)) \le \epsilon.$$

Assume the contrary. Then for all $j \in \{0, \dots, n_B\}$

$$\rho(x_j, x(A, \gamma)) > \epsilon.$$

Together with (12.175) and (12.209), this implies that for all integers $j \in [0, n_B]$,

$$\rho(x_j, x(A, \gamma)) - \rho(x_{j+1}, x(A, \gamma))$$

$$\geq \gamma\rho(x_j, x(A, \gamma)) - 2\delta(A, \gamma) > \gamma\epsilon - 2\delta(A, \gamma) > \gamma\epsilon/2.$$

Together with (12.206), this implies that

$$M(A, \gamma) \geq \rho(x_0, x(A, \gamma)) - \rho(x_{n_B}, x(A, \gamma))$$

$$= \sum_{j=0}^{n_B-1} (\rho(x_j, x(A, \gamma)) - \rho(x_{j+1}, x(A, \gamma))) \geq n_B\gamma\epsilon/2$$

and

$$n_B \leq 2(\gamma\epsilon)^{-1}M(A, \gamma).$$

This contradicts (12.201). The contradiction we have reached proves that there exists an integer j such that

$$j \in [0, n_B],\ \ \rho(x_j, x(A, \gamma)) \leq \epsilon. \tag{12.210}$$

Assume that an integer $i \geq 0$ satisfies

$$\rho(x_i, x(A, \gamma)) \leq \epsilon. \tag{12.211}$$

We show that

$$\rho(x_{i+1}, x(A, \gamma)) \leq \epsilon.$$

By (12.175), (12.209), and (12.211),

$$\rho(x_{i+1}, x(A, \gamma)) \leq (1 - \gamma)\epsilon + 2\delta(A, \gamma) \leq \epsilon. \tag{12.212}$$

We have shown that if an integer $i \geq 0$ satisfies (12.211), then (12.212) holds. Together with (12.210), this implies that (12.211) holds for all integers $i \geq n_B$. This completes the proof of Theorem 12.16.

12.10 Extensions of Theorems 12.6 and 12.8

Let (X, ρ) be a complete metric space, $T : X \to 2^X \setminus \{\emptyset\}$, $T(x)$ be closed for each $x \in X$, $c \in [0, 1)$, $X_0 \subset X$ be a nonempty closed set,

$$T(X_0) \subset X_0$$

and let for each $x \in X_0$ and each $y \in X$,

$$H(T(x), T(y)) \leq c\rho(x, y). \tag{12.213}$$

Theorem 12.1 implies that there exists

$$z_* \in X_0$$

such that

$$z_* \in T(z_*). \tag{12.214}$$

The following extension of Theorem 12.8 was obtained in [233].

Theorem 12.18 *Assume that* $\{x_i\}_{i=0}^{\infty} \subset X$, $\{r_i\}_{i=0}^{\infty} \subset (0, \infty)$,

$$\sum_{i=0}^{\infty} r_i < \infty \tag{12.215}$$

and that for each integer $i \geq 0$,

$$B(x_{i+1}, r_i) \cap \{\xi \in T(x_i) : \ \rho(\xi, x_i) \leq r_i + \rho(x_i, T(x_i))\} \neq \emptyset. \tag{12.216}$$

Then $\{x_i\}_{i=0}^{\infty}$ *converges to a fixed point* x_* *of* T *such that* $x_* \in T(x_*)$.

Proof By (12.216), for each integer $i \geq 0$, there exists

$$\xi_i \in B(x_{i+1}, r_i) \cap T(x_i) \tag{12.217}$$

such that

$$\rho(x_i, \xi_i) \leq \rho(x_i, T(x_i)) + r_i. \tag{12.218}$$

For each integer $i \geq 0$, there exists

$$u_i \in X_0 \tag{12.219}$$

such that

$$\rho(u_i, x_i) \leq \rho(x_i, X_0) + r_i. \tag{12.220}$$

Let $i \geq 0$ be an integer. By (12.213), (12.217), (12.219), (12.220), and the inclusion

$$T(X_0) \subset X_0, \tag{12.221}$$

we have

$$\begin{aligned}\rho(x_{i+1}, u_{i+1}) &\leq \rho(x_{i+1}, X_0) + r_{i+1} \\ &\leq \rho(x_{i+1}, \xi_i) + \rho(\xi_i, X_0) + r_{i+1} \\ &\leq \rho(\xi_i, X_0) + r_{i+1} + r_i \\ &\leq \rho(\xi_i, T(u_i)) + r_{i+1} + r_i \\ &\leq H(T(x_i), T(u_i)) + r_{i+1} + r_i \\ &c\rho(x_i, u_i) + +r_{i+1} + r_i.\end{aligned} \tag{12.222}$$

By (12.112), (12.217), and (12.219),

$$\xi_i \in T(x_i), \;\; \rho(T(u_i), T(x_i)) \leq c\rho(x_i, u_i). \tag{12.223}$$

In view of (12.223), there exists

$$v_i \in T(u_i) \tag{12.224}$$

such that

$$\rho(v_i, \xi_i) \leq c\rho(x_i, u_i) + r_i. \tag{12.225}$$

It follows from (12.217), (12.222), and (12.225) that

$$\begin{aligned}\rho(v_i, u_{i+1}) &\leq \rho(v_i, \xi_i) + \rho(\xi_i, u_{i+1}) \\ &\leq c\rho(x_i, u_i) + r_i + \rho(\xi_i, x_{i+1}) + \rho(x_{i+1}, u_{i+1}) \\ &\leq c\rho(x_i, u_i) + r_i + r_i + c\rho(x_i, u_i) + r_{i+1} + r_i \\ &= 2c\rho(x_i, u_i) + r_{i+1} + 3r_i.\end{aligned} \tag{12.226}$$

Equations (12.218) and (12.225) imply that

$$\begin{aligned}\rho(u_i, v_i) &\leq \rho(u_i, x_i) + \rho(x_i, \xi_i) + \rho(\xi_i, v_i) \\ &\leq \rho(x_i, u_i) + \rho(x_i, T(x_i)) + r_i + c\rho(x_i, u_i) + r_i.\end{aligned} \tag{12.227}$$

By (12.213), (12.216), and (12.219),

$$\rho(x_i, T(x_i)) \le \rho(x_i, u_i) + \rho(u_i, T(u_i)) + H(T(u_i), T(x_i))$$
$$\le \rho(u_i, T(u_i)) + (1 + c)\rho(x_i, u_i). \tag{12.228}$$

In view of (12.227) and (12.228),

$$\rho(u_i, v_i) \le (1 + c)\rho(u_i, x_i) + 2r_i + \rho(x_i, T(x_i))$$
$$\le (1 + c)\rho(u_i, x_i) + 2r_i + \rho(u_i, T(u_i)) + (1 + c)\rho(x_i, u_i)$$
$$\le 2(1 + c)\rho(u_i, x_i) + 2r_i + \rho(u_i, T(u_i)).$$

By the relation above, (12.123) and (12.125),

$$v_i \in T(u_i),\ \ \rho(v_i, u_{i+1}) \le 2c\rho(x_i, u_i) + r_{i+1} + 3r_i,$$
$$\rho(u_i, v_i) \le 2(1 + c)\rho(u_i, x_i) + 2r_i + \rho(u_i, T(u_i)). \tag{12.229}$$

We show that

$$\sum_{i=0}^{\infty} \rho(u_i, v_i) < \infty.$$

For each integer $i \ge 0$, set

$$\Delta_i = r_{i+1} + r_i. \tag{12.230}$$

By (12.212) and (12.230),

$$\sum_{i=0}^{\infty} \Delta_i < \infty.$$

In view of (12.222) and (12.230), for each integer $i \ge 0$,

$$\rho(x_{i+1}, u_{i+1}) \le c\rho(x_i, u_i) + \Delta_i. \tag{12.231}$$

By (12.231),

$$\rho(x_1, u_1) \le c\rho(x_0, u_0) + \Delta_0$$

and

$$\begin{aligned}\rho(x_2, u_2) &\le c\rho(x_1, u_1) + \Delta_1 \\ &\le c^2\rho(x_0, u_0) + c\Delta_0 + \Delta_1.\end{aligned} \tag{12.232}$$

By induction we show that for each integer $n \ge 1$,

$$\rho(x_n, u_n) \le c^n \rho(x_0, u_0) + \sum_{i=0}^{n-1} c^i \Delta_{n-i-1}. \tag{12.233}$$

In view of (12.232), inequality (12.233) is valid for $n = 1, 2$.

Assume that $k \ge 1$ is an integer and that (12.233) holds for $n = k$. When combined with (12.231), this implies that

$$\begin{aligned}\rho(x_{k+1}, u_{k+1}) &\le c\rho(x_k, u_k) + \Delta_k \\ &\le c^{k+1}\rho(x_0, u_0) + \sum_{i=0}^{k-1} c^{i+1} \Delta_{k-i-1} + \Delta_k \\ &= c^{k+1}\rho(x_0, u_0) + \sum_{i=0}^{k} c^i \Delta_{k-i}.\end{aligned}$$

Thus, (12.233) holds with $n = k + 1$ and therefore (12.233) holds for all integers $n \ge 1$. By (12.233),

$$\begin{aligned}\sum_{n=1}^{\infty} \rho(x_n, u_n) &\le \sum_{n=1}^{\infty} (c^n \rho(x_0, u_0) + \sum_{i=0}^{n-1} c^i \Delta_{n-1-i}) \\ &\le \rho(x_0, u_0) \sum_{n=1}^{\infty} c^n + \sum_{j=0}^{\infty} c^j \sum_{j=0}^{\infty} \Delta_j < \infty.\end{aligned} \tag{12.234}$$

For each integer $i \ge 0$, set

$$\gamma_i = 4\rho(x_i, u_i) + 3r_i + r_{i+1}. \tag{12.235}$$

In view of (12.212), (12.234), and (12.235),

$$\sum_{i=0}^{\infty} \gamma_i < \infty.$$

By (12.219) and (12.229), $\{u_n\}_{n=0}^{\infty} \subset X_0$ and for each integer $i \geq 0$,

$$v_i \in B(u_{i+1}, \gamma_i) \cap T(u_i),$$

$$\rho(u_i, v_i) \leq \rho(u_i, T(u_i)) + \gamma_i.$$

Applying Theorem 12.8 for the space X_0 and the restriction of T to X_0 with $\epsilon_i = r_i = \gamma_i$, $i = 0, 1, \ldots$ we obtain that $\{u_i\}_{i=0}^{\infty}$ converges to a point $x_* \in X_0$ such that $x_* \in T(x_*)$. By (12.234),

$$x_* = \lim_{n \to \infty} x_n.$$

Theorem 12.18 is proved.

The following extension of Theorem 12.6 was obtained in [234].

Theorem 12.19 *Let* $\theta \in X$, $\epsilon \in (0, 1)$, $M > 0$,

$$B(\theta, M) \cap X_0 \neq \emptyset, \tag{12.236}$$

$$M_0 > (1 - c)^{-1}(2M + 2)(10 + \rho(\theta, T(\theta))), \tag{12.237}$$

$$\delta = 360^{-1}\epsilon(1 - c)^2, \tag{12.238}$$

$$n_0 = \lfloor 4M_0(1 - c)^{-1}\epsilon^{-1} \rfloor + \lfloor 72(2M + 1)(1 - c)^{-2}\epsilon^{-1} \rfloor + 1. \tag{12.239}$$

Then for each sequence $\{x_i\}_{i=0}^{\infty} \subset X$ *such that*

$$\rho(x_0, \theta) \leq M$$

and that for each integer $n \geq 0$,

$$B(x_{n+1}, \delta) \cap \{\xi \in T(x_n) : \ \rho(\xi, x_n) \leq \delta + \rho(x_n, T(x_n))\} \neq \emptyset \tag{12.240}$$

the following relations hold for each integer $n \geq n_0$:

$$\rho(x_n, x_{n+1}) < \epsilon \text{ and } B(x_n, \epsilon) \cap T(x_n) \neq \emptyset.$$

Proof Set

$$\delta_1 = 36^{-1}(1 - c)\epsilon. \tag{12.241}$$

By (12.238) and (12.241),

$$\delta \leq (1 - c)\delta_1/10. \tag{12.242}$$

Set

$$n_1 = \lfloor 2(2M+1)\delta_1^{-1}(1-c)^{-1} \rfloor + 1, \tag{12.243}$$

$$n_2 = \lfloor 4M_0(1-c)^{-1}\epsilon^{-1} \rfloor. \tag{12.244}$$

By (12.239), (12.243), and (12.244),

$$n_0 = n_1 + n_2. \tag{12.245}$$

Assume that

$$\{x_i\}_{i=0}^{\infty} \subset X, \tag{12.246}$$

$$\rho(x_0, \theta) \le M \tag{12.247}$$

and that (12.240) holds for each integer $n \ge 0$. By (12.236), there exists

$$\theta_1 \in B(\theta, M) \cap X_0. \tag{12.248}$$

In view of (12.247) and (12.248),

$$\rho(x_0, \theta_1) \le \rho(x_0, \theta) + \rho(\theta, \theta_1) \le 2M. \tag{12.249}$$

Equations (12.213), (12.247), (12.248), and (12.249) imply that

$$\begin{aligned}
\rho(x_0, T(x_0)) &\le \rho(x_0, \theta_1) + \rho(\theta_1, T(\theta_1)) + H(T(\theta_1), T(x_0)) \\
&\le 2\rho(x_0, \theta_1) + \rho(\theta_1, T(\theta_1)) \\
&\le 4M + \rho(\theta_1, \theta) + \rho(\theta, T(\theta)) + H(T(\theta), T(\theta_1)) \\
&\le 4M + \rho(\theta, T(\theta)) + 2\rho(\theta, \theta_1) \le 6M + \rho(\theta, T(\theta)).
\end{aligned} \tag{12.250}$$

By (12.240), (12.242), and (12.250),

$$\rho(x_0, x_1) \le \rho(x_0, T(x_0)) + 2\delta \le 6M + \rho(\theta, T(\theta)) + 1. \tag{12.251}$$

In view of (12.240), for each integer $n \ge 0$, there exists

$$\xi_n \in B(x_{n+1}, \delta) \cap T(x_n) \tag{12.252}$$

such that

$$\rho(x_n, \xi_n) \le \delta + \rho(x_n, T(x_n)). \tag{12.253}$$

For each integer $n \geq 0$, there exists

$$u_n \in X_0 \tag{12.254}$$

such that

$$\rho(u_n, x_n) \leq \rho(x_n, X_0) + \delta. \tag{12.255}$$

Recall that

$$T(X_0) \subset X_0. \tag{12.256}$$

Let $n \geq 0$ be an integer. By (12.112), (12.252), (12.255), and (12.256),

$$\begin{aligned} \rho(u_{n+1}, x_{n+1}) &\leq \rho(x_{n+1}, X_0) + \delta \\ &\leq \rho(x_{n+1}, \xi_n) + \rho(\xi_n, X_0) + \delta \\ &\leq 2\delta + \rho(\xi_n, T(u_n)) \leq 2\delta + H(T(x_n), T(u_n)) \\ &\leq 2\delta + c\rho(x_n, u_n). \end{aligned} \tag{12.257}$$

Equations (12.213), (12.252), and (12.254) imply that

$$\xi_n \in T(x_n), \ \ \rho(T(u_n), T(x_n)) \leq c\rho(x_n, u_n). \tag{12.258}$$

In view of (12.258), there exists

$$v_n \in T(u_n) \tag{12.259}$$

such that

$$\rho(v_n, \xi_n) \leq c\rho(x_n, u_n) + \delta. \tag{12.260}$$

It follows from (12.252), (12.257), and (12.260) that

$$\begin{aligned} \rho(v_n, u_{n+1}) &\leq \rho(v_n, \xi_n) + \rho(\xi_n, u_{n+1}) \\ &\leq c\rho(x_n, u_n) + \delta + \rho(\xi_n, x_{n+1}) + \rho(x_{n+1}, u_{n+1}) \\ &\leq 2c\rho(x_n, u_n) + 4\delta. \end{aligned} \tag{12.261}$$

By (12.253) and (12.260),

$$\rho(u_n, v_n) \leq \rho(x_n, u_n) + \rho(x_n, \xi_n) + \rho(\xi_n, v_n)$$

$$\leq \rho(u_n, x_n) + \rho(x_n, T(x_n)) + \delta + c\rho(x_n, u_n) + \delta$$

$$= (1+c)\rho(x_n, u_n) + \rho(x_n, T(x_n)) + 2\delta. \tag{12.262}$$

Equation (12.213) implies that

$$\rho(x_n, T(x_n)) \leq \rho(x_n, u_n) + \rho(u_n, T(u_n)) + H(T(u_n), T(x_n))$$

$$\leq \rho(u_n, T(u_n)) + (1+c)\rho(x_n, u_n). \tag{12.263}$$

In view of (12.262) and (12.263),

$$\rho(u_n, v_n) \leq (1+c)\rho(u_n, x_n) + 2\delta + \rho(x_n, T(x_n))$$

$$\leq (1+c)\rho(x_n, u_n) + 2\delta + (1+c)\rho(x_n, u_n) + \rho(u_n, T(u_n))$$

$$\leq 2(1+c)\rho(x_n, u_n) + 2\delta + \rho(u_n, T(u_n)). \tag{12.264}$$

It follows from (12.259), (12.259), and (12.264) that

$$v_n \in T(u_n),\ \ \rho(v_n, u_{n+1}) \leq 2c\rho(x_n, u_n) + 4\delta,$$

$$\rho(u_n, v_n) \leq 2(1+c)\rho(x_n, u_n) + 2\delta + \rho(u_n, T(u_n)). \tag{12.265}$$

Let $n \geq 0$ be an integer. Equations (12.213), (12.254), and (12.265) imply that

$$\rho(u_{n+1}, u_{n+2}) \leq \rho(u_{n+1}, v_{n+1}) + \rho(v_{n+1}, u_{n+2})$$

$$\leq \rho(u_{n+1}, v_{n+1}) + 2c\rho(x_{n+1}, u_{n+1}) + 4\delta$$

$$\leq \rho(u_{n+1}, T(u_{n+1})) + 2\delta + 2(1+c)\rho(u_{n+1}, x_{n+1}) + 2c\rho(x_{n+1}, u_{n+1}) + 4\delta$$

$$\leq \rho(u_{n+1}, v_n) + \rho(v_n, T(u_{n+1})) + (2+4c)\rho(u_{n+1}, x_{n+1}) + 6\delta$$

$$\leq \rho(v_n, T(u_{n+1})) + 2c\rho(x_n, u_n) + 10\delta + \rho(u_{n+1}, x_{n+1})(2+4c)$$

$$\leq H(T(u_n), T(u_{n+1})) + 2c\rho(u_n, x_n) + (2+4c)\rho(x_{n+1}, u_{n+1}) + 10\delta$$

$$\leq c\rho(u_n, u_{n+1}) + 2c\rho(u_n, x_n) + (2+4c)\rho(x_{n+1}, u_{n+1}) + 10\delta. \tag{12.266}$$

We show that the following claim is true.

(A) There exists an integer $k \in \{0, \ldots, n_1\}$ such that

$$\rho(x_i, u_i) \leq \delta_1 \text{ for each integer } i \geq k_1 \tag{12.267}$$

and for each integer i satisfying $0 \le i < k_1$,

$$\rho(x_i, u_i) > \delta_1 \tag{12.268}$$

and

$$\rho(x_{i+1}, u_{i+1}) \le \rho(x_i, u_i) - (1 - c)\delta_1/2. \tag{12.269}$$

Assume that $i \ge 0$ is an integer and

$$\rho(x_i, u_i) > \delta_1.$$

By (12.242) and (12.257),

$$\rho(x_{i+1}, u_{i+1}) \le 2\delta + c\rho(x_i, u_i)$$

and

$$\rho(x_i, u_i) - \rho(x_{i+1}, u_{+1}) \ge (1 - c)\rho(x_i, u_i) - 2\delta$$

$$\ge (1 - c)\delta_1 - 2\delta \ge (1 - c)\delta_1/2.$$

Thus, we have shown that the following property holds:

(i) If $i \ge 0$ is an integer and $\rho(x_i, u_i) > \delta_1$, then

$$\rho(x_i, u_i) - \rho(x_{i+1}, u_{i+1}) \ge (1 - c)\delta_1/2.$$

Assume that $i \ge 0$ is an integer and

$$\rho(x_i, u_i) \le \delta_1.$$

By (12.242) and (12.257),

$$\rho(x_{i+1}, u_{i+1}) \le 2\delta + c\rho(x_i, u_i) \le 2\delta + c\delta_1 \le \delta_1.$$

Thus, the following property holds:

(ii) If $i \ge 0$ is an integer and

$$\rho(x_i, u_i) \le \delta_1,$$

then

$$\rho(x_{i+1}, u_{i+1}) \le \delta_1.$$

Assume that $k \geq 0$ is an integer and that

$$\rho(x_k, u_k) > \delta_1. \tag{12.270}$$

Property (ii) implies that

$$\rho(x_i, u_i) > \delta_1 \text{ for each } i \in \{0, \ldots, k\}. \tag{12.271}$$

Property (i) and (12.271) imply that for each $i \in \{0, \ldots, k\}$,

$$\rho(x_i, u_i) - \rho(x_{i+1}, u_{i+1}) \geq (1-c)\delta_1/2. \tag{12.272}$$

By (12.247), (12.248), and (12.255),

$$\rho(u_0, x_0) < \rho(x_0, X_0) + 1 \leq \rho(x_0, \theta) + 1 \leq 2M + 1. \tag{12.273}$$

In view of (12.243), (12.271), and (12.273),

$$2M + 1 \geq \rho(x_0, u_0) \geq \rho(x_0, u_0) - \rho(x_{k+1}, u_{k+1})$$

$$= \sum_{i=0}^{k} (\rho(x_i, u_i) - \rho(x_{i+1}, u_{i+1})) \geq (k+1)(1-c)\delta_1/2$$

and

$$k \leq 2(2M+1)(1-c)^{-1}\delta_1^{-1} - 1 \leq n_1 - 1.$$

This implies that claim (A) holds. In view of (12.254),

$$\{u_i\}_{i=k}^{\infty} \subset X_0.$$

Claim (A), (12.242), and (12.265) imply that for each integer $i \geq n_1$,

$$v_i \in T(u_i),\ \rho(v_i, u_{i+1}) \leq 2c\rho(x_i, u_i) + 4\delta \leq 2c\delta_1 + 4\delta \leq 2\delta_1, \tag{12.274}$$

$$\rho(u_i, v_i) \leq \rho(u_i, T(u_i)) + 2\delta + 2(1+c)\delta_1 \leq \rho(u_i, T(u_i)) + 6\delta_1. \tag{12.275}$$

By (12.247), (12.248), (12.255), (12.268), and (12.269), for each $i \in \{0, \ldots, k\} \setminus \{k\}$,

$$\rho(x_{i+1}, u_{i+1}) \leq \rho(x_i, u_i) \leq \rho(x_0, u_0) \leq \rho(x_0, X_0) + 1$$

$$\leq \rho(x_0, \theta_1) + 1 \leq 2M + 1$$

and

$$\rho(x_i, u_i) \le 2M + 1,\ i = 0, \dots, k. \tag{12.276}$$

Claim (A) and Eqs. (12.266) and (12.276) imply that

$$\rho(u_{n+1}, u_{n+2}) \le c\rho(u_n, u_{n+1}) + 8(M + 2). \tag{12.277}$$

It follows from (12.213), (12.265), and (12.276) that

$$\begin{aligned}\rho(u_0, u_1) &\le \rho(u_0, v_0) + \rho(v_0, u_1)\\ &\le \rho(u_0, v_0) + 2c\rho(x_0, u_0) + 4\delta\\ &\le \rho(u_0, T(u_0)) + 2\delta + 2(1 + c)\rho(u_0, x_0) + 2c\rho(x_0, u_0) + 4\delta\\ &\le \rho(u_0, T(u_0)) + 6(2M + 2)\\ &\le 6(2M + 2) + \rho(u_0, \theta) + \rho(\theta, T(\theta)) + H(T(\theta), T(u_0))\\ &\le 6(2M + 2) + 2\rho(u_0, \theta) + \rho(\theta, T(\theta))\\ &\le 10(2M + 2) + \rho(\theta, T(\theta)) < M_0.\end{aligned}$$

We show by induction that for each integer $i \ge 0$,

$$\rho(u_i, u_{i+1}) \le M_0. \tag{12.278}$$

We already showed that for $i = 0$, (12.278) holds. Assume that $i \ge 0$ is an integer and (12.278) is true. By (12.237), (12.277), and (12.278),

$$\rho(u_{i+1}, u_{i+2}) \le cM_0 + 8(M + 2) \le M_0.$$

Thus, our assumption holds for $i + 1$ too. Therefore, we showed by induction that (12.278) holds for each integer $i \ge 0$. Claim (A) and Eqs. (12.242) and (12.266) imply that for each integer $i \ge k$,

$$\rho(u_{i+1}, u_{i+2}) \le c\rho(u_i, u_{i+1}) + 8\delta_1 + 10\delta \le c\rho(u_i, u_{i+1}) + 9\delta_1. \tag{12.279}$$

Assume that $i \ge k$ is an integer and that

$$\rho(u_i, u_{i+1}) \le \epsilon/2.$$

By (12.241) and (12.279),

$$\rho(u_{i+1}, u_{i+2}) \le c\epsilon/2 + 9\delta_1 \le \epsilon/2.$$

Thus, we have shown that the following property holds:

(iii) If $i \ge 0$ is an integer and $\rho(u_i, u_{i+1}) \le \epsilon/2$, then $\rho(u_{i+1}, u_{i+2}) \le \epsilon/2$.

We show that there exists

$$j \in \{k, \dots, k + n_2\}$$

such that

$$\rho(u_j, u_{j+1}) \le \epsilon/2.$$

Assume the contrary. Then for each $i \in \{k, \dots, k + n_2\}$,

$$\rho(u_i, u_{i+1}) > \epsilon/2. \tag{12.280}$$

In view of (12.241), (12.279), and (12.280), for each $i \in \{k, \dots, k + n_2\}$,

$$\begin{aligned} &\rho(u_i, u_{i+1}) - \rho(u_{i+1}, u_{i+2}) \\ &\ge (1 - c)\rho(u_i, u_{i+1}) - 9\delta_1 \\ &\ge (1 - c)\epsilon/2 - 9\delta_1 \ge (1 - c)\epsilon/4. \end{aligned} \tag{12.281}$$

By (12.278) and (12.281),

$$\begin{aligned} M_0 &\ge \rho(u_k, u_{k+1}) \ge \rho(u_k, u_{k+1}) - \rho(u_{k+n_2+2}, x_{n_2+k+1}) \\ &\ge \sum_{i=k}^{k+n_2} (\rho(u_i, u_{i+1}) - \rho(u_{i+1}, u_{i+2})) \ge (n_2 + 1)(1 - c)\epsilon/4 \end{aligned}$$

and

$$n_2 \le \lfloor 4M_0(1 - c)^{-1}\epsilon^{-1} \rfloor - 1.$$

This contradicts (12.244). The contradiction we have reached proves that there exists

$$p \in \{k, \dots, k + n_2\}$$

such that

$$\rho(u_p, u_{p+1}) \le \epsilon/2.$$

Together with Claim (A), property (iii), and (12.245), this implies that

$$\rho(u_i, u_{i+1}) \leq \epsilon/2$$

for each integer $i \geq n_0 \geq n_1 + n_2$. Combined with Claim (A) and (12.241), this implies that for each integer $i \geq n_0$,

$$\rho(x_i, x_{i+1}) \leq \rho(x_i, u_i) + \rho(u_i, u_{i+1}) + \rho(u_{i+1}, x_{i+1}) \leq \epsilon/2 + 2\delta_1 \leq \epsilon$$

and

$$\xi_n \in B(x_{n+1}, \epsilon/2 - 2\delta_1) \subset B(x_n, \epsilon) \cap T(x_n).$$

Theorem 12.19 is proved.

Theorem 12.2 implies the following result.

Proposition 12.20 *Let $\epsilon > 0$. Then for each $x \in X_0$ satisfying*

$$\rho(x, T(x)) \leq 6^{-1}(1 - c)\epsilon,$$

there exists $\bar{x} \in X_0$ such that

$$\bar{x} \in T(\bar{x}) \cap B(x, \epsilon).$$

Theorem 12.21 *Let $\epsilon \in (0, 1)$ and*

$$\delta = 64^{-1}(1 - c)^{-2}\epsilon. \tag{12.282}$$

Then for each $x \in X$ satisfying $\rho(x, T(x)) \leq \delta$, there exists $\bar{x} \in X_0$ such that

$$\bar{x} \in T(\bar{x}) \cap B(x, \epsilon).$$

Proof Assume that $x \in X$ and

$$\rho(x, T(x)) \leq \delta. \tag{12.283}$$

We show that there exists

$$\bar{x} \in T(\bar{x}) \cap B(x, \epsilon).$$

In view of Proposition 12.20, we may assume without loss of generality that

$$x \notin X_0.$$

Let $\gamma > 0$. There exists

$$x_0 \in X_0 \tag{12.284}$$

such that

$$\rho(x, x_0) < \rho(x, X_0) + \gamma. \tag{12.285}$$

By (12.213), the relation

$$T(X_0) \subset X_0$$

and the relations (12.283)–(12.285),

$$\rho(x, X_0) \le \rho(x, T(x_0)) \le \rho(x, T(x)) + H(T(x), T(x_0))$$

$$\le \delta + c\rho(x, x_0) \le \delta + c\rho(x, X_0) + \gamma$$

and

$$(1 - c)\rho(x, X_0) \le \delta + \gamma.$$

Since γ is any positive number, we conclude that

$$\rho(x, X_0) \le \delta(1 - c)^{-1}. \tag{12.286}$$

By (12.286), there exists $x_0 \in X_0$ such that

$$\rho(x, X_0) \le 2\delta(1 - c)^{-1}. \tag{12.287}$$

Equations (12.213), (12.283), and (12.287) imply that

$$\rho(x_0, T(x_0)) \le \rho(x_0, x) + \rho(x, T(x)) + H(T(x), T(x_0))$$

$$\le 2\rho(x_0, x) + \delta \le \delta(4(1 - c)^{-1} + 1).$$

Together with Proposition 12.20, this implies that there exists

$$\bar{x} \in X_0$$

such that

$$\bar{x} \in T(\bar{x}),$$

$$\rho(\bar{x}, x_0) \le 6\delta(4(1 - c)^{-1} + 1)(1 - c)^{-1}. \tag{12.288}$$

By (12.287),

$$\rho(x, \bar{x}) \le \rho(x, x_0) + \rho(x_0, \bar{x})$$

$$\le 2\delta(1-c)^{-1} + 6\delta(4(1-c)^{-1} + 1)(1-c)^{-1}$$

$$\le \delta((1-c)^{-1}(8 + 24(1-c)^{-1})) \le 32(1-c)^{-2}\delta \le \epsilon.$$

Theorem 12.21 is proved.

Theorems 12.19 and 12.21 imply the following result.

Theorem 12.22 *Let $\theta \in X$, $\epsilon \in (0,1)$, $M > 0$,*

$$\epsilon_0 = 64^{-1}\epsilon(1-c)^{-2},$$

$$\delta = 360^{-1}\epsilon_0(1-c)^2,$$

$$B(\theta, M) \cap X_0 \ne \emptyset,$$

$$M_0 > (1-c)^{-1}((2M+2)10 + \rho(\theta, T(\theta))),$$

$$n_0 = \lfloor 4M_0(1-c)^{-1}\epsilon^{-1} \rfloor + \lfloor 2 \cdot 36^2(2M+1)(1-c)^{-2}\epsilon^{-1} \rfloor + 1.$$

Assume that $\{x_i\}_{i=0}^{\infty} \subset X$,

$$\rho(x_0, \theta) \le M$$

and that for each integer $n \ge 0$,

$$B(x_{n+1}, \delta) \cap \{\xi \in T(x_n) : \ \rho(\xi, x_n) \le \delta + \rho(x_n, T(x_n))\} \ne \emptyset.$$

Then for each integer $n \ge n_0$,

$$\rho(x_n, x_{n+1}) < \epsilon_0, \ \ B(x_n, \epsilon_0) \cap T(x_n) \ne \emptyset$$

and there exists $y_n \in X_0$ such that

$$y_n \in T(y_n) \cap B(x_n, \epsilon).$$

Example 12.23 Let $n \ge 2$ be an integer and let $X_i \subset X, i = 1, \dots, m$ be nonempty closed sets,

$$X_0 = \cap_{i=1}^{n} X_i \ne \emptyset,$$

$$X_{n+1} = X_1.$$

$$T : X \to 2^X \setminus \{\emptyset\},$$

$c \in [0, 1)$, for each $i \in \{1, \dots, n\}$

$$T(X_i) \subset X_{i+1}$$

and let for each $x, y \in X_i$,

$$H(T(x), T(y)) \le c\rho(x, y).$$

The map T is called cyclical [109]. Clearly, it belongs to the class of quasi-contractions studied in the paper.

12.11 A Contractive Type Set-Valued Mapping

In this section, we study an iterative process induced by a contractive type set-valued mapping in a complete metric space, show its convergence under the presence of computational errors, and prove the results obtained in [230].

Assume that (X, ρ) is a complete metric space, Assume that $\phi : [0, \infty) \to [0, 1]$ is a decreasing function such that

$$\phi(t) < 1 \text{ for each } t \in (0, \infty),$$

$$\phi(0) = 1,$$

$T : X \to 2^X \setminus \{\emptyset\}$, $T(x)$ is closed for each $x \in X$, and $X_0 \subset X$ is a nonempty compact set such that

$$T(X_0) \subset X_0 \tag{12.289}$$

and that for each $x \in X_0$ and each $y \in X$,

$$H(T(x), T(y)) \le \phi(\rho(x, y))\rho(x, y). \tag{12.290}$$

Clearly, T is a contractive type set-valued mapping [156, 168]. We are interested to solve the problem

$$\text{Find } z \in X \text{ such that } z \in T(z).$$

In practice we can obtain only an approximate solution of this problem $z \in X$ such that $\rho(z, T(z))$ is small. In order to meet this goal, we use the following algorithm:

Initialization: select an arbitrary point $x_0 \in X$ and a small positive constant δ.

Iterative step: given a current iteration point x_n, calculate the next iteration point x_{n+1} such that

$$B(x_{n+1}, \delta) \cap \{\xi \in T(x_n) : \rho(\xi, x_n) \le \delta + \rho(x_n, T(x_n))\} \ne \emptyset.$$

Clearly, at the each iterative step, x_{n+1} is an approximate solution of the problem

$$\rho(\xi, x_n) \to \min, \ \xi \in T(x_n).$$

Since the space X is not compact, a solution of the problem above does not exist in general. Note that we only assume the existence of the compact set X_0 but do not assume that it is given.

We show that if δ is small enough, then our algorithm generates approximate solutions of the fixed point problem.

Proposition 12.24 *There exists $x_T \in X_0$ such that*

$$x_T \in T(x_T).$$

Proof We show that

$$\inf\{\rho(x, T(x)) : x \in X_0\} = 0.$$

Assume the contrary. Then

$$\Delta := \inf\{\rho(x, T(x)) : x \in X_0\} > 0. \tag{12.291}$$

There is a positive number ϵ satisfying

$$\epsilon < \Delta(1 - \phi(\Delta))/2. \tag{12.292}$$

In view of (12.291), there exists

$$x_0 \in X_0 \tag{12.293}$$

such that

$$\rho(x_0, T(x_0)) < \Delta + \epsilon. \tag{12.294}$$

By (12.294), there exists

$$x_1 \in T(x_0) \tag{12.295}$$

such that

$$\rho(x_0, x_1) < \Delta + \epsilon. \tag{12.296}$$

It follows from (12.290), (12.291), (12.293), (12.295), and (12.296) that

$$\Delta \le \rho(x_1, T(x_1)) \le H(T(x_0), T(x_1)) \le \phi(\rho(x_0, x_1))\rho(x_0, x_1)$$
$$\le \phi(\Delta)\rho(x_0, x_1) \le \phi(\Delta)(\Delta + \epsilon)$$

and

$$\Delta(1 - \phi(\Delta)) \le \epsilon.$$

This contradicts (12.292). The contradiction we have reached proves that

$$\Delta = 0.$$

Together with (12.291), this implies that there exists a sequence $\{x_n\}_{n=0}^{\infty} \subset X_0$ such that

$$\lim_{n\to\infty} \rho(x_n, T(x_n)) = 0. \tag{12.297}$$

Since X_0 is compact, it has a convergent subsequence. We may assume without loss of generality that there exists

$$x_T = \lim_{n\to\infty} x_n. \tag{12.298}$$

By (12.290), (12.297), and (12.298), for each integer $n \ge 0$,

$$\rho(x_T, T(x_T)) \le \rho(x_T, x_n) + \rho(x_n, T(x_n)) + H(T(x_n), T(x_T))$$
$$\le 2\rho(x_T, x_n) + \rho(x_n, T(x_n)) \to 0 \text{ as } n \to +\infty.$$

This implies that $x_T \in T(x_T)$ and completes the proof of Proposition 12.24.

Proposition 12.25 *Let $\epsilon > 0$. Then there exists $\delta > 0$ such that for each $x \in X_0$ satisfying*

$$\rho(x, T(x)) \le \delta,$$

there exists $y \in X_0$ such that

$$y \in T(y) \cap B(x, \epsilon).$$

Proof Assume the contrary. Then for each natural number n, there exists

$$x_n \in X_0$$

such that

$$\rho(x_n, T(x_n)) \le n^{-1} \tag{12.299}$$

and

$$\{y \in B(x_n, \epsilon) \cap X_0 : \ y \in T(y)\} = \emptyset. \tag{12.300}$$

Since X_0 is compact, the sequence $\{x_n\}_{n=1}^{\infty}$ has a convergent subsequence. Because instead of the sequence, one can consider its convergent subsequence, we may assume without loss of generality that there exists

$$x_* = \lim_{n\to\infty} x_n. \tag{12.301}$$

In view of (12.290) and (12.299), for each natural number n,

$$\rho(x_*, T(x_*)) \le \rho(x_*, x_n) + \rho(x_n, T(x_n)) + H(T(x_n), T(x_*))$$

$$\le 2\rho(x_*, x_n) + 1/n \to 0 \text{ as } n \to +\infty.$$

This implies that

$$x_* \in T(x_*).$$

By (12.301), for all sufficiently large natural numbers n,

$$\rho(x_n, x_*) < \epsilon.$$

This contradicts (13). The contradiction we have reached completes the proof of Proposition 12.25.

We begin with the following result that shows that the approximate fixed points of T are closed to its fixed points.

Theorem 12.26 *Let $\epsilon > 0$. Then there exists $\delta > 0$ such that for each integer $x \in X$ satisfying*

$$\rho(x, T(x)) < \delta,$$

there exists

$$y \in X_0 \cap B(x, \epsilon) \cap T(y).$$

Proof Proposition 12.25 implies that there exists

$$\epsilon_1 \in (0, \epsilon/4)$$

such that the following property holds:

(i) For each $z \in X_0$ satisfying

$$\rho(z, T(z)) \leq 3\epsilon_1,$$

there exists

$$y \in X_0 \cap B(z, 2^{-1}\epsilon) \cap T(y).$$

Fix a positive number

$$\delta < 2^{-1}\epsilon_1(1 - \phi(\epsilon_1)).$$

Let $x \in X$ satisfy

$$\rho(x, T(x)) < \delta. \tag{12.302}$$

By (12.302), there exists

$$x_0 \in T(x) \tag{12.303}$$

such that

$$\rho(x, x_0) < \delta. \tag{12.304}$$

We show that

$$\rho(x, X_0) < \epsilon_1.$$

Assume the contrary. Then

$$\rho(x, X_0) \geq \epsilon_1. \tag{12.305}$$

There exists

$$x_1 \in X_0 \tag{12.306}$$

such that

$$\rho(x, x_1) \leq \rho(x, X_0) + \delta. \tag{12.307}$$

It follows from (12.290) and (12.303)–(12.307) that

$$\rho(x, X_0) - \delta \leq \rho(x_0, X_0) \leq \rho(x_0, T(x_1))$$

$$\le H(T(x), T(x_1)) < \phi(\rho(x, x_1))\rho(x, x_1)$$

$$\le \phi(\epsilon_1)\rho(x, x_1) \le \phi(\epsilon_1)\rho(x, X_0) + \delta$$

and

$$\rho(x, X_0) \le \phi(\epsilon_1)\rho(x, X_0) + 2\delta.$$

Together with (12.305), this implies that

$$\epsilon_1(1 - \phi(\epsilon_1)) \le \rho(x, X_0)(1 - \phi(\epsilon_1)) \le 2\delta.$$

This contradicts (12.11). The contradiction we have reached proves that

$$\rho(x, X_0) < \epsilon_1$$

and there exists

$$z \in X_0 \tag{12.308}$$

such that

$$\rho(x, z) < \epsilon_1. \tag{12.309}$$

By (12.290), (12.11), (12.302), and (12.309),

$$\rho(z, T(z)) \le \rho(z, x) + \rho(x, T(x)) + H(T(x), T(z))$$

$$\le 2\rho(z, x) + \delta \le 2\epsilon_1 + \delta \le 3\epsilon_1. \tag{12.310}$$

Property (i) and Eqs. (12.308) and (12.310) imply that there exists

$$y \in X_0 \cap T(y) \cap B(z, 2^{-1}\epsilon).$$

In view of the inclusion above and (12.309),

$$\rho(x, y) \le \rho(x, z) + \rho(z, y) \le \epsilon_1 + \epsilon/2 \le \epsilon.$$

Theorem 12.26 is proved.

Fix

$$\theta \in X.$$

Set

$$\mathrm{diam}(X_0) = \sup\{\rho(z_1, z_2) : z_1, z_2 \in X_0\} < \infty.$$

Theorem 12.27 *Let* $\epsilon \in (0, 1)$, $M > 0$,

$$M_0 \geq 8(1 - \phi(1))^{-1}(2M + 3) + \rho(\theta, T(\theta)) + 4M + 1, \tag{12.311}$$

$$B(\theta, M) \cap X_0 \neq \emptyset, \tag{12.312}$$

a positive number δ_1 *satisfy*

$$\delta_1 \leq 72^{-1}(1 - \phi(\epsilon/4))\epsilon, \tag{12.313}$$

$$n_1 = \lfloor 8(2M + 1)\delta_1^{-1}(1 - \phi(4^{-1}\delta_1))^{-1} \rfloor + 2, \tag{12.314}$$

$$n_2 = \lfloor 8M_0\epsilon_1^{-1}(1 - \phi(4^{-1}\epsilon))^{-1} \rfloor + 1, \tag{12.315}$$

$$n_0 = n_1 + n_2 \tag{12.316}$$

and let a positive number δ *satisfy*

$$\delta \leq 16^{-1}(1 - \phi(\delta_1/4))\delta_1. \tag{12.317}$$

Then for each sequence $\{x_i\}_{i=0}^{\infty} \subset X$ *such that*

$$\rho(x_0, \theta) \leq M$$

and for each integer $n \geq 0$,

$$B(x_{n+1}, \delta) \cap \{\xi \in T(x_n) : \rho(\xi, x_n) \leq \delta + \rho(x_n, T(x_n))\} \neq \emptyset \tag{12.318}$$

the relations

$$\rho(x_n, x_{n+1}) < \epsilon, \ B(x_n, \epsilon) \cap T(x_n) \neq \emptyset$$

hold for every integer $n \geq n_0$.

Proof Assume that a sequence $\{x_n\}_{n=0}^{\infty} \subset X$ satisfies

$$\rho(x_0, \theta) \leq M$$

and that for each integer $n \geq 0$, (12.318) holds. In view of (12.312), there exists

$$\theta_1 \in B(\theta, M) \cap X_0. \tag{12.319}$$

By (12.318) and (12.319),

$$\rho(x_0, \theta_1) \leq \rho(x_0, \theta) + \rho(\theta, \theta_1) \leq 2M. \tag{12.320}$$

Equations (12.290) and (12.319) imply that

$$\rho(x_0, T(x_0)) \leq \rho(x_0, \theta_1) + \rho(\theta_1, T(\theta_1)) + H(T(\theta_1), T(x_0))$$

$$\leq 2\rho(x_0, \theta_1) + \mathrm{diam}(X_0) \leq 4M + \mathrm{diam}(X_0). \tag{12.321}$$

It follows from (12.321) and (12.318) that

$$\rho(x_0, x_1) \leq \rho(x_0, T(x_0)) + 2\delta \leq 4M + \mathrm{diam}(X_0) + 1. \tag{12.322}$$

By (12.318), for each integer $n \geq 0$, there exists

$$\xi_n \in B(x_{n+1}, \delta) \cap T(x_n) \tag{12.323}$$

such that

$$\rho(x_n, \xi_n) \leq \delta + \rho(x_n, T(x_n)). \tag{12.324}$$

For each integer $n \geq 0$, there exists

$$u_n \in X_0 \tag{12.325}$$

such that

$$\rho(u_n, x_n) \leq \rho(x_n, X_0) + \delta. \tag{12.326}$$

Let $n \geq 0$ be an integer. By (12.290), (12.323), (12.325), and (12.326),

$$\rho(u_{n+1}, x_{n+1}) \leq \rho(x_{n+1}, X_0) + \delta$$

$$\leq \rho(x_{n+1}, \xi_n) + \rho(\xi_n, X_0) + \delta$$

$$\leq 2\delta + \rho(\xi_n, T(u_n)) \leq 2\delta + H(T(x_n), T(u_n))$$

$$\leq 2\delta + \phi(\rho(x_n, u_n))\rho(x_n, u_n). \tag{12.327}$$

It follows from (12.290), (12.323), and (12.325) that

$$\xi_n \in T(x_n),\ \ H(T(x_n), T(u_n)) \leq \phi(\rho(x_n, u_n))\rho(x_n, u_n). \tag{12.328}$$

In view of (12.328), there exists

$$v_n \in T(u_n) \tag{12.329}$$

such that

$$\rho(v_n, \xi_n) \le \phi(\rho(x_n, u_n))\rho(x_n, u_n) + \delta. \tag{12.330}$$

Equations (12.323), (12.327), and (12.330) imply that

$$\begin{aligned}
\rho(v_n, u_{n+1}) &\le \rho(v_n, \xi_n) + \rho(\xi_n, u_{n+1}) \\
&\le \phi(\rho(x_n, u_n))\rho(x_n, u_n) + \delta + \rho(x_{n+1}, \xi_n) + \rho(x_{n+1}, u_{n+1}) \\
&\le \phi(\rho(x_n, u_n))\rho(x_n, u_n) + 2\delta + 2\delta + \phi(\rho(x_n, u_n))\rho(x_n, u_n) \\
&\le 2\phi(\rho(x_n, u_n))\rho(x_n, u_n) + 4\delta.
\end{aligned} \tag{12.331}$$

By (12.324) and (12.330),

$$\begin{aligned}
\rho(u_n, v_n) &\le \rho(u_n, x_n) + \rho(x_n, \xi_n) + \rho(\xi_n, v_n) \\
&\le \rho(u_n, x_n) + \rho(x_n, T(x_n)) + \delta + \phi(\rho(u_n, x_n))\rho(u_n, x_n) + \delta \\
&= (1 + \phi(\rho(u_n, x_n))\rho(u_n, x_n) + 2\delta + \rho(x_n, T(x_n)).
\end{aligned} \tag{12.332}$$

Equations (12.290) and (12.325) imply that

$$\begin{aligned}
\rho(x_n, T(x_n)) &\le \rho(x_n, u_n) + \rho(u_n, T(u_n)) + H(T(u_n), T(x_n)) \\
&\le \rho(x_n, u_n) + \rho(u_n, T(u_n)) + \phi(\rho(u_n, x_n))\rho(u_n, x_n) \\
&= (1 + \phi(\rho(u_n, x_n)))\rho(u_n, x_n) + \rho(u_n, T(u_n)).
\end{aligned} \tag{12.333}$$

In view of (12.332) and (12.333),

$$\begin{aligned}
\rho(u_n, v_n) &\le (1 + \phi(\rho(u_n, x_n)))\rho(u_n, x_n) + 2\delta + \rho(x_n, T(x_n)) \\
&\le 2(1 + \phi(\rho(u_n, x_n)))\rho(u_n, x_n) + 2\delta + \rho(u_n, T(u_n)).
\end{aligned} \tag{12.334}$$

It follows from (12.329), (12.331), and (12.334) that

$$\begin{aligned}
&v_n \in T(u_n),\ \ \rho(v_n, u_{n+1}) \le 2\phi(\rho(x_n, u_n))\rho(x_n, u_n) + 4\delta, \\
&\rho(u_n, v_n) \le 2(1 + \phi(\rho(u_n, x_n)))\rho(u_n, x_n) + 2\delta + \rho(u_n, T(u_n)).
\end{aligned} \tag{12.335}$$

Let $n \geq 0$ be an integer. Equations (12.290), (12.325), and (12.335) imply that

$$\rho(u_{n+1}, u_{n+2}) \leq \rho(u_{n+1}, v_{n+1}) + \rho(v_{n+1}, u_{n+2})$$

$$\leq \rho(u_{n+1}, v_{n+1}) + 2\phi(\rho(x_{n+1}, u_{n+1}))\rho(x_{n+1}, u_{n+1}) + 4\delta$$

$$\leq \rho(u_{n+1}, T(u_{n+1})) + 2\delta + 2(1 + \phi(\rho(x_{n+1}, u_{n+1})))\rho(x_{n+1}, u_{n+1})$$

$$+ 2\phi(\rho(x_{n+1}, u_{n+1}))\rho(x_{n+1}, u_{n+1}) + 4\delta$$

$$\leq \rho(u_{n+1}, v_n) + \rho(v_n, T(u_{n+1}))$$

$$+ (2 + 4\phi(\rho(x_{n+1}, u_{n+1})))\rho(x_{n+1}, u_{n+1}) + 6\delta$$

$$\leq H(T(u_n), T(u_{n+1})) + 2\phi(\rho(x_n, u_n))\rho(x_n, u_n) + 4\delta$$

$$+ (2 + 4\phi(\rho(x_{n+1}, u_{n+1})))\rho(x_{n+1}, u_{n+1}) + 6\delta$$

$$\leq \phi(\rho(u_n, u_{n+1}))\rho(u_n, u_{n+1}) + 6\rho(x_{n+1}, u_{n+1}) + 10\delta + 2\rho(x_n, u_n). \quad (12.336)$$

By (12.318), (12.319), and (12.326),

$$\rho(u_0, x_0) \leq \rho(x_0, X_0) + 1 \leq \rho(x_0, \theta_1) + 1 \leq 2M + 1. \quad (12.337)$$

We show that there exists an integer $k \in \{0, \ldots, n_1\}$ such that the following property holds:

(a)

$$\rho(x_i, u_i) \leq \delta_1 \text{ for each integer } i \geq k \quad (12.338)$$

and $\rho(x_i, u_i) > \delta_1$ for each nonnegative integer $i < k$.

Clearly, it is sufficient to show that there exists an integer

$$k \in \{0, \ldots, n_1\}$$

such that (12.338) holds.

Assume that $i \geq 0$ is an integer and that

$$\rho(x_i, u_i) > \delta_1/4. \quad (12.339)$$

By (12.317), (12.327), and (12.329),

$$\rho(x_{i+1}, u_{i+1}) \leq 2\delta + \phi(\rho(x_i, u_i))\rho(u_i, u_i) \leq 2\delta + \phi(4^{-1}\delta_1)\rho(x_i, u_i)$$

and

$$\rho(x_i, u_i) - \rho(x_{i+1}, u_{i+1}) \geq (1 - \phi(4^{-1}\delta_1))\rho(x_i, u_i) - 2\delta$$

$$\geq (1 - \phi(4^{-1}\delta_1))\delta_1/4 - 2\delta \geq (1 - \phi(4^{-1}\delta_1))\delta_1/8.$$

Thus, the following property holds:

(b) For each integer $i \geq 0$ satisfying $\rho(x_i, u_i) > \delta_1/4$, we have

$$\rho(x_i, u_i) - \rho(x_{i+1}, u_{i+1}) \geq (1 - \phi(4^{-1}\delta_1))\delta_1/8.$$

Assume that $i \geq 0$ is an integer and

$$\rho(x_i, u_i) \leq \delta_1. \tag{12.340}$$

There are two cases:

$$\rho(x_i, u_i) \leq \delta_1/4; \tag{12.341}$$

$$\rho(x_i, u_i) > \delta_1/4. \tag{12.342}$$

If (12.342) holds, then property (b) and (12.340) imply that

$$\rho(x_{i+1}, u_{i+1}) \leq \rho(x_i, u_i) \leq \delta_1.$$

Assume that (12.341) holds. Then by (12.317) and (12.327),

$$\rho(x_{i+1}, u_{i+1}) \leq \rho(x_i, u_i) + 2\delta \leq 2\delta + \delta_1/4 \leq \delta_1.$$

Thus, in both cases, we obtained

$$\rho(x_{i+1}, u_{i+1}) \leq \delta_1.$$

Therefore, the following property holds:

(c) If an integer $i \geq 0$ and $\rho(x_i, u_i) \leq \delta_1$, then $\rho(x_{i+1}, u_{i+1}) \leq \delta_1$.

Property (c) implies that in order to show that there exists $k \in \{0, \ldots, n_1\}$ such that (a) holds, it is sufficient to show that there exists $k \in \{0, \ldots, n_1\}$ such that

$$\rho(x_i, u_i) \leq \delta_1.$$

Assume the contrary. Then for each $i \in \{0, \ldots, n_1\}$, we have

$$\rho(x_i, u_i) > \delta_1$$

and in view of property (b),

$$\rho(x_i, u_i) - \rho(x_{i+1}, u_{i+1}) \geq (1 - \phi(4^{-1}\delta_1))\delta_1/8.$$

Together with (12.337), this implies that

$$2M + 1 \geq \rho(u_0, x_0) \geq \rho(u_0, x_0) - \rho(u_{n_1}, x_{n_1})$$

$$= \sum_{i=0}^{n_1-1} (\rho(u_i, x_i) - \rho(u_{i+1}, x_{i+1})) \geq 8^{-1}n_1\delta_1(1 - \phi(4^{-1}\delta_1))$$

and

$$n_1 \leq 8(2M + 1)\delta_1^{-1}(1 - \phi(4^{-1}\delta_1))^{-1}.$$

This contradicts (12.314). The contradiction we have reached proves that there exists $i \in \{0, \ldots, n_1\}$ such that

$$\rho(x_i, u_i) \leq \delta_1$$

and therefore there is

$$k \in \{0, \ldots, n_1\}$$

such that property (a) holds.

Properties (a) and (b) and (12.337) imply that for each $i \in \{0, \ldots, k\}$,

$$\rho(x_i, u_i) \leq \rho(x_0, u_0) \leq 2M + 1$$

and that

$$\rho(x_i, u_i) \leq 2M + 1 \tag{12.343}$$

for each integer $i \geq 0$. By (12.336), for each integer $n \geq 0$,

$$\begin{aligned}\rho(u_{n+1}, u_{n+2}) &\leq \phi(\rho(u_n, u_{n+1}))\rho(u_n, u_{n+1}) \\ &\quad + 6\rho(x_{n+1}, u_{n+1}) + 10\delta + 2\rho(x_n, u_n) \\ &\leq \phi(\rho(u_n, u_{n+1}))\rho(u_n, u_{n+1}) + 8(2M + 2).\end{aligned} \tag{12.344}$$

By (12.290), (12.311), (12.317), (12.318), (12.335), and (12.337),

$$\rho(u_0, u_1) \leq \rho(u_0, v_0) + \rho(v_0, u_1)$$

$$\leq \rho(u_0, v_0) + 2\phi(\rho(x_0, u_0))\rho(x_0, u_0) + 4\delta$$

$$\le \rho(u_0, T(u_0)) + 2\delta + 4\rho(x_0, u_0) + 2\rho(x_0, u_0) + 4\delta$$

$$\le \rho(u_0, T(u_0)) + 6\rho(x_0, u_0) + 1$$

$$\le \rho(u_0, T(u_0)) + 6(2M + 1) + 1$$

$$\le \rho(u_0, \theta) + \rho(\theta, T(\theta)) + H(T(\theta), T(u_0)) + 6(2M + 2)$$

$$\le 6(2M + 2) + \rho(\theta, T(\theta))$$

$$\le \rho(\theta, T(\theta)) + 6(2M + 2) + 3M + 1 < M_0. \tag{12.345}$$

We show that

$$\rho(u_k, u_{k+1}) \le M_0. \tag{12.346}$$

If $k = 0$, then in view of (12.345), inequality (12.346) is true. Consider the case

$$k > 0.$$

Assume that $i \ge 0$ is an integer and that

$$\rho(u_i, u_{i+1}) \le M_0. \tag{12.347}$$

There are two cases:

$$\rho(u_i, u_{i+1}) \le 1; \;\; \rho(u_i, u_{i+1}) > 1.$$

If

$$\rho(u_i, u_{i+1}) \le 1,$$

then in view of (12.311), (12.336), and (12.343),

$$\rho(u_{i+1}, u_{i+2}) \le \rho(u_i, u_{i+1}) + 16M + 16 < 16M + 17 < M_0.$$

If

$$\rho(u_i, u_{i+1}) > 1,$$

then by (12.311), (12.336), (12.343), and (12.347),

$$\rho(u_{i+1}, u_{i+2}) \le \phi(\rho(u_i, u_{i+1}))\rho(u_i, u_{i+1}) + 8(2M + 2)$$

$$\le \phi(1)\rho(u_i, u_{i+1}) + 8(2M + 2)$$

$$\leq \phi(1)M_0 + 8(2M+2) \leq M_0.$$

Thus, we have shown that if a nonnegative integer i satisfies

$$\rho(u_i, u_{i+1}) \leq M_0,$$

then

$$\rho(u_{i+1}, u_{i+2}) \leq M_0.$$

Together with (12.345), this implies that $\rho(u_i, u_{i+1}) \leq M_0$ for all integers $i \geq 0$ and, in particular, (12.346) holds.
Assume that $i \geq k$ in an integer and that

$$\rho(u_i, u_{i+1}) \geq \epsilon/4. \tag{12.348}$$

By (12.313), (12.326), (12.336), (12.347), property (a), and the definition of k,

$$\rho(u_{i+1}, u_{i+2}) \leq \phi(\rho(u_i, u_{i+1}))\rho(u_i, u_{i+1}) + 6\rho(x_i, u_i) + 10\delta + 2\rho(x_i, u_i)$$

$$\leq \phi(\epsilon/4)\rho(u_i, u_{i+1}) + 8\delta_1 + 10\delta$$

$$\leq \phi(\epsilon/4)\rho(u_i, u_{i+1}) + 9\delta_1$$

and

$$\rho(u_i, u_{i+1}) - \rho(u_{i+1}, u_{i+2})$$

$$\geq (1 - \phi(\epsilon/4))\rho(u_i, u_{i+1}) - 9\delta_1$$

$$\geq (1 - \phi(\epsilon/4))\epsilon/4 - 9\delta_1 \geq (1 - \phi(\epsilon/4))\epsilon/8.$$

Thus, we showed that the following property holds:

(d) If $i \geq k_1$ is an integer and $\rho(u_i, u_{i+1}) \geq \epsilon/4$, then

$$\rho(u_i, u_{i+1}) - \rho(u_{i+1}, u_{i+2}) \geq (1 - \phi(\epsilon/4))\epsilon/8.$$

We show that there exists $j \in \{k, \ldots, k + n_2\}$ such that

$$\rho(u_j, u_{j+1}) \leq \epsilon/4.$$

Assume the contrary. Then for each $i \in \{k, \ldots, k + n_2\}$,

$$\rho(u_i, u_{i+1}) > \epsilon/4$$

and by property (d),

$$\rho(u_i, u_{i+1}) - \rho(u_{i+1}, u_{i+2}) \geq (1 - \phi(\epsilon/4))\epsilon/8. \tag{12.349}$$

Property (d) and (12.346) and (12.349) imply that

$$M_0 \geq \rho(u_k, u_{k+1}) \geq \rho(u_k, u_{k+1}) - \rho(u_{k+n_2-1}, u_{k+n_2})$$

$$= \sum_{i=k}^{k+n_2-1} (\rho(u_i, u_{i+1}) - \rho(u_{i+1}, u_{i+2})) \geq 8^{-1}\epsilon n_2(1 - \phi(\epsilon/4))$$

and

$$n_2 \leq 8M_0\epsilon^{-1}(1 - \phi(\epsilon/4))^{-1}.$$

This contradicts (12.315). The contradiction we have reached proves that there exists

$$p \in \{k, \dots, k + n_2\} \tag{12.350}$$

such that

$$\rho(u_p, u_{p+1}) \leq \epsilon/4. \tag{12.351}$$

Assume that $i \geq p$ is an integer and

$$\rho(u_i, u_{i+1}) \leq \epsilon/2. \tag{12.352}$$

We show that

$$\rho(u_{i+1}, u_{i+2}) \leq \epsilon/2.$$

There are two cases:

$$\rho(u_i, u_{i+1}) \leq \epsilon/4; \tag{12.353}$$

$$\rho(u_i, u_{i+1}) > \epsilon/4. \tag{12.354}$$

Assume that (12.353) holds. Then in view of (12.316), (12.317), (12.336), (12.350), (12.353), and property (a),

$$\rho(u_{i+1}, u_{i+2}) \leq \rho(u_i, u_{i+1}) + 10\delta + 8\delta_1 \leq \epsilon/4 + 9\delta_1 \leq \epsilon/2.$$

Assume that (12.354) holds. Then by (12.352), (12.354), and property (b),

$$\rho(u_{i+1}, u_{i+2}) \leq \rho(u_i, u_{i+1}) - 8^{-1}\epsilon(1 - \phi(\epsilon/4)) \leq \epsilon/2.$$

Thus, in both cases,

$$\rho(u_{i+1}, u_{i+2}) \leq \epsilon/2.$$

Thus, we have shown that if $i \geq p$ is an integer and (12.352) holds, then the relation above is true. Combined with (12.351), this implies that for each integer $i \geq p$,

$$\rho(u_i, u_{i+1}) \leq \epsilon/2. \tag{12.355}$$

Property (a) and (12.313), (12.350), and (12.355) imply that for each integer $i \geq p$,

$$\rho(x_i, x_{i+1}) \leq \rho(x_i, u_i) + \rho(u_i, u_{i+1}) + \rho(u_{i+1}, x_{i+1})$$

$$\leq \epsilon/2 + 2\delta_1 \leq 3\epsilon/4. \tag{12.356}$$

Let $i \geq p$ be an integer. By (12.313), (12.317), (12.323), and (12.356)

$$\xi_i \in T(x_i) \cap B(x_{i+1}, \delta),$$

$$\rho(x_i, \xi_i) \leq \delta + 3\epsilon/4 \leq \epsilon,$$

$$\xi_i \in B(x_i, \epsilon) \cap T(x_i).$$

This completes the proof of Theorem 12.27.

Theorems 12.26 and 12.27 easily imply the following result.

Theorem 12.28 *Let* $\epsilon \in (0, 1)$, $M > 0$, *and*

$$B(\theta, M) \cap X_0 \neq \emptyset.$$

Then there exist a number $\delta \in (0, \epsilon)$ *and a natural number* n_0 *such that for each sequence* $\{x_i\}_{i=0}^{\infty} \subset X$ *such that*

$$\rho(x_0, \theta) \leq M$$

and for each integer $n \geq 0$,

$$B(x_{n+1}, \delta) \cap \{\xi \in T(x_n) :$$

$$\rho(\xi, x_n) \leq \delta + \rho(x_n, T(x_n))\} \neq \emptyset$$

there exists

$$y_n \in X_0 \cap B(x_n, \epsilon) \cap T(y_n)$$

for every integer $n \geq n_0$.

Example 12.29 Assume that X is a nonempty closed set in a Banach space $(E, \|\cdot\|)$, $\rho(x, y) = \|x - y\|$, $x, y \in X$, and $\theta \in X$ and that for each $x \in X$,

$$tx + (1 - t)\theta \in X, \; t \in [0, 1].$$

Assume that $\phi : [0, \infty) \to [0, 1]$ is a decreasing function such that

$$\phi(0) = 1,$$

$$\phi(t) < 1, \; t \in (0, 1),$$

$H : X \to 2^X \setminus \{\emptyset\}$, $T(x)$ is compact for each $x \in X$,

$$H(T(x), T(y)) \leq \phi(\rho(x, y))\rho(x, y), \; x, y \in X$$

and that there exists a nonempty bounded set $K \subset X$ such that

$$T(K) \subset K.$$

Then there exists a unique compact set $X_T \subset X$ [171] such that

$$T(A_T) = A_T.$$

12.12 Uniformly Locally Strict Contractions

Let (X, ρ) be a complete metric space. Assume that $T : X \to 2^X \setminus \{\emptyset\}$, $T(x)$ is closed for each $x \in X$, $c \in [0, 1)$, $\Delta \in (0, 1]$ and that for each $x, y \in X$ satisfying $\rho(x, y) \leq \Delta$,

$$H(T(x), T(y)) \leq c\rho(x, y). \tag{12.357}$$

We are interested to solve the problem

Find $z \in X$ such that $z \in T(z)$.

In practice we can obtain only an approximate solution of this problem $z \in X$ such that $\rho(z, T(z))$ is small.

The following result provides some simple iterative process that generates approximate fixed points of T.

Theorem 12.30 *Assume that* $x_0 \in X$,

$$\{\epsilon_i\}_{i=1}^{\infty} \subset (0, (1-c)\Delta], \ \sum_{i=0}^{\infty} \epsilon_i < \infty, \tag{12.358}$$

$q \geq 1$ *is an integer,* $x_i \in X$, $i = 1, \dots, q$,

$$\rho(x_i, x_{i+1}) \leq \Delta, \ i = 0, \dots, q-1 \tag{12.359}$$

and

$$x_q \in T(x_0,). \tag{12.360}$$

Assume that $\{x_i\}_{i=q+1}^{\infty} \subset X$ *and that for each integer* $n \geq 1$ *and each* $i \in \{0, \dots, q\}$,

$$x_{nq+i} \in T(x_{(n-1)q+i}) \tag{12.361}$$

and for each $j \in \{0, \dots, q-1\}$,

$$\rho(x_{nq+j}, x_{nq+j+1}) \leq \rho(x_{nq+j}, T(x_{(n-1)q+j+1})) + \epsilon_n. \tag{12.362}$$

Then for each integer $j \geq 0$,

$$\rho(x_j, x_{j+1}) \leq \Delta,$$

for each integer $n \geq 1$ *and each* $j \in \{0, \dots, q-1\}$,

$$\rho(x_{nq+j}, x_{inq+j+1}) \leq c\rho(x_{(n-1)q+j}, x_{(n-1)q+j+1}) + \epsilon_n \tag{12.363}$$

and the sequence $\{x_i\}_{i=0}^{\infty}$ *converges to a fixed point of* T.

Proof Clearly, the sequence $\{x_i\}_{i=0}^{\infty}$ is well defined by (12.361) and (12.362). Assume that $n \geq 1$ is an integer and

$$\rho(x_i, x_{i+1}) \leq \Delta, \ i = 0, \dots, nq-1. \tag{12.364}$$

(In view of (12.359), relation (12.363) holds for $n = 1$.) By (12.357), (12.361), (12.362), and (12.364), for each $j \in \{0, \dots, q-1\}$,

$$\rho(x_{nq+j}, x_{nq+j+1}) \leq \rho(x_{nq+j}, T(x_{(n-1)q+j+1})) + \epsilon_n$$

$$\leq H(T(x_{(n-1)q+j}), T(x_{(n-1)q+j+1})) + \epsilon_n$$

$$\leq c\rho(x_{(n-1)q+j}, x_{(n-1)q+j+1}) + \epsilon_n. \tag{12.365}$$

In view of (12.358), (12.364), and (12.365),

$$\rho(x_{nq+j}, x_{nq+j+1}) \leq c\Delta + \epsilon \leq c\Delta + (1-c)\Delta \leq \Delta.$$

Thus, for each $j \in \{0, \ldots, q-1\}$,

$$\rho(x_{nq+j}, x_{nq+j+1}) \leq c\rho(x_{(n-1)q+j}, x_{(n-1)q+j+1}) + \epsilon_n,$$

$$\rho(x_{nq+j}, x_{nq+j+1}) \leq \Delta.$$

By induction this implies that for each integer $n \geq 1$ and each $j \in \{0, \ldots, q-1\}$, (12.363) holds and that

$$\rho(x_i, x_{i+1}) \leq \Delta, \; i = 0, 1, \ldots.$$

Let $n \geq 1$ be an integer. By (12.363),

$$\sum_{i=0}^{q-1} \rho(x_{nq+i}, x_{nq+i+1}) \leq c \sum_{i=0}^{q-1} \rho(x_{(n-1)q+i}, x_{(n-1)q+i+1}) + q\epsilon_n. \tag{12.366}$$

We show by induction that for each integer $n \geq 1$,

$$\sum_{i=0}^{q-1} \rho(x_{nq+i}, x_{nq+i+1}) \leq c^n \sum_{i=0}^{q-1} \rho(x_i, x_{i+1}) + q \sum_{i=0}^{n-1} c^i \epsilon_{n-i}. \tag{12.367}$$

In view of (12.366), inequality (12.367) is valid for $n = 1$.

Assume that $k \geq 1$ is an integer and that (12.367) holds for $n = k$. When combined with (12.366), this implies that

$$\sum_{i=0}^{q-1} \rho(x_{(k+1)q+i}, x_{(k+1)q+i+1}) \leq c \sum_{i=0}^{q-1} \rho(x_{kq+i}, x_{kq+i+1}) + q\epsilon_{k+1}$$

$$\leq c^{k+1} \sum_{i=0}^{q-1} \rho(x_i, x_{i+1}) + q \sum_{i=0}^{k-1} c^{i+1} \epsilon_{k-i} + \epsilon_{k+1} q$$

$$\leq c^{k+1} \sum_{i=0}^{q-1} \rho(x_i, x_{i+1}) + q \sum_{i=0}^{k} c^i \epsilon_{k+1-i}$$

and (12.367) holds for $k+1$ too. Therefore (12.367) holds for each integer $n \geq 1$ and this implies that

$$\sum_{n=1}^{\infty}(\sum_{i=0}^{q-1}\rho(x_{nq+i}, x_{nq+i+1})) \leq \sum_{n=1}^{\infty}(c^n\sum_{i=0}^{q-1}\rho(x_i, x_{i+1}) + q\sum_{i=0}^{n-1}c^i\epsilon_{n-i})$$

$$\leq (\sum_{n=0}^{\infty}c^n)\sum_{i=0}^{q-1}\rho(x_i, x_{i+1}) + q\sum_{n=0}^{\infty}\epsilon_n(\sum_{n=0}^{\infty}c^n) < \infty.$$

Thus, $\{x_n\}_{n=0}^{\infty}$ is indeed a Cauchy sequence and there exists

$$x_* = \lim_{n\to\infty} x_n. \tag{12.368}$$

We claim that $x_* \in T(x_*)$. By (12.357), (12.361), and (12.368), for each sufficiently large natural number i,

$$\rho(x_i, x_*) \leq \Delta, \ H(T(x_i), T(x_*)) \leq \rho(x_i, x_*)$$

and

$$\rho(x_*, T(x_*)) \leq \rho(x_*, x_{(i+1)q}) + H(T(x_{iq}), T(x_*))$$

$$\leq \rho(x_*, x_{(i+1)q}) + \rho(x_{iq}, x_*) \to 0$$

as $i \to \infty$. Since $T(x_*)$ is closed, we conclude that

$$x_* \in T(x_*),$$

as claimed. Theorem 12.30 is proved.

The next result shows the well-posedness of the fixed point problem.

Theorem 12.31 *Let*

$$\epsilon \in (0, \Delta], \tag{12.369}$$

be given and let

$$\delta \in (0, 4^{-1}(1-c)\epsilon]. \tag{12.370}$$

Then if $x \in X$ and $\rho(x, T(x)) < \delta$, then there is $\bar{x} \in X$ such that $\bar{x} \in T(\bar{x})$ and $\rho(x, \bar{x}) \leq \epsilon$.

Proof Consider

$$x \in X \text{ such that } \rho(x, T(x)) < \delta. \tag{12.371}$$

Set

$$x_0 = x. \tag{12.372}$$

By (12.371) and (12.372), there is

$$x_1 \in T(x_0) \tag{12.373}$$

such that

$$\rho(x_0, x_1) < \delta. \tag{12.374}$$

For each integer $n \geq 1$, choose

$$x_{n+1} \in T(x_n) \tag{12.375}$$

such that

$$\rho(x_{n+1}, x_n) \leq \rho(x_n, T(x_n))(1 + c)/(2c). \tag{12.376}$$

Assume that $n \geq 1$ is an integer and that

$$\rho(x_n, x_{n-1}) \leq \Delta. \tag{12.377}$$

(In view of (12.374), our assumption holds for $n = 1$.) By (12.357), (12.373), and (12.375)–(12.377),

$$\rho(x_n, x_{n+1}) \leq (1 + c)(2c)^{-1}\rho(x_n, T(x_n))$$

$$\leq (1 + c)H(T(x_{n-1}), T(x_n))/(2c) \leq ((1 + c)/2)\rho(x_n, x_{n-1})$$

and

$$\rho(x_n, x_{n+1}) \leq \Delta.$$

Thus, by induction we showed that for each integer $n \geq 1$,

$$\rho(x_n, x_{n-1}) \leq \Delta,$$

$$\rho(x_n, x_{n+1}) \leq ((1 + c)/2)\rho(x_n, x_{n-1}).$$

When combined with (12.374), this implies that for each integer $n \geq 1$,

$$\rho(x_n, x_{n+1}) \leq [(1+c)/2]^n \rho(x_0, x_1) \leq [(1+c)/2]^n \delta. \quad (12.378)$$

Therefore, in view of (12.378),

$$\sum_{n=0}^{\infty} \rho(x_n, x_{n+1}) < \infty,$$

$\{x_n\}_{n=0}^{\infty}$ is a Cauchy sequence and there exists $\bar{x} \in X$ such that

$$\bar{x} = \lim_{n\to\infty} x_n. \quad (12.379)$$

By (12.379), for each integer $n \geq 0$,

$$\rho(\bar{x}, T(\bar{x})) \leq \rho(\bar{x}, x_{n+1}) + H(T(x_n), T(\bar{x}))$$

$$\leq \rho(\bar{x}, x_{n+1}) + \rho(x_n, \bar{x}) \to 0$$

as $n \to \infty$ and

$$\bar{x} \in T(\bar{x}).$$

By (12.370), (12.378), and (12.379),

$$\rho(x_0, \bar{x}) = \lim_{n\to\infty} \rho(x_0, x_n) \leq \sum_{n=0}^{\infty} \rho(x_n, x_{n+1})$$

$$\leq \sum_{i=0}^{\infty} [(1+c)/2]^i \delta = 2\delta/(1-c) < \epsilon/2.$$

This completes the proof of Theorem 12.31.

Theorem 12.32 *Let*

$$\epsilon \in (0, \Delta], \quad (12.380)$$

$q \geq 1$ *be an integer, a positive number* δ *satisfy*

$$\delta \leq 2^{-1}\epsilon(1-c)q^{-1}, \quad (12.381)$$

$x_i \in X$, $i = 0, 1, \ldots, q$,

$$\rho(x_i, x_{i+1}) \le \Delta,\ i = 0, \dots, q-1, \tag{12.382}$$

$$x_q \in T(x_0) \tag{12.383}$$

and a natural number n_0 satisfy

$$c^{n_0} < (2q)^{-1}\epsilon. \tag{12.384}$$

Assume that $\{x_i\}_{i=q+1}^{\infty} \subset X$ and that for each integer $n \ge 1$ and each $i \in \{0, \dots, q\}$,

$$x_{nq+i} \in T(x_{(n-1)q+i}) \tag{12.385}$$

and for each $j \in \{0, \dots, q-1\}$,

$$\rho(x_{nq+j}, x_{nq+j+1}) \le \rho(x_{(n-1)q+j}, T(x_{(n-1)q+j+1})) + \delta. \tag{12.386}$$

Then for each integer $i \ge 0$,

$$\rho(x_i, x_{i+1}) \le \Delta$$

and for each integer $n \ge n_0$,

$$\rho(x_{nq}, x_{(n+1)q}) \le \sum_{i=0}^{q-1} \rho(x_{nq+i}, x_{nq+i+1}) < \epsilon.$$

Proof Clearly, the sequence $\{x_i\}_{i=0}^{\infty}$ is well defined. Assume that $n \ge 1$ is an integer and

$$\rho(x_i, x_{i+1}) \le \Delta,\ i = 0, \dots, nq-1. \tag{12.387}$$

(In view of (12.382), relation (12.387) holds for $n = 1$.) By (12.357) and (12.385)–(12.387), for each $j \in \{0, \dots, q-1\}$,

$$\begin{aligned}\rho(x_{nq+j}, x_{nq+j+1}) &\le \rho(x_{nq+j}, T(x_{(n-1)q+j+1})) + \delta \\ &\le H(T(x_{(n-1)q+j}), T(x_{(n-1)q+j+1})) + \delta \\ &\le c\rho(x_{(n-1)q+j}, x_{(n-1)q+j+1}) + \delta. \end{aligned} \tag{12.388}$$

In view of (12.369), (12.370), (12.387), and (12.388),

$$\rho(x_{nq+j}, x_{nq+j+1}) \le c\Delta + \delta < \Delta.$$

Thus, we showed by induction that for each integer $i \ge 0$,

$$\rho(x_i, x_{i+1}) \le \Delta \tag{12.389}$$

and that for each integer $n \ge 1$ and each $j \in \{0, \dots, q-1\}$,

$$\rho(x_{nq+j}, x_{nq+j+1}) \le c\rho(x_{(n-1)q+j}, x_{(n-1)q+j+1}) + \delta. \tag{12.390}$$

By (12.390), for each integer $n \ge 1$,

$$\sum_{i=0}^{q-1} \rho(x_{nq+i}, x_{nq+i+1}) \le c \sum_{i=0}^{q-1} \rho(x_{(n-1)q+i}, x_{(n-1)q+i+1}) + q\delta. \tag{12.391}$$

We show by induction that for each integer $n \ge 1$,

$$\sum_{i=0}^{q-1} \rho(x_{nq+i}, x_{nq+i+1}) \le c^n \sum_{i=0}^{q-1} \rho(x_i, x_{i+1}) + q\delta \sum_{i=0}^{n-1} c^i. \tag{12.392}$$

In view of (12.391), inequality (12.392) is valid for $n = 1$.

Assume that $k \ge 1$ is an integer and that (12.392) holds for $n = k$. Equations (12.391) and (12.392) imply that

$$\sum_{i=0}^{q-1} \rho(x_{(k+1)q+i}, x_{(k+1)q+i+1}) \le c \sum_{i=0}^{q-1} \rho(x_{kq+i}, x_{kq+i+1}) + q\delta$$

$$\le c^{k+1} \sum_{i=0}^{q-1} \rho(x_i, x_{i+1}) + q\delta \sum_{i=0}^{k-1} c^{i+1} + \delta q$$

$$= c^{k+1} \sum_{i=0}^{q-1} \rho(x_i, x_{i+1}) + q\delta \sum_{i=0}^{k} c^i.$$

Thus, (12.392) holds for $n = k + 1$ too. Therefore, we showed by induction that (12.392) holds for each integer $n \ge 1$. By (12.382) and (12.392), for each natural number n,

$$\sum_{i=0}^{q-1} \rho(x_{nq+i}, x_{nq+i+1}) \le q\delta \sum_{i=0}^{n-1} c^i + c^n q\Delta < q\delta(1-c)^{-1} + c^n q\Delta. \tag{12.393}$$

It follows from (12.381), (12.384), and (12.393) that for each integer $n \ge n_0$,

$$\rho(x_{nq}, x_{(n+1)q}) \le \sum_{i=0}^{q-1} \rho(x_{nq+i}, x_{nq+i+1}) < q\delta(1-c)^{-1} + c^{n_0} q\Delta < \epsilon.$$

Theorem 12.32 is proved.

Theorems 12.31 and 12.32 imply the following additional result.

Theorem 12.33 *Let*

$$\epsilon \in (0, \Delta],$$

$q \geq 1$ *be an integer, a positive number* δ *satisfy*

$$\delta \leq 8^{-1}\epsilon(1-c)^2 q^{-1},$$

and a natural number n_0 *satisfy*

$$c^{n_0} < (8q)^{-1}(1-c)\epsilon,$$

$x_i \in X$, $i = 0, 1, \ldots, q$,

$$\rho(x_i, x_{i+1}) \leq \Delta, \; i = 0, \ldots, q-1$$

and

$$x_q \in T(x_0).$$

Assume that $\{x_i\}_{i=q+1}^{\infty} \subset X$ *and that for each integer* $n \geq 1$ *and each* $i \in \{0, \ldots, q\}$,

$$x_{nq+i} \in T(x_{(n-1)q+i})$$

and for each $j \in \{0, \ldots, q-1\}$,

$$\rho(x_{nq+j}, x_{nq+j+1}) \leq \rho(x_{(n-1)q+j}, T(x_{(n-1)q+j+1})) + \delta.$$

Then for each integer $i \geq 0$,

$$\rho(x_i, x_{i+1}) \leq \Delta$$

and for each integer $n \geq n_0$, *there is*

$$y \in B(x_{nq}, \epsilon) \cap T(y).$$

12.13 Extension of Theorems 12.30 and 12.32

In this section, we prove extensions of Theorems 12.30 and 12.32 in which computational errors are taken into account.

Let (X, ρ) be a complete metric space. We use all the notations, definitions, and assumptions introduced and assumed in Sect. 12.12.

The following result is an extension of Theorem 12.30.

Theorem 12.34 *Assume that $q \geq 1$ is an integer, $x_i \in X$, $i = 0, 1, \dots, q$*

$$\rho(x_i, x_{i+1}) \leq \Delta,\ i = 0, \dots, q-1, \tag{12.394}$$

$$x_q \in T(x_0,), \tag{12.395}$$

$$\epsilon_i \in (0, 4^{-1}(1-c)\Delta],\ i = 1, 2, \dots, \tag{12.396}$$

$$\sum_{i=1}^{\infty} \epsilon_i < \infty. \tag{12.397}$$

Let $\{x_i\}_{i=q+1}^{\infty} \subset X$, for each integer $n \geq 1$ and each $i \in \{0, \dots, q\}$,

$$B(x_{nq+i}, \epsilon) \cap T(x_{(n-1)q+i}) \neq \emptyset \tag{12.398}$$

and for each $j \in \{0, \dots, q-1\}$,

$$B(x_{nq+j+1}, \epsilon_n) \cap \{\xi \in T(x_{(n-1)q+j+1}) :$$

$$\rho(x_{nq+j}, \xi) \leq \rho(x_{nq+j}, T(x_{(n-1)q+j+1})) + \epsilon_n\} \neq \emptyset. \tag{12.399}$$

Set $\epsilon_0 = 0$,

$$\widehat{\epsilon}_n = \max\{\epsilon_n,\ \epsilon_{n+1}\},\ n = 0, 1, \dots.$$

Then for each integer $i \geq 0$,

$$\rho(x_i, x_{i+1}) \leq \Delta$$

and for each integer $n \geq 1$ and each $j \in \{0, \dots, q-1\}$,

$$\rho(x_{nq+j}, x_{inq+j+1}) \leq c\rho(x_{(n-1)q+j}, x_{(n-1)q+j+1}) + 3\widehat{\epsilon}_n$$

and the sequence $\{x_i\}_{i=0}^{\infty}$ converges to a fixed point of T.

Proof In view of (12.399), for each integer $n \geq 1$ and each $j \in \{0, \dots, q-1\}$, there exists

$$\xi_{nq+j+1} \in T(x_{(n-1)q+j+1}) \tag{12.400}$$

such that

$$\rho(x_{nq+j+1}, \xi_{nq+j+1}) \leq \epsilon_n, \tag{12.401}$$

$$\rho(x_{nq+j}, \xi_{nq+j+1}) \leq \rho(x_{nq+j}, T(x_{(n-1)q+j+1})) + \epsilon_n. \tag{12.402}$$

Assume that $n \geq 1$ is an integer and

$$\rho(x_i, x_{i+1}) \leq \Delta, \; i = 0, \ldots, nq - 1. \tag{12.403}$$

(In view of (12.394), relation (12.403) holds for $n = 1$.) By (12.401) and (12.402), for each $j \in \{0, \ldots, q-1\}$,

$$\begin{aligned} \rho(x_{nq+j}, x_{nq+j+1}) &\leq \rho(x_{nq+j}, \xi_{nq+j+1}) + \rho(\xi_{nq+j+1}, x_{nq+j+1}) \\ &\leq \rho(x_{nq+j}, T(x_{(n-1)q+j+1})) + 2\epsilon_n. \end{aligned} \tag{12.404}$$

Let $j \in \{0, \ldots, q-1\}$. There are two cases: $j \geq 1$; $j = 0$. Assume that $j \geq 1$. By (12.357), (12.400), (12.401), and (12.403),

$$\begin{aligned} \rho(x_{nq+j}, T(x_{(n-1)q+j+1})) &\leq \rho(x_{nq+j}, \xi_{nq+j}) + \rho(\xi_{nq+j}, T(x_{(n-1)q+j+1})) \\ &\leq H(T(x_{(n-1)q+j}), T(x_{(n-1)q+j+1})) + \epsilon_n \\ &\leq c\rho(x_{(n-1)q+j}, x_{(n-1)q+j+1}) + \epsilon_n. \end{aligned} \tag{12.405}$$

In view of (12.404) and (12.405),

$$\rho(x_{nq+j}, x_{nq+j+1}) \leq 3\epsilon_n + c\rho(x_{(n-1)q+j}, x_{(n-1)q+j+1}).$$

Assume that $j = 0$. There are two cases: $n \geq 2$; $n = 1$. Assume that $n \geq 2$. By (12.357), (12.400), (12.401), and (12.403),

$$\rho(x_{(n-1)q+q}, \xi_{(n-1)q+q}) \leq \epsilon_{n-1},$$

$$\xi_{(n-1)q+q} \in T(x_{(n-2)q+q}),$$

$$\begin{aligned} \rho(x_{nq}, T(x_{(n-1)q+1})) &\leq \rho(x_{nq}, \xi_{nq}) + \rho(\xi_{nq}, T(x_{(n-1)q+1})) \\ &\leq H(T(x_{(n-1)q}), T(x_{(n-1)q+1})) + \epsilon_{n-1} \\ &\leq c\rho(x_{(n-1)q}, x_{(n-1)q+1}) + \epsilon_{n-1}. \end{aligned}$$

Together with (12.404), this implies that

$$\rho(x_{nq}, x_{nq+1}) \le c\rho(x_{(n-1)q}, x_{(n-1)q+1}) + \epsilon_{n-1} + 2\epsilon_n.$$

Now assume that $j = 0$ and $n = 1$. By (12.357), (12.394), (12.395), and (12.404),

$$\rho(x_q, T(x_1)) \le H(T(x_0), T(x_1)) \le c\rho(x_0, x_1)$$

and

$$\rho(x_q, x_{q+1}) \le 2\epsilon_1 + \rho(x_q, T(x_1)) \le c\rho(x_0, x_1) + 2\epsilon_1.$$

Therefore, in all these cases for each $j \in \{0, \dots, q-1\}$,

$$\rho(x_{nq+j}, x_{nq+j+1}) \le 3\widehat{\epsilon}_n + c\rho(x_{(n-1)q+j}, x_{(n-1)q+j+1}). \tag{12.406}$$

In view of (12.396), (12.399), (12.403), and (12.406), $j \in \{0, \dots, q-1\}$,

$$\rho(x_{nq+j}, x_{nq+j+1}) \le c\Delta + (1-\epsilon)\Delta \le \Delta.$$

Hence, our assumption holds for $n+1$ too, and by induction we showed that it holds for each integer $n \ge 1$ and that for each integer $n \ge 1$ and each $j \in \{0, \dots, q-1\}$ (12.406) is true. By (12.406), for each integer $n \ge 1$,

$$\sum_{i=0}^{q-1} \rho(x_{nq+i}, x_{nq+i+1}) \le c \sum_{i=0}^{q-1} \rho(x_{(n-1)q+i}, x_{(n-1)q+i+1}) + 3q\widehat{\epsilon}_n. \tag{12.407}$$

We show by induction that for each integer $n \ge 1$,

$$\sum_{i=0}^{q-1} \rho(x_{nq+i}, x_{nq+i+1}) \le c^n \sum_{i=0}^{q-1} \rho(x_i, x_{i+1}) + 3q \sum_{i=0}^{n-1} c^i \widehat{\epsilon}_{n-i}. \tag{12.408}$$

In view of (12.407), inequality (12.408) is valid for $n = 1$.

Assume that $k \ge 1$ is an integer and that (12.408) holds for $n = k$. When combined with (12.407), this implies that

$$\sum_{i=0}^{q-1} \rho(x_{(k+1)q+i}, x_{(k+1)q+i+1}) \le c \sum_{i=0}^{q-1} \rho(x_{kq+i}, x_{kq+i+1}) + 3q\widehat{\epsilon}_{k+1}$$

$$\le c^{k+1} \sum_{i=0}^{q-1} \rho(x_i, x_{i+1}) + 3q \sum_{i=0}^{k-1} c^{i+1} \widehat{\epsilon}_{k-i} + 3q\widehat{\epsilon}_{k+1}$$

$$\leq c^{k+1} \sum_{i=0}^{q-1} \rho(x_i, x_{i+1}) + 3q \sum_{i=0}^{k} c^i \widehat{\epsilon}_{k+1-i}$$

and (12.408) holds for $k + 1$ too. Therefore (12.408) holds for each integer $n \geq 1$. By (12.408),

$$\sum_{n=1}^{\infty} (\sum_{i=0}^{q-1} \rho(x_{nq+i}, x_{nq+i+1})) \leq \sum_{n=1}^{\infty} (c^n \sum_{i=0}^{q-1} \rho(x_i, x_{i+1}) + 3q \sum_{i=0}^{n-1} c^i \widehat{\epsilon}_{n-i})$$

$$\leq (\sum_{n=0}^{\infty} c^n) \sum_{i=0}^{q-1} \rho(x_i, x_{i+1}) + 3q \sum_{n=0}^{\infty} \widehat{\epsilon}_n (\sum_{n=0}^{\infty} c^n) < \infty.$$

Thus, $\{x_n\}_{n=0}^{\infty}$ is indeed a Cauchy sequence and there exists

$$x_* = \lim_{n \to \infty} x_n. \tag{12.409}$$

By (12.357), (12.397), (12.400), (12.401), and (12.409), for each sufficiently large natural number n,

$$\rho(x_n, x_*) \leq \Delta, \;\; H(T(x_n), T(x_*)) \leq \rho(x_n, x_*)$$

and

$$\rho(x_*, T(x_*)) \leq \rho(x_*, x_{(n+1)q}) + \rho(x_{(n+1)q}, \xi_{(n+1)q}) + H(T(x_{nq}), T(x_*))$$

$$\leq \rho(x_*, x_{(n+1)q}) + \rho(x_{nq}, x_*) + \epsilon_{n-1} \to 0$$

as $n \to \infty$ and

$$x_* \in T(x_*).$$

Theorem 12.34 is proved.

The following result is an extension of Theorem 12.32.

Theorem 12.35 *Let*

$$\epsilon \in (0, \Delta], \tag{12.410}$$

$q \geq 1$ is an integer, $x_i \in X$, $i = 0, 1, \ldots, q$,

$$x_q \in T(x_0), \tag{12.411}$$

$$\rho(x_i, x_{i+1}) \le \Delta,\ i = 0, \dots, q-1, \tag{12.412}$$

a positive number δ *satisfy*

$$\delta \le 6^{-1}\epsilon(1-c)q^{-1}, \tag{12.413}$$

and a natural number n_0 *satisfy*

$$c^{n_0} < (2q)^{-1}\epsilon. \tag{12.414}$$

Assume that $\{x_i\}_{i=q+1}^{\infty} \subset X$, *for each integer* $n \ge 1$ *and each* $i \in \{0, \dots, q\}$,

$$B(x_{nq+i}, \delta) \cap T(x_{(n-1)q+i}) \ne \emptyset \tag{12.415}$$

and that for each $j \in \{0, \dots, q-1\}$,

$$B(x_{nq+j+1}, \delta) \cap \{\xi \in T(x_{(n-1)q+j+1}) :$$

$$\rho(x_{nq+j}, \xi) \le \rho(x_{nq+j}, T(x_{(n-1)q+j+1})) + \delta\} \ne \emptyset. \tag{12.416}$$

Then for each integer $i \ge 0$,

$$\rho(x_i, x_{i+1}) \le \Delta$$

and for each integer $n \ge n_0$,

$$\rho(x_{nq}, x_{(n+1)q}) \le \sum_{i=0}^{q-1} \rho(x_{nq+i}, x_{nq+i+1}) < \epsilon.$$

Proof In view of (12.416), for each integer $n \ge 1$ and each $j \in \{0, \dots, q-1\}$, there exists

$$\xi_{nq+j+1} \in T(x_{(n-1)q+j+1}) \tag{12.417}$$

such that

$$\rho(x_{nq+j+1}, \xi_{nq+j+1}) \le \delta, \tag{12.418}$$

$$\rho(x_{nq+j}, \xi_{nq+j+1}) \le \rho(x_{nq+j}, T(x_{(n-1)q+j+1})) + \delta. \tag{12.419}$$

In view of (12.418) and (12.419), for each $j \in \{0, \dots, q-1\}$,

$$\rho(x_{q+j}, x_{q+j+1}) \le \rho(x_{q+j}, \xi_{q+j+1}) + \rho(\xi_{q+j+1}, x_{q+j+1})$$

$$\leq \delta + \rho(x_{q+j}, T(x_{j+1})) + 2\delta. \tag{12.420}$$

If $j = 0$, then by (12.357), (12.411), (12.412), and (12.420),

$$\rho(x_{q+j}, x_{q+j+1}) = \rho(x_q, x_{q+1}) \leq 2\delta + H(T(x_0), T(x_1))$$

$$\leq 2\delta + c\rho(x_0, x_1) = 2\delta + c\rho(x_j, x_{j+1}).$$

If $j \in \{0, \ldots, q-1\} \setminus \{0\}$, then in view of (12.357), (12.412), (12.417), and (12.418),

$$\rho(x_{q+j}, T(x_{j+1})) \leq \rho(x_{q+j}, \xi_{q+j}) + \rho(\xi_{q+j}, T(x_{q+j}))$$

$$\leq \delta + H(T(x_j), T(x_{j+1})) \leq \delta + \rho(x_j, x_{j+1})c.$$

Thus, in both cases for each $j \in \{0, \ldots, q-1\}$,

$$\rho(x_{q+j}, x_{q+j+1}) \leq 3\delta + c\rho(x_j, x_{j+1}). \tag{12.421}$$

By (12.412), (12.413), and (12.421), for each $j \in \{0, \ldots, q-1\}$,

$$\rho(x_{q+j}, x_{q+j+1}) \leq 3\delta + c\Delta. \tag{12.422}$$

Assume that $n \geq 2$ is an integer and

$$\rho(x_i, x_{i+1}) \leq \Delta, \ i = 0, \ldots, nq-1. \tag{12.423}$$

(In view of (12.412) and (12.422), our assumption holds for $n = 2$.) By (12.418) and (12.419), for each $j \in \{0, \ldots, q-1\}$,

$$\rho(x_{nq+j}, x_{nq+j+1}) \leq \rho(x_{nq+j}, \xi_{nq+j+1}) + \rho(\xi_{nq+j+1}, x_{nq+j+1})$$

$$\leq 2\delta + \rho(x_{nq+j}, T(x_{(n-1)q+j+1})). \tag{12.424}$$

Assume that $j = 0$. By (12.357), (12.417), (12.418), and (12.423),

$$\rho(x_{nq}, T(x_{(n-1)q+1})) \leq \rho(x_{nq}, \xi_{nq}) + \rho(\xi_{(n-1)q+q}, T(x_{(n-1)q+1})),$$

$$\leq \delta + H(T(x_{(n-1)q}), T(x_{(n-1)q+1}))$$

$$\leq c\rho(x_{(n-1)q}, x_{(n-1)q+1}) + \delta.$$

Together with (12.424), this implies for $j = 0$ that

$$\rho(x_{nq+j}, x_{nq+j+1}) \leq 3\delta + c\rho(x_{(n-1)q+j}, x_{(n-1)q+j+1}).$$

Assume that $q - 1 \geq j \geq 1$. By (12.357), (12.417), (12.418), and (12.423),

$$\rho(x_{nq+j}, T(x_{(n-1)q+j+1})) \leq \rho(x_{nq+j}, \xi_{nq+j}) + \rho(\xi_{nq+j}, T(x_{(n-1)q+j+1}))$$

$$\leq H(T(x_{(n-1)q+j}), T(x_{(n-1)q+j+1})) + \delta$$

$$\leq c\rho(x_{(n-1)q+j}, x_{(n-1)q+j+1}) + \delta.$$

Together with (12.424), this implies that

$$\rho(x_{nq+j}, x_{nq+j+1}) \leq 3\delta + c\rho(x_{(n-1)q+j}, x_{(n-1)q+j+1}). \tag{12.425}$$

Therefore, in all the cases for each $j \in \{0, \dots, q-1\}$, (12.425) holds. In view of (12.410), (12.417), (12.423), and (12.425), for each $j \in \{0, \dots, q-1\}$,

$$\rho(x_{nq+j}, x_{nq+j+1}) \leq c\Delta + 3\delta \leq \Delta.$$

Hence, by induction we showed that for each integer $i \geq 0$,

$$\rho(x_i, x_{i+1}) \leq \Delta$$

and that for each integer $n \geq 1$ and each $j \in \{0, \dots, q-1\}$,

$$\rho(x_{nq+j}, x_{nq+j+1}) \leq 3\delta + c\rho(x_{(n-1)q+j}, x_{(n-1)q+j+1}).$$

This implies that for each integer $n \geq 1$,

$$\sum_{i=0}^{q-1} \rho(x_{nq+i}, x_{nq+i+1}) \leq c \sum_{i=0}^{q-1} \rho(x_{(n-1)q+i}, x_{(n-1)q+i+1}) + 3q\delta. \tag{12.426}$$

We show by induction that for each integer $n \geq 1$,

$$\sum_{i=0}^{q-1} \rho(x_{nq+i}, x_{nq+i+1}) \leq c^n \sum_{i=0}^{q-1} \rho(x_i, x_{i+1}) + 3q\delta \sum_{i=0}^{n-1} c^i. \tag{12.427}$$

In view of (12.426), inequality (12.427) is valid for $n = 1$.

Assume that $k \geq 1$ is an integer and that (12.427) holds for $n = k$. Equations (12.426) and (12.427) imply that

$$\sum_{i=0}^{q-1} \rho(x_{(k+1)q+i}, x_{(k+1)q+i+1}) \leq c \sum_{i=0}^{q-1} \rho(x_{kq+i}, x_{kq+i+1}) + 3q\delta$$

$$\leq c^{k+1}\sum_{i=0}^{q-1}\rho(x_i, x_{i+1}) + 3q\delta\sum_{i=0}^{k-1}c^{i+1} + 3\delta q$$

$$= c^{k+1}\sum_{i=0}^{q-1}\rho(x_i, x_{i+1}) + 3q\delta\sum_{i=0}^{k}c^{i}.$$

Thus, (12.427) holds for $n = k + 1$ too. Therefore, we showed by induction that (12.427) holds for each integer $n \geq 1$. By (12.422) and (12.427), for each natural number n,

$$\sum_{i=0}^{q-1}\rho(x_{nq+i}, x_{nq+i+1}) \leq 3q\delta\sum_{i=0}^{n-1}c^i + c^n q\Delta < 3q\delta(1-c)^{-1} + c^n\Delta. \quad (12.428)$$

It follows from (12.413), (12.4122), and (12.428) that for each integer $n \geq n_0$,

$$\rho(x_{nq}, x_{(n+1)q}) \leq \sum_{i=0}^{q-1}\rho(x_{nq+i}, x_{nq+i+1}) \leq 3q\delta(1-c)^{-1} + c^{n_0}\Delta < \epsilon.$$

Theorem 12.35 is proved.

Theorems 12.31 and Theorem 12.35 imply the following result.

Theorem 12.36 *Let*

$$\epsilon \in (0, \Delta],$$

$q \geq 1$ is an integer, $x_i \in X$, $i = 0, 1, \ldots, q$,

$$x_q \in T(x_0),$$

$$\rho(x_i, x_{i+1}) \leq \Delta,\ i = 0, \ldots, q-1$$

a positive number δ satisfy

$$\delta \leq 24^{-1}\epsilon(1-c)^2 q^{-1},$$

and a natural number n_0 satisfy

$$c^{n_0} < (8q)^{-1}\epsilon(1-c).$$

Assume that $\{x_i\}_{i=q+1}^{\infty} \subset X$, for each integer $n \geq 1$ and each $i \in \{0, \ldots, q\}$,

$$B(x_{nq+i}, \delta) \cap T(x_{(n-1)q+i}) \neq \emptyset$$

and that for each $j \in \{0, \dots, q-1\}$,

$$B(x_{nq+j+1}, \delta) \cap \{\xi \in T(x_{(n-1)q+j+1}) :$$

$$\rho(x_{nq+j}, \xi) \le \rho(x_{nq+j}, T(x_{(n-1)q+j})) + \delta\} \ne \emptyset.$$

Then for each integer $i \ge 0$,

$$\rho(x_i, x_{i+1}) \le \Delta$$

and for each integer $n \ge n_0$, *there exist*

$$y \in B(x_{nq}, \epsilon) \cap T(y).$$

12.14 The First Asymptotic Regularity Result

Assume that (X, ρ) is a complete metric space. Assume that $\phi : [0, \infty) \to [0, 1]$ is a decreasing function such that

$$\phi(t) < 1 \text{ for each } t \in (0, \infty), \tag{12.429}$$

$$\phi(0) = 1,$$

$T : X \to 2^X \setminus \{\emptyset\}$ and that for each $x, y \in X$,

$$H(T(x), T(y)) \le \phi(\rho(x, y))\rho(x, y). \tag{12.430}$$

Clearly, T is a contractive type set-valued mapping [156, 168]. We are interested to find approximate solutions of the problem

$$\text{Find } z \in X \text{ such that } z \in T(z)$$

under the presence of computational errors. We show that if they are small enough, then our algorithm generates approximate solutions of the fixed point problem.

Theorem 12.37 *Let* $\epsilon \in (0, 1)$, $M > 0$,

$$\delta = 8^{-1}(1 - \phi(\epsilon/2))\epsilon, \tag{12.431}$$

$$n_0 = \lfloor 4(M+1)\epsilon^{-1}(1 - \phi(2^{-1}\epsilon))^{-1} \rfloor. \tag{12.432}$$

Assume that $\{x_i\}_{i=0}^{\infty} \subset X$,

$$\rho(x_0, T(x_0)) \le M \tag{12.433}$$

and that for each integer $n \ge 0$,

$$B(x_{n+1}, \delta) \cap \{\xi \in T(x_n) : \ \rho(\xi, x_n) \le \delta + \rho(x_n, T(x_n))\} \ne \emptyset. \tag{12.434}$$

Then the relations

$$\rho(x_n, x_{n+1}) < \epsilon, \ B(x_n, \epsilon) \cap T(x_n) \ne \emptyset$$

hold for every integer $n \ge n_0$.

Proof Let $n \ge 0$ be an integer. By (12.434), there exists

$$\xi_n \in B(x_{n+1}, \delta) \cap T(x_n) \tag{12.435}$$

such that

$$\rho(x_n, \xi_n) \le \delta + \rho(x_n, T(x_n)). \tag{12.436}$$

In view of (12.430), (12.435), and (12.436), for each integer $n \ge 0$,

$$\begin{aligned} \rho(x_{n+1}, x_{n+2}) &\le \rho(x_{n+1}, \xi_{n+1}) + \rho(\xi_{n+1}, x_{n+2}) \\ &\le \delta + \rho(x_{n+1}, T(x_{n+1})) \\ &\le \delta + \rho(x_{n+1}, \xi_n) + \rho(\xi_n, T(x_{n+1})) \\ &\le 2\delta + H(T(x_n), T(x_{n+1})) \\ &\le 2\delta + \phi(\rho(x_n, x_{n+1}))\rho(x_n, x_{n+1}). \end{aligned} \tag{12.437}$$

Equations (12.431), (12.433), (12.435), and (12.436) imply that

$$\begin{aligned} &\rho(x_0, x_1) \le \rho(x_0, \xi_0) + \rho(\xi_0, x_1) \\ &\le \delta + \rho(x_0, \xi_0) \le 2\delta + \rho(x_0, T(x_0)) \le M + 1. \end{aligned} \tag{12.438}$$

Assume that an integer $n \ge 0$ and

$$\rho(x_{n+1}, x_n) \ge \epsilon/2. \tag{12.439}$$

By (12.431), (12.437), and (12.439),

$$\rho(x_{n+1}, x_{n+2}) \le 2\delta + \phi(\epsilon/2)\rho(x_{n+1}, x_n),$$

$$\rho(x_n, x_{n+1}) - \rho(x_{n+1}, x_{n+2})$$

$$\geq (1 - \phi(\epsilon/2))\rho(x_{n+1}, x_n) - 2\delta \geq (1 - \phi(\epsilon/2))\epsilon/4. \tag{12.440}$$

Thus, we have shown that the following property holds:

(a) For each integer $n \geq 0$ for which $\rho(x_{n+1}, x_n) \geq \epsilon/2$, we have

$$\rho(x_n, x_{n+1}) - \rho(x_{n+1}, x_{n+2}) \geq (1 - \phi(\epsilon/2))\epsilon/4.$$

We show that there exists $i \in \{0, \ldots, n_0\}$ such that $\rho(x_{n+1}, x_n) \leq \epsilon/2$. Assume the contrary. Then,

$$\rho(x_{n+1}, x_n) > \epsilon/2, \ i = 0, \ldots, n_0. \tag{12.441}$$

Property (a) and (12.441) imply that for each $i \in \{0, \ldots, n_0\}$,

$$\rho(x_i, x_{i+1}) - \rho(x_{i+1}, x_{i+2}) \geq (1 - \phi(\epsilon/2))\epsilon/4. \tag{12.442}$$

By (12.438) and (12.442),

$$M + 1 \geq \rho(x_0, x_1) \geq \rho(x_0, x_1) - \rho(x_{n_0}, x_{n_0+1})$$

$$= \sum_{i=0}^{n_0}(\rho(x_i, x_{i+1}) - \rho(x_{i+1}, x_{i+2})) \geq 4^{-1}(n_0 + 1)\epsilon(1 - \phi(\epsilon/2)),$$

$$n_0 \leq 4(M + 1)\epsilon^{-1}(1 - \phi(\epsilon/2)) - 1.$$

This contradicts (12.432). The contradiction we have reached proves that there exists $k \in \{0, \ldots, n_0\}$ such that

$$\rho(x_{k+1}, x_k) \leq \epsilon/2. \tag{12.443}$$

We show that $\rho(x_i, x_{i+1}) \leq 3\epsilon/4$ for each integer $i \geq n_0$.

Assume the contrary. Then there exists $p > k$ such that

$$\rho(x_p, x_{p+1}) > 3\epsilon/4. \tag{12.444}$$

In view of (12.443), we may assume without loss of generality that

$$\rho(x_i, x_{i+1}) \leq 3\epsilon/4$$

for each integer $i \in \{k, \ldots, p - 1\}$. In particular,

$$\rho(x_{p-1}, x_p) \leq 3\epsilon/4. \tag{12.445}$$

If $\rho(x_{p-1}, x_p) > \epsilon/2$, then by (a) and (12.445),

$$\rho(x_{p+1}, x_p) \le \rho(x_{p-1}, x_p) \le 3\epsilon/4.$$

If $\rho(x_{p-1}, x_p) \le \epsilon/2$, then by (12.437),

$$\rho(x_{p+1}, x_p) \le \rho(x_{p-1}, x_p) + 2\delta \le 3\epsilon/4.$$

Clearly, in the both cases, we have a contradiction that proves that

$$\rho(x_i, x_{i+1}) \le 3\epsilon/4$$

for each integer $i \ge n_0$. Combined with (12.431) and (12.434), this completes the proof of Theorem 12.37.

12.15 The Second Asymptotic Regularity Result

Assume that (X, ρ) is a complete metric space. Assume that $\phi : [0, \infty) \to [0, 1]$ is a decreasing function such that

$$\phi(t) < 1 \text{ for each } t \in (0, \infty),$$

$$\phi(0) = 1,$$

$T : X \to 2^X \setminus \{\emptyset\}$, $X_0 \subset X$ is a nonempty set such that

$$T(X_0) \subset X_0 \tag{12.446}$$

and that for each $x \in X_0$ and each $y \in X$,

$$H(T(x), T(y)) \le \phi(\rho(x, y))\rho(x, y). \tag{12.447}$$

Clearly, T is a contractive type set-valued mapping [156, 168].

Theorem 12.38 *Let* $\epsilon \in (0, 1)$, $M > 0$,

$$M_0 = (16M + 10)(1 - \phi(1))^{-1}, \tag{12.448}$$

a positive number δ_1 *satisfy*

$$\delta_1 = 80^{-1}(1 - \phi(\epsilon/4))\epsilon, \tag{12.449}$$

$$\delta = 16^{-1}(1 - \phi(\delta_1/4))\delta_1, \tag{12.450}$$

$$n_1 = \lfloor 8(M+1)\delta_1^{-1}(1-\phi(4^{-1}\delta_1))^{-1} \rfloor + 1, \tag{12.451}$$

$$n_2 = \lfloor 8(M_0+2)\epsilon_1^{-1}(1-\phi(4^{-1}\epsilon))^{-1} \rfloor, \tag{12.452}$$

$$n_0 = n_1 + n_2 \tag{12.453}$$

and let a sequence $\{x_i\}_{i=0}^{\infty} \subset X$ *satisfy*

$$\rho(x_0, T(x_0)) \le M, \tag{12.454}$$

$$\rho(x_0, X_0) \le M \tag{12.455}$$

and for each integer $n \ge 0$,

$$B(x_{n+1}, \delta) \cap \{\xi \in T(x_n) : \ \rho(\xi, x_n) \le \delta + \rho(x_n, T(x_n))\} \ne \emptyset \tag{12.456}$$

the relations

$$\rho(x_n, x_{n+1}) < \epsilon, \ \ B(x_n, \epsilon) \cap T(x_n) \ne \emptyset \tag{12.457}$$

hold for every integer $n \ge n_0$.

Proof By (12.457), for each integer $n \ge 0$, there exists

$$\xi_n \in B(x_{n+1}, \delta) \cap T(x_n) \tag{12.458}$$

such that

$$\rho(x_n, \xi_n) \le \delta + \rho(x_n, T(x_n)). \tag{12.459}$$

For each integer $n \ge 0$, there exists

$$u_n \in X_0 \tag{12.460}$$

such that

$$\rho(u_n, x_n) \le \rho(x_n, X_0) + \delta. \tag{12.461}$$

By Eqs. (12.450), (12.454), (12.458), and (12.459),

$$\begin{aligned} \rho(x_0, x_1) &\le \rho(x_0, \xi_0) + \rho(\xi_0, x_1) \\ &\le \delta + \rho(x_0, T(x_0)) \le M + 1. \end{aligned} \tag{12.462}$$

In view of (12.455) and (12.461),

$$\rho(u_0, x_0) \le \rho(x_0, X_0) + \delta \le M + 1. \tag{12.463}$$

Let $n \ge 0$ be an integer. By (12.446), (12.447), (12.458), (12.460), and (12.461),

$$\rho(u_{n+1}, x_{n+1}) \le \rho(x_{n+1}, X_0) + \delta$$

$$\le \rho(x_{n+1}, \xi_n) + \rho(\xi_n, X_0) + \delta$$

$$\le 2\delta + \rho(\xi_n, T(u_n)) \le 2\delta + H(T(x_n), T(u_n))$$

$$\le 2\delta + \phi(\rho(x_n, u_n))\rho(x_n, u_n). \tag{12.464}$$

It follows from (12.430), (12.458), and (12.460) that

$$\xi_n \in T(x_n), \;\; H(T(x_n), T(u_n)) \le \phi(\rho(x_n, u_n))\rho(x_n, u_n). \tag{12.465}$$

In view of (12.465), there exists

$$v_n \in T(u_n)$$

such that

$$\rho(v_n, \xi_n) \le \phi(\rho(x_n, u_n))\rho(x_n, u_n) + \delta. \tag{12.466}$$

Equations (12.458), (12.464), and (12.466) imply that

$$\rho(v_n, u_{n+1}) \le \rho(v_n, \xi_n) + \rho(\xi_n, u_{n+1})$$

$$\le \phi(\rho(x_n, u_n))\rho(x_n, u_n) + \delta + \rho(x_{n+1}, \xi_n) + \rho(x_{n+1}, u_{n+1})$$

$$\le \phi(\rho(x_n, u_n))\rho(x_n, u_n) + 2\delta + 2\delta + \phi(\rho(x_n, u_n))\rho(x_n, u_n)$$

$$\le 2\phi(\rho(x_n, u_n))\rho(x_n, u_n) + 4\delta. \tag{12.467}$$

By (12.459) and (12.466),

$$\rho(u_n, v_n) \le \rho(u_n, x_n) + \rho(x_n, \xi_n) + \rho(\xi_n, v_n)$$

$$\le \rho(u_n, x_n) + \rho(x_n, T(x_n)) + \delta + \phi(\rho(u_n, x_n))\rho(u_n, x_n) + \delta$$

$$= (1 + \phi(\rho(u_n, x_n))\rho(u_n, x_n) + 2\delta + \rho(x_n, T(x_n)). \tag{12.468}$$

Equation (12.465) implies that

$$\rho(x_n, T(x_n)) \le \rho(x_n, u_n) + \rho(u_n, T(u_n)) + H(T(u_n), T(x_n))$$

$$\leq \rho(x_n, u_n) + \rho(u_n, T(u_n)) + \phi(\rho(u_n, x_n))\rho(u_n, x_n)$$

$$= (1 + \phi(\rho(u_n, x_n)))\rho(u_n, x_n) + \rho(u_n, T(u_n)). \quad (12.469)$$

In view of (12.468) and (12.469),

$$\rho(u_n, v_n) \leq (1 + \phi(\rho(u_n, x_n)))\rho(u_n, x_n) + 2\delta + \rho(x_n, T(x_n))$$

$$\leq 2(1 + \phi(\rho(u_n, x_n)))\rho(u_n, x_n) + 2\delta + \rho(u_n, T(u_n)). \quad (12.470)$$

It follows from (12.465), (12.470), and (12.467) that

$$v_n \in T(u_n), \ \ \rho(v_n, u_{n+1}) \leq 2\phi(\rho(x_n, u_n))\rho(x_n, u_n) + 4\delta,$$

$$\rho(u_n, v_n) \leq 2(1 + \phi(\rho(u_n, x_n)))\rho(u_n, x_n) + 2\delta + \rho(u_n, T(u_n)). \quad (12.471)$$

We show that there exists an integer $k \in \{0, \dots, n_1\}$ such that the following property holds:

(a)

$$\rho(x_i, u_i) \leq \delta_1 \text{ for each integer } i \geq k \quad (12.472)$$

and $\rho(x_i, u_i) > \delta_1$ for each nonnegative integer $i < k$.
Clearly, it is sufficient to show that there exists an integer

$$k \in \{0, \dots, n_1\}$$

such that (12.472) holds.
Assume that $i \geq 0$ is an integer and that

$$\rho(x_i, u_i) > \delta_1/4. \quad (12.473)$$

By (12.450), (12.464)), and (12.473),

$$\rho(x_{i+1}, u_{i+1}) \leq 2\delta + \phi(\rho(x_i, u_i))\rho(u_i, u_i) \leq 2\delta + \phi(4^{-1}\delta_1)\rho(x_i, u_i)$$

and

$$\rho(x_i, u_i) - \rho(x_{i+1}, u_{i+1}) \geq (1 - \phi(4^{-1}\delta_1))\rho(x_i, u_i) - 2\delta$$

$$\geq (1 - \phi(4^{-1}\delta_1))\delta_1/4 - 2\delta \geq (1 - \phi(4^{-1}\delta_1))\delta_1/8.$$

Thus, the following property holds:

(b) For each integer $i \geq 0$ satisfying $\rho(x_i, u_i) > \delta_1/4$, we have

$$\rho(x_i, u_i) - \rho(x_{i+1}, u_{i+1}) \geq (1 - \phi(4^{-1}\delta_1))\delta_1/8.$$

Assume that $i \geq 0$ is an integer and

$$\rho(x_i, u_i) \leq \delta_1.$$

There are two cases:

$$\rho(x_i, u_i) \leq \delta_1/4; \;\; \rho(x_i, u_i) > \delta_1/4.$$

If $\rho(x_i, u_i) > \delta_1/4$, then property (b) implies that

$$\rho(x_{i+1}, u_{i+1}) \leq \rho(x_i, u_i) \leq \delta_1.$$

If $\rho(x_i, u_i) \leq \delta_1/4$, then (12.450) and (12.458) imply that

$$\rho(x_{i+1}, u_{i+1}) \leq \rho(x_i, u_i) + 2\delta \leq 2\delta + \delta_1/4 \leq \delta_1.$$

Thus, in both cases, we obtained

$$\rho(x_{i+1}, u_{i+1}) \leq \delta_1.$$

Therefore, the following property holds:

if an integer $i \geq 0$ and $\rho(x_i, u_i) \leq \delta_1$, then $\rho(x_{i+1}, u_{i+1}) \leq \delta_1$.

This property implies that in order to show that there exists $k \in \{0, \dots, n_1\}$ such that (a) holds, it is sufficient to show that there exists $k \in \{0, \dots, n_1\}$ such that

$$\rho(x_k, u_k) \leq \delta_1.$$

Assume the contrary. Then for each $i \in \{0, \dots, n_1\}$, we have

$$\rho(x_i, u_i) > \delta_1$$

and in view of property (b),

$$\rho(x_i, u_i) - \rho(x_{i+1}, u_{i+1}) \geq (1 - \phi(4^{-1}\delta_1))\delta_1/8.$$

Together with (12.463), this implies that

$$M + 1 \geq \rho(u_0, x_0) \geq \rho(u_0, x_0) - \rho(u_{n_1}, x_{n_1})$$

$$= \sum_{i=0}^{n_1-1} (\rho(u_i, x_i) - \rho(u_{i+1}, x_{i+1})) \ge 8^{-1} n_1 \delta_1 (1 - \phi(4^{-1}\delta_1))$$

and

$$n_1 \le 8(2M+1)\delta_1^{-1}(1 - \phi(4^{-1}\delta_1))^{-1}.$$

This contradicts (12.451). The contradiction we have reached proves that there exists $i \in \{0, \ldots, n_1\}$ such that

$$\rho(x_i, u_i) \le \delta_1$$

and therefore there is

$$k \in \{0, \ldots, n_1\}$$

such that property (a) holds.

Properties (a) and (b) and (12.463) imply that for each $i \in \{0, \ldots, k\}$,

$$\rho(x_i, u_i) \le \rho(x_0, u_0) \le M + 1$$

and that

$$\rho(x_i, u_i) \le M + 1, \ i = 0, 1, \ldots. \tag{12.474}$$

Let $n \ge 0$ be an integer. By (12.447), (12.460), and (12.471),

$$\rho(u_{n+1}, u_{n+2}) \le \rho(u_{n+1}, v_{n+1}) + \rho(v_{n+1}, u_{n+2})$$

$$\le \rho(u_{n+1}, T(u_{n+1})) + 2\delta + 4\rho(x_{n+1}, u_{n+1}) + 2\rho(x_{n+1}, u_{n+1}) + 4\delta$$

$$\le \rho(u_{n+1}, v_n) + \rho(v_n, T(u_{n+1})) + 6\rho(x_{n+1}, u_{n+1}) + 6\delta$$

$$\le H(T(u_n), T(u_{n+1})) + 6\rho(x_{n+1}, u_{n+1}) + 2\rho(x_n, u_n) + 10\delta$$

$$\le \phi(\rho(u_n, u_{n+1}))\rho(u_n, u_{n+1}) + 6\rho(x_{n+1}, u_{n+1}) + 2\rho(x_n, u_n) + 10\delta. \tag{12.475}$$

In view of (12.450), (12.474), and (12.475),

$$\rho(u_{n+1}, u_{n+2}) \le \phi(\rho(u_n, u_{n+1}))\rho(u_n, u_{n+1}) + 8(2M+1) + 1. \tag{12.476}$$

We show that for each integer $n \ge 0$,

$$\rho(u_n, u_{n+1}) \le M_0. \tag{12.477}$$

It follows from (12.462)–(12.464) that

$$\rho(u_0, u_1) \le \rho(x_0, x_1) + \rho(u_0, x_0) + \rho(u_1, x_1) \le 3M + 4$$

and (12.477) holds for $n = 0$. Assume that $n \ge 0$ is an integer and that (12.477) holds. There are two cases:

$$\rho(u_n, u_{n+1}) \le 1; \ \rho(u_n, u_{n+1}) > 1.$$

If

$$\rho(u_n, u_{n+1}) \le 1,$$

then in view of (12.448) and (12.477),

$$\rho(u_{n+1}, u_{n+2}) \le 8M + 10 < M_0.$$

If

$$\rho(u_n, u_{n+1}) > 1,$$

then by (12.448), (12.476), and (12.477),

$$\rho(u_{n+1}, u_{n+2}) \le \phi(1)M_0 + 8(2M + 1) + 1 < M_0.$$

Thus, we have shown by induction that (12.477) holds for each integer $n \ge 0$. It follows from (12.446), (12.460), (12.465), (12.471), (12.477), and property (a) that for each integer $n \ge k$,

$$v_n, u_n \in X_0, \ \rho(u_n, u_{n+1}) \le M_0, \ v_n \in T(u_n),$$

$$\rho(v_n, u_{n+1}) \le 3\delta_1, \ \rho(u_n, v_n) \le \rho(u_n, T(u_n)) + 5\delta_1. \tag{12.478}$$

By (12.449), (12.477), and (12.478),

$$\rho(u_k, T(u_k)) \le \rho(u_k, v_k) \le \rho(u_k, u_{k+1}) + \rho(u_{k+1}, v_k) \le M_0 + 1.$$

Now we apply Theorem 12.37 with $X = X_0$, $M = M_0 + 1$, $\delta = 5\delta_1$, $x_i = u_{k+i}$, and $i = 0, 1, \ldots$ and obtain that for each integer $n \ge n_1 + n_2$,

$$\rho(u_n, u_{n+1}) < \epsilon/2, \ B(u_n, \epsilon/2) \cap T(u_n) \ne \emptyset. \tag{12.479}$$

Property (a) and (12.449), (12.450), (12.456), and (12.479) imply that for each integer $n \geq n_0$,

$$\rho(x_n, x_{n+1}) \leq \rho(x_n, u_n) + \rho(u_n, u_{n+1}) + \rho(u_{n+1}, x_{n+1}) \leq \epsilon/2 + 2\delta_1 \leq (3/4)\epsilon,$$

$$B(x_{n+1}, \delta) \subset B(x_n, \epsilon), \ \ B(x_n, \epsilon) \cap T(x_n) \neq \emptyset.$$

Theorem 12.38 is proved.

Chapter 13
Nonexpansive Mappings in Spaces with Graphs

In this chapter, we study nonexpansive mappings in spaces with graphs. We consider single-valued operators as well as set-valued maps. Our goals are to show the existence of a fixed point and construct iteration processes associated with a given mapping which generate its approximate fixed point and in some cases converge to a fixed point of the mapping. Our results are proved for various classes of mappings of contractive type. The results of Sects. 13.11–13.14 are new.

13.1 Preliminaries

Let (X, ρ) be a complete metric space and let G be a (directed) graph. Let $V(G)$ be the set of its vertices and let $E(G)$ be the set of its edges. We identify the graph G with the pair $(V(G), E(G))$.

Denote by $\mathcal{M}_{ne}$ the set of all mappings $T : X \to X$ such that for each $x, y \in X$ satisfying $(x, y) \in E(G)$, we have

$$(T(x), T(y)) \in E(G) \text{ and } \rho(T(x), T(y)) \leq \rho(x, y). \tag{13.1}$$

A mapping $T \in \mathcal{M}_{ne}$ is called G-nonexpansive. If $T \in \mathcal{M}_{ne}$, $\alpha \in (0, 1)$, and for each $x, y \in X$ satisfying $(x, y) \in E(G)$, we have

$$\rho(T(x), T(y)) \leq \alpha\rho(x, y), \tag{13.2}$$

then T is called a G-strict contraction.

Our goal is to show that analogs of many results obtained in our previous research for nonexpansive mappings hold for nonexpansive operators in metric spaces with graphs. For these operators, relation (13.1) is true only if $(x, y) \in E(G)$. This makes our study more difficult and less understood than in the case

A. J. Zaslavski, *Approximate Fixed Points of Nonexpansive Mappings*, Developments in Mathematics 80, https://doi.org/10.1007/978-3-031-70710-0_13

of nonexpansive operators. To overcome this difficulty, we introduced the quasi-metric ρ_1 (see (13.3)). It turns out that nonexpansive operators in the metric spaces with graph G with respect to the metric ρ are nonexpansive operators in their usual sense with respect to the metric ρ_1, and this fact plays a crucial role in our study.

Fix $\theta \in X$. For each $x \in X$ and each $r > 0$, set

$$B_\rho(x,r) := \{y \in X : \rho(x,y) \le r\}.$$

We may assume without loss of generality that if $x, y \in X$ satisfies $(x, y) \in E(G)$, then

$$(y, x) \in E(G).$$

We assume that the following assumption holds.

(P) For each $x, y \in X$, there exist an integer $q \ge 1$ and points $x_i \in X$, $i = 0, \dots, q$, such that

$$x_0 = x,\ x_q = y,\ (x_i, x_{i+1}) \in E(G),\ i = 0, \dots, q-1.$$

Thus, $V(G) = X$ and the graph G is connected.

Let $T \in \mathcal{M}_{ne}$ be a G-strict contraction. It is known [92] that under certain mild assumptions, the mapping T has a unique fixed point that attracts all the iterates of T. We begin this chapter with presenting the results of [174] where it was provided a very simple proof of this fact by using a certain metric on X which is defined below. In [174] it was also shown uniform convergence of the iterates of T on bounded subsets of X and that this convergence is stable under small perturbations of these iterates.

For each $x, y \in X$, define

$$\rho_1(x,y) := \inf\{\sum_{i=0}^{q-1} \rho(x_i, x_{i+1}) : q \ge 1 \text{ is an integer},$$

$$x_i \in X,\ i = 0, \dots, q,\ x_0 = x,\ x_q = y,\ (x_i, x_{i+1}) \in E(G),\ i = 0, \dots, q-1\}. \tag{13.3}$$

It is easy to see that for each $x, y, z \in X$, $\rho_1(x, y)$ is finite,

$$\rho_1(x,y) \ge \rho(x,y), \tag{13.4}$$

$$\rho_1(x,y) = \rho_1(y,x),$$

$$\rho_1(x,z) \le \rho_1(x,y) + \rho_1(y,z),$$

and if $\rho_1(x, y) = 0$, then $x = y$. However, ρ_1 is a metric only if $(x, x) \in E(G)$ for all $x \in X$. This pseudometric ρ_1 plays an important role in our study.

For each $x \in X$ and each $r > 0$, we set

$$B_{\rho_1}(x, r) := \{y \in X : \rho_1(x, y) \leq r\}.$$

13.2 Strict Contractions

The results of this section were obtained in [174].

Theorem 13.1 *Let $T \in \mathcal{M}_{ne}$, $\alpha \in (0, 1)$ and assume that for each $x, y \in X$ satisfying $(x, y) \in E(G)$, inequality (13.2) holds. Then for each $x, y \in X$, we have*

$$\rho_1(T(x), T(y)) \leq \alpha\rho_1(x, y).$$

If T is continuous as a self-mapping of (X, ρ), then there exists a unique point $x_* \in X$ satisfying $T(x_*) = x_*$ and for each $x \in X$,

$$\lim_{i\to\infty} T^i(x) = x_* \text{ in } (X, \rho).$$

Proof Assume that $x, y \in X$, that q is an integer, and that

$$x_i \in X,\ i = 0, \ldots, q,\ x_0 = x,\ x_q = y,\ (x_i, x_{i+1}) \in E(G),\ i = 0, \ldots, q-1. \tag{13.5}$$

By (13.2) and (13.5), we have

$$T(x) = T(x_0),\ T(y) = T(x_q),\ (T(x_i), T(x_{i+1})) \in E(G),\ i = 0, \ldots, q-1,$$

and

$$\rho(T(x_i), T(x_{i+1})) \leq \alpha\rho(x_i, y_i),\quad i = 0, \ldots, q-1.$$

When combined with (13.3), these relations imply that

$$\rho_1(T(x), T(y)) \leq \sum_{i=0}^{q-1} \rho(T(x_i), T(x_{i+1})) \leq \alpha \sum_{i=0}^{q-1} \rho(x_i, x_{i+1}).$$

Since the above inequalities hold for each integer $q \geq 1$ and every finite sequence $\{x_i\}_{i=0}^{q}$ satisfying (13.5), we conclude that

$$\rho_1(T(x), T(y)) \leq \alpha\rho_1(x, y). \tag{13.6}$$

Assume now that T is continuous, $x \in X$ and consider the sequence of iterates $\{T^i(x)\}_{i=0}^\infty$. By (13.6), for each integer $i \geq 0$, we have

$$\rho_1(T^{i+1}(x), T^{i+2}(x)) \leq \alpha\rho_1(T^i(x), T^{i+1}(x))$$

and

$$\sum_{i=0}^{\infty} \rho_1(T^i(x), T^{i+1}(x)) < \infty.$$

When combined with (13.4), this implies that

$$\sum_{i=0}^{\infty} \rho(T^i(x), T^{i+1}(x)) < \infty$$

and that $\{T^i(x)\}_{i=0}^\infty$ is a Cauchy sequence in (X, ρ). Thus, there exists

$$x_* = \lim_{i \to \infty} T^i(x) \tag{13.7}$$

in (X, ρ). Since T is assumed to be continuous, it follows that $T(x_*) = x_*$. If a point $y_* \in X$ also satisfies $T(y_*) = y_*$, then in view of (13.6),

$$\rho_1(x_*, y_*) = \rho_1(T(x_*), T(y_*)) \leq \alpha\rho_1(x_*, y_*)$$

and

$$\rho_1(x_*, y_*) = 0.$$

When combined with (13.4), this equality implies that $x_* = y_*$. This completes the proof of Theorem 13.1.

The next result shows the uniform convergence of iterates of the mapping T to its unique fixed point.

Theorem 13.2 *Let all the assumptions of Theorem 13.1 hold, let the mapping T be continuous as a self-mapping of (X, ρ), and let $x_* \in X$ be as guaranteed by Theorem 13.1 and satisfy*

$$x_* = T(x_*).$$

Suppose that the following assumption holds.

(A1) For each $M_0 > 0$, there exists $M_1 > 0$ such that for each point $x \in B_\rho(\theta, M_0)$, we have

$$\rho_1(x, \theta) \le M_1.$$

Then $T^i(x) \to x_*$ as $i \to \infty$ in (X, ρ), uniformly on all bounded subsets of (X, ρ).

Proof Let $\epsilon, M_0 > 0$ be given. By (A1), there exists $M_1 > 0$ such that

$$B_\rho(\theta, M_0) \subset B_{\rho_1}(\theta, M_1). \tag{13.8}$$

We may assume without any loss of generality that

$$x_* \in B_\rho(\theta, M_0). \tag{13.9}$$

Choose a natural number n_0 such that

$$2M_1\alpha^{n_0} < \epsilon. \tag{13.10}$$

Next, assume that

$$x \in B_\rho(\theta, M_0). \tag{13.11}$$

By (13.9) and (13.11),

$$\rho_1(x, \theta) \le M_1, \ \ \rho_1(x_*, \theta) \le M_1,$$

$$\rho_1(x, x_*) \le 2M_1. \tag{13.12}$$

It now follows from Theorem 13.1, (13.2), (13.10), and (13.12) that for all natural numbers $n \ge n_0$, we have

$$\rho(T^n(x), x_*) \le \rho_1(T^n(x), x_*) \le \alpha^n \rho_1(x, x_*) \le 2\alpha^{n_0} M_1 < \epsilon.$$

This completes the proof of Theorem 13.2.

Theorem 13.3 *Let all the assumptions of Theorem 13.2 hold and let $x_* \in X$ be as guaranteed by Theorem 13.1 and satisfy*

$$x_* = T(x_*).$$

Suppose that T is uniformly continuous and bounded on bounded sets as a self-mapping of (X, ρ) and let $\epsilon, M > 0$ be given. Then there exist $\delta > 0$ and a natural number n_0 such that for each integer $n \ge n_0$ and every finite sequence $\{x_i\}_{i=0}^n \subset X$ satisfying

$$x_0 \in B_\rho(\theta, M)$$

and

$$\rho(x_{i+1}, T(x_i)) \le \delta, \ i = 0, \dots, n-1,$$

the inequality $\rho(x_i, x_*) \le \epsilon$ *holds for all* $i = n_0, \dots, n$.

Theorem 13.3 follows from Theorem 13.2 and Theorem 2.65 of [168], which was obtained in [31].

Denote by $\mathcal{M}$ the set of all mappings $S \in \mathcal{M}_{ne}$ which are uniformly continuous and bounded on bounded sets as self-mappings of (X, ρ). We equip the space $\mathcal{M}$ with the uniformity which has the base

$$\mathcal{E}(\epsilon, M) = \{(S_1, S_2) \in \mathcal{M} \times \mathcal{M} : \ \rho(S_1(x), S_2(x)) \le \epsilon \text{ for all } x \in B(\theta, M)\},$$

where $M, \epsilon > 0$. It is not difficult to see that the uniform space $\mathcal{M}$ is metrizable and complete (by a metric d).

Our next theorem follows from Theorem 13.2 and Theorem 2.68 of [168], which was obtained in [31].

Theorem 13.4 *Let all the assumptions of Theorem 13.2 hold and let* $x_* \in X$ *be as guaranteed by Theorem 13.1 and satisfy*

$$x_* = T(x_*).$$

Suppose that $T \in \mathcal{M}$ *and that* $\epsilon, M > 0$ *are given. Then there exist a neighborhood* $\mathcal{U}$ *of* T *in* $\mathcal{M}$ *and a natural number* n_0 *such that for each* $C \in \mathcal{U}$, *each* $x \in B_\rho(\theta, M)$, *and each integer* $n \ge n_0$, *we have*

$$\rho(C^n(x), x_*) \le \epsilon.$$

13.3 Contractive Single-Valued Mappings

Now our goal is to show that the results of the previous sections obtained [174] also hold for mappings on complete metric spaces with graphs which are merely contractive under certain additional assumptions on the graphs. These results were obtained in [179].

Let (X, ρ) be a complete metric space and let G be a (directed) graph. Let $V(G)$ be the set of its vertices and let $E(G)$ be the set of its edges. We identify the graph G with the pair $(V(G), E(G))$. Denote by $\mathcal{M}_{ne}$ the set of all mappings $T : X \to X$ such that for each $x, y \in X$ satisfying $(x, y) \in E(G)$, we have

$$(T(x), T(y)) \in E(G), \quad \rho(T(x), T(y)) \leq \rho(x, y) \tag{13.13}$$

and such that

$$\{(x, T(x)) : \ x \in X\} \text{ is a closed set in the product space } X \times X. \tag{13.14}$$

Recall that a mapping $T \in \mathcal{M}_{ne}$ is called G-nonexpansive. If $T \in \mathcal{M}_{ne}$, $\alpha \in (0, 1)$ and for each $x, y \in X$ satisfying $(x, y) \in E(G)$, we have

$$\rho(T(x), T(y)) \leq \alpha\rho(x, y), \tag{13.15}$$

then T is called a G-strict contraction.

A mapping $T \in \mathcal{M}_{ne}$ is called G-contractive (or G-Rakotch contraction [156]) if there exists a decreasing function $\phi : [0, \infty) \to [0, 1]$ such that

$$\phi(t) < 1, \ t \in [0, \infty), \tag{13.16}$$

and for each $x, y \in X$ satisfying $(x, y) \in E(G)$, we have

$$\rho(T(x), T(y)) \leq \phi(\rho(x, y))\rho(x, y). \tag{13.17}$$

In the sequel, we assume that the infimum over the empty set is ∞, $\infty + \infty = \infty$, and $a + \infty = \infty$ for each $a \in R^1$.

For each point $x \in X$ and each number $r > 0$, set

$$B_\rho(x, r) := \{y \in X : \ \rho(x, y) \leq r\}.$$

Recall that for each $x, y \in X$,

$$\rho_1(x, y) := \inf\{\sum_{i=0}^{q-1} \rho(x_i, x_{i+1}) : \ q \geq 1 \text{ is an integer,}$$

$$x_i \in X, \ i = 0, \dots, q, \ x_0 = x, \ x_q = y, \ (x_i, x_{i+1}) \in E(G), \ i = 0, \dots, q-1\}. \tag{13.18}$$

For each point $x \in X$ and each number $r > 0$, we set

$$B_{\rho_1}(x, r) := \{y \in X : \ \rho_1(x, y) \leq r\}.$$

We assume that there exists a point $\bar{x} \in X$ such that the following assumption holds:

(A1) For each $x \in X$, there exist an integer $q \geq 1$ and points $x_i \in X, i = 0, \dots, q$, such that

$$x_0 = \bar{x},\ x_q = x,\ (x_i, x_{i+1}) \in E(G),\ i = 0, \dots, q-1.$$

(Note that (A1) holds if and only if $\rho_1(\bar{x}, x) < \infty$ for all $x \in X$.) We also assume that there exists a number $\bar{\Delta} > 0$ such that the following assumption holds:

(A2) If $(x_0, x_1), (x_1, x_2) \in E(G)$ satisfy $\rho(x_0, x_1) \le \bar{\Delta},\ \rho(x_1, x_2) \le \bar{\Delta}$, then $(x_0, x_2) \in E(G)$.

It turns out that these assumptions hold for many important graphs; see, for instance, the examples below.

Example 13.5 Assume that $\Delta > 0$, for each $x, y \in X$, $(x, y) \in E(G)$ if and only if $\rho(x, y) \le \Delta$, and that there exists a point $\bar{x} \in X$ such that for each $x \in X$, there exist an integer $q \ge 1$ and points $x_i \in X$, $i = 0, \dots, q$, satisfying $x_0 = \bar{x}$, $x_q = x$ and $\rho(x_i, x_{i+1}) \le \Delta$, $i = 0, \dots, q$. Clearly, (A1) holds and (A2) holds with $\bar{\Delta} = \Delta/2$. Note that in this case, (A1) means that (X, ρ) is Δ-chainable.

Example 13.6 Let X be a closed set in a Banach space $(E, \|\cdot\|)$ ordered by a closed and convex cone E_+ such that $\|x\| \le \|y\|$ for each $x, y \in E_+$ satisfying $x \le y$, $\rho(x, y) = \|x - y\|,\ x, y \in X$, and $(x, y) \in E(G)$ if and only if $x \le y$. Assume that $\bar{x} \in X$ and $\bar{x} \le x$ for each $x \in X$. Clearly, both (A1) and (A2) hold. It is not difficult to see that every partially ordered metric space with a smallest element satisfies both (A1) and (A2).

13.4 Existence and Convergence Results

Assume that $T \in \mathcal{M}_{ne}$. It is easy to see that for each $x, y \in X$,

$$\rho_1(T(x), T(y)) \le \rho_1(x, y). \tag{13.19}$$

Note that if $\alpha \in (0, 1)$ and for each $x, y \in E(G)$ satisfying $(x, y) \in E(G)$ the inequality

$$\rho(T(x), T(y)) \le \alpha \rho(x, y)$$

holds, then for each $x, y \in X$,

$$\rho_1(T(x), T(y)) \le \alpha \rho_1(x, y).$$

Assume that $\phi : [0, \infty) \to [0, 1]$ is a decreasing function satisfying

$$\phi(t) < 1,\ t \in [0, \infty),$$

and that for each $x, y \in X$ satisfying $(x, y) \in E(G)$, the inequality

$$\rho(T(x), T(y)) \le \phi(\rho(x, y))\rho(x, y) \tag{13.20}$$

holds. In other words, T is G-contractive.

The following results were obtained in [179].

Theorem 13.7 *Assume that $\rho_1(T(\bar{x}), \bar{x}) < \infty$. Then the following assertions hold:*

1. *There exists*

$$x_* = \lim_{i \to \infty} T^i(\bar{x})$$

in (X, ρ) and $T(x_*) = x_*$.
2. *For every $M_0 > 0$, $\lim_{n\to\infty} T^n(x) = x_*$ in (X, ρ), uniformly on $B_{\rho_1}(\bar{x}, M_0)$.*
3. *Assume that for each $M_0 > 0$, there exists $M_1 > 0$ such that*

$$B_\rho(\bar{x}, M_0) \subset B_{\rho_1}(\bar{x}, M_1).$$

Then $\lim_{n\to\infty} T^n(x) = x_*$ in (X, ρ), uniformly on all bounded subsets of (X, ρ).

Theorem 13.8 *Assume that $\rho_1(T(\bar{x}), \bar{x}) < \infty$ and that for each $M_0 > 0$, there exists $M_1 > 0$ such that*

$$B_\rho(\bar{x}, M_0) \subset B_{\rho_1}(\bar{x}, M_1)$$

and that T is uniformly continuous and bounded on all bounded sets as a self-mapping of (X, ρ). Let $M, \epsilon > 0$. Then there exist a number $\delta > 0$ and a natural number n_0 such that for each integer $n \ge n_0$ and each sequence $\{x_i\}_{i=0}^n \subset X$ which satisfies

$$x_0 \in B_\rho(\bar{x}, M)$$

and

$$\rho(x_{i+1}, T(x_i)) \le \delta, \ i = 0, 1, \dots, n-1,$$

the inequality $\rho(x_i, x_*) \le \epsilon$ holds for all $i = n_0, \dots, n$.

This result easily follows from Theorem 13.7 and Theorem 2.65 of [168].

Denote by $\mathcal{M}$ the set of all $S \in \mathcal{M}_{ne}$ which are uniformly continuous and bounded on bounded sets as a self-mapping of (X, ρ). Denote by $\mathcal{M}_c$ the set of all continuous mappings $T \in \mathcal{M}_{ne}$. Fix a point

$$\theta \in X.$$

The space $\mathcal{M}_{ne}$ is equipped with the uniformity determined by the base

$$\mathcal{E}(\epsilon, M) = \{(S_1, S_2) \in \mathcal{M}_{ne} \times \mathcal{M}_{ne} : \ \rho(S_1(x), S_2(x)) \le \epsilon$$

$$\text{for all } x \in B(\theta, M)\},$$

where $M, \epsilon > 0$. The uniform space $\mathcal{M}_{ne}$ is metrizable (by a metric d) and complete. It is complete if the set $\{(x, y) \in X \times X : \ (x, y) \in E(G)\}$ is closed in the product topology. It is clear that $\mathcal{M}$ is a closed subset of $\mathcal{M}_{ne}$. We equip it with the relative topology. The next result easily follows from Theorem 13.7 and Theorem 2.68 of [168].

We say that a set $C \subset X$ is bounded in (X, ρ_1) if $\sup\{\rho_1(x, y) : \ x, y \in C\} < \infty$.

Theorem 13.9 *Assume that $\rho_1(T(\bar{x}), \bar{x}) < \infty$ and let x_* be as guaranteed by Theorem 13.7. Assume further that each bounded set in (X, ρ) is also bounded in (X, ρ_1) and let $\epsilon, M > 0$. Then there exist a neighborhood $\mathcal{U}$ of T in $\mathcal{M}$ and a natural number n_0 such that for each $x \in B(\theta, M)$, each integer $n \ge n_0$ and each $\{B_i\}_{i=1}^n \subset \mathcal{U}$, we have*

$$\rho(B_q \cdots, B_1(x), x_*) \le \epsilon, \ q = n_0, \ldots, n.$$

13.5 Auxiliary Results for Theorem 13.7

Lemma 13.10 *Let $0 < \epsilon < \bar{\Delta}$, $x, y \in X$ satisfy*

$$\epsilon \le \rho_1(x, y), \tag{13.21}$$

$q \ge 1$ be an integer and let $\{x_i\}_{i=0}^q \subset X$ satisfy

$$x_0 = x, \ x_q = y, \tag{13.22}$$

$$(x_i, x_{i+1}) \in E(G), \ i = 0, \ldots, q-1, \tag{13.23}$$

$$\rho(x_i, x_{i+1}) < \epsilon, \ i = 0, \ldots, q-1. \tag{13.24}$$

Then there exist an integer $\widehat{q} \in [1, q)$ and $\{\widehat{x}_i\}_{i=0}^{\widehat{q}} \subset X$ such that

$$\widehat{x}_0 = x, \ \widehat{x}_{\widehat{q}} = y,$$

$$(\widehat{x}_i, \widehat{x}_{i+1}) \in E(G), \ i = 0, \ldots, \widehat{q}-1,$$

$$\sum_{i=0}^{\widehat{q}-1} \rho(\widehat{x}_i, \widehat{x}_{i+1}) \le \sum_{i=0}^{q-1} \rho(x_i, x_{i+1}).$$

Proof In view of (13.21) and (13.24), $q \geq 2$,

$$\rho(x_0, x_1),\ \rho(x_1, x_2) < \epsilon.$$

Together with assumption (A2) and (13.22), this implies that

$$(x_0, x_2) \in E(G).$$

Set

$$\widehat{q} = q - 1,\ \widehat{x}_0 = x_0,\ \widehat{x}_i = x_{i+1},\ i = 1, \ldots, q - 1.$$

It is not difficult to see that this completes the proof of Lemma 13.10.

Applying Lemma 13.10 and induction, we also obtain the following auxiliary result:

Lemma 13.11 *Let* $0 < \epsilon < \bar{\Delta}/2$, $x, y \in X$ *satisfy*

$$\epsilon \leq \rho_1(x, y),$$

$q \geq 1$ *be an integer and let* $\{x_i\}_{i=0}^{q} \subset X$ *satisfy*

$$x_0 = x,\ x_q = y,$$

$$(x_i, x_{i+1}) \in E(G),\ i = 0, \ldots, q - 1,$$

$$\rho(x_i, x_{i+1}) < \epsilon/4,\ i = 0, \ldots, q - 1.$$

Then there exist an integer $\widehat{q} \in [1, q)$ *and points* $\{\widehat{x}_i\}_{i=0}^{\widehat{q}} \subset X$ *such that*

$$\widehat{x}_0 = x,\ \widehat{x}_{\widehat{q}} = y,$$

$$(\widehat{x}_i, \widehat{x}_{i+1}) \in E(G),\ i = 0, \ldots, \widehat{q} - 1,$$

$$\sum_{i=0}^{\widehat{q}-1} \rho(\widehat{x}_i, \widehat{x}_{i+1}) \leq \sum_{i=0}^{q-1} \rho(x_i, x_{i+1})$$

and

$$\max\{\rho(\widehat{x}_i, \widehat{x}_{i+1}) :\ i = 0, \ldots, \widehat{q} - 1\} \geq \epsilon/4.$$

Lemma 13.11 implies our next auxiliary result.

Lemma 13.12 *Let* $0 < \epsilon < \bar{\Delta}/2$ *and let* $x, y \in X$ *satisfy*

$$\epsilon \leq \rho_1(x, y).$$

Then,

$$\rho_1(x, y) = \inf\{\sum_{i=0}^{q-1} \rho(x_i, x_{i+1}) : \; q \geq 1 \textit{ is an integer},$$

$$x_i \in X, \; i = 0, \ldots, q, \; x_0 = x, \; x_q = y, \; (x_i, x_{i+1}) \in E(G), \; i = 0, \ldots, q-1,$$

$$\max\{\rho(x_i, x_{i+1}) : \; i = 0, \ldots, q-1\} \geq \epsilon/4\}.$$

Lemma 13.13 *Assume that* $0 < \epsilon < \bar{\Delta}/2$ *and let* $x, y \in X$ *satisfy*

$$\epsilon \leq \rho_1(x, y) < \infty.$$

Then,

$$\rho_1(T(x), T(y)) \leq \rho_1(x, y) - 4^{-1}\epsilon(1 - \phi(\epsilon/4)).$$

Proof Assume that $q \geq 1$ is an integer and $x_i \in X, \; i = 0, \ldots, q$ satisfy

$$x_0 = x, \; x_q = y, \tag{13.25}$$

$$(x_i, x_{i+1}) \in E(G), \; i = 0, \ldots, q-1, \tag{13.26}$$

$$\max\{\rho(x_i, x_{i+1}) : \; i = 0, \ldots, q-1\} \geq \epsilon/4. \tag{13.27}$$

By (13.13), (13.25), and (13.27),

$$\rho_1(T(x), T(y)) \leq \sum_{i=0}^{q-1} \rho(T(x_i), T(x_{i+1})) \tag{13.28}$$

and for $i = 0, \ldots, q-1$,

$$\rho(T(x_i), T(x_{i+1})) \leq \rho(x_i, x_{i+1}). \tag{13.29}$$

In view of (13.27), there exists

$$j \in \{0, \ldots, q-1\} \tag{13.30}$$

such that

$$\rho(x_j, x_{j+1}) \geq \epsilon/4. \tag{13.31}$$

It follows from (13.20), (13.26), and (13.31) that

$$\rho(T(x_j), T(x_{j+1})) \leq \phi(\rho(x_j, x_{j+1}))\rho(x_j, x_{j+1}) \leq \phi(\epsilon/4)\rho(x_j, x_{j+1}). \tag{13.32}$$

Inequalities (13.31) and (13.32) imply that

$$\begin{aligned}\rho(x_j, x_{j+1}) - \rho(T(x_j), T(x_{j+1})) &\geq (1 - \phi(\epsilon/4))\rho(x_j, x_{j+1}) \\ &\geq (1 - \phi(\epsilon/4))\epsilon/4.\end{aligned} \tag{13.33}$$

By (13.28)–(13.30) and (13.33),

$$\rho_1(T(x), T(y)) \leq \sum_{i=0}^{q-1} \rho(x_i, x_{i+1}) - (1 - \phi(\epsilon/4))\epsilon/4. \tag{13.34}$$

In view of (13.23) and (13.34),

$$\begin{aligned}\rho_1(T(x), T(y)) &\leq -(1 - \phi(\epsilon/4))\epsilon/4 \\ &+ \inf\{\sum_{i=0}^{q-1} \rho(x_i, x_{i+1}) : \ q \geq 1 \text{ is an integer, } \{x_i\}_{i=0}^{q} \subset X, \\ &x_0 = x, \ x_q = y, \\ &(x_i, x_{i+1}) \in E(G), \ i = 0, \dots, q-1, \\ &\max\{\rho(x_i, x_{i+1}) : \ i = 0, \dots, q-1\} \geq \epsilon/4\} \\ &= \rho_1(x, y) - (1 - \phi(\epsilon/4))\epsilon/4.\end{aligned}$$

This completes the proof of Lemma 13.13.

13.6 Proof of Theorem 13.7

By (A1), for each $x \in X$,

$$\rho_1(\bar{x}, x) < \infty.$$

When combined with our assumptions and (13.19), this implies that for each integer $i \geq 0$ and each integer $j > 0$, we have

$$\rho_1(\bar{x}, T^i(\bar{x})) < \infty,$$

$$\rho_1(T^j(\bar{x}), T^{j+i}(\bar{x})) \le \rho_1(\bar{x}, T^i(\bar{x})) < \infty, \; \rho_1(T^{i+1}(\bar{x}), T^i(\bar{x})) < \infty. \tag{13.35}$$

Let $y, z \in X$ satisfy $\rho_1(y, z) < \infty$ and

$$\epsilon \in (0, \bar{\Delta}/2). \tag{13.36}$$

We claim that there exists an integer $p \ge 0$ such that

$$\rho_1(T^p(y), T^p(z)) < \epsilon. \tag{13.37}$$

Suppose to the contrary that no such integer exists. Then for each integer $i \ge 0$, we have

$$\rho_1(T^i(y), T^i(z)) \ge \epsilon. \tag{13.38}$$

Lemma 13.13 and (13.38) imply that for each integer $i \ge 0$, we have

$$\rho_1(T^{i+1}(y), T^{i+1}(z)) \le \rho_1(T^i(y), T^i(z)) - 4^{-1}\epsilon(1 - \phi(\epsilon/4))$$

and

$$\lim_{i\to\infty} \rho_1(T^i(y), T^i(z)) = -\infty.$$

The contradiction we have reached proves that there indeed exists an integer $p \ge 0$ such that (13.37) is true. By (13.19) and (13.37),

$$\rho_1(T^i(y), T^i(z)) < \epsilon \text{ for all integers } i \ge p.$$

This implies that the following property holds:

(a) For each $\epsilon \in (0, \bar{\Delta}/2)$, there exists an integer $p \ge 0$ such that

$$\rho_1(T^i(\bar{x}), T^{i+1}(\bar{x})) < \epsilon \text{ and } \rho_1(T^{i+1}(\bar{x}), T^i(\bar{x})) < \epsilon \text{ for all integers } i \ge p.$$

Let

$$\epsilon \in (0, \bar{\Delta}/2).$$

Choose a positive number

$$\delta < 8^{-1}\epsilon(1 - \phi(\epsilon/4)). \tag{13.39}$$

Property (a) implies that there exists an integer $p \geq 0$ such that

$$\rho_1(T^{i+1}(\bar{x}), T^i(\bar{x})) < \delta \text{ and } \rho_1(T^i(\bar{x}), T^{i+1}(\bar{x})) < \delta \text{ for all integers } i \geq p. \tag{13.40}$$

Let $i, j \geq p$ be integers. We claim that

$$\rho_1(T^i(\bar{x}), T^j(\bar{x})) \leq \epsilon. \tag{13.41}$$

Suppose to the contrary that this is not true. Then

$$\rho_1(T^i(\bar{x}), T^j(\bar{x})) > \epsilon. \tag{13.42}$$

Lemma 13.13 and (13.42) imply that

$$\rho_1(T^{i+1}(\bar{x}), T^{j+1}(\bar{x})) \leq \rho_1(T^i(\bar{x}), T^j(\bar{x})) - 4^{-1}\epsilon(1 - \phi(\epsilon/4)). \tag{13.43}$$

It follows from (13.40) and (13.43) that

$$\rho_1(T^i(\bar{x}), T^j(\bar{x}))$$

$$\leq \rho_1(T^i(\bar{x}), T^{i+1}(\bar{x})) + \rho_1(T^{i+1}(\bar{x}), T^{j+1}(\bar{x})) + \rho_1(T^{j+1}(\bar{x}), T^j(\bar{x}))$$

$$\leq 2\delta + \rho_1(T^i(\bar{x}), T^j(\bar{x})) - 4^{-1}\epsilon(1 - \phi(\epsilon/4))$$

and

$$\delta \geq 8^{-1}\epsilon(1 - \phi(\epsilon/4)).$$

This contradicts (13.39). The contradiction we have reached proves that (13.41) holds for all integers $j > i \geq p$. Therefore, $\{T^i(\bar{x})\}_{i=0}^{\infty}$ is a Cauchy sequence in (X, ρ). Therefore, there exists

$$x_* = \lim_{i \to \infty} T^i(\bar{x})$$

in (X, ρ). Since the graph of T is closed (see (13.14)), we obtain $T(x_*) = x_*$. Assertion 1 is proved.

Next, we prove Assertion 2. Assume that $M_0 > 1$ and let

$$\epsilon \in (0, \bar{\Delta}/4)$$

be given. Assertion 1 implies that there exists an integer $n_1 \geq 1$ such that

$$\rho(T^i(\bar{x}), x_*) \leq \epsilon/4 \text{ for all integers } i \geq n_1. \tag{13.44}$$

Choose a positive number

$$\delta < 8^{-1}\epsilon(1 - \phi(\epsilon/4)) \tag{13.45}$$

and an integer

$$n_0 > 8M_0\epsilon^{-1}(1 - \phi(\epsilon/8))^{-1} + n_1. \tag{13.46}$$

Assume that $x \in X$ satisfies

$$\rho_1(\bar{x}, x) \le M_0. \tag{13.47}$$

Assumption (A1) and (13.19) imply that for each integer $i \ge 0$, we have

$$\rho_1(T^{i+1}(\bar{x}), T^{i+1}(x)) \le \rho_1(T^i(\bar{x}), T^i(x)) < \infty. \tag{13.48}$$

We claim that there exists an integer $j \in \{0, \dots, n_0\}$ such that

$$\rho_1(T^j(\bar{x}), T^j(x)) \le \epsilon/2.$$

Suppose to the contrary that no such integer exists. Then for each $i \in \{0, \dots, n_0\}$,

$$\rho_1(T^i(\bar{x}), T^i(x)) > \epsilon/2. \tag{13.49}$$

Lemma 13.3 and (13.49) imply that for each $i \in \{0, \dots, n_0\}$, we have

$$\rho_1(T^{i+1}(\bar{x}), T^{i+1}(x)) \le \rho_1(T^i(\bar{x}), T^i(x)) - 8^{-1}\epsilon(1 - \phi(\epsilon/8)). \tag{13.50}$$

In view of (13.47) and (13.50), we have

$$M_0 \ge \rho_1(\bar{x}, x) \ge \rho_1(\bar{x}, x) - \rho_1(T^{n_0}(\bar{x}), T^{n_0}(x))$$

$$= \sum_{i=0}^{n_0-1} (\rho_1(T^i(\bar{x}), T^i(x)) - \rho_1(T^{i+1}(\bar{x}), T^{i+1}(x))$$

$$\ge 8^{-1}\epsilon n_0(1 - \phi(\epsilon/8))$$

and

$$n_0 \le 8M_0\epsilon^{-1}(1 - \phi(\epsilon/8)).$$

This, however, contradicts (13.46). The contradiction we have reached proves that there exists $j \in \{0, \dots, n_0\}$ such that

$$\rho_1(T^j(\bar{x}), T^j(x)) \le \epsilon/2.$$

It follows from (13.19) and the relation above that

$$\rho_1(T^i(\bar{x}), T^i(x)) \le \epsilon/2 \text{ for every integer } i \ge n_0.$$

It follows from (13.44), (13.46), and the relation above that for each integer $i \ge n_0$,

$$\rho(T^i(x), \bar{x}_*) \le \rho(T^i(x), T^i(\bar{x})) + \rho(T^i(\bar{x}), x_*) \le \epsilon/2 + \epsilon/4.$$

Assertion 2 is proved. Assertion 3 follows from Assertion 2. This completes the proof of Theorem 13.7.

13.7 The First Extension of Theorem 13.7

Assume that (X, d) is a complete metric space equipped with a graph G. We denote by $V(G)$ the set of its vertices and by $E(G)$ be the set of its edges. We assume that $(x, x) \in E(G)$ for any point $x \in X$ and that $\phi : [0, \infty) \to [0, 1]$ is a decreasing function such that

$$\phi(t) < 1, \; t \in [0, \infty) \tag{13.51}$$

and

$$\phi(0) = 1. \tag{13.52}$$

Assume that a mapping $T : X \to X$ satisfies the following assumption:

(A1) For every pair of points $x, y \in X$ satisfying $(x, y) \in E(G)$, the inequality

$$d(T(x), T(y)) \le \phi(d(x, y))d(x, y)$$

holds and

$$(T(x), T(y)) \in E(G).$$

Theorem 13.7 shows that T possesses a unique fixed point under the assumption that there exists a number $\bar{\Delta} > 0$ such that if $(x_0, x_1), (x_1, x_2) \in E(G)$ satisfy $\rho(x_0, x_1) \le \bar{\Delta}$, $\rho(x_1, x_2) \le \bar{\Delta}$, then $(x_0, x_2) \in E(G)$. In this section, we prove the following existence result obtained in [231] without this assumption.

Theorem 13.14 *Assume that $\bar{x} \in X$ and that there exist a natural number q and $M > 1$ such that for each integer $n \ge 1$, there exist $x_i^{(n)} \in X$, $i = 0, \dots, q$ such that*

$$x_0^{(n)} = \bar{x},\ x_q^{(n)} = T^n(\bar{x}) \tag{13.53}$$

and that for each $i \in \{0, \dots, n-1\}$,

$$d(x_i^{(n)}, x_{i+1}^{(n)}) \le M \tag{13.54}$$

and at least one of the following relations hold:

$$(x_i^{(n)}, x_{i+1}^{(n)}) \in E(G),\ (x_{i+1}^{(n)}, x_i^{(n)}) \in E(G). \tag{13.55}$$

Then there exists

$$x_* = \lim_{i\to\infty} T^i(x)$$

and if T is continuous at x_*, then $T(x_*) = x_*$.

Proof Let $\epsilon \in (0, 1)$. Choose a natural number p such that

$$p > q^2 M\epsilon^{-1}(1 - \phi(\epsilon q^{-1}))^{-1}. \tag{13.56}$$

Assume that $n \ge 1$ is an integer. We show that

$$d(T^p(\bar{x}), T^{n+p}(\bar{x})) \le \sum_{i=0}^{q-1} d(T^p(x_i^{(n)}), T^p(x_{i+1}^{(n)})) \le \epsilon. \tag{13.57}$$

Assume the contrary. Then

$$\sum_{i=0}^{q-1} d(T^p(x_i^{(n)}), T^p(x_{i+1}^{(n)})) > \epsilon. \tag{13.58}$$

It follows from (A1), (13.55), and (13.58) that for each integer $l \in \{0, \dots, p\}$,

$$\sum_{i=0}^{q-1} d(T^l(x_i^{(n)}), T^l(x_{i+1}^{(n)})) > \epsilon. \tag{13.59}$$

Let $l \in \{0, \dots, p\}$. Then in view of (A1), for each $i \in \{0, \dots, q-1\}$,

$$d(T^{l+1}(x_i^{(n)}), T^{l+1}(x_{i+1}^{(n)})) \le d(T^l(x_i^{(n)}), T^l(x_{i+1}^{(n)})). \tag{13.60}$$

By (13.59), there exists $j \in \{0, \dots, q-1\}$ such that

$$d(T^l(x_j^{(n)}), T^l(x_{j+1}^{(n)})) > \epsilon/q. \tag{13.61}$$

Assumption (A1) and Eqs. (13.55) and (13.61) imply that

$$d(T^{l+1}(x_j^{(n)}), T^{l+1}(x_{j+1}^{(n)})) \le \phi(d(T^l(x_j^{(n)}), T^l(x_{j+1}^{(n)})))d(T^l(x_j^{(n)}), T^l(x_{j+1}^{(n)}))$$

$$\le \phi(\epsilon q^{-1})d(T^l(x_j^{(n)}), T^l(x_{j+1}^{(n)})).$$

Together with (13.61), this implies that

$$d(T^l(x_j^{(n)}), T^l(x_{j+1}^{(n)})) - d(T^{l+1}(x_j^{(n)}), T^{l+1}(x_{j+1}^{(n)}))$$

$$\ge (1 - \phi(\epsilon/4))d(T^l(x_j^{(n)}), T^l(x_{j+1}^{(n)}))$$

$$\ge (1 - \phi(\epsilon q^{-1}))\epsilon q^{-1}. \tag{13.62}$$

Equations (13.60) and (13.62) imply that

$$\sum_{i=0}^{q-1} d(T^l(x_i^{(n)}), T^l(x_{i+1}^{(n)})) - \sum_{i=0}^{q-1} d(T^{l+1}(x_i^{(n)}), T^{l+1}(x_{i+1}^{(n)}))$$

$$\ge (1 - \phi(\epsilon q^{-1}))\epsilon q^{-1}. \tag{13.63}$$

It follows from (13.54) and (13.63) that

$$qM \ge \sum_{i=0}^{q-1} d(x_i^{(n)}, x_{i+1}^{(n)})$$

$$\ge \sum_{i=0}^{q-1} d(x_i^{(n)}), x_{i+1}^{(n)}) - \sum_{i=0}^{q-1} d(T^{p+1}(x_i^{(n)}), T^{p+1}(x_{i+1}^{(n)}))$$

$$= \sum_{l=0}^{p}(\sum_{i=0}^{q-1} d(T^l(x_i^{(n)}), T^l(x_{i+1}^{(n)})) - \sum_{i=0}^{q-1} d(T^{l+1}(x_i^{(n)}), T^{l+1}(x_{i+1}^{(n)}))$$

$$\ge p(1 - \phi(\epsilon q^{-1}))\epsilon q^{-1}$$

and

$$p \le Mq^2(1 - \phi(\epsilon q^{-1}))^{-1}\epsilon^{-1}.$$

This contradicts (13.56) and proves that (13.57) holds. Since ϵ is any number from $(0, 1)$ $\{T^n(\bar{x})\}_{n=0}^{\infty}$ is a Cauchy sequence. Therefore, there exists

$$x_* = \lim_{n\to\infty} T^n(\bar{x}).$$

Combined with Eq. (13.57), this implies that

$$d(T^p(\bar{x}), x_*) \le \epsilon$$

and

$$d(T^{n+p}(\bar{x}), x_*) \le 2\epsilon$$

for each integer $n \ge 0$. Therefore,

$$x_* = \lim_{i\to\infty} T^i(\bar{x}).$$

Evidently, if T is continuous at x_*, then $T(x_*) = x_*$. Theorem 13.14 is proved.

13.8 The Second Extension of Theorem 3.7

We continue to use the notation and definitions and assume the assumptions introduced in Sect. 13.7. In particular, we assume that $T : X \to X$ satisfies (A1).

We prove the following convergence result obtained in [231] which is also an extension of Theorem 3.7:

Theorem 13.15 *Assume that $x_* \in X$ satisfies*

$$T(x_*) = x_*, \tag{13.64}$$

q is a natural number, and $M > 0$ and that

$$X_{M,q} = \{x \in X : \text{ there exist } x_i \in X,\ i = 0, \dots, q$$

$$\text{such that } x_0 = x,\ x_q = x_* \text{ and for each } i \in \{0, \dots, q-1\},$$

$$d(x_i, x_{i+1}) \le M \text{ and at least one of the following relations holds:}$$

$$(x_i, x_{i+1}) \in E(G);\ (x_{i+1}, x_i) \in E(G)\}. \tag{13.65}$$

Then $d(T^n(x), x_*) \to 0$ as $n \to \infty$ uniformly on $X_{q,M}$.

Proof Let $\epsilon \in (0, 1)$. Choose an integer p such that

$$p > q^2(M+1)\epsilon^{-1}(1 - \phi(\epsilon q^{-1}))^{-1}. \tag{13.66}$$

Assume that

$$x \in X(M, q). \tag{13.67}$$

In view of (13.65) and (13.67), there exist $x_i \in X,\ i = 0, \ldots, q$ such that

$$x_0 = x,\ x_q = x_* \tag{13.68}$$

and for each $i \in \{0, \ldots, q-1\}$,

$$d(x_i, x_{i+1}) \le M \tag{13.69}$$

and at least one of the following relations holds:

$$(x_i, x_{i+1}) \in E(G);\ (x_{i+1}, x_i) \in E(G). \tag{13.70}$$

By (A1) and (13.70), for each integer $i \in \{0, \ldots, q-1\}$ and each integer $n \ge 0$,

$$d(T^{n+1}(x_i), T^{n+1}(x_{i+1})) \le \phi(d(T^n(x_i), T^n(x_{i+1}))d(T^n(x_i), T^n(x_{i+1})) \tag{13.71}$$

and at least one of the following relations holds:

$$(T^n(x_i), T^n(x_{i+1})) \in E(G);\ (T^n(x_{i+1}), T^n(x_i)) \in E(G). \tag{13.72}$$

Assume that an integer $n \ge 0$ and

$$\sum_{i=0}^{q-1} d(T^n(x_i), T^n(x_{i+1})) > \epsilon. \tag{13.73}$$

In view of (13.73), there exists an integer $j \in \{0, \ldots, q-1\}$ such that

$$d(T^n(x_j), T^n(x_{j+1})) > \epsilon/q. \tag{13.74}$$

By (A1), (13.72), and (13.74),

$$d(T^{n+1}(x_j), T^{n+1}(x_{j+1})) \le \phi(d(T^n(x_j), T^n(x_{j+1})))d(T^n(x_j), T^n(x_{j+1}))$$
$$\le \phi(q^{-1}\epsilon)d(T^n(x_j), T^n(x_{j+1}))$$

and

$$d(T^n(x_j), T^n(x_{j+1})) - d(T^{n+1}(x_j), T^{n+1}(x_{j+1}))$$

$$\geq (1 - \phi(q^{-1}\epsilon))d(T^n(x_j), T^n(x_{j+1})) \geq (1 - \phi(q^{-1}\epsilon))\epsilon/q. \tag{13.75}$$

Equations (13.71) and (13.75) imply that

$$\sum_{i=0}^{q-1} d(T^n(x_i), T^n(x_{i+1})) - \sum_{i=0}^{q-1} d(T^{n+1}(x_i), T^{n+1}(x_{i+1}))$$

$$\geq d(T^n(x_j), T^n(x_{j+1})) - d(T^{n+1}(x_j), T^{n+1}(x_{j+1})) \geq (1 - \phi(\epsilon/q))\epsilon/q.$$

Thus, we have shown that the following property holds:

(i) If $n \geq 0$ is an integer and (13.73) holds, then

$$\sum_{i=0}^{q-1} d(T^n(x_i), T^n(x_{i+1})) - \sum_{i=0}^{q-1} d(T^{n+1}(x_i), T^{n+1}(x_{i+1})) \geq (1 - \phi(\epsilon/q))\epsilon/q.$$

We show that there exists an integer $n \in \{0, \dots, p\}$ such that

$$\sum_{i=0}^{q-1} d(T^n(x_i), T^n(x_{i+1})) \leq \epsilon.$$

Assume the contrary. Then for each $n \in \{0, \dots, p\}$,

$$\sum_{i=0}^{q-1} d(T^n(x_i), T^n(x_{i+1})) > \epsilon$$

and in view of property (i),

$$\sum_{i=0}^{q-1} d(T^n(x_i), T^n(x_{i+1})) - \sum_{i=0}^{q-1} d(T^{n+1}(x_i), T^{n+1}(x_{i+1})) \geq (1 - \phi(\epsilon/q))\epsilon/q. \tag{13.76}$$

It follows from (13.69) and (13.76) that

$$qM \geq \sum_{i=0}^{q-1} d(x_i, x_{i+1})$$

$$\geq \sum_{i=0}^{q-1} d(x_i, x_{i+1}) - \sum_{i=0}^{q-1} d(T^p(x_i), T^p(x_{i+1}))$$

$$= \sum_{n=0}^{p} \sum_{i=0}^{q-1} (d(T^n(x_i), T^n(x_{i+1})) - d(T^{n+1}(x_i), T^{n+1}(x_{i+1})))$$

$$\geq p(1 - \phi(\epsilon q^{-1}))\epsilon q^{-1}$$

and

$$p \leq Mq^2(1 - \phi(\epsilon q^{-1}))^{-1}\epsilon^{-1}.$$

This contradicts (13.66). The contradiction we have reached proves that there exists an integer $n \in \{0, \dots, p\}$ such that

$$d(T^n(x), T^n(x_*)) \leq \sum_{i=0}^{q-1} d(T^n(x_i), T^n(x_{i+1})) \leq \epsilon.$$

By (13.68), (13.71), and the relation above, for each integer $s > n$,

$$d(T^s(x), x_*) \leq d(T^s(x), T^s(x_*)) \leq \sum_{i=0}^{q-1} d(T^s(x_i), T^s(x_{i+1}))$$

$$\leq \sum_{i=0}^{q-1} d(T^n(x_i), T^n(x_{i+1})) \leq \epsilon.$$

Theorem 13.15 is proved.

13.9 Set-Valued Mappings

We continue to use the notation and definitions introduced in Sect. 3.7 and prove the following result for set-valued mappings obtained in [231]. It shows that if the mapping has a fixed point, then for a given initial point, there exists a sequence of iterates converging to this fixed point.

Theorem 13.16 *Let $T : X \to 2^X \setminus \{\emptyset\}$, $x_* \in X$ satisfy*

$$x_* \in T(x_*) \tag{13.77}$$

$q \geq 2$ be natural number, $M > 0$, and let

$$X_{M,q} = \{x \in X : \text{ there exist } y_i \in X, \ i = 1, \dots, q \text{ such that}$$

$$y_1 = x_*,\ y_q = x,\ d(y_i, y_{i+1}) \le M \text{ and } (y_i, y_{i+1}) \in E(G),\ i \in \{1, \dots, q-1\}\}. \tag{13.78}$$

Assume that the following assumption holds:

(A2) For each $(x, y) \in E(G)$ and each $z \in T(x)$, the set

$$\{\xi \in T(y) :\ (z, \xi) \in E(G)\}$$

is nonempty and

$$d(z, \{\xi \in T(y) :\ (z, \xi) \in E(G)\}) \le \phi(d(x, y))d(x, y).$$

Then for each $x \in X(M, q)$, there exists a sequence $\{x_i\}_{i=1}^{\infty}$ such that

$$x_1 = x,\ x_{i+1} \in T(x_i),\ i = 1, 2, \dots,\ \lim_{i \to \infty} x_i = x_*$$

and for each $\epsilon > 0$, there exists a natural number $p(\epsilon)$ depending only on ϵ such that $d(x_i, x_*) \le \epsilon$ for each integer $i \ge p(\epsilon)$.

Proof Let

$$x \in X_{M,q}.$$

There exist $y_i \in X,\ i = 1, \dots, q$ such that (13.78) holds. Assume that $s \ge 1$ is an integer and we defined a sequence $\{y_i\}_{i=1}^{sq} \subset X$ such that

$$y_{(s-1)q+1} = x_*, \tag{13.79}$$

$$(y_i, y_{i+1}) \in E(G),\ i = (s-1)q + 1, \dots, sq - 1. \tag{13.80}$$

(Clearly, for $s = 1$ our assumption holds.) By induction, using (13.77), (13.79), and (13.80), we define $y_i \in X, i = sq + 1, \dots, s(q + 1)$ such that

$$y_{sq+1} = x_* \tag{13.81}$$

and that for each $i \in \{(s-1)q + 1, \dots, sq - 1\}$,

$$y_{i+q} \in T(y_i), \tag{13.82}$$

$$(y_{i+q}, y_{i+q+1}) \in E(G) \tag{13.83}$$

and

$$d(y_{i+q}, y_{i+q+1})$$

$$\leq d(y_{i+q}, \{\xi \in T(y_{i+1}) : (y_{i+q}, \xi) \in E(G)\})(1 + \phi(d(y_i, y_{i+1}))^{-1})/2. \tag{13.84}$$

(Note that if $y_i = y_{i+1}$, then $y_{i+q} = y_{i+q+1}$.) Thus, by induction, we defined the sequence $\{y_i\}_{i=0}^{\infty} \subset X$ such that for each integer $s \geq 1$ relations (13.79), (13.80) hold and that for each $i \in \{(s-1)q+1, \ldots, sq-1\}$, Eqs. (13.82)–(13.84) are true. For each natural number s, set

$$x_s = y_{sq}. \tag{13.85}$$

Assumption (A2) and Eqs. (13.80), (13.82), and (13.85) imply and that for each integer $s \geq 1$ and each $i \in \{(s-1)q+1, \ldots, sq-1\}$,

$$x_{s+1} \in T(x_s) \tag{13.86}$$

and

$$d(y_{i+q}, y_{i+q+1}) \leq \phi(d(y_i, y_{i+1}))d(y_i, y_{i+1})(1 + \phi(d(y_i, y_{i+1}))^{-1})/2$$
$$\leq 2^{-1}(1 + \phi(d(y_i, y_{i+1})))d(y_i, y_{i+1}). \tag{13.87}$$

Let s be a natural number. By (13.87),

$$\sum_{i=sq+1}^{(s+1)q-1} d(y_i, y_{i+1}) = \sum_{i=(s-1)q+1}^{sq-1} d(y_{i+q}, y_{i+q+1})$$
$$\leq \sum_{i=(s-1)q+1}^{sq-1} d(y_i, y_{i+1})(1 + \phi(d(y_i, y_{i+1})))/2 \tag{13.88}$$

and

$$\sum_{i=sq+1}^{(s+1)q-1} d(y_i, y_{i+1}) \leq \sum_{i=(s-1)q+1}^{sq-1} d(y_i, y_{i+1}). \tag{13.89}$$

Let $\epsilon \in (0, 1)$ and

$$p(\epsilon) > 2q^2 M\epsilon^{-1}(1 - \phi(\epsilon q^{-1}))^{-1} \tag{13.90}$$

be an integer. Assume that s is a natural number and

$$\sum_{i=(s-1)q+1}^{sq-1} d(y_i, y_{i+1}) > \epsilon. \tag{13.91}$$

By (13.91), there exists $j \in \{(s-1)q+1, \ldots, sq-1\}$ such that

$$d(y_i, y_{i+1}) > \epsilon/q. \tag{13.92}$$

By (13.87) and (13.92),

$$d(y_{j+q}, y_{j+q+1}) \le 2^{-1}(1+\phi(d(y_j, y_{j+1})))d(y_j, y_{j+1})$$
$$\le 2^{-1}(1+\phi(\epsilon/q))d(y_j, y_{j+1})$$

and

$$d(y_j, y_{j+1}) - d(y_{j+q}, y_{j+q+1})$$
$$\ge 2^{-1}d(y_j, y_{j+1})(1-\phi(\epsilon/q)) \ge 2^{-1}\epsilon q^{-1}(1-\phi(\epsilon/q)).$$

Together with (13.87), this implies that

$$\sum_{i=(s-1)q+1}^{sq-1} d(y_i, y_{i+1}) - \sum_{i=sq+1}^{(s+1)q-1} d(y_i, y_{i+1})$$
$$\ge 2^{-1}\epsilon q^{-1}(1-\phi(\epsilon/q)). \tag{13.93}$$

Thus, we have shown that the following property holds:

(i) If s is a natural number, then (13.91) implies (13.93).

We show that there exists $s \in \{0, \ldots, p(\epsilon)\}$ such that

$$\sum_{i=(s-1)q+1}^{sq-1} d(y_i, y_{i+1}) \le \epsilon. \tag{13.94}$$

Assume the contrary. Then for each $s \in \{0, \ldots, p(\epsilon)\}$,

$$\sum_{i=(s-1)q+1}^{sq-1} d(y_i, y_{i+1}) \ge \epsilon. \tag{13.95}$$

Property (i) and (13.95) imply that for each $s \in \{0, \ldots, p(\epsilon)\}$,

$$\sum_{i=(s-1)q+1}^{sq-1} d(y_i, y_{i+1}) - \sum_{i=sq+1}^{(s+1)q-1} d(y_i, y_{i+1})$$

$$\geq 2^{-1}\epsilon q^{-1}(1 - \phi(\epsilon/q)). \tag{13.96}$$

In view of (13.78) and (13.96),

$$Mq \geq \sum_{i=1}^{q-1} d(y_i, y_{i+1}) \geq \sum_{i=1}^{q-1} d(y_i, y_{i+1}) - \sum_{i=p(\epsilon)q+1}^{(p(\epsilon)+1)q-1} d(y_i, y_{i+1})$$

$$= \sum_{s=0}^{p(\epsilon)} (\sum_{i=(s-1)q+1}^{sq-1} d(y_i, y_{i+1}) - \sum_{i=sq+1}^{(s+1)q-1} d(y_i, y_{i+1}))$$

$$\geq 2^{-1}p(\epsilon)\epsilon q^{-1}(1 - \phi(\epsilon/q))$$

and

$$p(\epsilon) \leq 2Mq^2\epsilon^{-1}(1 - \phi(\epsilon/q))^{-1}.$$

This contradicts (13.90). The contradiction we have reached proves that there exists $s \in \{0, \ldots, p(\epsilon)\}$ such that (13.94) holds. By (79), (85), (89), and (94), for each integer $m \geq s$,

$$d(x_*, x_m) = d(y_{(m-1)q+1}, y_{mq}) \leq \sum_{i=(m-1)q+1}^{mq-1} d(y_i, y_{i+1}) \leq \epsilon.$$

Theorem 13.16 is proved.

13.10 Set-Valued Strict Contractions

We continue to use the notation and definitions introduced in Sects. 3.7 and 3.9 and prove the following results for set-valued strict contractions obtained in [231].

Assume that $E(G)$ is a closed set in $X \times X$, $c \in (0, 1)$, $T : X \to 2^X \setminus \{\emptyset\}$, $T(x)$ is closed for each $x \in X$ and that the following assumption holds:

(A3) For each $(x, y) \in E(G)$ and each $\xi \in T(x)$,

$$d(\xi, \{u \in T(y) : (\xi, u) \in E(G)\}) \leq cd(x, y).$$

Theorem 13.17 *Assume that $x_0 \in X$:*

$$\{u \in T(x_0) : (x_0, u) \in E(G)\} \neq \emptyset \tag{13.97}$$

and that $\{\epsilon_i\}_{i=0}^{\infty} \subset (0, \infty)$ *satisfies*

$$\sum_{i=0}^{\infty} \epsilon_i < \infty. \tag{13.98}$$

Then there exists a sequence $\{x_i\}_{i=1}^{\infty}$ *such that for each integer* $i \geq 0$,

$$(x_i, x_{i+1}) \in E(G),\ x_{i+1} \in T(x_i) \tag{13.99}$$

and

$$d(x_i, x_{i+1}) \leq d(x_i, \{u \in T(x_i) : (x_i, u) \in E(G)\}) + \epsilon_i. \tag{13.100}$$

Moreover, if a sequence $\{x_i\}_{i=1}^{\infty}$ *satisfies* (13.99) *and* (13.100) *for each integer* $i \geq 0$, *then it converges to a point* $x_* \in X$ *and if, in addition, the graph of* T *is closed, then* $x_* \in T(x_*)$.

Proof By (13.97), there exists

$$x_1 \in T(x_0), \tag{13.101}$$

such that

$$(x_0, x_1) \in E(G), \tag{13.102}$$

$$d(x_0, x_1) \leq d(x_0, \{u \in T(x_0) : (x_0, x_1) \in E(G)\}) + \epsilon_0. \tag{13.103}$$

Assume that $n \geq 1$ is an integer and that a finite sequence $\{x_i\}_{i=1}^{n}$ was defined such that (13.99) and (13.100) are valid for each integer $i \in \{0, \ldots, n-1\}$. (Note that for $n = 1$ our assumption holds.) By (13.99),

$$(x_{n-1}, x_n) \in E(G),\ x_n \in T(x_{n-1}). \tag{13.104}$$

In view of (13.104), there exists

$$x_{n+1} \in T(x_n)$$

such that

$$(x_n, x_{n+1} \in E(G)$$

and

$$d(x_n, x_{n+1}) \le d(x_n, \{u \in T(x_n) : (x_n, u) \in E(G)\}) + \epsilon_n.$$

Thus, by induction, we constructed the sequence $\{x_i\}_{i=1}^{\infty}$ that satisfies (13.99) and (13.100) for each integer $i \ge 0$.

Assume that a sequence $\{x_i\}_{i=0}^{\infty} \subset X$ satisfies (13.99) and (13.100) for each integer $i \ge 0$. We show that it is a Cauchy sequence.

Let $i \ge 1$ be an integer. Assumption (A3), (13.99), and (13.100) imply that

$$d(x_{i+1}, x_{i+2}) \le d(x_{i+1}, \{u \in T(x_{i+1}) : (x_{i+1}, u) \in E(G)\}) + \epsilon_{i+1}$$

$$\le cd(x_i, x_{i+1}) + \epsilon_{i+1}. \tag{13.105}$$

In view of (13.105),

$$d(x_1, x_2) \le cd(x_0, x_1) + \epsilon_1,$$

$$d(x_2, x_3) \le cd(x_1, x_2) + \epsilon_2 \le c^2 d(x_0, x_1) + c\epsilon_1 + \epsilon_2.$$

Now we show by induction that for every integer $n \ge 1$,

$$d(x_n, x_{n+1}) \le c^n d(x_0, x_1) + \sum_{i=0}^{n-1} c^i \epsilon_{n-i}. \tag{13.106}$$

(Note that by the relations above, (13.106) holds for $n = 1, 2$.)

Assume that $k \ge 1$ is an integer and that (13.106) holds for $n = k$. When combined with (13.105), this implies that

$$d(x_{k+1}, x_{k+2}) \le cd(x_k, x_{k+1}) + \epsilon_{k+1}$$

$$\le c^{k+1} d(x_0, x_1) + \sum_{i=0}^{k-1} c^{i+1} \epsilon_{k-i} + \epsilon_{k+1}$$

$$= c^{k+1} d(x_0, x_1) + \sum_{i=0}^{k} c^i \epsilon_{k+1-i}.$$

Thus, (13.106) holds with $n = k + 1$ too. Thus, we have showed by induction that (13.106) holds for all integers $n \ge 1$. By (13.98) and (13.106),

$$\sum_{n=1}^{\infty} d(x_n, x_{n+1}) \le \sum_{n=1}^{\infty} (c^n d(x_0, x_1) + \sum_{i=1}^{n} c^{n-i} \epsilon_i)$$

$$\leq d(x_0, x_1) \sum_{n=1}^{\infty} c^n + \sum_{i=1}^{\infty} (\sum_{j=0}^{\infty} c^j) \epsilon_i$$

$$\leq (\sum_{n=0}^{\infty} c^n)[d(x_0, x_1) + \sum_{n=1}^{\infty} \epsilon_n] < \infty.$$

Thus, $\{x_n\}_{n=0}^{\infty}$ is indeed a Cauchy sequence and there exists

$$x_* = \lim_{n \to \infty} x_n.$$

Clearly, if the graph of T is closed, then $x_* \in T(x_*)$. Theorem 13.17 is proved.

The next result shows the well-posedness of the fixed point problem.

Theorem 13.18 *Assume that the graph of T is closed and that $\epsilon > 0$. Then there exists $\delta > 0$ such that if $x \in X$ and*

$$d(x, \{u \in T(x) : (x, u) \in E(G)\}) < \delta, \tag{13.107}$$

then there is $\bar{x} \in X$ such that $\bar{x} \in T(\bar{x})$ and $d(x, \bar{x}) \leq \epsilon$.

Proof Choose

$$\delta \in (0, 4^{-1}(1 - c)\epsilon) \tag{13.108}$$

Let $x \in X$ and (13.107) hold. Set

$$x_0 = x. \tag{13.109}$$

By (13.107) and (13.109), there is

$$x_1 \in T(x_0) \tag{13.110}$$

such that

$$(x_0, x_1) \in E(G),\ d(x_0, x_1) < \delta, \tag{13.111}$$

$$d(x_0, x_1) \leq d(x_0, \{u \in T(x_0) : (x_0, u) \in E(G)\})(1 + c)(2c)^{-1}. \tag{13.112}$$

By induction, we construct a sequence $\{x_n\}_{n=2}^{\infty} \subset X$ such that for each integer $n \geq 0$,

$$x_{n+1} \in T(x_n),\ (x_n, x_{n+1}) \in E(G), \tag{13.113}$$

$$d(x_{n+1}, x_n) \le d(x_n, \{u \in T(x_n) : (x_n, u) \in E(G)\})(1+c)(2c)^{-1}. \quad (13.114)$$

(Note that by (13.111) and (13.112), the relations above hold for $n = 0$.)

Assume that k is a natural number and (13.113) and (13.114) hold for all integers $n = 0, \dots, k-1$. Then,

$$(x_{k-1}, x_k) \in E(G), \ x_k \in T(x_{k-1}).$$

By (A3),

$$\{u \in T(x_k) : (x_k, u) \in E(G)\} \neq \emptyset$$

and there is

$$x_{k+1} \in \{u \in T(x_k) : (x_k, u) \in E(G)\}$$

such that

$$d(x_k, x_{k+1}) \le d(x_{k+1}, \{u \in T(x_k) : (x_k, u) \in E(G)\})(1+c)(2c)^{-1}.$$

Thus, by induction, we constructed the sequence $\{x_n\}_{n=0}^{\infty} \subset X$ such that for each integer $n \ge 0$, relations (13.113) and (13.114) hold. Assumption (A3), (13.113), and (13.114) imply that for each integer $n \ge 1$,

$$d(x_n, x_{n+1}) \le ((1+c)/2) d(x_n, x_{n-1})$$

and

$$d(x_n, x_{n+1}) \le [(1+c)/2]^n d(x_0, x_1) \le [(1+c)/2]^n \delta. \quad (13.115)$$

Therefore,

$$\sum_{n=0}^{\infty} d(x_n, x_{n+1}) < \infty,$$

$\{x_n\}_{n=0}^{\infty}$ is a Cauchy sequence and there exists $\bar{x} \in X$ such that

$$\bar{x} = \lim_{n \to \infty} x_n.$$

Since the graph of T is closed, we have

$$\bar{x} \in T(\bar{x}).$$

By (13.108) and (13.115),

$$d(x_0, \bar{x}) = \lim_{n\to\infty} d(x_0, x_n) \le \sum_{n=0}^{\infty} d(x_n, x_{n+1})$$

$$\le \sum_{i=0}^{\infty} [(1+c)/2]^i \delta = 2\delta/(1-c) < \epsilon.$$

This completes the proof of Theorem 13.18.

The next result shows the behavior of inexact iterates under the presence of nonsummable computational errors.

Theorem 13.19 *Let* $\epsilon \in (0, 1)$, $M > 0$,

$$\delta \in (0, 2^{-1}(1-c)\epsilon) \tag{13.116}$$

and a natural number n_0 *satisfy*

$$c^{n_0}(2M+1) < \epsilon/2. \tag{13.117}$$

Then for each sequence $\{x_i\}_{i=0}^{\infty} \subset X$ *such that*

$$\{u \in T(x_0) : (x_0, u) \in E(G)\} \neq \emptyset,$$

$$d(x_0, \{u \in T(x_0) : (x_0, u) \in E(G)\}) \le M \tag{13.118}$$

and that for each integer $n \ge 0$,

$$x_{n+1} \in T(x_n), \ (x_n, x_{n+1}) \in E(G), \tag{13.119}$$

$$d(x_n, x_{n+1}) \le d(x_n, \{u \in T(x_n) : (x_n, u) \in E(G)\}) + \delta \tag{13.120}$$

the inequality

$$d(x_{n+1}, x_n) \le \epsilon \text{ for all integers } n \ge n_0.$$

Proof Let $\{x_n\}_{n=0}^{\infty} \subset X$ satisfy (13.118) and (13.119) and (13.120) for each integer $i \ge 0$. By (13.118) and (13.119),

$$d(x_0, x_1) \le \delta + d(x_0, \{u \in T(x_0) : (x_0, u) \in E(G)\}) \le M + 1. \tag{13.121}$$

Assumption (A3), (13.119), and (13.120) imply that for each integer $n \ge 0$,

$$d(x_{n+1}, x_{n+2})$$

$$\leq d(x_{n+1}, \{u \in T(x_{n+1}) : (x_{n+1}, u) \in E(G)\}) + \delta \leq cd(x_n, x_{n+1}) + \delta. \tag{13.122}$$

By induction we show that for each integer $n \geq 1$,

$$d(x_{n+1}, x_n) \leq \delta \sum_{i=0}^{n-1} c^i + c^n d(x_0, x_1). \tag{13.123}$$

Clearly, for $n = 1$, (13.123) holds. Assume that $k \geq 1$ is an integer and that (13.123) holds with $n = k$. Together with (13.122), this implies that

$$d(x_{k+2}, x_{k+1}) \leq cd(x_k, x_{k+1}) + \delta \leq \delta \sum_{i=0}^{k} c^i + c^{k+1} d(x_0, x_1)$$

and (13.123) holds for $n = k + 1$. Thus, (13.123) holds for each integer $n \geq 0$. By (13.116), (13.117), and (13.121), for each integer $n \geq n_0$,

$$d(x_n, x_{n+1}) \leq \delta(1 - c)^{-1} + c^n(2M + 1) \leq \epsilon.$$

Theorem 13.19 is proved.

Theorems 13.18 and 13.19 imply the following additional result:

Theorem 13.20 *Assume that the graph of T is closed and let positive numbers ϵ and M be given. Then there exist $\delta > 0$ and an integer $n_0 \geq 1$ such that if a sequence $\{x_i\}_{i=0}^{\infty} \subset X$ satisfies*

$$d(x_0, \{u \in T(x_0) : (x_0, u) \in E(G)\}) \leq M$$

and that for each integer $n \geq 0$,

$$(x_n, x_{n+1}) \in E(G), \ x_{n+1} \in T(x_n),$$

$$d(x_n, x_{n+1}) \leq d(x_n, \{u \in T(x_n) : (x_n, u) \in E(G)\}) + \delta$$

for each integer $n \geq n_0$ there is a point $y \in X$ such that $y \in T(y)$ and $\rho(y, x_n) \leq \epsilon$.

13.11 The Mizoguchi-Takahashi Class of Mappings

Assume that (X, d) is a complete metric space equipped with a graph G, $V(G) \subset X$ is the set of its vertices, and $E(G) \subset X \times X$ is the closed set of its edges.

Assume that a function $\phi : [0, \infty) \to [0, 1)$ satisfies for each $t \in [0, \infty)$,

$$\limsup_{r \to t^+} \phi(r) < 1, \ \phi(0) = 0. \tag{13.124}$$

$$(x, x) \in E(G), \ x \in X \tag{13.125}$$

and that a mapping $T : X \to 2^X \setminus \{\emptyset\}$ satisfies the following assumption:

(A) For each $x, y \in X$ satisfying $(x, y) \in E(G)$ and each $\xi \in T(x)$,

$$d(\xi, \{u \in T(y) : \ (\xi, u) \in E(G)\}) \leq \phi(d(x, y))d(x, y).$$

The mapping T is of the Mizoguchi-Takahashi type [132].

In this section, we study this mapping and analyze an iterative scheme that allows us, for a given initial state, to construct a sequence of iterates which converge to a fixed point of the mapping or generates approximate fixed points.

Theorem 13.21 *Let the graph of T be closed, $x_0 \in X$, a sequence $\{\Delta_i\}_{i=0}^{\infty} \subset (0, \infty)$ satisfy*

$$\sum_{i=0}^{\infty} \Delta_i < \infty, \tag{13.126}$$

$$\{u \in T(x_0)) : \ (x_0, u) \in E(G)\} \neq \emptyset, \tag{13.127}$$

and let a sequence $\{x_i\}_{i=0}^{\infty} \subset X$ satisfy for each integer $i \geq 0$,

$$(x_i, x_{i+1}) \in E(G), \ x_{i+1} \in T(x_i), \tag{13.128}$$

$$\textit{if } x_i \in T(x_i), \textit{ then } x_{i+1} = x_i, \tag{13.129}$$

if $x_i \notin T(x_i)$, then

$$d(x_i, x_{i+1}) \leq d(x_i, \{u \in T(x_i) : \ (x_i, u) \in E(G)\}) + \Delta_i \tag{13.130}$$

and have the following property:

(i) If $i \geq 1$ is an integer and

$$d(x_i, \{u \in T(x_i) : \ (x_i, u) \in E(G)\}) < d(x_{i-1}, x_i),$$

then

$$d(x_i, x_{i+1}) < d(x_{i-1}, x_i).$$

Then the sequence $\{x_i\}_{i=0}^{\infty}$ converges to a fixed point x_* of T satisfying $x_* \in T(x_*)$.

Proof Note that it is not difficult to see that for every $x_0 \in X$ satisfying (13.127) and every sequence $\{\Delta_i\}_{i=0}^{\infty} \subset (0, 1)$ satisfying (13.126), there exists a sequence $\{x_i\}_{i=0}^{\infty} \subset X$ that satisfies all the assumptions of our theorem.

We may assume without loss of generality that for all integers $i \geq 0$,

$$x_i \notin T(x_i). \tag{13.131}$$

This implies that for all integers $i \geq 0$,

$$x_i \neq x_{i+1}. \tag{13.132}$$

In view of (13.130) and (13.131),

$$d(x_0, x_1) \leq d(x_0, \{u \in T(x_0) : (x_0, u) \in E(G)\}) + \Delta_0. \tag{13.133}$$

By (A), (13.128), (13.130), and (13.132), for each integer $i \geq 1$,

$$\begin{aligned} d(x_i, x_{i+1}) &\leq d(x_i, \{u \in T(x_i) : (x_i, u) \in E(G)\}) + \Delta_i \\ &\leq \phi(d(x_{i-1}, x_i))d(x_{i-1}, x_i) + \Delta_i. \end{aligned} \tag{13.134}$$

By (A), (13.128), and (13.132), for each integer $i \geq 1$,

$$d(x_i, \{u \in T(x_i) : (x_i, u) \in E(G)\}) \leq \phi(d(x_{i-1}, x_i))d(x_{i-1}, x_i) < d(x_{i-1} x_i). \tag{13.135}$$

It follows from (13.135) and property (i) that for each integer $i \geq 1$,

$$d(x_i, x_{i+1}) < d(x_{i-1}, x_i). \tag{13.136}$$

In view of (13.136), there exists

$$\begin{aligned} \tau &= \lim_{i \to \infty} \rho(x_i, x_{i+1}) \\ &= \inf\{\rho(x_i, x_{i+1}) : i \geq 0 \text{ is an integer}\}. \end{aligned} \tag{13.137}$$

Set

$$\lambda = \limsup_{r \to \tau^+} \phi(r). \tag{13.138}$$

In view of (13.134), (13.137), and (13.138),

$$0 \leq \lambda < 1. \tag{13.139}$$

By (13.138) and (13.139), there exists a natural number p_0 such that for each integer $i \geq p_0$,

$$\phi(d(x_i, x_{i+1})) \leq 2^{-1}(1+\lambda). \tag{13.140}$$

It follows from (13.134) and (13.140) that for each integer $i \geq p_0 + 1$,

$$d(x_i, x_{i+1}) \leq \phi(d(x_{i-1}, x_i))d(x_{i-1}, x_i) + \Delta_i$$
$$\leq 2^{-1}(1+\lambda)d(x_{i-1}, x_i) + \Delta_i \tag{13.141}$$

and

$$d(x_{p_0+1}, x_{p_0+2}) \leq 2^{-1}(1+\lambda)d(x_{p_0}, x_{p_0+1}) + \Delta_{p_0+1}. \tag{13.142}$$

In view of (13.141) and (13.142),

$$\rho(x_{p_0+2}, x_{p_0+3}) \leq 2^{-1}(1+\lambda)d(x_{p_0+1}, x_{p_0+2}) + \Delta_{p_0+2}$$
$$\leq (2^{-1}(1+\lambda))^2 d(x_{p_0}, x_{p_0+1}) + 2^{-1}(1+\lambda)\Delta_{p_0+1} + \Delta_{p_0+2}.$$

We show by induction that for each integer $n \geq 1$,

$$d(x_{p_0+n}, x_{p_0+n+1})$$
$$\leq (2^{-1}(1+\lambda))^n d(x_{p_0}, x_{p_0+1}) + \sum_{i=0}^{n-1}(2^{-1}(1+\lambda))^i \Delta_{p_0+n-i}. \tag{13.143}$$

In view of (13.142) and the relation above, inequality (13.143) holds for $n = 1, 2$.

Assume that $k \geq 1$ is an integer and that (13.143) holds for $n = k$. Together with (13.141), this implies that

$$d(x_{p_0+k+1}, x_{p_0+k+2}) \leq 2^{-1}(1+\lambda)d(x_{p_0+k}, x_{p_0+k+1}) + \Delta_{p_0+k+1}$$
$$\leq (2^{-1}(1+\lambda))^{k+1} d(x_{p_0}, x_{p_0+1}) + \sum_{i=0}^{k-1}(2^{-1}(1+\lambda))^{i+1}\Delta_{p_0+k-i} + \Delta_{p_0+k+1}$$
$$= (2^{-1}(1+\lambda))^{k+1}\rho(x_{p_0}, x_{p_0+1}) + \sum_{i=0}^{k}(2^{-1}(1+\lambda))^i \Delta_{p_0+k+1-i}.$$

Thus, (13.143) holds for $n = k + 1$. Therefore, we have shown that (13.143) holds for all integers $n \geq 1$.

By (13.126) and (13.143),

$$\sum_{n=1}^{\infty} \rho(x_{p_0+n}, x_{p_0+n+1})$$

$$\leq \sum_{n=1}^{\infty}((2^{-1}(1+\lambda))^n \rho(x_{p_0}, x_{p_0+1}) + \sum_{i=1}^{n}(2^{-1}(1+\lambda))^{n-i} \Delta_{p_0+i})$$

$$\leq d(x_{p_0}, x_{p_0+1}) \sum_{n=1}^{\infty}(2^{-1}(1+\lambda))^n + \sum_{i=1}^{\infty}(\sum_{j=0}^{\infty}(2^{-1}(1+\lambda))^j)\Delta_{p_0+i}$$

$$\leq (\sum_{n=0}^{\infty}(2^{-1}(1+\lambda))^n)[d(x_{p_0}, x_{p_0+1}) + \sum_{i=1}^{\infty} \Delta_{p_0+i}] < \infty.$$

Thus, $\{x_n\}_{n=0}^{\infty}$ is a Cauchy sequence and there exists

$$x_* = \lim_{n\to\infty} x_n.$$

Since the graph of T is closed $x_* \in T(x_*)$. Theorem 13.21 is proved.

The next result shows the well-posedness of the fixed point problem.

Theorem 13.22 *Let the graph of T be closed and $\epsilon > 0$. Then there exists $\delta > 0$ such that for each $x \in X$ satisfying*

$$d(x, \{u \in T(x) : (x, u) \in E(G)\}) < \delta, \tag{13.144}$$

there exists $\bar{x} \in X$ such that $\bar{x} \in T(\bar{x})$ and $d(x, \bar{x}) < \epsilon$.

Proof By (13.124), there exist $\delta_0 > 0$ and $\lambda_0 \in (0, 1)$ such that

$$\phi(t) < \lambda_0 \text{ for all } t \in (0, \delta_0]. \tag{13.145}$$

Fix

$$\lambda_1 \in (\lambda_0, 1). \tag{13.146}$$

Choose a positive number δ such that

$$\delta < \delta_0 \text{ and } \delta(1-\lambda_1)^{-1} < \epsilon. \tag{13.147}$$

Let $x \in X$ satisfy (13.144). Set

$$x_0 = x. \tag{13.148}$$

By (13.144) and (13.148), there exists $x_1 \in X$ such that

$$(x_0, x_1) \in E(G),\ \ x_1 \in T(x_0),\ \ d(x_0, x_1) < \delta. \tag{13.149}$$

$$\text{if } x_0 \in T(x_0),\ \text{ then } x_1 = x_0. \tag{13.150}$$

By induction, (A), (13.149), and (13.150), we obtain a sequence $\{x_n\}_{n=0}^{\infty} \subset X$ such that for each integer $n \geq 1$, the following properties hold:

$$\text{if } x_n \in T(x_n),\ \text{ then } x_{n+1} = x_n; \tag{13.151}$$

$$\text{if } x_n \notin T(x_n),\ \text{ then } x_{n+1} \in T(x_n),\ (x_n, x_{n+1}) \in E(G), \tag{13.152}$$

$$d(x_n, x_{n+1}) < d(x_n, \{u \in T(x_n) :\ (x_n, u) \in E(G)\})\lambda_1\lambda_0^{-1}. \tag{13.153}$$

and the following property holds:

(i) If

$$d(x_n, \{u \in T(x_n) :\ (x_n, u) \in E(G)\}) < d(x_n, x_{n-1}),$$

then

$$d(x_n, x_{n+1}) < d(x_n, x_{n-1}).$$

If $x_1 \in T(x_1)$, then in view of (13.149), the assertion of the theorem holds. Thus, we may assume without loss of generality that

$$x_1 \notin T(x_1). \tag{13.154}$$

By (13.152) and (13.154),

$$x_2 \neq x_1$$

and

$$x_1 \neq x_0.. \tag{13.155}$$

Assume that an integer $k \geq 2$ and

$$x_k \neq x_{k-1},\ x_k \in T(x_{k-1}),\ (x_{k-1}, x_k) \in E(G). \tag{13.156}$$

(Note that in view of (13.152)–(13.154), (13.156) is true for $k = 2$.) By (13.151), (13.152), and (13.156),

$$x_i \neq x_{i-1},\ x_{i-1} \notin T(x_{i-1}),\ x_i \in T(x_{i-1}),\ (x_{i-1}, x_i) \in E(G). \tag{13.157}$$

for all integers i satisfying $1 \le i \le k$.

Let an integer i satisfy $1 \le i \le k$. By (13.157), property (i), and (A),

$$d(x_i, \{u \in T(x_i) : (x_i, u) \in E(G)\}) \le \phi(d(x_i, x_{i-1}))d(x_i, x_{i-1}), \tag{13.158}$$

$$d(x_i, \{u \in T(x_i) : (x_i, u) \in E(G)\}) < d(x_i, x_{i-1}), \tag{13.159}$$

$$d(x_i, x_{i+1}) < d(x_i, x_{i-1}). \tag{13.160}$$

Since (13.167) holds for all integers i satisfying $1 \le i \le k$, we conclude using (13.149) that

$$d(x_k, x_{k+1}) \le d(x_{k-1}, x_k) \le \cdots \le d(x_0, x_1) < \delta. \tag{13.161}$$

By (13.145), (13.153), (13.157), (13.158), and (13.161), for each integer i satisfying $1 \le i < k$,

$$d(x_i, x_{i+1}) \le d(x_i, \{u \in T(x_i) : (x_i, u) \in E(G)\})\lambda_1\lambda_0^{-1}$$
$$\le \lambda_1\lambda_0^{-1}\phi(d(x_i, x_{i-1}))d(x_i, x_{i-1})$$
$$\le \lambda_1\lambda_0^{-1}d(x_i, x_{i-1}) \le \lambda_1 d(x_{i-1}, x_i).$$

Together with (13.149), this implies that for each integer i satisfying $1 \le i < k$,

$$d(x_i, x_{i+1}) \le \lambda_1^i d(x_0, x_1) \le \lambda_1^i \delta.$$

Thus, we have shown that the following property holds:

(ii) If an integer $k \ge 2$ satisfies $x_k \neq x_{k-1}$, then for each integer i satisfying $1 \le i < k$, the inequality

$$d(x_i, x_{i+1}) \le \lambda_1^i \delta$$

holds.

There are two cases:

(a) There is an integer $k \ge 2$ such that

$$x_k \neq x_{k-1},\ x_{k+1} = x_k;$$

(b) $x_k \neq x_{k-1}$ for all integers $k \geq 2$.

Assume that the case (a) holds and let an integer $k \geq 2$ be as guaranteed in the case (a). Then by property (ii) and (13.151),

$$d(x_i, x_{i+1}) \leq \lambda_1^i \delta, \; i = 1, \dots, k-1, \tag{13.162}$$

$$x_i = x_k \text{ for all integers } i \geq k. \tag{13.163}$$

It follows from (13.147), (13.149), and (13.162) that

$$d(x_0, x_k) \leq \sum_{i=0}^{k-1} d(x_i, x_{i+1}) \leq \sum_{i=0}^{\infty} \lambda_1^i \delta = \delta(1-\lambda_1)^{-1} < \epsilon$$

and in the case (a) the assertion of the theorem is proved.

Assume that the case (b) holds. By property (ii), for all integers $i \geq 1$,

$$d(x_i, x_{i+1}) \leq \lambda_1^i \delta,$$

$$\sum_{i=0}^{\infty} d(x_i, x_{i+1}) < \infty.$$

Thus, $\{x_i\}_{i=0}^{\infty}$ is a Cauchy sequence. Let

$$\bar{x} = \lim_{i \to \infty} x_i.$$

Since the graph of T is closed, we have $\bar{x} \in T(\bar{x})$. By (13.147), (13.48), and the relation above,

$$d(x, \bar{x}) = d(x_0, \bar{x}) = \lim_{n \to \infty} d(x_0, x_n)$$

$$\leq \sum_{i=0}^{\infty} d(x_i, x_{i+1}) \leq \sum_{i=0}^{\infty} \lambda_1^i \delta = \delta(1-\lambda_1)^{-1} < \epsilon.$$

This completes the proof of Theorem 13.22.

Theorem 13.23 *Let $x_0 \in X$ and a sequence $\{\Delta_i\}_{i=0}^{\infty} \subset (0, \infty)$ satisfy*

$$\lim_{i \to \infty} \Delta_i = 0 \tag{13.164}$$

and let a sequence $\{x_i\}_{i=0}^{\infty} \subset X$ satisfy

$$\{u \in T(x_0) : \; (x_0, u) \in E(G)\} \neq \emptyset, \tag{13.165}$$

for each integer $i \geq 0$,

$$(x_i, x_{i+1}) \in E(G), \ \ x_{i+1} \in T(x_i), \tag{13.166}$$

$$\text{if } x_i \in T(x_i), \ \text{ then } x_{i+1} = x_i, \tag{13.167}$$

if $x_i \notin T(x_i)$, *then*

$$d(x_i, x_{i+1}) \leq d(x_i, \{u \in T(x_i) : \ (x_i, u) \in E(G)\}) + \Delta_i, \tag{13.168}$$

and let the following property hold:

(i) *If* $i \geq 1$ *is an integer and*

$$d(x_i, \{u \in T(x_i) : \ (x_i, u) \in E(G)\}) < d(x_{i-1}, x_i),$$

then

$$d(x_i, x_{i+1}) < d(x_{i-1}, x_i).$$

Then $\lim_{i \to \infty} \rho(x_i, x_{i+1}) = 0$.

Proof We may assume without loss of generality that for all integers $i \geq 0$,

$$x_i \notin T(x_i). \tag{13.169}$$

In view of (13.169), for all integers $i \geq 0$,

$$x_i \neq x_{i+1}. \tag{13.170}$$

By (13.168),

$$d(x_0, x_1) \leq d(x_0, \{u \in T(x_0) : \ (x_0, u) \in E(G)\}) + \Delta_0. \tag{13.171}$$

By (A), (13.165), and (13.170), for each integer $i \geq 1$,

$$d(x_i, \{u \in T(x_i) : \ (x_i, u) \in E(G)\}) \leq \phi(d(x_{i-1}, x_i))d(x_{i-1}, x_i), \tag{13.172}$$

$$d(x_i, \{u \in T(x_i) : \ (x_i, u) \in E(G)\}) < d(x_{i-1} x_i). \tag{13.173}$$

By (13.170) and (13.173), for each integer $i \geq 1$,

$$d(x_i, x_{i+1}) < d(x_{i-1}, x_i). \tag{13.174}$$

In view of (13.174), there exists

$$\lambda_0 = \lim_{i\to\infty} d(x_i, x_{i+1})$$

$$= \inf\{d(x_i, x_{i+1}) :\ i \geq 0 \text{ is an integer}\}. \quad (13.175)$$

It is sufficient to show that

$$\lambda_0 = 0.$$

Assume the contrary. Then,

$$\lambda_0 > 0. \quad (13.176)$$

By (13.168), (13.169), (13.172), and (13.176), for each integer $i \geq 1$,

$$d(x_i, x_{i+1}) \leq d(x_i, \{u \in T(x_i) :\ (x_i, u) \in E(G)\}) + \Delta_i$$

$$\leq \phi(d(x_{i-1}, x_i))d(x_{i-1}, x_i) + \Delta_i. \quad (13.177)$$

In view of (13.124) and (13.176), there exist

$$\lambda_1 > \lambda_0,\ \gamma_0 \in (0, 1)$$

such that

$$\phi(t) < \gamma_0 \text{ for all } t \in (\lambda_0, \lambda_1]. \quad (13.178)$$

It follows from (13.174), (13.175), and (13.177) that there exists a natural number p_0 such that for every integer $i \geq p_0$,

$$d(x_i, x_{i+1}) \in (\lambda_0, \lambda_1]. \quad (13.179)$$

Together with (13.178), this implies that for every integer $i \geq p_0$,

$$\phi(d(x_i, x_{i+1})) < \gamma_0. \quad (13.180)$$

By (13.176), (13.179), and (13.180) that for every integer $i \geq p_0 + 1$,

$$d(x_i, x_{i+1}) \leq \phi(d(x_{i-1}, x_i))d(x_{i-1}, x_i) + \Delta_i$$

$$\leq \gamma_0 d(x_{i-1}, x_i) + \Delta_i$$

$$\leq d(x_{i-1}, x_i) - (1 - \gamma_0)\lambda_0 + \Delta_i. \quad (13.181)$$

In view of (13.164), there exists an integer $p_1 > p_0$ such that for each integer $i \geq p_1$,

$$\Delta_i < \lambda_0(1 - \gamma_0)/4.$$

Together with (13.181), this implies that for each integer $i \geq p_1$,

$$d(x_i, x_{i+1}) \leq d(x_{i-1}, x_i) - \lambda_0(1 - \gamma_0)/2.$$

This implies that

$$\lim_{i \to \infty} \rho(x_i, x_{i+1}) = -\infty.$$

The contradiction we have reached proves Theorem 13.23.

13.12 Extensions of Theorems 13.17–13.20

Let (X, d) be a complete metric space equipped with a graph G, $V(G) \subset X$ is the set of its vertices, and $E(G) \subset X \times X$ is the closed set of its edges.

We say that $x, y \in X$ are G-connected if there exist a natural number q and $x_i \in X, i = 0, \ldots, q$ such that

$$x_0 = x, \ x_q = y,$$

$$(x_i, x_{i+1}) \in E(G), \ i = 0, \ldots, q - 1.$$

Assume that $T : X \to 2^X \setminus \{\emptyset\}$, $c \in [0, 1)$, the graph

$$\{(x, y) : \in X, \ y \in T(x)\}$$

is closed, for each $(x, y) \in E(G)$ and each $\xi \in T(x)$,

$$d(\xi, \{u \in T(y) : \ (\xi, u) \in E(G)\}) \leq cd(x, y)d(x, y). \tag{13.182}$$

In Sect. 13.10 we studied this mapping and analyzes an iterative scheme that allows us, for a given initial state, to construct a sequence of iterates $\{x_i\}_{i=0}^{\infty}$ which converges to a fixed point of the mapping or generates approximate fixed points. It was done under assumption that

$$x_1 \in T(x_0), \ (x_0, x_1) \in E(G).$$

In this section, we extend the results of Sect. 13.10 for the case when $x_1 \in T(x_0)$ and x_0, x_1 are merely G-connected.

Theorem 13.24 *Assume that* $x^{(0)} \in X$,

$$x^{(1)} \in T(x^{(0)}),$$

$q \geq 1$ *is an integer,* $x_i \in X$, $i = 0, \ldots, q$,

$$x_0 = x^{(0)},\ x_q = x^{(1)},\ (x_i, x_{i+1}) \in E(G),\ i = 0, \ldots, q-1, \tag{13.183}$$

$$\{\epsilon_i\}_{i=0}^{\infty} \subset (0, \infty),\ \sum_{i=0}^{\infty} \epsilon_i < \infty. \tag{13.184}$$

Then there exists a sequence $\{x_i\}_{i=q+1}^{\infty} \subset X$ *such that for each integer* $i \geq 0$,

$$(x_i, x_{i+1}) \in E(G), \tag{13.185}$$

$$x_{i+q} \in T(x_i) \tag{13.186}$$

and that for each integer $i \geq 0$ *and each* $j \in \{0, \ldots, q-1\}$,

$$d(x_{(i+1)q+j}, x_{(i+1)q+j+1})$$

$$\leq d(x_{(i+1)q+j}, \{u \in T(x_{iq+j+1}) :\ (x_{(i+1)q+j}, u) \in E(G)\}) + \epsilon_i.$$

Moreover, if a sequence $\{x_i\}_{i=q+1}^{\infty} \subset X$ *satisfies* (13.185) *and* (13.186) *for each integer* $i \geq 0$ *and* (13.187) *for each integer* $i \geq 0$ *and each* $j \in \{0, \ldots, q-1\}$, *then*

$$\sum_{i=0}^{\infty} d(x_i, x_{i+1}) \leq ((1-c)^{-1} + 1) \sum_{i=0}^{q-1} d(x_i, x_{i+1}) + q(1-c)^{-1} \sum_{i=0}^{\infty} \epsilon_i \tag{13.187}$$

and it converges to a point $x_* \in X$ *such that* $x_* \in T(x_*)$.

Proof We may consider only the case when $q \geq 2$ because the case $q = 1$ is covered by Theorem (13.10). By (13.183),

$$x_q \in T(x_0). \tag{13.188}$$

By (A), (13.183), and (13.188), there exists

$$x_{1+q} \in T(x_1) \tag{13.189}$$

such that

$$(x_q, x_{q+1}) \in E(G), \tag{13.190}$$

$$d(x_q, x_{q+1}) \leq d(x_q, \{u \in T(x_q) : (x_q, u) \in E(G)\}) + \epsilon_0. \tag{13.191}$$

Assume that $p \in \{1, \dots, q\} \setminus \{q\}$, $x_i \in X$, $i = q, \dots, p+q$ are defined such that

$$(x_i, x_{i+1}) \in E(G), \tag{13.192}$$

$i = 0, \dots, p+q-1$,

$$x_{i+q} \in T(x_i), \tag{13.193}$$

$i = 0, \dots, p$ and that that for each $i \in \{0, \dots, p-1\}$,

$$d(x_{i+q}, x_{i+q+1}) \leq d(x_{i+q}, \{u \in T(x_{i+q}) : (x_{i+q}, u) \in E(G)\}) + \epsilon_0. \tag{13.194}$$

(Clearly, for $p = 1$ our assumption holds.) By (13.182), (13.183), and (13.193), there exists

$$x_{p+q+1} \in T(x_{p+1})$$

such that

$$(x_{q+p}, x_{q+p+1}) \in E(G)$$

and

$$d(x_{q+p}, x_{q+p+1}) \leq d(x_{q+p}, \{u \in T(x_{p+1}) : (x_{q+p}, u) \in E(G)\}) + \epsilon_0.$$

Thus, our assumption holds for $p + 1$ too and by induction we have constructed $\{x_i\}_{i=0}^{2q} \subset X$ such that (13.192) holds for $i = 0, \dots, 2q-1$, (13.193) holds for $i = 0, \dots, q$, and (13.194) holds for $i = 0, \dots, q-1$.

Assume that $n \geq 2$ is an integer and that $\{x_i\}_{i=0}^{qn} \subset X$ is defined, (13.191) holds for $i = 0, \dots, qn-1$, (13.193) holds for $i = 0, \dots, q(n-1)$, and

$$d(x_{i+q}, x_{i+q+1}) \leq d(x_{i+q}, \{u \in T(x_{i+q}) : (x_{i+q}, u) \in E(G)\}) + \epsilon_{\lfloor i/q \rfloor - 1} \tag{13.195}$$

for each $i \in \{0, \dots, (n-1)q - 1\}$. (Clearly, for $n = 2$ our assumption holds.) This implies that

$$x_{nq} \in T(x_{(n-1)q}), \ (x_{(n-1)q}, x_{(n-1)q+1}) \in E(G). \tag{13.196}$$

By (13.182) and (13.196), there exists

$$x_{nq+1} \in T(x_{(n-1)q+1})$$

such that

$$(x_{nq}, x_{nq+1}) \in E(G),$$

$$d(x_{nq}, x_{nq+1}) \le d(x_{nq}, \{u \in T(x_{(n-1)q+1}) : (x_{nq}, u) \in E(G)\}) + \epsilon_{n-1}.$$

Assume that $s \in \{1, \dots, q\} \setminus \{q\}$, $x_{nq+j} \in X$, $j = 1, \dots, s$ are defined such that

$$(x_{nq+s-1}, x_{nq+s}) \in E(G), \tag{13.197}$$

$$x_{nq+s} \in T(x_{(n-1)q+s}). \tag{13.198}$$

(Clearly, our assumption holds for $s = 1$.) By (13.182), (13.197), and (13.198), there exists

$$x_{nq+s+1} \in T(x_{(n-1)q+s+1})$$

such that

$$(x_{nq+s}, x_{nq+s+1}) \in E(G)$$

and

$$d(x_{nq+s}, x_{nq+s+1})$$

$$\le d(x_{nq+s}, \{u \in T(x_{(n-1)q+s+1}) : (x_{nq+s}, u) \in E(G)\}) + \epsilon_{n-1}. \tag{13.199}$$

Thus, our assumption holds for $s + 1$ too and by induction we have constructed $x_{nq+j} \in X$, $j = 1, \dots, q$ such that for each $s = 1, \dots, q$ relations (13.197) and (13.198) hold and for each $s = 0, \dots, q - 1$ relation (13.199) holds. Thus, our assumption holds for $n + 1$ too and by induction we have constructed $\{x_i\}_{i=0}^{\infty} \subset X$ such that (13.192), (13.193), and (13.195) hold for all integers $i \ge 0$.

Assume that $\{x_i\}_{i=q+1}^{\infty} \subset X$ satisfies (12.185) and (12.186) for each integer $i \ge 0$ and (12.187) for each integer $i \ge 0$ and each $j \in \{0, \dots, q - 1\}$.

Let $n \ge 0$ be an integer. By (13.181) and (13.284)–(13.286), for each $i \in \{0, \dots, q - 1\}$,

$$d(x_{(n+1)q+i}, x_{(n+1)q+i+1})$$

$$\le d(x_{(n+1)q+i}, \{u \in T(x_{nq+i+1}) : (x_{(n+1)q+i}, u) \in E(G)\}) + \epsilon_n$$

$$\le cd(x_{nq+i}, x_{nq+i+1}) + \epsilon_n$$

and

$$\sum_{i=0}^{q-1} d(x_{(n+1)q+i}, x_{(n+1)q+i+1}) \le c \sum_{i=0}^{q-1} d(x_{nq+i}, x_{nq+i+1}) + q\epsilon_n. \tag{13.200}$$

We show by induction that for each integer $n \ge 1$,

$$\sum_{i=0}^{q-1} d(x_{nq+i}, x_{nq+i+1}) \le c^n \sum_{i=0}^{q-1} d(x_i, x_{i+1}) + q \sum_{i=0}^{n-1} c^i \epsilon_{n-i-1}. \tag{13.201}$$

In view of (13.200), inequality (13.201) is valid for $n = 1$.

Assume that $n \ge 1$ is an integer and that (13.201) holds. In view of (13.200) and (13.201),

$$\sum_{i=0}^{q-1} d(x_{(n+1)q+i}, x_{(n+1)q+i+1}) \le c^{n+1} \sum_{i=0}^{q-1} d(x_i, x_{i+1}) + q \sum_{i=0}^{n-1} c^{i+1} \epsilon_{n-i-1} + \epsilon_n q$$

$$= c^{n+1} \sum_{i=0}^{q-1} d(x_i, x_{i+1}) + q \sum_{i=0}^{n} c^i \epsilon_{n-i}.$$

Thus, by induction, we showed that (13.201) holds for each integer $n \ge 1$. By (13.201),

$$\sum_{n=0}^{\infty} \sum_{i=0}^{q-1} d(x_{nq+i}, x_{nq+i+1})$$

$$\le \sum_{i=0}^{q-1} d(x_i, x_{i+1}) + \sum_{n=1}^{\infty} (c^n \sum_{i=0}^{q-1} d(x_i, x_{i+1}) + q \sum_{i=0}^{n-1} c^i \epsilon_{n-i-1})$$

$$\le (\sum_{n=0}^{\infty} c^n + 1) \sum_{i=0}^{q-1} d(x_i, x_{i+1}) + q \sum_{i=0}^{\infty} \epsilon_i (\sum_{i=0}^{\infty} c^i)$$

$$= ((1-c)^{-1} + 1) \sum_{i=0}^{q-1} d(x_i, x_{i+1}) + q(1-c)^{-1} \sum_{i=0}^{\infty} \epsilon_i.$$

Thus, $\{x_n\}_{n=0}^{\infty}$ is indeed a Cauchy sequence and there exists

$$x_* = \lim_{n \to \infty} x_n.$$

Since the graph of T is closed, we conclude that $x_* \in T(x_*)$. This completes the proof of Theorem 13.24.

Theorem 13.25 *Assume that* $\epsilon \in (0, 1]$, $x^{(0)} \in X$,

$$x^{(1)} \in T(x^{(0)}),$$

$q \geq 1$ *is an integer,* $x_i \in X$, $i = 0, \dots, q$,

$$x_0 = x^{(0)},\ x_q = x^{(1)},\ (x_i, x_{i+1}) \in E(G),\ i = 0, \dots, q-1,$$

$$((1-c)^{-1} + 1) \sum_{i=0}^{q-1} d(x_i, x_{i+1}) < \epsilon.$$

Then there exists $x_* \in X$ *such that* $x_* \in T(x_*)$ *and* $d(x^{(0)}, x_*) \leq \epsilon$.

Proof There exists a sequence $\{\epsilon_i\}_{i=0}^{\infty} \subset (0, 1)$ such that

$$((1-c)^{-1} + 1)(\sum_{i=0}^{q-1} d(x_i, x_{i+1}) + \sum_{i=0}^{\infty} \epsilon_i) < \epsilon.$$

Let a sequence $\{x_i\}_{i=q+1}^{\infty} \subset X$ be as guaranteed by Theorem 13.24 which implies that $\sum_{i=0}^{\infty} d(x_i, x_{i+1}) < \epsilon$. Thus, $\{x_i\}_{i=0}^{\infty}$ is a Cauchy sequence and it converges to a point $x_* \in X$ such that $x_* \in T(x_*)$. Clearly, $d(x^{(0)}, x_*) \leq \epsilon$. Theorem 13.25 is proved.

Theorem 13.26 *Assume that* $\epsilon \in (0, 1]$, $q \geq 1$ *is an integer,*

$$\delta \in (0, 2^{-1}\epsilon(1-c)q^{-1}), \tag{13.202}$$

a natural number n_0 *satisfy*

$$c^{n_0} \sum_{i=0}^{q-1} d(x_i, x_{i+1}) < 2^{-1}\epsilon, \tag{13.203}$$

$$x^{(0)} \in X,\ x^{(1)} \in T(x^{(0)}), \tag{13.204}$$

q *is natural number,* $x_i \in X$, $i = 0, \dots, q$,

$$x_0 = x^{(0)},\ x_q = x^{(1)},$$

$$(x_i, x_{i+1}) \in E(G),\ i = 0, \dots, q-1,$$

and that for each integer $i \geq 0$ *and each* $j \in \{0, \ldots, q-1\}$,

$$(x_i, x_{i+1}) \in E(G), \; x_{i+q} \in T(x_i), \tag{13.205}$$

$$d(x_{(i+1)q+j}, x_{(i+1)q+j+1})$$

$$\leq d(x_{(i+1)q+j}, \{u \in T(x_{iq+j+1}) : \; (x_{(i+1)q+j}, u) \in E(G)\}) + \delta. \tag{13.206}$$

Then for each integer $n \geq n_0$,

$$d(x_{qn}, x_{q(n+1)}) \leq \sum_{i=qn}^{q(n+1)-1} d(x_i, x_{i+1}) \leq \epsilon.$$

Proof Clearly, the sequence $\{x_i\}_{i=0}^{\infty} \subset X$ is well-defined by the relations above. Assume that $k \geq 0$ is an integer. By (13.182), (13.205), and (13.206),

$$\sum_{i=kq}^{(k+1)q-1} d(x_{i+q}, x_{i+1+q}) \leq \sum_{i=0}^{q-1} d(x_{(k+1)q+i}, x_{(k+1)q+i+1})$$

$$\leq \sum_{i=0}^{q-1} (d(x_{(k+1)q+i}, \{u \in T(x_{kq+i+1}) : \; (x_{(k+1)q+i}, u) \in E(G)\}) + \delta)$$

$$\leq q\delta + c \sum_{i=0}^{q-1} d(x_{kq+i}, x_{kq+i+1}). \tag{13.207}$$

We show by induction that for each integer $k \geq 1$,

$$\sum_{i=0}^{q-1} d(x_{kq+i}, x_{kq+i+1}) \leq c^k \sum_{i=0}^{q-1} d(x_i, x_{i+1}) + q\delta \sum_{i=0}^{k-1} c^i. \tag{13.208}$$

In view of (13.207), relation (13.208) holds for $k = 1$.

Assume that $k \geq 1$ is an integer and (13.208) holds. By (13.207) and (13.208),

$$\sum_{i=0}^{q-1} d(x_{(k+1)q+i}, x_{(k+1)q+i+1}) \leq q\delta + c \sum_{i=0}^{q-1} d(x_{kq+i}, x_{kq+i+1})$$

$$\leq c^{k+1} \sum_{i=0}^{q-1} d(x_i, x_{i+1}) + \delta q \sum_{i=0}^{k-1} c^{i+1} + \delta q$$

$$= c^{k+1}\sum_{i=0}^{q-1} d(x_i, x_{i+1}) + \delta q \sum_{i=0}^{k} c^i.$$

Thus, (13.208) holds for $k + 1$ too and by induction we showed that (13.208) holds for each integer $k \geq 1$. If an integer $k \geq n_0$, then by (13.202), (13.203), and (13.208),

$$\sum_{i=0}^{q-1} d(x_{kq+i}, x_{kq+i+1}) \leq c^k \sum_{i=0}^{q-1} d(x_i, x_{i+1}) + q\delta(1-c)^{-1} < \epsilon.$$

This completes the proof of Theorem 13.26.

Theorems 13.25 and 13.26 imply the following result:

Theorem 13.27 *Assume that* $\epsilon \in (0, 1]$,

$$\epsilon_0 \in (0, 2^{-1}(1-c)\epsilon),$$

$q \geq 1$ *is an integer,*

$$\delta \in (0, 2^{-1}\epsilon_0(1-\epsilon_0)q^{-1}),$$

$$x^{(0)} \in X,\ x^{(1)} \in T(x^{(0)}),$$

$x_i \in X$, $i = 0, 1, \ldots,$,

$$x_0 = x^{(0)},\ x_q = x^{(1)},$$

$$x_{i+q} \in T(x_i),\ (x_i, x_{i+1}) \in E(G),\ i = 0, 1, \ldots,$$

$$(1 + (1-c)^{-1}) \sum_{i=0}^{q-1} d(x_i, x_{i+1}) < 2^{-1}\epsilon$$

and a natural number n_0 *satisfy*

$$c^{n_0} \sum_{i=0}^{q-1} d(x_i, x_{i+1}) < 2^{-1}\epsilon_0,$$

and that for each integer $i \geq 0$ *and each* $j \in \{0, \ldots, q-1\}$,

$$d(x_{(i+1)q+j}, x_{(i+1)q+j+1})$$

$$\leq d(x_{(i+1)q+j}, \{u \in T(x_{iq+j+1}) :\ (x_{(i+1)q+j}, u) \in E(G)\}) + \delta.$$

Then for each integer $n \geq n_0$,

$$\sum_{i=qn}^{q(n+1)-1} d(x_i, x_{i+1}) \leq \epsilon_0$$

and

$$B(x_{qn}, \epsilon) \cap \{y \in X : \ y \in T(y)\} \neq \emptyset.$$

13.13 Extensions of Theorems 13.21–13.23

Assume that (X, d) is a complete metric space equipped with a graph G, $V(G) \subset X$ is the set of its vertices, $E(G) \subset X \times X$ is the closed set of its edges, and

$$(x.x) \in E(G), \ \ x \in X.$$

Assume that a function $\phi : [0, \infty) \to [0, 1)$ satisfies for each $t \in [0, \infty)$,

$$\limsup_{r \to t^+} \phi(r) < 1, \ \ \phi(0) = 0 \tag{13.209}$$

and that $T : X \to 2^X \setminus \{\emptyset\}$, the graph

$$\{(x, y) \in X \times X : \ y \in T(x)\}$$

of T is closed. Assume that the following assumption holds:

(P) For each $x, y \in X$ satisfying $(x, y) \in E(G)$ and each $\xi \in T(x)$,

$$d(\xi, \{u \in T(y) : \ (\xi, u) \in E(G)\}) \leq \phi(d(x, y))d(x, y). \tag{13.210}$$

The mapping T is of the Mizoguchi-Takahashi type [132]. In Sect. 13.11 we studied this mapping and analyzed an iterative scheme that allows us, for a given initial state, to construct a sequence of iterates $\{x_i\}_{i=0}^{\infty}$ which converge to a fixed point of the mapping or generates approximate fixed points. It was done under assumption that

$$x_1 \in T(x_0), \ \ (x_0, x_1) \in E(G).$$

In this section, we extend the results of Sect. 13.11 for the case when $x_1 \in T(x_0)$ and x_0, x_1 are merely G-connected.

Theorem 13.28 *Let*

$$x^{(0)} \in X,\ x^{(1)} \in T(x^{(0)}), \tag{13.211}$$

$q \geq 2$ *is an integer,* $x_i \in X$, $i = 0, 1, \dots,$

$$(x_i, x_{i+1}) \in E(G),\ i = 0, \dots, q-1, \tag{13.212}$$

$$x_0 = x^{(0)},\ x_q = x^{(1)}, \tag{13.213}$$

$$x_i \neq x_j \text{ for each } i, j \in \{0, \dots, q\} \text{ such that } i \neq j, \tag{13.214}$$

a sequence $\{\Delta_i\}_{i=0}^{\infty} \subset (0, \infty)$ *satisfy*

$$\lim_{i \to \infty} \Delta_i = 0,$$

for each integer $i \geq 0$,

$$(x_i, x_{i+1}) \in E(G),\ x_{i+q} \in T(x_i), \tag{13.215}$$

and that for each integer $i \geq 0$ *and each* $j \in \{0, \dots, q-1\}$,

$$d(x_{(i+1)q+j}, x_{(i+1)q+j+1})$$

$$\leq d(x_{(i+1)q+j}, \{u \in T(x_{iq+j+1}) : (x_{(i+1)q+j}, u) \in E(G)\}) + \Delta_i, \tag{13.216}$$

$$\text{if } x_{iq+j} = x_{iq+j+1}, \text{ then } x_{(i+1)q+j} = x_{(i+1)q+j+1}; \tag{13.217}$$

otherwise,

$$d(x_{(i+1)q+j}, x_{(i+1)q+j+1}) < d(x_{iq+j}, x_{iq+j+1}). \tag{13.218}$$

Then,

$$\lim_{i \to \infty} d(x_i, x_{i+1}) = 0$$

and if

$$\sum_{i=0}^{\infty} \Delta_i < \infty,$$

then the sequence $\{x_i\}_{i=0}^{\infty}$ *converges to a fixed point* x_* *of* T *satisfying* $x_* \in T(x_*)$.

Proof It is not difficult to see that there exists a sequence $\{x_i\}_{i=0}^{\infty} \subset X$ that satisfies all the assumptions of our theorem if (13.211)–(13.214) are true.

Let $n \geq 0$ be an integer. By (P) and (13.215)–(13.217), for each integer $i \in \{0, \ldots, q-1\}$,

$$d(x_{(n+1)q+i}, x_{(n+1)q+i+1})$$

$$\leq d(x_{(n+1)q+i}, \{u \in T(x_{nq+i+1}) : (x_{(n+1)q+i}, u) \in E(G)\}) + \Delta_n$$

$$\leq \phi(d(x_{nq+i}, x_{nq+i+1}))d(x_{nq+i}, x_{nq+i+1}) + \Delta_n. \tag{13.219}$$

$$\text{if } x_{nq+i} = x_{nq+i+1}, \text{ then } x_{(n+1)q+i} = x_{(n+1)q+i+1}; \tag{13.220}$$

otherwise,

$$d(x_{(n+1)q+i}, x_{(n+1)q+i+1}) < d(x_{nq+i}, x_{nq+i+1}). \tag{13.221}$$

Fix $i \in \{0, \ldots, q-1\}$. There are two cases:

(a) $x_{nq+i} \neq x_{nq+i+1}$ for each integer $n \geq 0$.
(b) There is a natural number n_i such that $x_{nq+i} = x_{nq+i+1}$ for each integer $n \geq n_i$.

Set

$$E_1 = \{i \in \{0, \ldots, q-1\} : \text{ (a) holds}\}, \tag{13.222}$$

$$E_2 = \{0, \ldots, q-1\} \setminus E_1. \tag{13.223}$$

It is easy to see that if $E_1 = \emptyset$, then the assertion of the theorem holds. Therefore, we may assume without loss of generality that

$$E_1 \neq \emptyset. \tag{13.224}$$

Let

$$p \in E_1. \tag{13.225}$$

Set

$$\lambda_p = \lim_{i \to \infty} d(x_{iq+p}, x_{iq+p+1})$$

$$= \inf\{d(x_{iq+p}, x_{iq+p+1}) : i \geq 0 \text{ is an integer}\}. \tag{13.226}$$

We show that $\lambda_p = 0$. Assume the contrary. Then

$$\lambda_p > 0 \tag{13.227}$$

and in view of (13.221), (13.225), and (13.226),

$$d(x_{iq+p}, x_{iq+p+1}) > \lambda_p, \ i = 0, 1, \ldots. \tag{13.228}$$

By (13.209), there exist

$$\widehat{\lambda}_p > \lambda_p, \ \gamma_p \in (0, 1) \tag{13.229}$$

such that

$$\phi(t) < \gamma_p \text{ for each } t \in (\lambda_p, \widehat{\lambda}_p). \tag{13.230}$$

It follows from (13.226)–(13.229) that there exists a natural number n_p such that for each integer $i \geq n_p$,

$$d(x_{iq+p}, x_{iq+p+1}) \in (\lambda_p, \widehat{\lambda}_p). \tag{13.231}$$

Equations (13.230) and (13.231) imply that for each integer $i \geq n_p$,

$$\phi(d(x_{iq+p}, x_{iq+p+1})) < \gamma_p. \tag{13.232}$$

In view of (13.219), (13.228), and (13.232),

$$d(x_{(i+1)q+p}, x_{(i+1)q+p+1})) \leq \phi(d(x_{iq+p}, x_{iq+p+1}))d(x_{iq+p}, x_{iq+p+1}) + \Delta_i$$

$$\leq \gamma_p d(x_{iq+p}, x_{iq+p+1}) + \Delta_i \leq d(x_{iq+p}, x_{iq+p+1}) - (1-\gamma_p)\lambda_p + \Delta_i. \tag{13.233}$$

Since

$$\lim_{i\to\infty} \Delta_i = 0,$$

there exists an integer $m_p > n_p$ such that

$$\Delta_i < \lambda_p(1 - \gamma_p)/4 \tag{13.234}$$

for each integer $i \geq m_p$. By (13.233) and (13.234), for each integer $i \geq m_p$,

$$d(x_{(i+1)q+p}, x_{(i+1)q+p+1})) \leq d(x_{iq+p}, x_{iq+p+1}) - 2^{-1}(1 - \gamma_p)\lambda_p$$

and this implies that

$$\lim_{i\to\infty} d(x_{iq+p}, x_{iq+p+1}) = -\infty.$$

The contradiction we have reached proves that

$$\lambda_p = 0, \ \ p \in E_1. \tag{13.235}$$

This implies that

$$\lim_{i\to\infty} d(x_i, x_{i+1}) = 0.$$

Assume that

$$\sum_{i=0}^{\infty} \Delta_i < \infty.$$

Set

$$\lambda = \limsup_{r\to 0^+} \phi(r). \tag{13.236}$$

In view of (13.209) and (13.236),

$$0 \le \lambda < 1. \tag{13.237}$$

Let

$$p \in E_1.$$

By (13.221), (13.226), (13.235), (13.236), and the inclusion above, there exists an integer $n_p \ge 1$ such that for each integer $i \ge n_p$,

$$\phi(d(x_{iq+p}, x_{iq+p+1})) \le 2^{-1}(1+\lambda).$$

It follows from (13.219), (13.232), and the inclusion above that for each integer $i \ge n_p$,

$$\begin{aligned} d(x_{(i+1)q+p}, x_{(i+1)q+p+1}) &\le \phi(d(x_{iq+p}, x_{iq+p+1}))d(x_{iq+p}, x_{iq+p+1}) + \Delta_i \\ &\le 2^{-1}(1+\lambda)d(x_{iq+p}, x_{iq+p+1}) + \Delta_i. \end{aligned} \tag{13.238}$$

We show by induction that for each integer $n \ge 1$,

$$\begin{aligned} &d(x_{(n_p+n)q+p}, x_{(n_p+n)q+p+1}) \\ &\le (2^{-1}(1+\lambda))^n d(x_{n_pq+p}, x_{n_pq+p+1}) + \sum_{i=0}^{n-1}(2^{-1}(1+\lambda))^i \Delta_{n_p+n-i-1}. \end{aligned} \tag{13.239}$$

In view of (13.238), inequality (13.239) holds for $n = 1$.

Assume that $n \geq 1$ is an integer and that (13.239) holds. By (13.238) and (13.239),

$$d(x_{(n_p+n+1)q+p}, x_{(n_p+n+1)q+p+1})$$

$$\leq \Delta_{n_p+n} + (2^{-1}(1+\lambda))[(2^{-1}(1+\lambda))^n d(x_{n_pq+p}, x_{n_pq+p+1})$$

$$+\sum_{i=0}^{n-1}(2^{-1}(1+\lambda))^i \Delta_{n_p+n-i-1}]$$

$$= (2^{-1}(1+\lambda))^{n+1} d(x_{n_pq+p}, x_{n_pq+p+1}) + \sum_{i=0}^{n}(2^{-1}(1+\lambda))^i \Delta_{n_p+n-i}.$$

Thus, we showed by induction that (13.239) holds for each integer $n \geq 1$.

By (13.239),

$$\sum_{n=1}^{\infty} d(x_{(n_p+n)q+p}, x_{(n_p+n)q+p+1})$$

$$\leq \sum_{n=1}^{\infty}((2^{-1}(1+\lambda))^n d(x_{n_pq+p}, x_{n_pq+p+1})) + \sum_{i=1}^{n-1}(2^{-1}(1+\lambda))^i \Delta_{n_p+n-i-1})$$

$$\leq \sum_{n=1}^{\infty}((2^{-1}(1+\lambda))^n d(x_{n_pq+p}, x_{n_pq+p+1})) + \sum_{j=n_p}^{\infty} \Delta_j \sum_{j=0}^{\infty}((2^{-1}(1+\lambda))^j) < \infty.$$

This implies that $\sum_{i=0}^{\infty} d(x_i, x_{i+1})) < \infty$ and there exists

$$x_* = \lim_{n\to\infty} x_n.$$

Since the graph of T is closed, $x_* \in T(x_*)$. Theorem 13.28 is proved.

Theorem 13.29 *Let $\epsilon > 0$. Then there exists $\delta > 0$ such that if*

$$x^{(0)} \in X,\ x^{(1)} \in T(x^{(0)}), \tag{13.240}$$

$q \geq 1$ is an integer, $x_i \in X$, $i = 1, \dots, q$,

$$x_0 = x^{(0)},\ x_q = x^{(1)}, \tag{13.241}$$

$$(x_i, x_{i+1}) \in E(G),\ i = 0, \dots, q-1, \tag{13.242}$$

$$x_i \neq x_j \text{ for each } i, j \in \{0, \dots, q\} \text{ such that } i \neq j \tag{13.243}$$

and

$$\sum_{i=0}^{q-1} d(x_i, x_{i+1}) < \delta, \tag{13.244}$$

then there exists $\bar{x} \in X$ *such that*

$$\bar{x} \in T(\bar{x}), \ d(x, \bar{x}) < \epsilon.$$

Proof By (13.209), there exist $\delta_0 > 0$ and $\lambda_0 \in (0, 1)$ such that

$$\phi(t) < \lambda_0 \text{ for all } t \in (0, \delta_0]. \tag{13.245}$$

Fix

$$\lambda_1 \in (\lambda_0, 1). \tag{13.246}$$

Choose a positive number δ such that

$$\delta < \delta_0 \text{ and } \delta(1 - \lambda_1)^{-1} < \epsilon. \tag{13.247}$$

Let $x^{(0)}, x^{(1)} \in X$ satisfy (13.240), $q \geq 1$ is an integer, $x_i \in X$, $i = 1, \dots, q$ and (13.241)–(13.244) hold. By induction, we define a sequence $\{x_i\}_{i=q+1}^{\infty} \subset X$ such that for each integer $i \geq 0$,

$$(x_i, x_{i+1}) \in E(G), \ x_{i+q} \in T(x_i), \tag{13.248}$$

for each integer $i \geq 0$ and each $j \in \{0, \dots, q-1\}$,

$$\text{if } x_{iq+j} = x_{iq+j+1}, \text{ then } x_{(i+1)q+j} = x_{(i+1)q+i+1}; \tag{13.249}$$

otherwise,

$$d(x_{(i+1)q+j}, x_{(i+1)q+j+1}) < d(x_{iq+j}, x_{iq+j+1}),$$

$$d(x_{(i+1)q+j}, x_{(i+1)q+j+1})$$

$$\leq d(x_{(i+1)q+j}, \{u \in T(x_{iq+j+1}) : (x_{(i+1)q+j}, u) \in E(G)\})\lambda_1\lambda_0^{-1}. \tag{13.250}$$

Let $p \in \{0, \dots, q-1\}$ and $n \geq 0$ be an integer. In view of the relation above and (13.244),

$$d(x_{(n+1)q+p}, x_{(n+1)q+p+1}) \le d(x_{nq+p}, x_{nq+p+1}) \le d(x_p, x_{p+1}) < \delta, \tag{13.251}$$

if

$$x_{nq+p} \ne x_{nq+p+1},$$

then by (P) and (13.245), (13.248), (13.250), and (13.251),

$$d(x_{(n+1)q+p}, x_{(n+1)q+p+1}) \le \lambda_1\lambda_0^{-1}\phi(d(x_{nq+p}, x_{nq+p+1}))d(x_{nq+p}, x_{nq+p+1})$$

$$\le \lambda_1\lambda_0^{-1}\lambda_0 d(x_{nq+p}, x_{nq+p+1}) = \lambda_1 d(x_{nq+p}, x_{nq+p+1}),$$

$$d(x_{nq+p}, x_{(n+1)q+p+1}) \le \lambda_1^n d(x_p, x_{p+1}). \tag{13.252}$$

Then, $\sum_{i\to\infty} d(x_i, x_{i+1}) < \infty$, and there exists

$$x_* = \lim_{i\to\infty} x_i$$

and $x_* \in T(x_*)$. By (13.247) and (13.252),

$$d(x_0, x_*) \le \sum_{i=0}^{\infty} d(x_i, x_{i+1}) \le (\sum_{i=0}^{q-1} d(x_i, x_{i+1})) \sum_{i=0}^{\infty} \lambda_1^i \le \delta(1-\lambda_1)^{-1} < \epsilon.$$

Theorem 13.29 is proved.

13.14 Set-Valued Rakotch Contractions

Assume that (X, d) is a metric space equipped with a graph G, $V(G) \subset X$ is the set of its vertices, and $E(G) \subset X \times X$ is a set of its edges. Assume that a function $\phi : [0, \infty) \to [0, 1]$ is decreasing and satisfies

$$\phi(t) < 1, \ t \in (0, \infty), \tag{13.253}$$

$$\phi(0) = 1,$$

$T : X \to 2^X \setminus \{\emptyset\}$ and that for each $(x, y) \in E(G)$ and each $u \in T(x)$,

$$\{v \in T(y) : \ (u, v) \in E(G)\} \ne \emptyset,$$

$$d(u, \{v \in T(y) : \ (u, v) \in E(G)\}) \le \phi(d(x, y))d(x, y). \tag{13.254}$$

Theorem 13.30 *Let $M > 1$, $\epsilon \in (0, 1)$, $q \geq 1$ is an integer,*

$$\delta = 16^{-1}(1 - \phi(\epsilon(4q)^{-1}))\epsilon q^{-2}, \tag{13.255}$$

$$n_0 = \lfloor 16Mq^2\epsilon^{-1}(1 - \phi(4^{-1}\epsilon))^{-1} \rfloor, \tag{13.256}$$

$x_i \in X$, $i = 1, \dots, q$, for each $i \in \{0, \dots, q-1\}$,

$$(x_i, x_{i+1}) \in E(G),\ d(x_i, x_{i+1}) \leq M, \tag{13.257}$$

$$x_q \in T(x_0).$$

By induction define $x_i \in X$, $i = q+1, \dots$ such that for each integer $i \geq 0$,

$$(x_i, x_{i+1}) \in E(G),\ x_{i+q} \in T(x_i), \tag{13.258}$$

$$d(x_{i+q}, x_{i+q+1}) \leq d(x_{i+q}, \{v \in T(x_{i+1}) :\ (x_{i+1}, v) \in E(G)\}) + \delta. \tag{13.259}$$

Then for each integer $n \geq n_0 q$,

$$d(x_n, x_{n+1}) \leq \epsilon/q,\ d(x_n, x_{n+q}) \leq \epsilon,\ d(x_n, T(x_n)) \leq \epsilon.$$

Proof It is not difficult to see that the sequence $\{x_i\}_{i=0}^{\infty} \subset X$ is well-defined.

Let $n \geq 0$ be an integer and

$$d(x_n, x_{n+1}) > \epsilon(4q)^{-1}. \tag{13.260}$$

By (13.254), (13.255), (13.259), and (13.260),

$$d(x_{n+q}, x_{n+q+1}) \leq \delta + d(x_{n+q}, \{v \in T(x_{n+1}) :\ (x_{n+1}, v) \in E(G)\})$$

$$\leq \delta + \phi(d(x_n, x_{n+1}))d(x_n, x_{n+1}) \leq \delta + \phi((4q)^{-1}\epsilon)d(x_n, x_{n+1}),$$

$$d(x_n, x_{n+1}) - d(x_{n+q}, x_{n+1+q}) \geq (1 - \phi((4q)^{-1}\epsilon))d(x_n, x_{n+1}) - \delta$$

$$\geq (1 - \phi((4q)^{-1}\epsilon))(4q)^{-1}\epsilon - \delta = (1 - \phi((4q)^{-1}\epsilon))(8q)^{-1}\epsilon.$$

Thus, we showed that the following property holds:

(a) For each integer $n \geq 0$ satisfying $d(x_n, x_{n+1}) > \epsilon(4q)^{-1}$, we have

$$d(x_n, x_{n+1}) - d(x_{n+q}, x_{n+1+q}) \geq (1 - \phi((4q)^{-1}\epsilon))(8q)^{-1}\epsilon.$$

Assume that $n \geq 0$ is an integer. By (13.254) and (13.259),

$$d(x_{n+q}, x_{n+q+1}) \leq \delta + d(x_{n+q}, \{v \in T(x_{n+1}) : (x_{n+1}, v) \in E(G)\})$$

$$\leq \delta + d(x_n, x_{n+1}). \tag{13.261}$$

Assume that $n \geq 0$ is an integer and that

$$\max\{d(x_i, x_{i+1}) : i \in \{nq, \ldots, (n+1)q - 1\}\} > \epsilon(4q)^{-1}. \tag{13.262}$$

By (13.261) and (13.262), for each $i \in \{nq, \ldots, (n+1)q - 1\}$,

$$d(x_{i+q}, x_{i+q+1}) \leq \delta + d(x_i, x_{i+1}). \tag{13.263}$$

In view of (13.262), there exists

$$j \in \{nq, \ldots, (n+1)q - 1\}$$

such that

$$d(x_j, x_{j+1}) > \epsilon(4q)^{-1}. \tag{13.264}$$

Property (a) and (13.264) imply that

$$d(x_j, x_{j+1}) - d(x_{j+q}, x_{j+1+q}) \geq (1 - \phi((4q)^{-1}\epsilon))(8q)^{-1}\epsilon. \tag{13.265}$$

It follows from (13.256), (13.263), and (13.265) that

$$\sum_{i=nq}^{(n+1)q-1} (d(x_i, x_{i+1}) - d(x_{i+q}, x_{i+1+q}))$$

$$\geq (1 - \phi((4q)^{-1}\epsilon))(8q)^{-1}\epsilon - q\delta \geq (1 - \phi((4q)^{-1}\epsilon))(16q)^{-1}\epsilon.$$

Thus, we have shown that the following property holds:

(b) If $n \geq 0$ is an integer and (13.262) holds, then

$$\sum_{i=nq}^{(n+1)q-1} (d(x_i, x_{i+1}) - d(x_{i+q}, x_{i+1+q}))$$

$$\geq (1 - \phi((4q)^{-1}\epsilon))(16q)^{-1}\epsilon. \tag{13.266}$$

We show that there exists $k \in \{0, \ldots, n_0\}$ such that

$$d(x_i, x_{i+1}) \leq \epsilon(4q)^{-1}, \ i \in \{kq, \ldots, (k+1)q - 1\}.$$

Assume the contrary. Then for each $k \in \{0, \dots, n_0\}$,

$$\max\{d(x_i, x_{i+1}) : i \in \{kq, \dots, (k+1)q - 1\}\} > \epsilon(4q)^{-1}. \tag{13.267}$$

Property (b) and (13.267) imply that for each integer $k \in \{0, \dots, n_0\}$ relation (13.266) holds. It follows from (13.257) and (13.266) that

$$Mq \geq \sum_{i=0}^{q-1} d(x_i, x_{i+1}) \geq \sum_{i=0}^{q-1} d(x_i, x_{i+1}) - \sum_{i=n_0q}^{(n_0+1)q-1} d(x_i, x_{i+1})$$

$$= \sum_{k=0}^{n_0} (\sum_{i=kq}^{(k+1)q-1} d(x_i, x_{i+1}) - \sum_{i=kq}^{(k+1)q-1} d(x_{i+q}, x_{i+1+q}))$$

$$\geq (n_0 + 1)(1 - \phi((4q)^{-1}\epsilon))(16q)^{-1}\epsilon$$

and

$$n_0 \leq 16Mq^2\epsilon^{-1}(1 - \phi((4q)^{-1}\epsilon))^{-1} - 1.$$

This contradicts (13.256). The contradiction we have reached proves that there exists $k \in \{0, \dots, n_0\}$ such that

$$d(x_i, x_{i+1}) \leq \epsilon(4q)^{-1}, \; i \in \{kq, \dots, (k+1)q - 1\}\}. \tag{13.268}$$

Assume that $n \geq 0$ is an integer and that

$$d(x_n, x_{n+1}) \leq \epsilon/q.$$

We show that

$$d(x_{n+q}, x_{n+q+1}) \leq \epsilon/q. \tag{13.269}$$

If

$$d(x_n, x_{n+1}) \leq \epsilon(2q)^{-1},$$

then in view of (13.255) and (13.261),

$$d(x_{n+q}, x_{n+q+1}) \leq \delta + d(x_n, x_{n+1}) \leq \delta + \epsilon(2q)^{-1} \leq \epsilon/q.$$

If

$$d(x_n, x_{n+1}) > \epsilon(2q)^{-1},$$

then property (a) implies that

$$d(x_{n+q}, x_{n+1+q}) \le d(x_n, x_{n+1}) \le \epsilon/q.$$

Thus, in both cases, (13.269) holds. Together with (13.268), this implies that for each $i \ge n_0 q$,

$$d(x_i, x_{i+1}) \le \epsilon/q, \ d(x_i, x_{i+q}) \le \epsilon, \ B(x_i, \epsilon) \cap T(x_i) \ne \emptyset.$$

Theorem 13.30 is proved.

References

1. Agarwal, R.P., Karapinar, E., O'Regan, D., Roldán-López-de-Hierro, A.F.: Fixed Point Theory in Metric Type Spaces. Springer, Cham (2015)
2. Agarwal, R.P., O'Regan, D., Soori, E.: Existence of group nonexpansive retractions and ergodic theorems in topological groups. Fixed Point Theory **22**, 455–463 (2021)
3. Alber, Y.I., Li, J.L.: The connection between the metric and generalized projection operators in Banach spaces. Acta Math. Sin. (Engl. Ser.) **23**, 1109–1120 (2007)
4. Alber, Y.I., Verma, R.U.: Fixed point problems with operators given by total asymptotically nonexpansive approximations. Comm. Appl. Nonlinear Anal. **24**, 1–28 (2017)
5. Alber, Y.I., Yao, J.C.: Another version of the proximal point algorithm in a Banach space. Nonlinear Anal. **70**, 3159–3171 (2009)
6. Aleyner, A., Reich, S.: Block-iterative algorithms for solving convex feasibility problems in Hilbert and Banach spaces. J. Math. Anal. Appl. **343**, 427–435 (2008)
7. Alfuraidan, M.R., Khamsi, M.A.: A fixed point theorem for monotone asymptotically nonexpansive mappings. Proc. Am. Math. Soc. **146**, 2451–2456 (2018)
8. Alfuraidan, M.R., Khamsi, M.A., Kozlowski, W.M.: On Monotone Mappings in Modular Function Spaces. Advances in Metric Fixed Point Theory and Applications, pp. 217–240. Springer, Singapore (2021)
9. Alsulami, S.M., Takahashi, W.: Iterative methods for the split feasibility problem in Banach spaces. J. Nonlinear Convex Anal. **16**, 585–596 (2015)
10. Ariza–Ruiz, D., Leustean, L., Lopez—Acedo, G.: Firmly nonexpansive mappings in classes of geodesic spaces. Trans. Am. Math. Soc. **366**, 4299–4322 (2024)
11. Assad, N.A., Kirk, W.A.: Fixed point theorems for set-valued mappings of contractive type. Pac. J. Math. **43**, 553–562 (1972)
12. Bacak, M.: Proximal point algorithm in metric spaces. Israel J. Math. **160**, 1–13 (2012)
13. Bacak, M.: Convex Analysis and Optimization in Hadamard Spaces. De Gruyter Series in Nonlinear Analysis and Applications. Berlin (2014)
14. Baillon, J.B., Bruck, R.E., Reich, S.: On the asymptotic behavior of nonexpansive mappings and semigroups in Banach spaces. Houston J. Math. **4**, 1–9 (1978)
15. Banach, S.: Sur les opérations dans les ensembles abstraits et leur application aux équations intégrales. Fund. Math. **3**, 133–181 (1922)
16. Bargetz, C., Klemenc, J., Reich, S., Skorokhod, N.: On angles, projections and iterations. Linear Algebra Appl. **603**, 41–56 (2020)
17. Bargetz, C., Medjic, E.: On the rate of convergence of iterated Bregman projections and of the alternating algorithm. J. Math. Anal. Appl. **481**, 123482, 1–23 (2020)

A. J. Zaslavski, *Approximate Fixed Points of Nonexpansive Mappings*,
Developments in Mathematics 80, https://doi.org/10.1007/978-3-031-70710-0

18. Bargetz, C., Reich, S., Thimm, D.: Generic properties of nonexpansive mappings on unbounded domains. J. Math. Anal. Appl. **526**, Paper No. 127179, 1–33 (2023)
19. Bauschke, H.H., Combettes, P.L.: Convex Analysis and Monotone Operator Theory in Hilbert Spaces. CMS Books in Mathematics/Ouvrages de Mathématiques de la SMC. Springer, Cham (2017)
20. Betiuk-Pilarska, A., Domínguez Benavides, T.: Fixed points for nonexpansive mappings and generalized nonexpansive mappings on Banach lattices. Pure Appl. Func. Anal. **1**, 343–359 (2016)
21. Bin Dehaish, B.A., Khamsi, M.A.: Approximating common fixed points of semigroups in metric spaces. Fixed Point Theory Appl. 2015–2051 (2015). https://doi.org/10.1186/s13663-015-0291-8
22. Blum, E., Oettli, W.: From optimization and variational inequalities to equilibrium problems. Math. Student **63**, 123–145 (1994)
23. Boyd, D.W., Wong, J.S.W.: On nonlinear contractions. Proc. Am. Math. Soc. **20**, 458–464 (1969)
24. Bruck, R.E.: On the random product of orthogonal projections in Hilbert space II. Contemp. Math. **513**, 65–98 (2010)
25. Bruck, R.E., Kirk, W.A., Reich, S.: Strong and weak convergence theorems for locally nonexpansive mappings in Banach spaces. Nonlinear Anal. **6**, 151–155 (1982)
26. Bruck, R.E., Reich, S.: Nonexpansive projections and resolvents of accretive operators in Banach spaces. Houston J. Math. **3**, 459–470 (1977)
27. Butnariu, D., Censor, Y., Reich, S. (eds.): Inherently Parallel Algorithms in Feasibility and Optimization and Their Applications. Elsevier Science Publishers, Amsterdam (2001)
28. Butnariu, D., Davidi, R., Herman, G.T., Kazantsev, I.G.: Stable convergence behavior under summable perturbations of a class of projection methods for convex feasibility and optimization problems. IEEE J. Sel. Top. Sign Process **1**, 540–547 (2007)
29. Butnariu, D., Iusem, A.N.: Totally Convex Functions for Fixed Points Computation and Infinite Dimensional Optimization. Kluwer, Dordrecht (2000)
30. Butnariu, D., Reich, S., Zaslavski, A.J.: Convergence to fixed points of inexact orbits of Bregman-monotone and of nonexpansive operators in Banach spaces. In: Fixed Point Theory and Its Applications, pp. 11–32. Yokohama Publishers, Mexico (2006)
31. Butnariu, D., Reich, S., Zaslavski, A.J.: Asymptotic behavior of inexact orbits for a class of operators in complete metric spaces. J. Appl. Anal. **13**, 1–11 (2007)
32. Butnariu, D., Reich, S., Zaslavski, A.J.: Stable convergence theorems for infinite products and powers of nonexpansive mappings. Numer. Funct. Anal. Optim. **29**, 304–323 (2008)
33. Butnariu, D., Resmerita, E.: The Outer Bregman Projection Method for Stochastic Feasibility Problems in Banach Spaces. Studies in Computational Mathematics, vol. 8, pp. 69–86. North-Holland Publishing Co., Amsterdam (2001)
34. Cegielski, A.: Iterative Methods for Fixed Point Problems in Hilbert Spaces. Lecture Notes in Mathematics, vol. 2057. Springer, Berlin/Heidelberg (2012)
35. Cegielski, A., Gibali, A., Reich, S., Zalas, R.: An algorithm for solving the variational inequality problem over the fixed point set of a quasi-nonexpansive operator in Euclidean space. Numer. Funct. Anal. Optim. **34**, 1067–1096 (2013)
36. Ceng, L.C., Ansari, Q.H., Perrusel, A., Yao, J.C.: Approximation methods for triple hierarchical variational inequalities. Fixed Point Theory **16**, 67–90 (2015)
37. Ceng, L.C., Hadjisavvas, N., Wong, N.C.: Strong convergence theorem by a hybrid extragradient-like approximation method for variational inequalities and fixed point problems. J. Glob. Optim. **46**, 635–646 (2010)
38. Ceng, L.C., Mordukhovich, B.S., Yao, J.C.: Hybrid approximate proximal method with auxiliary variational inequality for vector optimization. J. Optim. Theory Appl. **146**, 267–303 (2010)
39. Ceng, L.C., Petrusel, A., Qin, X., Yao, J.-C.: Pseudomonotone variational inequalities and fixed points. Fixed Point Theory **22**, 543–558 (2021)

40. Ceng, L.C., Plubtieng, S., Wong, M.M., Yao, J.C.: System of variational inequalities with constraints of mixed equilibria, variational inequalities, and convex minimization and fixed point problems. J. Nonlinear Convex Anal. **16**, 385–421 (2015)
41. Ceng, L.C., Wong, N.C., Yao, J.C.: Regularized hybrid iterative algorithms for triple hierarchical variational inequalities. J. Inequalities Appl. **490**, 1–26 (2014)
42. Ceng, L.C., Wong, N.C., Yao, J.C.: Hybrid extragradient methods for finding minimum norm solutions of split feasibility problems. J. Nonlinear Convex Anal. **16**, 1965–1983 (2015)
43. Censor, Y., Cegielski, A.: Projection methods: an annotated bibliography of books and reviews. Optimization **64**, 2343–2358 (2015)
44. Censor, Y., Chen, W., Combettes, P.L., Davidi, R., Herman, G.T.: On the effectiveness of projection methods for convex feasibility problems with linear inequality constraints. Comput. Optim. Appl. **51**, 1065–1088 (2012)
45. Censor, Y., Davidi, R, Herman, G.T.: Perturbation resilience and superiorization of iterative algorithms. Inverse Problems **26**, 1–12 (2010)
46. Censor, Y., Elfving, T., Herman, G.T.: Averaging strings of sequential iterations for convex feasibility problems. In: Butnariu, D., Censor, Y., Reich, S. (eds.) Inherently Parallel Algorithms in Feasibility and Optimization and Their Applications, pp. 101–113. North-Holland, Amsterdam (2001)
47. Censor, Y., Elfving, T., Herman, G.T., Nikazad, T.: Diagonally-relaxed orthogonal projection methods. SIAM J. Sci. Comput. **30**, 473–504 (2008)
48. Censor, Y., Lent, A.: Cyclic subgradient projections. Math. Program **24**, 233–235 (1982)
49. Censor, Y., Segal, A.: On the string averaging method for sparse common fixed point problems. Inter. Trans. Oper. Res. **16**, 481–494 (2009)
50. Censor, Y., Segal, A.: The split common fixed point problem for directed operators. J. Convex Anal. **16**, 587–600 (2009)
51. Censor, Y., Zaslavski, A.J.: Convergence and perturbation resilience of dynamic string-averaging projection methods. Comput. Optim. Appl. **54**, 65–76 (2013)
52. Censor, Y., Zaslavski, A.J.: Strict Fejer monotonicity by superiorization of feasibility-seeking projection methods. J. Optim. Theory Appl. **165**, 172–187 (2015)
53. Censor, Y., Zenios, S.: Parallel Optimization: Theory, Algorithms and Applications. Oxford University Press, New York (1997)
54. Cheval, H., Kohlenbach, U., Leustean, L.: On modified Halpern and Tikhonov-Mann iterations. J. Optim. Theory Appl. **197**, 233–251 (2023)
55. Cheval, H., Leustean, L.: Quadratic rates of asymptotic regularity for the Tikhonov–Mann iteration. Optim. Methods Softw. **37**, 2225–2240 (2022)
56. Cheval, H., Leustean, L.: Linear rates of asymptotic regularity for Halpern-type iterations, arXiv:2303.05406v2 [math.OC]. https://doi.org/10.48550/arXiv.2303.05406
57. Colao, V., Leustean, L., Lopez-Acedo, G., Martin-Marquez, V.: Alternative iterative methods for nonexpansive mappings, rates of convergence and applications. J. Convex Anal. **18**, 465–487 (2011)
58. Combettes, P.L.: The convex feasibility problem in image recovery. Adv. Imag. Electron Phys. **95**, 155–270 (1996)
59. Combettes, P.L.: Hilbertian convex feasibility problems: convergence of projection methods. Appl. Math. Optim. **35**, 311–330 (1997)
60. Combettes, P.L., Hirstoaga, S.A.: Equilibrium problems in Hilbert spaces. J. Nonlinear Convex Anal. **63**, 117–136 (2005)
61. de Blasi, F.S., Myjak, J.: Sur la convergence des approximations successives pour les contractions non linéaires dans un espace de Banach. C. R. Acad. Sci. Paris **283**, 185–187 (1976)
62. de Blasi, F.S., Myjak, J., Reich, S., Zaslavski, A.J.: Generic existence and approximation of fixed points for nonexpansive set-valued maps. Set-Valued Var. Anal. **17**, 97–112 (2009)
63. Dinis, B., Pinto, P.: Strong convergence for the alternating Halpern-Mann iteration in CAT(0) spaces. SIOPT **33**, 785–815 (2023)

64. Djafari-Rouhani, B.: Asymptotic behaviour of almost nonexpansive sequences in a Hilbert space. J. Math. Anal. Appl. **151**, 226–235 (1990)
65. Djafari-Rouhani, B.: Ergodic theorems for hybrid sequences in a Hilbert space with applications. J. Math. Anal. Appl. **409**, 205–211 (2014)
66. Djafari-Rouhani, B.: Remarks on asymptotically nonexpansive mappings in a Hilbert space. Nonlinear Anal. **49**, 1099–1104 (2002)
67. Djafari-Rouhani, B., Farid, M., Kazmi, K.R.: Common solution to generalized mixed equilibrium problem and fixed point problem for a nonexpansive semigroup in Hilbert space. J. Korean Math. Soc. **53**, 89–114 (2016)
68. Djafari-Rouhani, B., Kazmi, K.R., Farid, M.: Common solutions to some systems of variational inequalities and fixed point problems. Fixed Point Theory **18**, 167–190 (2017)
69. Djafari-Rouhani, B., Kazmi, K.R., Moradi, S., Ali, R., Khan, S.A.: Solving the split equality hierarchical fixed point problem. Fixed Point Theory **23**, 351–369 (2022)
70. Djafari-Rouhani, B., Khatibzadeh, H.: Nonlinear Evolution and Difference Equations of Monotone Type in Hilbert Spaces. CRC Press/Taylor & Francis Group, Boca Raton (2019)
71. Djafari Rouhani, B., Khatibzadeh, H., Mohebbi, V.: Asymptotic behaviour of Φ-nonexpansive sequences and mappings in Banach spaces. Numer. Funct. Anal. Optim. **43**, 860–875 (2022)
72. Djafari-Rouhani, B., Mohebbi, V.: Proximal point method for quasi-equilibrium problems in Banach spaces. Numer. Funct. Anal. Optim. **41**, 1007–1026 (2020)
73. Djafari-Rouhani, B., Rahimi Piranfar, M.: Recent results on expansive-type evolution and difference equations: a survey. Axioms **12**, 1–15 (2023)
74. Du, W.S.: Some generalizations of fixed point theorems of Caristi type and Mizoguchi–Takahashi type under relaxed conditions. Bull. Braz. Math. Soc. (N.S.) **50**, 603–624 (2019)
75. Edelstein, M.: An extension of Banach's contraction principle. Proc. Am. Math. Soc. **12**, 7–10 (1961)
76. Eldred, A.A., Veeramani, P.: Existence and convergence of best proximity points. J. Math. Anal. Appl. **323**, 1001–1006 (2006)
77. Espínola, R., Wiśnicki, A.: The Knaster-Tarski theorem versus monotone nonexpansive mappings. Bull. Pol. Acad. Sci. Math. **66**, 1–7 (2018)
78. Flam, S.D., Zowe, J.: Relaxed outer projections, weighted averages and convex feasibility. BIT **30**, 289–300 (1990)
79. Fukhar-ud-din, H., Khamsi, M.A.: Approximating common fixed points in hyperbolic spaces. Fixed Point Theory Appl. **13**, 1–15 (2014)
80. Gerhardy, P., Kohlenbach, U.: General logical metatheorems for functional analysis. Trans. Am. Math. Soc. **360**, 2615–2660 (2008)
81. Gibali, A.: A new split inverse problem and an application to least intensity feasible solutions. Pure Appl. Funct. Anal. **2**, 243–258 (2017)
82. Gibali, A., Reich, S., Zalas, R.: Iterative methods for solving variational inequalities in Euclidean space. J. Fixed Point Theory Appl. **17**, 775–811 (2015)
83. Gibali, A., Reich, S., Zalas, R.: Outer approximation methods for solving variational inequalities in Hilbert space. Optimization **66**, 417–437 (2017)
84. Goebel, K., Kirk, W.A.: Topics in Metric Fixed Point Theory. Cambridge University Press, Cambridge (1990)
85. Goebel, K., Reich, S.: Uniform Convexity, Hyperbolic Geometry, and Nonexpansive Mappings. Marcel Dekker, New York/Basel (1984)
86. Granas, A., Dugundji, J.: Fixed Point Theory. Springer, New York (2003)
87. Gwinner, J., Jadamba, B., Khan, A.A., Sama, M.: Identification in variational and quasi-variational inequalities. J. Convex Anal. **25**, 545–569 (2018)
88. Gwinner, J., Raciti, F.: On monotone variational inequalities with random data. J. Math. Inequal. **3**, 443–453 (2009)
89. He, H., Ling, C., Xu, H.K.: A relaxed projection method for split variational inequalities. J. Optim. Theory Appl. **166**, 213–233 (2015)

90. He, H., Ling, C., Xu, H.K.: A projection-based splitting method for structured variational inequalities. J. Nonlinear Convex Anal. **16**, 1539–1556 (2015)
91. Iusem, A., Resmerita, E.: A proximal point method in nonreflexive Banach spaces. Set-Valued Var. Anal. **18**, 109–120 (2010)
92. Jachymski, J.: The contraction principle for mappings on a metric space with a graph. Proc. Am. Math. Soc. **136**, 1359–1373 (2008)
93. Jachymski, J.: Extensions of the Dugundji-Granas and Nadler's theorems on the continuity of fixed points. Pure Appl. Funct. Anal. **2**, 657–666 (2017)
94. Jadamba, B., Khan, A.A., Sama, M.: Generalized solutions of quasi-variational inequalities. Optim. Lett. **6**, 1221–1231 (2017)
95. Karapinar, E., Agarwal, R.P.: Fixed Point Theory in Generalized Metric Spaces. Synthesis Lectures on Mathematics and Statistics. Springer, Cham (2022)
96. Karapinar, E., Agarwal, R.P., Yesilkaya, S.S.: Perov type mappings with a contractive iterate. J. Nonlinear Convex Anal. **22**, 2531–2541 (2021)
97. Karapinar, E., Mitrovic, Z., Ozturk, A., Radenovic, S.: On a theorem of Ciric in b-metric spaces. Rend. Circ. Mat. Palermo **70**, 217–225 (2021)
98. Khamsi, M.A.: Generalized metric spaces: a survey. J. Fixed Point Theory Appl. **17**, 455–475 (2015)
99. Khamsi, M.A., Khan, A.R.: Inequalities in metric spaces with applications. Nonlinear Anal. **74**(12), 4036–4045 (2011)
100. Khamsi, M.A., Kirk, W.A.: An Introduction to Metric Spaces and Fixed Point Theory. Pure and Applied Mathematics (New York). Wiley-Interscience, New York (2001)
101. Khamsi, M.A., Kozlowski, W.M.: On asymptotic pointwise nonexpansive mappings in modular function spaces. J. Math. Anal. Appl. **380**, 697–708 (2011)
102. Khamsi, M.A., Kozlowski, W.M.: Fixed Point Theory in Modular Function Spaces. Birkhäuser/Springer, Cham (2015)
103. Khan, A.A., Li, J.: Characterizations of the metric and generalized metric projections on subspaces of Banach spaces. J. Math. Anal. Appl. **531**(2), part 1, Paper No. 127865, 1–23 (2024)
104. Khan, A.A., Li, J., Reich, S.: Generalized projections on general Banach spaces. J. Nonlinear Convex Anal. **24**, 1079–1112 (2023)
105. Khan, A.A., Motreanu, D.: Existence theorems for elliptic and evolutionary variational and quasi-variational inequalities. J. Optim. Theory Appl. **167**, 1136–1161 (2015)
106. Khan, A.A., Tammer, C., Zalinescu, C.: Regularization of quasi-variational inequalities. Optimization **64**, 1703–1724 (2015)
107. Kirk, W.A.: Contraction mappings and extensions. In: Handbook of Metric Fixed Point Theory, pp. 1–34. Kluwer, Dordrecht (2001)
108. Kirk, W.A., Shahzad, N.: Hyperbolic spaces and directional contractions. Bull. Math. Sci. **9**, 1–49 (2019)
109. Kirk, W.A., Srinivasan, P.S., Veeramani, P.: Fixed points for mappings satisfying cyclical contractive conditions. Fixed Point Theory **4**, 79–89 (2003)
110. Kohlenbach, U.: Some logical metatheorems with applications in functional analysis. Trans. Am. Math. Soc. **357**, 89–128 (2005)
111. Kohlenbach, U.: Applied Proof Theory: Proof Interpretations and Their Use in Mathematics. Springer Berlin, Heidelberg (2008)
112. Kohlenbach, U., Leustean, L.: Asymptotically nonexpansive mappings in uniformly convex hyperbolic spaces. J. Eur. Math. Soc. **12**, 71–92 (2010)
113. Konnov, I.V.: Combined Relaxation Methods for Variational Inequalities. Springer, Berlin/Heidelberg (2001)
114. Kopecka, E., Reich, S.: A note on the von Neumann alternating projections algorithm. J. Nonlinear Convex Anal. **5**, 379–386 (2004)
115. Kopecka, E., Reich, S.: A note on alternating projections in Hilbert space. J. Fixed Point Theory Appl. **12**, 41–47 (2012)

116. Kozlowski, W.M.: An Introduction to Fixed Point Theory in Modular Function Spaces, pp. 159–222. Springer, Cham (2014)
117. Kubota, R., Takahashi, W., Takeuchi, Y.: Extensions of Browder's demiclosedness principle and Reich's lemma and their applications. Pure Appl. Func. Anal. **1**, 63–84 (2016)
118. Leustean, L.: A quadratic rate of asymptotic regularity for CAT(0)-spaces. J. Math. Anal. Appl. **325**, 386–399 (2007)
119. Leustean, L.: Nonexpansive iterations in uniformly convex W–hyperbolic spaces. In: Leizarowitz, A., Mordukhovich, B.S., Shafrir, I., Zaslavski, A. (eds.) Nonlinear Analysis and Optimization I: Nonlinear Analysis, pp. 193–209. American Mathematical Society, Providence, Rhode Island (2010)
120. Leustean, L., Nicolae, A.: A note on an alternative iterative method for nonexpansive mappings. J. Convex Anal. **24**, 501–503 (2017)
121. Leustean, L., Pinto, P.: Rates of asymptotic regularity for the alternating Halpern-Mann iteration. Optim. Lett. (2023). https://doi.org/10.1007/s11590-023-02002-y
122. Li, J.: Iterative fixed point theorems and their applications to ordered variational inequalities on vector lattices. Fixed Point Theory **17**, 401–411 (2016)
123. Li, J.: Several fixed point theorems on partially ordered Banach spaces and applications. J. Nonlinear Convex Anal. **20**, 2095–2108 (2019)
124. Li, L., Xu, H.K.: Further convergence analysis of iterative methods for generalized split feasibility problems in Hilbert spaces. J. Nonlinear Convex Anal. **22**, 2575–2589 (2021)
125. Lopez, G., Martin, V., Xu, H.K.: Halpern's iteration for nonexpansive mappings. Contemp. Math. **513**, 211–230 (2010)
126. Marino, G., Xu, H.K.: Convergence of generalized proximal point algorithms. Commun. Pure Appl. Anal. **3**, 791–808 (2004)
127. Masad, E., Reich, S.: A note on the multiple-set split convex feasibility problem in Hilbert space. J. Nonlinear Convex Anal. **8**, 367–371 (2007)
128. Matkowski, J.: Integrable solutions of functional equations. Dissertationes Math. **127**, 1–68 (1975)
129. Mitrovic, Z.D., Parvaneh, V., Mlaiki N, Hussain, N., Radenovic, S.: On some new generalizations of Nadler contraction in b-metric spaces. Cogent Math. Stat. **7**(1), Art. ID 1760189, 1–13 (2020)
130. Mitrovic, Z.D., Radenovic, S.: The Banach and Reich contractions in bv(s)–metric spaces. J. Fixed Point Theory Appl. **19**, 3087–3095 (2017)
131. Mitrovic, Z.D., Radenovic, S.: On Meir-Keeler contraction in Branciari b-metric spaces. Trans. A. Razmadze Math. Inst. **173**, 83–90 (2019)
132. Mizoguchi, N., Takahashi, W.: Fixed point theorems for multivalued mappings on complete metric spaces. J. Math. Anal. Appl. **141**, 177–188 (1989)
133. Moudafi, A.: Viscosity approximation methods for fixed-point problems. J. Math. Anal. Appl. **241**, 46–55 (2000)
134. Moudafi, A.: Alternating CQ-algorithms for convex feasibility and split fixed-point problems. J. Nonlinear Convex Anal. **15**, 809–818 (2014)
135. Moudafi, A., Gibali, A.: $l_1 - l_2$ regularization of split feasibility problems. Numer. Algorithms **78**, 739–757 (2018)
136. Moudafi, A., Thakur, B.S.: Solving proximal split feasibility problems without prior knowledge of operator norms. Optim. Lett. **8**, 2099–2110 (2014)
137. Nadler, Jr S.B.: Multi–valued contraction mappings. Pac. J. Math. **30**, 475–488 (1969)
138. Nicolae, A., O'Regan, D., Petruşel, A.: Fixed point theorems for singlevalued and multivalued generalized contractions in metric spaces endowed with a graph. Georgian Math. J. **18**, 307–327 (2011)
139. ODHara, J.G., Pillay, P., Xu, H.K.: Iterative approaches to convex feasibility problems in Banach spaces. Nonlinear Anal. **64**, 2022–2042 (2006)
140. Qin, X., Cho, S.Y., Yao, J.C.: Weak and strong convergence of splitting algorithms in Banach spaces. Optimization **69**, 243–267 (2020)

141. Qin, X., Petrusel, A., Yao, J.C.: CQ iterative algorithms for fixed points of nonexpansive mappings and split feasibility problems in Hilbert spaces. J. Nonlinear Convex Anal. **19**, 157–165 (2018)
142. Qin, X., Yao, J.C.: Weak convergence of a Mann-like algorithm for nonexpansive and accretive operators. J. Inequal. Appl. **2016**, Paper No. 232, 1–9 (2016)
143. Orouji, B., Soori, E., O'Regan, D., Agarwal, R.P.: A strong convergence theorem for a finite family of Bregman demimetric mappings in a Banach space under a new shrinking projection method. J. Funct. Spaces, Art. ID 9551162, 1–11 (2021)
144. Peng, L., Li, C.: Porosity and fixed points of nonexpansive set-valued maps. Set-Valued Var. Anal. **22**, 333–348 (2014)
145. Peng, L., Li, C., Yao, J.C.: Porosity results on fixed points for nonexpansive set-valued maps in hyperbolic spaces. J. Math. Anal. Appl. **428**, 989–1004 (2015)
146. Perov, A.I.: A multidimensional version of M. A. Krasnosel'skii's generalized contraction principle. Funktsional. Anal. i Prilozhen. **44**, 83–87 (2010)
147. Perov, A.I., Kibenko, A.V.: On a certain general method for investigation of boundary value problems. Izv. Akad. Nauk SSSR **30**, 249–264 (1966)
148. Petrusel, A., Petrusel, G., Yao, J.C.: Coupled fixed point theorems for symmetric contractions in b-metric spaces with applications to operator equation systems. Fixed Point Theory **17**, 457–475 (2016)
149. Petrusel, A., Petrusel, G., Yao, J.C.: Perov type theorems for orbital contractions. J. Nonlinear Convex Anal. **21**, 759–769 (2020)
150. Petrusel, A., Petrusel, G., Yao, J.C.: Multi-valued graph contraction principle with applications. Optimization **69**, 1541–1556 (2020)
151. Petrusel, A., Petrusel, G., Yao, J.C.: Graph contractions in vector-valued metric spaces and applications. Optimization **70**, 763–775 (2021)
152. Petruşel, A., Rus, I.A., Serban, M.A.: Fixed points, fixed sets and iterated multifunction systems for nonself multivalued operators. Set-Valued Var. Anal. **23**, 223–237 (2015)
153. Planiden, C., Wang, X.: Most convex functions have unique minimizers. J. Convex Anal. **23**, 877–892 (2016)
154. Planiden, C., Wang, X.: Strongly convex functions, Moreau envelopes, and the generic nature of convex functions with strong minimizers. SIAM J. Optim. **26**, 1341–1364 (2016)
155. Rakotch, E.: On ϵ-contractive mappings. Bull. Res. Council Israel **10F**, 53–58 (1961)
156. Rakotch, E.: A note on contractive mappings. Proc. Am. Math. Soc. **13**, 459–465 (1962)
157. Reich, S.: Some remarks concerning contraction mappings. Can. Math. Bull. **14**, 121–124 (1971)
158. Reich, S.: Fixed points of contractive functions. Boll. Un. Mat. Ital. **5**, 26–42 (1972)
159. Reich, S.: Some fixed point problems. Atri. Acad. Nuz. Lincei **57**, 194–198 (1974)
160. Reich, S.: Approximate selections, best approximations, fixed points, and invariant sets. J. Math. Anal. Appl. **62**, 104–113 (1978)
161. Reich, S.: A limit theorem for projections. Linear Multilinear Algorithm **13**, 281–290 (1983)
162. Reich, S., Shafrir, I.: Nonexpansive iterations in hyperbolic spaces. Nonlinear Anal. **15**, 537–558 (1990)
163. Reich, S., Tuyen, T.M.: Projection algorithms for solving the split feasibility problem with multiple output sets. J. Optim. Theory Appl. **190**, 861–878 (2021)
164. Reich, S., Zalas, R.: A modular string averaging procedure for solving the common fixed point problem for quasi-nonexpansive mappings in Hilbert space. Numer. Algorithms **72**, 297–323 (2016)
165. Reich, S., Zaslavski, A.J.: Generic aspects of metric fixed point theory. In: Handbook of Metric Fixed Point Theory, pp. 557–575. Kluwer, Dordrecht (2001)
166. Reich, S., Zaslavski, A.J.: The set of noncontractive mappings is σ-porous in the space of all nonexpansive mappings. C. R. Acad. Sci. Paris **333**, 539–544 (2001)
167. Reich, S., Zaslavski, A.J.: Convergence of iterates of nonexpansive set-valued mappings. In: Set Valued Mappings with Applications in Nonlinear Analysis, pp. 411–420. Taylor & Francis, London (2002)

168. Reich, S., Zaslavski, A.J.: Genericity in Nonlinear Analysis. Springer, New York (2014)
169. Reich, S., Zaslavski, A.J.: Approximate fixed points of nonexpansive set-valued mappings in unbounded sets. J. Nonlinear Convex Anal. **16**, 1707–1717 (2015)
170. Reich, S., Zaslavski, A.J.: Monotone contractive mappings. J. Nonlinear Var. Anal. **1**, 391–401 (2017)
171. Reich, S., Zaslavski, A.J.: Contractive mappings on unbounded sets. Set-Valued Var. Anal. **26**, 27–47 (2018)
172. Reich, S., Zaslavski, A.J.: Well-posedness of fixed point problems for monotone nonexpansive mappings. Linear Nonlinear Anal. **4**, 1–8 (2018)
173. Reich, S., Zaslavski, A.J.: Generic well-posedness of the fixed point problem for monotone nonexpansive mappings. In: Mathematics Almost Everywhere, pp. 169–179. World Scientific Publication, Hackensack (2018)
174. Reich, S., Zaslavski, A.J.: Contractive mappings on metric spaces with graphs. Mathematics **9**, 2774 (2021). https://doi.org/10.3390/math9212774
175. Reich, S., Zaslavski, A.J.: Convergence and well-posedness properties of uniformly locally contractive mappings. Topol. Methods Nonlinear Anal. **61**(2), 761–773 (2022). https://doi.org/10.12775/TMNA.2022.035
176. Reich, S., Zaslavski, A.J.: Fixed point and well-posedness properties of uniformly locally contractive mappings. J. Nonlinear Convex Anal. **24**, 2543–2550 (2023)
177. Reich, S., Zaslavski, A.J.: Convergence of inexact iterates of monotone nonexpansive mappings with summable errors. Axioms **12**(1), 15 (2023). https://doi.org/10.3390/axioms12010015
178. Reich, S., Zaslavski, A.J.: A porosity result regarding uniformly locally contractive mappings. Pure Appl. Funct. Anal. **8**, 1781–1789 (2023)
179. Reich, S., Zaslavski, A.J.: Existence of a fixed point and stability results for contractive mappings on metric spaces with graphs. Topol. Methods Nonlinear Anal. (2023), in press
180. Rockafellar, R.T.: Monotone operators and the proximal point algorithm. SIAM J. Control Optim. **14**, 877–898 (1976)
181. Sabach, S., Shtern, S.: A first order method for solving convex bilevel optimization problems. SIAM J. Optim. **27**, 640–660 (2017)
182. Sahu, D.R., Wong, N.C., Yao, J.C.: A generalized hybrid steepest-descent method for variational inequalities in Banach spaces. Fixed Point Theory Appl. **2011**, 28 (2011)
183. Sahu, D.R., Wong, N.C., Yao, J.C.: A unified hybrid iterative method for solving variational inequalities involving generalized pseudocontractive mappings. SIAM J. Control Optim. **50**, 2335–2354 (2012)
184. Suparatulatorn, R., Cholamjiak, W., Suantai, S.: A modified S-iteration process for G-nonexpansive mappings in Banach spaces with graphs. Numer. Algorithms **77**, 479–490 (2018)
185. Suzuki, T.: Mizoguchi-Takahashi's fixed point theorem is a real generalization of Nadler's. J. Math. Anal. Appl. **340**, 752–755 (2008)
186. Takahashi, S., Takahashi, W.: Viscosity approximation methods for equilibrium problems and fixed point problems in Hilbert spaces. J. Math. Anal. Appl. **331**, 506–515 (2007)
187. Takahashi, W.: A convexity in metric space and nonexpansive mappings. I. Kodai Math. Sem. Rep. **22**, 142–149 (1970)
188. Takahashi, W.: The split common fixed point problem and the shrinking projection method for new nonlinear mappings in two Banach spaces. Pure Appl. Funct. Anal. **2**, 685–699 (2017)
189. Takahashi, W.: A general iterative method for split common fixed point problems in Hilbert spaces and applications. Pure Appl. Funct. Anal. **3**, 349–369 (2018)
190. Takahashi, W., Iiduka, H.: Weak convergence of a projection algorithm for variational inequalities in a Banach space. J. Math. Anal. Appl. **339**, 668–679 (2008)
191. Takahashi, W., Wen, C.F., Yao, J.C.: A strong convergence theorem by Halpern type iteration for a finite family of generalized demimetric mappings in a Hilbert space. Pure Appl. Funct. Anal. **4**, 407–426 (2019)

192. Takahashi, W., Xu, H.K., Yao, J.C.: Iterative methods for generalized split feasibility problems in Hilbert spaces. Set-Valued Var. Anal. **23**, 205–221 (2015)
193. Takahashi, W., Yao, J.C.: A strong convergence theorem by the hybrid method for a new class of nonlinear operators in a Banach space and applications. Appl. Anal. Optim. **1**, 1–17 (2017)
194. Takahashi, W., Yao, J.C.: Strong convergence theorems under shrinking projection methods for split common fixed point problems in two Banach spaces. J. Convex Anal. **28**, 1097–1118 (2021)
195. Tam, M.K.: Algorithms based on unions of nonexpansive maps. Optim. Lett. **12**, 1019–1027 (2018)
196. Tan, B., Qin, X., Yao, J.C.: Strong convergence of self-adaptive inertial algorithms for solving split variational inclusion problems with applications. J. Sci. Comput. **8**, 34 (2021)
197. Vanderwerff, J.: On the residuality of certain classes of convex functions. Pure Appl. Funct. Anal. **5**, 791–806 (2020)
198. Vetro, F., Radenovic, S.: Some results of Perov type in rectangular cone metric spaces. J. Fixed Point Theory Appl. **20**, 16 (2018)
199. Wang, F., Xu, H.K.: Cyclic algorithms for split feasibility problems in Hilbert spaces. Nonlinear Anal. **74**, 4105–4111 (2011)
200. Wang, X.: Most maximally monotone operators have a unique zero and a super-regular resolvent. Nonlinear Anal. **87**, 69–82 (2013)
201. Wang, X., Yang, X.: On the existence of minimizers of proximity functions for split feasibility problems. J. Optim. Theory Appl. **166**, 861–888 (2015)
202. Wlodarczyk, K., Plebaniak, R., Banach, A.: Best proximity points for cyclic and noncyclic set-valued relatively quasi-asymptotic contractions in uniform spaces. Nonlinear Anal. **70**, 3332–3341 (2009)
203. Wlodarczyk, K., Plebaniak, R., Obczynski, C.: Covergence theorems, best approximation and best proximity for set-valued dynamic systems of relatively quasi-asymptotic contractions in cone uniform spaces. Nonlinear Anal. **72**, 794–805 (2010)
204. Xu, H.K.: Viscosity approximation methods for nonexpansive mappings. J. Math. Anal. Appl. **298**, 279–291 (2004)
205. Xu, H.K.: A regularization method for the proximal point algorithm. J. Global Optim. **36**, 115–125 (2006)
206. Xu, H.K.: Iterative methods for the split feasibility problem in infinite-dimensional Hilbert spaces. Inver. Probl **26**, 1–17 (2010)
207. Xu, H.K.: An alternative regularization method for nonexpansive mappings with applications. In: Mordukhovich, B.S., Shafrir, I., Zaslavski, A. (eds.) Nonlinear Analysis and Optimization I: Nonlinear Analysis. Contemporary Mathematics, vol. 513, pp. 239–263. American Mathematical Society, Providence, Rhode Island (2010)
208. Xu, H.K., Cegielski, A.: The Landweber operator approach to the split equality problem. SIAM J. Optim. **31**, 626–652 (2021)
209. Yao, Y., Liou, Y.C., Yao, J.C.: Split common fixed point problem for two quasi-pseudocontractive operators and its algorithm construction. In: Fixed Point Theory and Applications, **2015**, (2015). Article number 127. https://doi.org/10.1186/s13663-015-0376-4
210. Zaslavski, A.J.: Optimization on metric and normed spaces. In: Springer Optimization and Its Applications. Springer, New York (2010)
211. Zaslavski, A.J.: Maximal monotone operators and the proximal point algorithm in the presence of computational errors. J. Optim. Theory Appl. **150**, 20–32 (2011)
212. Zaslavski, A.J.: Proximal point algorithm for finding a common zero of a finite family of maximal monotone operators in the presence of computational errors. Nonlinear Anal. **75**, 6071–6087 (2012)
213. Zaslavski, A.J.: Solving for (approximate) convex feasibility under finite precision. Nonlinear Stud. **19**, 653–660 (2012)
214. Zaslavski, A.J.: Approximate solutions of common fixed point problems. In: Springer Optimization and Its Applications. Springer, New York (2016)

215. Zaslavski, A.J.: Uniform convergence in inexact viscosity approximation methods for fixed–point problems. PanAm. Math. J. **26**, 61–70 (2016)
216. Zaslavski, A.J.: Inexact viscosity approximation methods for equilibrium problems and fixed point problems. Adv. Nonlinear Var. Inequal. **20**, 1–10 (2017)
217. Zaslavski, A.J.: Algorithms for solving common fixed point problems. In: Springer Optimization and Its Applications. Springer, Cham (2018)
218. Zaslavski, A.J.: Generic well-posedness of the fixed point problem for monotone nonexpansive mappings. In: Mathematics Almost Everywhere, pp. 169–179. World Scientific Publication, Hackensack (2018)
219. Zaslavski, A.J.: Approximate fixed points of nonexpansive mappings on hyperbolic spaces. Linear Nonlinear Anal. **5**, 517–524 (2019)
220. Zaslavski, A.J.: Convergence of inexact iterates of Perov nonexpansive mappings (2021). Preprint
221. Zaslavski, A.J.: An iterative method for solving the fixed point problem for a set-valued mapping. Linear Nonlinear Anal. **7**, 355–364 (2021)
222. Zaslavski, A.J.: Existence of a fixed point for a cyclical contractive mapping. PanAm. Math. J. **32**, 57–64 (2022)
223. Zaslavski, A.J.: Asymptotic behavior of iterates of a generic cyclical nonexpansive mapping. Numer. Funct. Anal. Optim. (2022). https://doi.org/10.1080/01630563.2022.2029484
224. Zaslavski, A.J.: Existence and convergence results for contractive mappings of Perov type. J. Nonlinear Convex Anal. **23**, 2677–2685 (2022)
225. Zaslavski, A.J.: Two convergence results for inexact iterates of cyclic nonexpansive mappings. Adv. Nonlinear Var. Inequal. **25**, 69–76 (2022)
226. Zaslavski, A.J.: Existence and convergence results for Perov contraction mappings. Fixed Point Theory **24**, 419–429 (2023). https://doi.org/10.24193/fpt-ro.2023.1.XX
227. Zaslavski, A.J.: Cyclical contractive mappings in hyperbolic spaces. Minimax Theory Appl. **8**, 285–300 (2023)
228. Zaslavski, A.J.: Existence and stability of approximate fixed points of nonexpansive set-valued mappings. Appl. Anal. Optim. **7**, 209–221 (2023)
229. Zaslavski, A.J.: Generic existence of fixed points for a class of cyclical nonlinear mappings. J. Nonlinear Convex Anal. accepted
230. Zaslavski, A.J.: Convergence results for contractive type set-valued mappings. Axioms **13**(2), 112 (2024). https://doi.org/10.3390/axioms13020112 (registering DOI)
231. Zaslavski, A.J.: Fixed point and convergence results for contractive-type self-mappings of metric spaces with graphs. Symmetry **16**(1), 119 (2024); https://doi.org/10.3390/sym16010119 (registering DOI)
232. Zaslavski, A.J.: Asymptotic regularity results for a viscosity version of Halpern-type iterations, submitted
233. Zaslavski, A.J.: Convergence results for set-valued strict contractions. Commun. Optim. Theory, accepted
234. Zaslavski, A.J.: Approximate fixed points for a set-valued quasi-contraction. Preprint
235. Zhou, H., Qin, X.: Fixed points of nonlinear operators. De Gruyter STEM, De Gruyter. BerlinNational Defense Industry Press, Beijing (2020)

Index

A. J. Zaslavski, *Approximate Fixed Points of Nonexpansive Mappings*, Developments in Mathematics 80, https://doi.org/10.1007/978-3-031-70710-0

The manufacturer's authorised representative in the EU is Springer Nature Customer Service Centre GmbH, Europaplatz 3, 69115 Heidelberg, Germany. If you have any concerns regarding our products, please contact ProductSafety@springernature.com

Printed and bound by CPI Group (UK) Ltd, Croydon, CR0 4YY
28/09/2026
02224061-0008